AF716925

Streptomyces

Molecular Biology and Biotechnology

Edited by

Paul Dyson

Institute of Life Science
School of Medicine
Swansea University
Swansea
UK

Caister Academic Press

Copyright © 2011

Caister Academic Press
Norfolk, UK

www.caister.com

British Library Cataloguing-in-Publication Data
A catalogue record for this book is available from the British Library

ISBN: 978-1-904455-77-6

Description or mention of instrumentation, software, or other products in this book does not imply endorsement by the author or publisher. The author and publisher do not assume responsibility for the validity of any products or procedures mentioned or described in this book or for the consequences of their use.

All rights reserved. No part of this publication may be reproduced, stored in a retrieval system, or transmitted, in any form or by any means, electronic, mechanical, photocopying, recording or otherwise, without the prior permission of the publisher. No claim to original U.S. Government works.

Cover image adapted from 'Leaving', an original silkscreen print by Keith Chater.

Printed and bound in Great Britain

Contents

Contributors

Bertrand Aigle
Laboratoire de Génétique et Microbiologie
Faculté des Sciences et Techniques
Université Henri Poincaré
Vandœuvre-lès-Nancy
France

Bertrand.Aigle@scbiol.uhp-nancy.fr

Robert Bunet
Laboratoire de Génétique et Microbiologie
Faculté des Sciences et Techniques
Université Henri Poincaré
Vandœuvre-lès-Nancy
France

robert.bunet@biotica.com

Gregory L. Challis
Department of Chemistry
University of Warwick
Coventry
UK

g.l.challis@warwick.ac.uk

Keith F. Chater
John Innes Centre
Norwich
UK

keith.chater@bbsrc.ac.uk

Carton W. Chen
Department of Life Sciences and Institute of Genome Sciences
National Yang-Ming University
Taipei
Taiwan

cwchen@ym.edu.tw

Christophe Corre
Department of Chemistry
University of Warwick
Coventry
UK

c.corre@warwick.ac.uk

Lubbert Dijkhuizen
Groningen Biomolecular Sciences and Biotechnology Institute (GBB)
Department of Microbiology
University of Groningen
Haren
The Netherlands

l.dijkhuizen@rug.nl

Paul Dyson
Institute of Life Science
School of Medicine
Swansea University
Swansea
UK

p.j.dyson@swansea.ac.uk

Amélie Garénaux
Université Paris-Sud
Centre Universitaire d'Orsay
Institut de Génétique et Microbiologie
Orsay cedex
France

amelie.garenaux@igmors.u-psud.fr

Marco Gottelt
Department of Microbial Physiology
Groningen Biomolecular Sciences and Biotechnology Institute (GBB)
University of Groningen
Haren
The Netherlands

marco.gottelt@biotica.com

F. Peter Guengerich
Department of Biochemistry
Center in Molecular Toxicology
Vanderbilt University School of Medicine
Nashville, TN
USA

f.guengerich@vanderbilt.edu

David A. Hopwood
John Innes Centre
Norwich
UK

david.hopwood@bbsrc.ac.uk

Laurence Hotel
Laboratoire de Génétique et Microbiologie
Faculté des Sciences et Techniques
Université Henri Poincaré
Vandœuvre-lès-Nancy
France

laurence.hotel@scbiol.uhp-nancy.fr

Sheng Huang
Laboratoire de Génétique et Microbiologie
Faculté des Sciences et Techniques
Université Henri Poincaré
Vandœuvre-lès-Nancy
France

shuang@iastate.edu

Matthew I. Hutchings
School of Biological Sciences
University of East Anglia
Norwich
UK

m.hutchings@uea.ac.uk

Steven L. Kelly
Institute of Life Science
School of Medicine
Swansea University
Swansea
UK

s.l.kelly@swansea.ac.uk

Geertje van Keulen
Institute of Life Science
School of Medicine
Swansea University
Swansea
UK

g.van.keulen@swansea.ac.uk

Ralph Kirby
Department of Life Sciences and Institute of Genome Sciences
National Yang-Ming University
Taipei
Taiwan

rkirby@ym.edu.tw

Stefan Kol
Department of Microbial Physiology
Groningen Biomolecular Sciences and Biotechnology Institute (GBB)
University of Groningen
Haren
The Netherlands

stef.kol@gmail.com

David C. Lamb
Institute of Life Science
School of Medicine
Swansea University
Swansea
UK

d.c.lamb@swansea.ac.uk

Luisa Laureti
Laboratoire de Génétique et Microbiologie
Faculté des Sciences et Techniques
Université Henri Poincaré
Vandœuvre-lès-Nancy
France

luisa.laureti@inserm.fr

Sylvie Lautru
Université Paris-Sud
Centre Universitaire d'Orsay
Institut de Génétique et Microbiologie
Orsay cedex
France

sylvie.lautru@igmors.u-psud.fr

Pierre Leblond
Laboratoire de Génétique et Microbiologie
Faculté des Sciences et Techniques
Université Henri Poincaré
Vandœuvre-lès-Nancy
France

leblond@nancy.inra.fr

Paloma Liras
Institute of Biotechnology of Leon (INBIOTEC);
Department of Microbiology
Faculty of Biological and Environmental Sciences
University of Leon
León
Spain

paloma.liras@unileon.es

Juan F. Martín
Institute of Biotechnology of Leon (INBIOTEC);
Department of Microbiology
Faculty of Biological and Environmental Sciences
University of Leon
León
Spain

jf.martin@unileon.es

Yvonne Mast
Interfaculty Institute of Microbiology and Infectious Diseases
Eberhard Karls University of Tübingen
Microbiology/Biotechnology
Tübingen
Germany

yvonne.mast@biotech.uni-tuebingen.de

Marta Vaz Mendes
Laboratoire de Génétique et Microbiologie
Faculté des Sciences et Techniques
Université Henri Poincaré
Vandœuvre-lès-Nancy
France

mvm@ibmc.up.pt

Carmen Méndez
Functional Biology Department and Institute of Oncology of the Principality of Asturias (I.U.O.P.A)
University of Oviedo
Oviedo

cmendezf@uniovi.es

Günther Muth
Microbiology/Biotechnology
Interfaculty Institute of Microbiology and Infectious Diseases
Eberhard Karls University Tübingen
Tübingen
Germany

gmuth@biotech.uni-tuebingen.de

Šárka Nezbedová
Université Paris-Sud
Centre Universitaire d'Orsay
Institut de Génétique et Microbiologie
Orsay cedex
France;
Institute of Microbiology
Academy of Sciences of the Czech Republic
Prague
Czech Republic

sarka@biomed.cas.cz

Hoang Chuong Nguyen
Université Paris-Sud
Centre Universitaire d'Orsay
Institut de Génétique et Microbiologie
Orsay cedex
France

nhchuong@hcmuns.edu.vn

Carlos Olano
Functional Biology Department and Institute of Oncology of the Principality of Asturias (I.U.O.P.A)
University of Oviedo
Oviedo
Spain

olanocarlos@uniovi.es

Tracy Palmer
Division of Molecular Microbiology
College of Life Sciences
University of Dundee
Dundee
UK

t.palmer@dundee.ac.uk

Rosario Pérez-Redondo
Institute of Biotechnology of Leon (INBIOTEC);
Department of Microbiology
Faculty of Biological and Environmental Sciences
University of Leon
León
Spain

r.perez.redondo@unileon.es

Jean-Luc Pernodet
Université Paris-Sud
Centre Universitaire d'Orsay
Institut de Génétique et Microbiologie
Orsay cedex
France

jean-luc.pernodet@igmors.u-psud.fr

Jens Reuther
Interfaculty Institute of Microbiology and Infectious Diseases
Eberhard Karls University of Tübingen
Microbiology/Biotechnology
Tübingen
Germany

jens.reuther@biotech.uni-tuebingen.de

Antonio Rodríguez-García
Institute of Biotechnology of Leon (INBIOTEC);
Department of Microbiology
Faculty of Biological and Environmental Sciences
University of Leon
León
Spain

arodg@unileon.es

José A. Salas
Functional Biology Department and Institute of Oncology of the Principality of Asturias (I.U.O.P.A)
University of Oviedo
Oviedo
Spain

jasalas@uniovi.es

Irene Santamarta
Institute of Biotechnology of Leon (INBIOTEC);
Department of Microbiology
Faculty of Biological and Environmental Sciences
University of Leon
León
Spain

isanh@unileon.es

Fernando Santos-Beneit
Institute of Biotechnology of Leon (INBIOTEC)
León
Spain

fsanb@unileon.es

Jeroen Siebring
Groningen Biomolecular Sciences and Biotechnology Institute (GBB)
Department of Microbiology
University of Groningen
Haren
The Netherlands

j.siebring@rug.nl

Alberto Sola-Landa
Institute of Biotechnology of Leon (INBIOTEC)
León
Spain

alberto.sola@unileon.es

Lijiang Song
Department of Chemistry
University of Warwick
Coventry
UK

lijiang.song@warwick.ac.uk

Eriko Takano
Department of Microbial Physiology
Groningen Biomolecular Sciences and Biotechnology Institute (GBB)
University of Groningen
Haren
The Netherlands

e.takano@rug.nl

Jutta Vogelmann
Microbiology/Biotechnology
Interfaculty Institute of Microbiology and Infectious Diseases
Eberhard Karls University Tübingen
Tübingen
Germany

juttav2000@yahoo.de

Michael R. Waterman
Department of Biochemistry;
Center in Molecular Toxicology
Vanderbilt University School of Medicine
Nashville, TN
USA

michael.waterman@vanderbilt.edu

Jaroslav Weiser
Institute of Microbiology
Academy of Sciences of the Czech Republic
Prague
Czech Republic

weiser@biomed.cas.cz

Wolfgang Wohlleben
Microbiology/Biotechnology
Interfaculty Institute of Microbiology and Infectious Diseases
Eberhard Karls University of Tübingen
Tübingen
Germany

wolfgang.wohlleben@uni.fuebingen.de

Bin Zhao
Department of Biochemistry
Vanderbilt University School of Medicine
Nashville, TN
USA

bin.zhao@vanderbilt.edu

Current Books of Interest

Viruses and Interferon: Current Research	2011
Essentials of Veterinary Parasitology	2011
Hepatitis C: Antiviral Drug Discovery and Development	2011
Streptomyces: Molecular Biology and Biotechnology	2011
Alphaherpesviruses: Molecular Virology	2011
Recent Advances in Plant Virology	2011
Vaccine Design: Innovative Approaches and Novel Strategies	2011
Salmonella: From Genome to Function	2011
PCR Troubleshooting and Optimization: The Essential Guide	2011
Insect Virology	2010
Environmental Microbiology: Current Technology and Water Applications	2011
Sensory Mechanisms in Bacteria: Molecular Aspects of Signal Recognition	2010
Bifidobacteria: Genomics and Molecular Aspects	2010
Molecular Phylogeny of Microorganisms	2010
Nanotechnology in Water Treatment Applications	2010
Iron Uptake and Homeostasis in Microorganisms	2010
Caliciviruses: Molecular and Cellular Virology	2010
Epstein-Barr Virus: Latency and Transformation	2010
Anaerobic Parasitic Protozoa: Genomics and Molecular Biology	2010
Lentiviruses and Macrophages: Molecular and Cellular Interactions	2010
Microbial Population Genetics	2010
Borrelia: Molecular and Cellular Biology	2010
Influenza: Molecular Virology	2010
RNA Interference and Viruses: Current Innovations and Future Trends	2010
Retroviruses: Molecular Biology, Genomics and Pathogenesis	2010
Metagenomics: Theory, Methods and Applications	2010
Aspergillus: Molecular Biology and Genomics	2010
Environmental Molecular Microbiology	2010
Neisseria: Molecular Mechanisms of Pathogenesis	2010
Frontiers in Dengue Virus Research	2010
ABC Transporters in Microorganisms	2009
Pili and Flagella: Current Research and Future Trends	2009
Lab-on-a-Chip Technology: Biomolecular Separation and Analysis	2009
Lab-on-a-Chip Technology: Fabrication and Microfluidics	2009
Bacterial Polysaccharides: Current Innovations and Future Trends	2009
Microbial Toxins: Current Research and Future Trends	2009
Acanthamoeba	2009
Bacterial Secreted Proteins: Secretory Mechanisms and Role in Pathogenesis	2009
Lactobacillus Molecular Biology: From Genomics to Probiotics	2009
Mycobacterium: Genomics and Molecular Biology	2009
Real-Time PCR: Current Technology and Applications	2009

Preface

In 1954, I was looking for an interesting PhD project in microbial genetics at the Cambridge University Botany School. Lewis Frost suggested that I might work on *Streptomyces*, pointing out that they and their relatives in the actinomycetes were usually regarded as intermediate between bacteria and fungi; as such they would be of great interest, given the fundamental differences in genetic behaviour becoming apparent between the two groups of microbes. While the fungi had a typical eukaryotic alternation of haploid and diploid stages, separated by meiosis with its reassortment of whole chromosomes and recombination of linked genes by crossing over – essentially the same as in animals and plants – the three systems of gene exchange so far discovered in bacteria were fundamentally different from this generalization, and diverse among themselves. Transformation, transduction and the bizarre conjugation mechanism recently discovered in *Escherichia coli* had in common the transfer of only a portion of the genes of a donor into a recipient, but there the resemblances ended. One depended on the transfer of naked DNA, the second harnessed bacteriophages as gene vectors, and the third was clearly a mating system requiring the contact of donor and recipient bacteria but resulted in the gradual spooling of DNA from male to female. Who knew what strange genetic phenomena awaited discovery in the actinomycetes?

In part through a fruitful collaboration with the electron microscopist Audrey Glauert, my own work soon placed *Streptomyces coelicolor* among the bacteria, but this did not induce me to abandon the search for genetic recombination. After all, in the words of Bill Hayes, one of the pioneers of bacterial genetics, the actinomycetes could be regarded as 'senior bacteria' compared with the simple unicellular *E. coli, Salmonella typhimurium* and the pneumococcus, which had until then provided our limited knowledge of bacterial genetics. As such, the actinomycetes could be expected to reveal many further genetic novelties. In the half-century since, this has turned out to be true in spades. This book tells us about many of them.

From my earliest studies of gene exchange in *S. coelicolor* I was intrigued by the architecture of the genome and its mode of transfer. While my early crosses generated a circular genetic map, as in *E. coli*, I could not prove if this reflected physical circularity of the chromosome, according to the generalization emerging for bacteria. It was not until the 1990s that the chromosomes of streptomycetes and their relatives were shown to be linear, and with a unique mechanism to address the universal problem of replicating DNA ends. Again, my first crosses indicated that some form of conjugation must be responsible for gene exchange, since distantly located genetic markers were often co-inherited, but it was only in this century that the process was proved to differ fundamentally from the *E. coli* paradigm. Instead of depending on the transfer of single-stranded DNA via a conjugation apparatus requiring many specialized proteins, streptomycetes have harnessed double-stranded DNA motor proteins that are widespread in bacteria, where they move daughter chromosomes into progeny cells as part of the normal life cycle.

The *S. coelicolor* genome sequence at the turn of the millennium revealed many further chromosomal novelties, including the distinction between the 'core' of the chromosome and its 'arms', the latter rich in conditionally adaptive genes that fit the organism for life in the complex and variable soil environment. As other streptomycete genome sequences were completed, the arms were shown to be much more variable between species than the core, and especially in the clusters of secondary metabolic gene clusters they contain. Horizontal gene transfer, emerging as a major driver in bacterial evolution in general, has evidently been at work on a grand scale.

The mycelial, differentiating lifestyle of the actinomycetes, superficially resembling that of moulds, had been responsible for early confusion as to their phylogenetic affinities. Genetic control of this life-style has revealed a wealth of unique features, many reflecting the fact that the colonies of these organism, almost uniquely for bacteria, consist of separate 'tissues', with different patterns of gene expression. There are many evolutionary parallels between the two major mycelial groups of microbes – actinomycetes and fungi – even extending to the functional equivalence, though phylogenetic independence, of particular proteins, such as the hydrophobic surface structures of the aerial mycelium.

Streptomycetes were the cradle of combinatorial biosynthesis of secondary metabolites in the 1990s. By that time, chemistry and biochemistry had revealed many of the secrets of biosynthesis, but mysteries remained: the programming of polyketide synthases was a notable example of the limitation of these approaches. *Streptomyces* genetics provided the key and now it is difficult to imagine bioorganic chemistry without a crucial input from gene sequencing, bioinformatic analysis, gene disruption and heterologous gene expression. As part of these endeavours, genome mining and the waking up of sleeping gene clusters – first shown to be abundant in the genomes of actinomycetes and now extended to other groups of differentiating bacteria, the myxobacteria and Cyanobacteria, and to filamentous fungi – are well established as promising tools in the fight against the emergence of antibiotic-resistant pathogens and tumours. Whatever the eventual contribution of genetic approaches to the discovery of useful drug leads, the study of natural products as an academic discipline will never be the same.

This book showcases these examples, and others, of the wealth of new knowledge that has emerged from the study of *Streptomyces* genetics. It is a 'must-read', not only for students of the actinomycetes, but for other microbiologists who will learn much of general interest about this fascinating group of microbes.

David A. Hopwood

Introduction

Paul Dyson

As the editor, with almost 30 years dedicated to investigating aspects of the biology of *Streptomyces*, my opinion that this volume provides an excellent insight and overview of the current status of many aspects of research into these fascinating bacteria is undoubtedly prejudiced. Nevertheless, not only for those for whom this book may be a first taste of the topic or for scientists who only occasionally seek an update on the status of the field, but also for dedicated *Streptomyces* researchers, I am convinced that the selection of chapters can inspire a great deal of interest. There are so many intriguing facets to discover about *Streptomyces* and the focus of this volume concerns what we know about their genomes, their physiology, their development and their capacity to synthesize an extraordinary diversity of secondary metabolites. There is a deliberate biotechnological slant and a consequence is that equally fascinating aspects of their ecology are only briefly alluded to. The focus of the content is largely driven by how we have exploited genome sequences of representative species, in particular that of *Streptomyces coelicolor*, in the first decade of this century: I would argue that as yet this information has not provided a huge impact on understanding streptomycete ecology, at least in experimental detail, despite some important insight based on bioinformatic predictions.

Streptomycetes are bacteria that mostly inhabit the soil. Indeed, by synthesizing volatile compounds such as geosmin, they provide the soil a typical odour. They are unusual bacteria in many respects, not least because their modes of growth and dispersal mimic those of eukaryotic filamentous fungi. Adopting a filamentous hyphal form enables them to make a vital contribution to soil ecology as saprophytes, important for recycling natural polymers such as chitin and cellulose. And being non-motile, they rely on producing unigenomic spores for their reproductive dispersal. The commitment to sporulation is usually a trigger to produce secondary metabolites, many of which have medicinal properties. Many of these compounds are antibiotics whose function may be to inhibit other organisms in the competitive niche of the soil, allowing the streptomycete life-cycle to culminate in production of spores that can be dispersed to new areas rich in organic material. The ability to synthesize antibiotics has, without doubt, inspired a great deal of research into understanding streptomycete biology and, as a consequence of genome sequencing, is it now apparent that the diversity of these compounds is much greater than we previously imagined. 'Genome mining' is an important advance in antibiotic discovery, likely to supplant traditional screening approaches of the 20th century.

Evidently a lot of our understanding of streptomycete biology preceded genome sequencing. The transition from classical genetic analysis to molecular genetics was guided by Professor Sir David Hopwood and co-workers at the John Innes Centre, UK, enabled by the development of appropriate genetic tools. The most authoritative account of the exciting developments that occurred during the first 50 years of this research has been written by David himself (Hopwood, 2007), and I strongly encourage anyone looking for inspiration and insight into how a research area can develop to read this. Indeed, David was largely responsible for the choice of *S. coelicolor* as the model species, motivated by the fact that two of the secondary metabolites it produces are pigmented: red prodiginines and blue actinorhodin. Consequently antibiotic production by this organism can be easily visualized with the naked eye and genetic mutations affecting production readily recognized. David was also the prime mover in instigating the genome sequencing project, at a time when this required a great deal more financing and time compared with 'new-generation' technologies. He convinced the UK BBSRC of the value of funding this project in part and then persuaded the UK Wellcome Trust to add *S. coelicolor* to their list of genomes of important pathogenic bacteria to be sequenced (most *Streptomyces* species are not mammalian pathogens) to finance the project to its conclusion.

If this investment was based on the promise of utilizing the genome sequence to improve our ability to harness *Streptomyces* for antibiotic production, and hence combat pathogens, then I believe this has already been vindicated. Indeed, this volume is testimony to what has already been forthcoming as a result of exploiting the genome sequences of *S. coelicolor* and a handful of other species. There is an account by Carton Chen and Ralph Kirby (Chapter 1) of the unusual genome architecture of streptomycetes with insight into the dynamics of the large linear (both aspects being remarkable for prokaryotes) chromosome and how this can contribute to streptomycete diversity, at the level of the species and also the secondary metabolic potential. Jutta Vogelmann, Wolfgang Wohlleben and Günther Muth (Chapter 2) have contributed a fascinating description of the topological diversity of streptomycete plasmids and the manner in which these elements interact with the chromosome. Importantly, they provide insight into how these plasmids can be transferred between different streptomycetes in a manner quite distinct to the paradigm of conjugative transfer of plasmids between enteric bacteria.

I've already alluded to the remarkable cell biology of streptomycetes. Keith Chater's chapter (Chapter 3) on the properties and programming of different cell types provides an extraordinarily valuable and detailed account of streptomycete developmental biology. The chapter emphasizes unique aspects of physiology and metabolism in the different cell types of a differentiating colony and highlights intriguing aspects of cell biology, for instance how cell division is uniquely inessential for viability and how growth and division are coordinated by processes that are specific to actinobacteria. In addition to describing what's going on inside different cell types, Keith introduces the complex extracellular biology of *Streptomyces*. Part of this involves secretion of signalling molecules (see below), but a great deal concerns export of proteins. Tracy Palmer and Matthew Hutchings (Chapter 4) detail the remarkable insight we have gained post-genome sequencing into systems employed by *Streptomyces* for protein secretion, including a vast array of proteins dependent on the non-conventional Tat pathway. The potential for metabolic versatility in streptomycetes is evident from scanning annotated genome sequences and functional genomics is now revealing how this plays out. It is true that these investigations are currently focussed on aspects of metabolism that potentially have a direct impact on antibiotic production, as elaborated on for central carbon metabolism by Geertje van Keulen, Jeroen Siebring and Lubbert Dijkhuizen (Chapter 5), for nitrogen assimilation by Wolfgang Wohlleben, Yvonne Mast and Jens Reuther (Chapter 6), and for phosphate supply and control of primary and secondary metabolisms by Juan Martín, Alberto Sola-Landa, Fernando Santos-Beneit and Antonio Rodríguez-García (Chapter 7). These chapters reveal that actinobacteria, and in particular streptomycetes, manage their metabolisms in quite distinct ways compared with other bacteria, including emerging insight into overlapping regulation influenced by nitrogen and phosphate availability.

Streptomyces species, with some of the largest prokaryotic genomes combined with a requirement as free-living organisms in the soil to adapt and differentiate in a range of environmental conditions, possess an enormous capacity for regulation. A total of 965 proteins encoded by the *S. coelicolor* genome are predicted to have regulatory function and recent research also indicates a significant role for regulatory RNA molecules (Pánek *et al.*, 2008; Swiercz *et al.*, 2008; D'Alia *et al.*, 2010). There are 65 predicted sigma factors, 85 sensor kinases and 79 response regulators (including 53 sensor–regulator pairs), 44 serine/threonine protein kinases and more than 100 likely transcription factors belonging to different regulator families. Unravelling the complexity of regulation is a current challenge that is employing global 'systems' approaches such as transcriptome and ChIP-on-chip analyses (Huang *et al.*, 2001; Bibb and Hesketh, 2009;

Bucca *et al.*, 2009; Nieselt *et al.*, 2010). As Keith Chater indicates, this regulatory complexity is defined by 'checkpoint multiplexes' or, as suggested by Charles Thompson (12th ISBA Meeting, Vancouver, 2001), interacting 'neural networks' that can be graphically represented in two dimensions by a spider's web. To date, most research has explored isolated linear regulatory pathways and hierarchies (the radials in a web) in *Streptomyces*, nevertheless revealing important new paradigms such as the response to oxidative stress mediated by the alternative sigma factor σ^R and its novel zinc-binding anti-sigma factor RsrA (Bae *et al.*, 2004; Li *et al.*, 2003; Newton and Fahey, 2008; Paget *et al.*, 2001; Zdanowski *et al.*, 2006). A next step in understanding regulatory complexity will be to investigate how different signals are integrated at regulatory nodes (where interconnecting strands connect to radials in the web), expanding the dimensionality of signal transduction and the resultant responses. Of biotechnological importance, of course, is the regulation of antibiotic production. Many gene clusters encoding antibiotic biosynthetic pathways encode pathway-specific regulatory genes. These can in turn be regulated by transcription factors whose activities are dependent on threshold concentrations of extracellular signalling molecules, the γ-butyrolactones. Marco Gottelt, Stefan Kol, and Eriko Takano explain exactly how these systems work and reveal how their manipulation can lead to the activation of an otherwise silent biosynthetic pathway in *S. coelicolor* (Chapter 8).

The 'golden era' of antibiotic discovery was 30 to 40 years ago. Research groups of large pharmaceutical companies assembled and screened large microbial (mostly streptomycete) strain collections for new compounds and a range of antimicrobials were discovered. Subsequently the law of diminishing returns applied, with increasing effort required to discover a reduced number of novel patentable drugs. One by one the large pharmaceutical companies disinvested in natural products research and development, so that by the end of the century there was only a fraction of the previous activity going into new antibiotic discovery at a time when multiple antibiotic resistant pathogenic 'superbugs' are increasingly causing clinical problems. I would argue that streptomycete genome sequencing, allied with our ability to genetically manipulate these organisms, should now catalyse a re-evaluation of these organisms as a source of vital new antibiotics. Genomics has motivated a rethink into how established antibiotics are produced: Paloma Liras, Irene Santamarta Rosario and Rosario Pérez-Redondo (Chapter 9) review our current understanding of how different β-lactams are synthesized by *S. clavuligerus* in which there are multiple dispersed clusters of the biosynthetic genes. But of greater significance has been recognition of an expanded biosynthetic potential of individual species as revealed by genome sequencing. As a rule of thumb, for a given well-characterized species, up to five or six different secondary metabolites may have been identified prior to genome sequencing. The sequence information now informs us that this is only approximately one quarter of the potential number of different secondary metabolites that can be produced. Moreover, the new products appear to be largely species-specific, implying a huge diversity of new compounds that can now be potentially accessed from existing strain collections. The synthesis of these new secondary metabolites is directed by 'silent' or 'cryptic' gene clusters identified by 'genome mining'. Bertrand Aigle and co-workers (Chapter 10) describe in detail their experiences of mining the *S. ambofaciens* genome and activating cryptic pathways therein, complementing the approach described by Marco Gottelt, Stefan Kol, and Eriko Takano for upregulating a silent pathway in *S. coelicolor*. These examples point towards a very exciting new era for antibiotic discovery, indicating that cryptic pathways are not redundant genetic relics but a source of new drugs.

Another strategy to extend the variety of secondary metabolites produced by the genus is to employ combinatorial biosynthesis. Essentially this means genetically manipulating biosynthetic pathways, based on precise knowledge of the roles of individual enzymes or active sites within these proteins. This strategy has been applied to change the aglycone skeletons of metabolites produced by an 'assembly-line' process – non-ribosomally synthesized peptides and type I polyketides – both very amenable to redesign by genetic intervention. The molecular biology underlying their assembly has been well reviewed in recent years (e.g. Marahiel and Essen, 2009; Weissman, 2009; and succeeding chapters of the same volume). Carlos Olano, Carmen Méndez and José A. Salas (Chapter 11) provide a comprehensive account including examples of combinatorial biosynthesis to change both aglycone skeletons and the patterns of subsequent decoration of these molecules by glycosylation. Lastly, critical steps in the synthesis of many secondary metabolites by streptomycetes depend on a plethora of different cytochrome P450 enzymes. David Lamb and co-workers, pioneers in elucidating the role of these enzymes, have written a timely review (Chapter 12) of the diversity of streptomycete cytochromes P450 and their roles, particularly in secondary metabolism, pointing to how they can exploited for combinatorial biosynthesis.

These developments provide the promise of a rewarding future for the development of new drugs. They underline the case that streptomycete-derived natural products will be the basis for future research and development for many years to come. More than half the content of this volume reports recent progress in research made during a 5-year European Union integrated project, ActinoGen, that I was fortunate enough to co-ordinate. For many, the prospect of managing a large project involving 19 partners would be daunting. It is true that the administrative burden was quite large. Nevertheless, I was inspired by the sense of cooperation existing between the participating partners and, above all, by the scientific outputs that are testimony to a very productive enterprise. Above all, they provide a basis for a new era of antibiotic discovery. I am very grateful to all ActinoGen partners and to all authors and co-authors who have contributed to this book.

References

Bae J.B., Park, J.H., Hahn, M.Y., Kim, M.S., and Roe, J.H. (2004). Redox-dependent changes in RsrA, an anti-sigma factor in *Streptomyces coelicolor*: zinc release and disulfide bond formation. J. Mol. Biol. *335*, 425–435.

Bibb M., and Hesketh, A. (2009). Analyzing the regulation of antibiotic production in streptomycetes. Methods Enzymol. *458*, 93–116.

Bucca G., Laing, E., Mersinias, V., Allenby, N., Hurd, D., Holdstock, J., Brenner, V., Harrison, M., and Smith, C.P. (2009). Development and application of versatile high density microarrays for genome-wide analysis of *Streptomyces coelicolor*: characterization of the HspR regulon. Genome Biol. *10*, R5.

D'Alia D., Nieselt, K., Steigele, S., Müller, J., Verburg, I., and Takano, E. (2010). Noncoding RNA of glutamine synthetase I modulates antibiotic production in *Streptomyces coelicolor* A3(2) J. Bacteriol. *192*, 1160–1164.

Hopwood, D.A. (2007). *Streptomyces* in Nature and Medicine: The Antibiotic Makers. (New York, N.Y., Oxford University Press).

Huang J., Lih, C.J., Pan, K.H., and Cohen, S.N. (2001). Global analysis of growth phase responsive gene expression and regulation of antibiotic biosynthetic pathways in *Streptomyces coelicolor* using DNA microarrays. Genes Dev. *15*, 3183–3192.

Li W., Bottrill, A.R., Bibb, M.J., Buttner, M.J., Paget, M.S., and Kleanthous, C. (2003). The role of zinc in the disulphide stress-regulated anti-sigma factor RsrA from *Streptomyces coelicolor*. J. Mol. Biol. *333*, 461–472.

Marahiel M.A., and Essen, L.O. (2009). Nonribosomal peptide synthetases mechanistic and structural aspects of essential domains. Methods Enzymol. *458*, 337–351.

Newton G.L., and Fahey, R.C. (2008). Regulation of mycothiol metabolism by sigma(R) and the thiol redox sensor anti-sigma factor RsrA. Mol. Microbiol. *68*, 805–809.

Nieselt K., Battke, F., Herbig, A., Bruheim, P., Wentzel, A., Jakobsen, Ø.M., Sletta, H., Alam, M.T., Merlo, M.E., Moore, J., *et al.* (2010). The dynamic architecture of the metabolic switch in *Streptomyces coelicolor*. BMC Genomics *11*, 10.

Paget M.S., Molle, V., Cohen, G., Aharonowitz, Y., and Buttner, M.J. (2001). Defining the disulphide stress response in *Streptomyces coelicolor* A3(2): identification of the sigmaR regulon. Mol. Microbiol. *42*, 1007–1020.

Pánek J., Bobek, J., Mikulík, K., Basler, M., and Vohradský, J. (2008). Biocomputational prediction of small non-coding RNAs in *Streptomyces*. BMC Genomics *9*, 217.

Swiercz J.P., Hindra, Bobek, J., Haiser, H.J., Di Berardo C., Tjaden, B., and Elliot, M.A. (2008). Small non-coding RNAs in *Streptomyces coelicolor*. Nucleic Acids Res. *36*, 7240–7251.

Weissman K.J. (2009). Introduction to polyketide biosynthesis. Methods Enzymol. *459*, 3–16.

Zdanowski K., Doughty, P., Jakimowicz, P., O'Hara L., Buttner, M.J., Paget, M.S., and Kleanthous, C. (2006). Assignment of the zinc ligands in RsrA, a redox-sensing ZAS protein from *Streptomyces coelicolor*. Biochemistry. *45*, 8294–8300.

Genome Architecture

1

Ralph Kirby and Carton W. Chen

Abstract

Linear replicons are relatively uncommon among bacteria and their preponderance among the Actinomycetales, and within the *Streptomyces* in particular, poses some interesting questions. These novel bacterial replicons are capped by terminal proteins that are covalently bound to the 5′ ends of the linear DNA and these terminal structures are directly involved in replicating and protecting the ends of the linear genome. In addition and perhaps related to their linear nature, these genomes are among the largest bacterial chromosomes. As far as can be ascertained at present, these large genomes have a specific organizational structure in terms of their genes. The genome structure can be divided into a core region that is present syntenously in most Actinomycetales, two terminal regions that are highly variable throughout the explored *Streptomyces* and two regions to the left and right of the core region that contain many syntenous genes specific to the *Streptomyces* and not found in other Actinomycetales. Genome dynamics seems to be important to the *Streptomyces* with plasmid–chromosome interactions, horizontal gene transfer and interspecific recombination probably playing important roles in how these genomes adapt to the diverse environment they reside in. Exploring the genome architecture of the *Streptomyces* helps our understanding of how and why the genus *Streptomyces* has a unique place in the evolution of the bacteria.

Introduction

Linear chromosomes

The chromosomes of *Streptomyces* species and those of some closely related actinobacteria, such as *Saccharopolyspora erythraea* (Reeves *et al.*, 1998), *Actinoplanes philippinensis*, *Micromonospora chalcea*, *Nocardia asteroides*, *Streptoverticillium abikoense*, *Streptoverticillium cinnamoneus* (Chen, 2000; Redenbach *et al.*, 2000) and *Rhodococcus* sp. RHA1 (McLeod *et al.*, 2006), represent a novel class of bacterial linear replicons capped by 'terminal proteins' (TPs) covalently bound at the 5′ ends (reviewed by Chaconas and Chen, 2005). Information on the genomes of actinobacteria was recently reviewed by Ventura *et al.* (2007).

The linear chromosomes of these bacteria are among the largest bacterial chromosomes. They range from about 8 Mb to over 10 Mb with a G+C content of about 72%. The extra length results from an increase in the number of genes rather than a greater average size of the genes. The large number of genes in turn reflects the generalist lifestyle and complex structure, the variability of the soil environment, and the rich secondary metabolism.

So far, there are published whole genome sequences of three *Streptomyces* species, *S. coelicolor* A3(2) (Bentley *et al.*, 2002), *S. avermitilis* (Ikeda *et al.*, 2003; Omura *et al.*, 2001), and *S. griseus* IFO 13350 (Ohnishi *et al.*, 2008). The genome sequence of *S. scabies* is partially annotated (http://www.sanger.ac.uk/

Projects/S_scabies/), but not yet published. All these chromosomal sequences may be divided into a 'core region' that houses the relatively conserved housekeeping genes, and two 'arms' that contain more divergent and horizontally transferred genes

The terminal regions of the chromosomes are highly unstable. Under laboratory conditions, spontaneous deletions of the terminal sequences (up to about 1 Mb) occur frequently, and can result in circularization of the chromosomes. Detection of such events by monitoring a terminally located genetic marker (e.g. chloramphenicol resistance, arginine prototrophy, tetracycline resistance) gives frequencies of about 10^{-2}–10^{-3} during one sporulation cycle. The actual frequencies certainly are higher.

In addition, unequal crossing over between the two arms of the chromosome or between one arm of the chromosome and a linear plasmid also occurs frequently, giving rise to gross rearrangements of the chromosome. It is not clear whether the high instability of the terminal regions is because they are intrinsically unstable (e.g. possibly because of the abundant transposable elements; see below) or because they are more tolerant of deletions and rearrangements than the core regions (because of the paucity of housekeeping genes). In any case, the dynamic nature of the arms is consistent with their high genetic diversity.

Circular and linear plasmids

Plasmids are commonly found in *Streptomyces* species, and linear plasmids appear to be as common as circular plasmids. The linear plasmids share the same general structural features as the linear chromosomes, i.e. (often) terminal inverted repeats capped by TP and a (usually) single internal replication origin (Hirochika *et al.*, 1985; Sakaguchi, 1990). The TPs of some linear plasmids are chromosomally encoded, while those of the others appear to be encoded by the plasmids. TP-capped linear plasmids are also found in other actinobacteria, including those with circular chromosomes.

Both linear and circular plasmids are generally conjugative, but rarely carry genes conferring resistance to toxic compounds or any other obvious selective advantage (Hopwood and Kieser, 1993). Only a small number of plasmids are known to contain one or more gene clusters for biosynthesis of secondary metabolites, such as SCP1 (Kirby *et al.*, 1975), pSV1 (Aguilar and Hopwood, 1982), and pSLA2-L (Mochizuki *et al.*, 2001). More detailed treatment of the biology of *Streptomyces* plasmids will be found elsewhere in this volume (Chapter 2).

Important issues regarding the linearity of chromosomes and plasmids

The linearity of *Streptomyces* chromosomes poses a number of challenges to some important classical biological concepts that have been based on circular replicons. The most important issue is replication.

In a manner similar to their circular counterparts, the replication of the linear chromosomes (and linear plasmids) of *Streptomyces* species is initiated from an internal origin, and then progresses bidirectionally towards the ends. Single-stranded gaps remain at each 3′ end of the daughter DNA molecules, about 250–320 nt long (Chang and Cohen, 1994; C.-H. Huang, unpublished results). These single-stranded gaps are patched to restore the double-stranded state, presumably by DNA synthesis primed by the TP. The use of such end patching to solve the classical end problem is very different from that used by eukaryotic chromosomes, where telomerase extends the 3′ ends to prevent shortening of the telomeres with each round of replication. It is also distinct from the TP-primed DNA synthesis during replication of other well-characterized TP-capped genomes such those of *Bacillus* phage ϕ29 and adenoviruses, which initiates on double-stranded DNA templates and copies the whole genome.

Another important issue is conjugal transfer. Based on the classical model, circular bacterial plasmids are transferred by rolling circle replication, in which one strand of the DNA is mobilized into the recipient. This model cannot be applied to linear plasmids unless one postulates that transfer begins from the end(s) of the DNA, because initiation of transfer from an internal site would not mobilize the complete plasmid DNA. Such a model of transfer has been proposed by Chen

(1996), in which the TP-capped 5′ strand leads the transfer in a manner similar to the leader protein-capped 5′ strand of a circular plasmid. There is also a possibility that linear plasmid DNA is transferred in a completely different fashion, such as in double-stranded form (Possoz *et al.*, 2001).

Finally, the evolutionary origin of linear chromosomes and plasmids must be addressed. Although, based on their phylogenetic distribution within the actinobacteria, it appears that the origin of linear chromosomes occurred after that of circular chromosomes, it is not certain how they arose. One reasonable hypothesis has been that they arose from an ancestral circular chromosome by recombination with linear plasmids (Chen *et al.*, 2002; Volff and Altenbuchner, 2000). However, it remains unclear how linear plasmids emerged and entered the scene because, among bacteria, TP-capped linear plasmids also seem to be limited to the actinobacteria, albeit with a wider distribution than linear chromosomes. During evolution, the linearity of chromosomes would certainly seem to have played a significant role in *Streptomyces* evolution in terms of the disposition and exchange of genes along the chromosome. With the availability of increasing numbers of sequenced genomes of *Streptomyces* and other actinobacteria, comparative analysis of these genomes should continue to shed light on this issue.

Replication origin and replication initiation

Initiation of replication of *Streptomyces* chromosomes takes place at *oriC*, which is approximately centrally located in most of the explored chromosomes (The *S. avermitilis* chromosome has an off-centre *oriC*). The basic structure of *oriC*, the action of the initiation protein DnaA, and the mechanism of replication initiation of *Streptomyces* chromosomes are similar to those of circular bacterial chromosomes, with some variations (Chen, 2007; Zawilak-Pawlik *et al.*, 2005).

The replication origins of linear plasmids are also internally located, but not necessarily near the centre of the molecule. In general, these replication origins have not been precisely characterized, but have been isolated together with one or more genes required for replication initiation. They are represented as a series of direct repeats (iterons) or AT-rich sequences, or both.

The combined replication origin (*oriP*) and replication gene(s) deserve the designation of autonomous replication sequence (ARS), since the unit is sufficient to support the replication of various artificially constructed circular or linear plasmids in *Streptomyces*. More than one ARS has been found in some larger linear plasmids, such as the 350-kb linear plasmid SCP1 (Redenbach *et al.*, 1999; Zhongjun Qin, personal communication) and the 210-kb linear plasmid pSLA2-L in *S. rochei* (Hiratsu *et al.*, 2000; Mochizuki *et al.*, 2001). Unlike the situation in the linear chromosomes, the required initiation proteins vary widely among different *Streptomyces* linear plasmids. Typically, the ARS includes at least one protein with a helicase domain, but these are often of different phylogenetic origins. For example, the first ARS identified by Shiffman and Cohen (1992) in pSCL1 of *S. clavuligerus* encodes a DnaB (replicative helicase) homologue and a *rep* gene homologue (Chang *et al.*, 1996).

The diversity of genetic constituents involved in the initiation of replication of *Streptomyces* linear plasmids indicates that there have been diverse origins of these linear plasmids. A recent study shows that novel initiation protein–*oriP* combinations constructed artificially using components from three linear plasmids, SLP2, SCP1 and pSLA2, are competent to support replication, but that transformation efficiencies are severely affected (Xu *et al.*, 2006). This indicates a relatively loose initiation protein-*oriP* specificity for these replication initiation systems. Like typical bacterial plasmids, *Streptomyces* plasmids generally do not encode a DNA polymerase, and presumably rely on chromosomally encoded polymerases (such as Pol III) for replication. An exception so far has been SCP1, which encodes the α and β subunits of Pol III (Bentley *et al.*, 2004), the importance of which has not been investigated.

Terminal DNA

Archetypal telomeres

The telomere sequences of the linear replicons of *Streptomyces* are more complex than those of eukaryotic chromosomes or other TP-capped linear replicons. The 3′ ends of the terminal 200–300 bp are exposed during replication and are involved in end patching. Thus, these 3′ terminal sequences and the secondary structures that they may form are very important not only for their functional roles but also for their resistance to nucleolytic attack.

The majority of *Streptomyces* telomere sequences may be grouped into a highly conservative 'archetypal' family. The remaining small fraction of *Streptomyces* telomeres are heterogeneous in sequence and size (designated 'atypical' telomeres), and they are found mostly in linear plasmids (such as SCP1, pRL1, and pRL2), but also in two chromosomes, those of *S. griseus* IFO 13350 (Ohnishi *et al.*, 2008) and *S. griseus* 2247 (Goshi *et al.*, 2002).

Archetypal telomeric sequences are highly packed with several palindromic sequences separated by zero to a few bp (Fig. 1.1). The first 13 bp form Palindrome I, which is most highly conserved (with the single possible exception of the *S. scabies* chromosome), and the remaining palindromes diverge increasingly from it. In many (but not all) cases, Palindrome I is complementary to an internal (usually the fourth) palindrome, and thus may form a complex secondary structure. However, the lack of the internal complementary palindrome in some archetypal telomeres, and the fact that palindrome IV of the *S. lividans* chromosome (shared by the right end of the SLP2 plasmid) can be deleted without affecting replication (H.-H. Tsai, unpublished results) indicate that the pairing of Palindrome I with an internal palindrome is not important. More likely, the palindrome folds back onto itself with a 3-nt loop.

Interestingly, most of the loops formed by the palindromes within the terminal 3′ sequences of the archetypal telomeres are of the GNA sequence (N being most often C). ^{13}C-NMR studies have shown that, in such hairpins, the G and A residues tend to form 'sheared purine–purine pairing', resulting in a single C-residue loop (Chou *et al.*, 1997; Zhu *et al.*, 1995), and that the GCA hairpins are the most stable among all the GNA hairpins. The fact that these single-nt hairpin loops are closed by sheared purine pairs makes them refractory to single-strand specific nuclease attack *in vitro* (Chou *et al.*, 1997), suggesting a similar protective role *in vivo* when the 3′ ends are exposed during replication. The minimum length of the telomere necessary for replication *in vivo* has been determined to be 144 bp for the linear plasmid pSLA2 (Qin and Cohen, 1998), spanning the first six palindromes. Of these, palindrome IV, but not palindromes II, III, and V, may be removed without affecting survival of the plasmid. In general, the telomere sequences of linear plasmids diverge more widely than those of linear chromosomes (Huang *et al.*, 1998), presumably

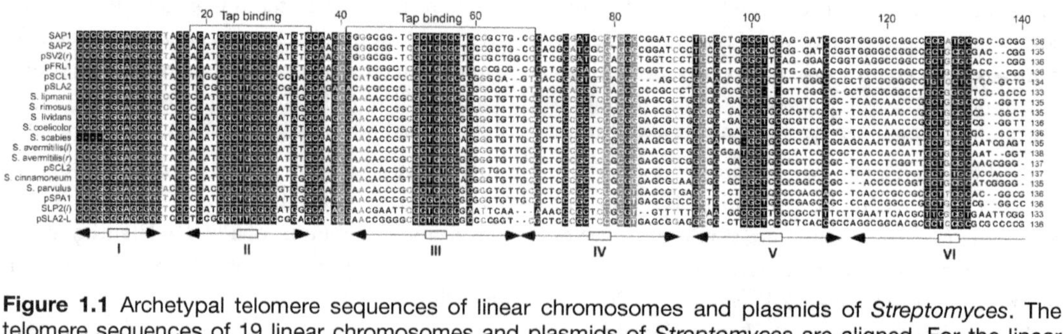

Figure 1.1 Archetypal telomere sequences of linear chromosomes and plasmids of *Streptomyces*. The telomere sequences of 19 linear chromosomes and plasmids of *Streptomyces* are aligned. For the linear plasmids, their designations are given, and for the linear chromosomes, the species name is given. In cases where the two telomeric sequences are different, *l* and *r* indicate the *left* and *right* telomeres, respectively. The fully conserved nucleotides are shown in black, and the less conserved nucleotides in grey shades of decreasing intensity. The first six palindromes are indicated by the double arrows at the bottom, with the central non-palindromic core depicted by the boxes. The indicated Tap binding site was determined on the pSAL2 telomere (Bao and Cohen, 2003).

reflecting their faster evolution and a larger latitude for evolutionary experimentation.

Atypical telomeres

The atypical telomeres differ from the archetypal telomeres as well as from one another in their sizes and their primary and predicted secondary structures. The telomeric sequence of SCP1 begins with a string of 4–6 Gs instead of three Cs in the archetypal telomeres (Bentley *et al.*, 2004; Kinashi *et al.*, 1991), and its predicted secondary structure contains an unpaired 3′ end (vs. fully duplexed 3′ ends formed by the archetypal telomeres) and 4-nt loops (vs. 3-nt loops of the archetypal telomeres). Most of the 4-nt loops may still undergo purine–purine sheared pairing. The minimum telomere length for SCP1 is 84 bp, spanning the first three palindromes (C.-H. Huang, unpublished results). Atypical telomere sequences were also found in two linear plasmids, pRL1 and pRL2 (Zhang *et al.*, 2006). These telomere sequences do not resemble, in primary or secondary structure, either the archetypal telomeres or the SCP1 telomere.

The atypical telomeres of the chromosomes of *S. griseus* 2247 (Goshi *et al.*, 2002) and IFO 13350 (Suzuki *et al.*, 2008) differ from each other. While the predicted secondary structures of the 3′ ends of the two telomeres contain 3-nt GNC loops and a free end, the terminal sequences are completely different. These distinct structural features of the atypical telomeres suggest that they are capped by very different TPs (see below).

Terminal inverted repeats

The telomeres on linear plasmids and chromosomes of *Streptomyces* are often part of longer terminal inverted repeats (TIRs), the sizes of which range from tens of bp to about one Mb. The initial discovery of such TIRs led to the designation 'invertron' by Sakaguchi (1990) for this type of replicon. However, with increasing studies of telomere sequences, it was discovered that such terminal inverted repeats are not an important requirement for survival of linear plasmids or chromosomes. Examples of very short TIRs are the 44-bp imperfect TIR of SLP2 (Chen *et al.*, 1993) and the 174-bp TIR of the *S. avermitilis* chromosome (Ikeda *et al.*, 2003).

The *S. coelicolor-S. lividans* hybrid chromosomes produced by recombination through conjugation (Wang *et al.*, 1999) have one telomere from each species and a TIR of less than 200 bp. At the most extreme, recombination between SCP1 and the *S. coelicolor* chromosome gave rise to two chimeric chromosomes, each containing an atypical (SCP1) and an archetypal (*S. coelicolor* chromosomal) telomere (Yamasaki and Kinashi, 2001). The telomeres of these chimeric chromosomes share no homology down to the terminal nt (C for *S. coelicolor* chromosomal telomere and G for the SCP1 telomere), and therefore they have no TIR. An artificially constructed linear plasmid containing this mixed pair of telomere can also readily replicate *in vivo* (Huang *et al.*, 2007). Notably, linear plasmids or chromosomes containing heterologous telomere sequences (even capped by different TPs) do not appear to suffer any obvious disadvantage in propagation or stability.

In contrast to the highly conserved telomere sequences of most *Streptomyces* chromosomes, the other sequences in the terminal regions are very heterogeneous, with no apparent constraints except for a terminally located helicase gene, *ttrA*, that is highly conserved and appears to be involved in conjugal transfer (Huang *et al.*, 2003). The sequence heterogeneity and length variations in the TIR reflect the highly dynamic nature of *Streptomyces* chromosomes (reviewed by Chen *et al.*, 2002) (see below).

Pseudotelomeres

Two 'pseudotelomeres', which markedly resemble archetypal telomeres in both primary and secondary structure (including the 3-nt GCA loops), were initially found in SLP2 (Huang *et al.*, 2003), and subsequently another two were found in SCP1 (Bentley *et al.*, 2004). These pseudotelomeres lie pairwise in the same orientation and within 4 kb from the end of both plasmids. The detection of these pseudotelomeres involved non-standard genomic sequence analysis procedures, and their presence in other *Streptomyces* replicons may have been overlooked. Presumably the pseudotelomeres represent the remnants of linear replicons that have been integrated during evolution. Both SLP2 and SCP1 have been proposed to have experienced more than one round of exchange with other plasmids during evolution.

Terminal proteins

Archetypal TPs

Terminal proteins that cap the archetypal telomere DNA are highly conserved in size (about 185 aa) and sequence (Fig. 1.2) (Bao and Cohen, 2001; Yang *et al.*, 2002). They are named 'archetypal TP' to distinguish from the heterologous TPs that cap atypical telomere DNA (see below). The archetypal TPs and the encoding gene bear the designations Tpg and *tpg*, respectively (Bao and Cohen, 2001). Most functional *tpg* form a bicistronic operon with an upstream gene, *tap*, which is also essential for replication of linear replicons (Bao and Cohen, 2003). All the *Streptomyces* chromosomes with archetypal telomeres contain the *tap-tpg* operon near one terminus in a position ranging from several kb to a few hundred kb from the end.

Streptomyces genomes often contain *tpg* pseudogenes, which are characterized by the aberrant (usually shortened) size of the encoded products and the lack of a typical *Streptomyces* codon preference. The first *tpg* pseudogene was found in the 15.3-kb terminal sequence in common between both arms of the *S. lividans* chromosome and the right arm of SLP2 (Huang *et al.*, 2003). This *tpg* pseudogene lies about 8 kb from the right telomere, and has been confirmed as defective (C.-C. Yang, unpublished results). Putative *tpg* pseudogenes were also found in the chromosome of *S. avermitilis* (Ikeda *et al.*, 2003), *S. ambofaciens* (Choulet *et al.*, 2006), and *S. griseus* (Ohnishi *et al.*, 2008; Suzuki *et al.*, 2008). The

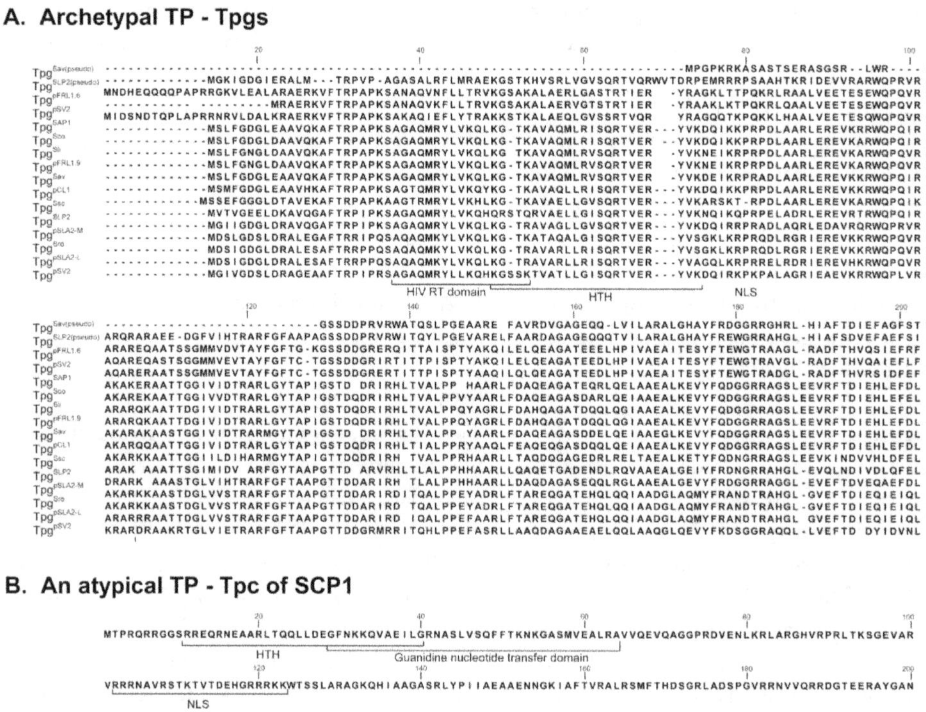

Figure 1.2 Archetypal and atypical TPs of *Streptomyces*. (A) The aligned amino acid sequences of the archetypal TPs proteins. The helix–turn–helix (HTH) domain, HIV reverse transcriptase (HIV RT) domain, and nuclear localization sequence (NLS) are indicated. (B) The atypical terminal protein, Tpc, of SCP1. The HTH, NLS, and guanidine nucleoside transfer domains are identified.

terminal location of the *tpg* pseudogenes hints at previous exchanges with other linear plasmids or chromosomes. The abundance of *tpg* pseudogenes suggests that such events are frequent and relatively recent in evolution.

While many smaller *Streptomyces* linear plasmids appear to rely on the host to supply the required TPs (Bao and Cohen, 2001), some larger linear plasmids contain a functional *tpg* (and *tpg* pseudogenes in some cases). The *tpg* homologues on the linear plasmids are often not accompanied by a *tap* gene. For example, the left arm of SLP2 contains a functional *tpg* gene (C.-C. Yang, personal communication) unaccompanied by a *tap* gene. In this case, the chromosomal *tap* gene is required for the replication (Lin, 1998).

Tpgs contain a putative helix–turn–helix domain similar to the DNA-binding 'thumb' domain of HIV reverse transcriptase (Fig. 1.2) (Bao and Cohen, 2001; Yang *et al.*, 2002). Interestingly, they all contain a typical monopartite nuclear localization signal (NLS) sequence or a similar sequence (Yang *et al.*, 2002)., which have been shown to be functional in targeting the TPs and the covalently linked DNA into human (Tsai *et al.*, 2008) and plant nuclei (H.-H. Tsai and J-S. Lai, unpublished results).

Atypical TPs — The TP of SCP1

Atypical telomeres would be expected to be capped by TPs different from the archetypal TPs. The first atypical TP identified is Tpc, which caps SCP1 and is strategically encoded by the plasmid itself (Huang *et al.*, 2007). Tpc (259 aa) is much longer than the archetypal TPs, and does not resemble Tpg or any other protein in the databases. Nevertheless, it also contains an HTH DNA binding domain in the N-terminal region and a bipartite NLS (Fig. 1.2), which has also been found to be functional (Tsai *et al.*, 2008) in nuclear targeting in human (Tsai *et al.*, 2008) and plant cells (H.-H. Tsai and J-S. Lai, unpublished results). The NLS domains in both Tpgs and Tpc are not required for replication of the linear plasmids, suggesting that the nuclear targeting function in these TPs has evolved independently of their replication function.

The discovery of functional NLSs of two different types in two different families of TPs suggests that their occurrence is not coincidental but is of biological significance. The TP-capped linear plasmids of *Streptomyces* are similar to the linear T-DNA capped by the pilot protein VirD2 during Ti plasmid-mediated conjugal transfer from *Agrobacterium tumefaciens* to plant cells. VirD2 also contains NLSs, which are necessary for nuclear targeting. The possibility that TP-mediated DNA transfer may also take place in nature, between *Streptomyces* and, for example, plants in soil, remains to be investigated. The *tac* gene, upstream of *tpc*, encoding a hypothetical protein, is also essential for replication of SCP1.

More recently, Suzuki *et al.* (2008) identified a protein encoded by SGR97t that appeared to cap the atypical telomere sequence of the *S. griseus* IFO 13350 chromosome. Although its translation product (GtpA) is of the same length (185 aa) as Tpgs, its amino acid sequence shows very low similarity to Tpgs (18% identity) or to other proteins in the databases (E-value >1.0e-01). Moreover, it does not contain a predicted NLS and HTH DNA-binding motif, and it has a low predicted pI (5.8) in contrast to the high predicted pI (above 10) of Tpgs.

End patching

The mechanisms

The internally initiated bidirectional replication of linear replicons of *Streptomyces* produces single-stranded 3′ overhangs of about 200–300 nt at the telomeres, which cannot be replicated by the replisomes. Since the telomeres of *Streptomyces* linear replicons have fixed lengths and sequences, these 3′ overhangs must be patched during each round of replication.

Several biochemical mechanisms for end patching were originally proposed (Chen, 1996) and, of these, the most favoured is a TP-primed DNA synthesis model, in which TP acts as a primer for DNA synthesis that fills the 3′ gaps. TP-primed DNA synthesis is used in replication of certain other protein-capped linear genomes, the best studied of which are ϕ29 *Bacillus* phage and adenoviruses. Unlike the *Streptomyces* linear replicons, these replicons use TP-primed synthesis for replication of the whole molecule.

In support of this model, *in vitro* deoxynucleotidylation of Tpg has been demonstrated (Yang *et al.*, 2006). In this system, dCMP is specifically incorporated onto a Thr residue present in Tpg. In contrast, in ϕ29 and adenoviruses, deoxynucleotidylation occurs at a Tyr residue of the TPs.

Both Tpg and Tap bind DNA *in vitro* (Bao and Cohen, 2003). Tpg binds non-specifically, whereas Tap binds strongly to the single-stranded telomeric 3′ overhangs at regions corresponding to Palindromes II and III. These results, together with the demonstration of Tpg–Tap interaction by immunoprecipitation and yeast two-hybrid studies, lead to a model in which Tap binds specifically at the 3′ overhang and recruits Tpg for end patching (Bao and Cohen, 2003).

This model indicates the importance of the structure of the 3′ overhang in initiating end patching. In contrast, the covalently attached TPs at the 5′ ends of the parental DNA appear not to be essential for end patching, because a functional linear plasmid can be generated, when its sequence (an ARS flanked by a pair of telomeres) embedded in adventitious DNA is introduced into *Streptomyces* (Shiffman and Cohen, 1992). This technique has been widely used to construct recombinant linear plasmids routinely. It indicates that end patching may be achieved on internally located telomere sequences in the absence of covalently bound TPs.

Bao and Cohen (2004) also isolated and identified several telomere-binding proteins from *S. lividans*, including DNA polymerase I (Pol I) and topoisomerase I. Interestingly, they discovered that both of these two proteins have a reverse transcriptase activity. While the finding is intriguing, its biological significance is not clear. Although it is tempting to speculate that Pol I is involved in TP-primed DNA synthesis, this appears to conflict with the finding of Huang *et al.* (2008) that linear chromosomes can replicate in *polA* deletion mutants of *Streptomyces*. One or both of two other polymerases appear to participate in end patching (see below).

DNA polymerase

While replication of *Streptomyces* chromosomes employs DNA polymerase III (Pol III) (Flett *et al.*, 1999; Flett *et al.*, 1992), the sequenced *Streptomyces* genomes contain duplicate copies of genes encoding some subunits of Pol III, including the catalytic subunit α (*dnaE1, dnaE2*). In addition to Pol III genes and the aforementioned *polA* gene, the *Streptomyces* genomes contain two genes (*dinB1, dinB2*) encoding the translesion repair DNA polymerase, Pol IV, but no *polB* homologue (DNA polymerase II) or *umuC* homologue (DNA polymerase V).

Recently, Tsai and Chen (unpublished results) discovered that either *dinB1* or *dinB2* is required for the replication of linear but not circular plasmids in *Streptomyces*. Deletion of both *dinB1* and *dinB2* is tolerated only in *Streptomyces* strains with a circularized mutant chromosome, but not with a linear chromosome. Moreover, the double mutants can support replication of circular plasmids, but not linear plasmids. These results indicate that *in vivo* one or both of the Pol IV enzymes are involved in end patching. Interestingly, both *dinB1*and *dinB2* lie in the '*Streptomyces*-specific region' (see below) on the left arm among sequenced *Streptomyces* chromosomes, and are essentially syntenous with the other DNA polymerase genes as well as *recA* and *tpg*, all of which are important for chromosome replication and end patching (Fig. 1.3). The overlap between *dinB2* and *polE2* also appears to be conserved among *Streptomyces* species.

TP–TP interactions

TP–TP interaction has been demonstrated previously in other TP-capped replicons, including ϕ29 phage (Ortin *et al.*, 1971), GA-1 phage (Arnberg and Arwert, 1976), and adenovirus (Keegstra *et al.*, 1977; Robinson *et al.*, 1973; Wong and Hsu, 1989). All these TP-capped linear replicons, despite their divergent evolutionary origins, were found to assume a circular configuration in the viral particles, suggesting the importance of such TP–TP interactions for this type of replicon.

Interactions between the TPs of the linear replicons were initially suggested by the fact that the TP–DNA complexes are immobilized during electrophoresis (for example, Lin *et al.*, 1993) and that the genetic maps of *Streptomyces* chromosomes are circular (Hopwood, 1967; Wang *et al.*, 1999). The TP–TP interaction at the telomeres *in vivo* was recently demonstrated by

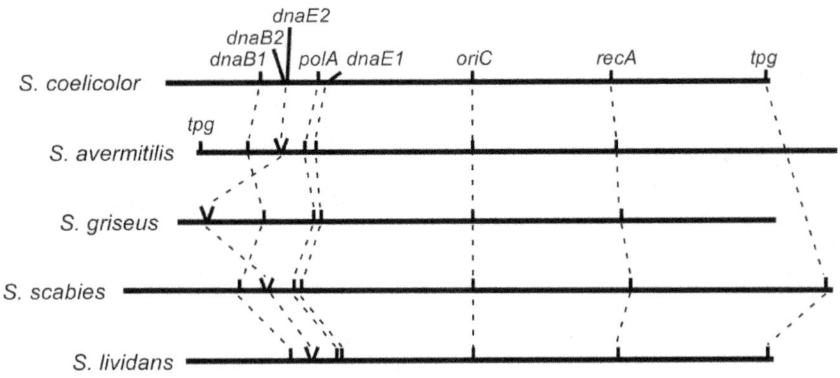

Figure 1.3 Location of selected genes involved in chromosome replication and end patching on *Streptomyces* chromosomes. The chromosomes are aligned to the central *oriC*. Synteny is preserved for the genes plotted, except for the overlapping *dnaE2* and *dnaB2* genes in *S. griseus* and *tpg* in *S. avermitilis*.

chemical cross-linking studies (Tsai *et al.*, unpublished results). The interactions may be either intramolecular or intermolecular. Intramolecular TP–TP interaction results in the formation of 'pseudocircular' replicons *in vivo* that are negatively supercoiled. This negative superhelicity in the linear DNA may be important for replication, transcription, and/or recombination. Supercoiling of adenovirus DNA has also been detected in the virus particles (Wong and Hsu, 1989).

The existence of linear *Streptomyces* replicons that possess heterologous telomeres capped by heterologous TPs raises the question whether intramolecular interaction between such heterologous TPs is still possible. At least for the Tpc (capping the SCP1 telomere) and Tpg (capping the *S. coelicolor* chromosome telomere), the answer is yes (Tsai and Chen, unpublished results). This, together with the finding of intermolecular TP–TP interactions, indicates that the telomeres of linear plasmids may be associated with those of the host chromosomes in *Streptomyces*, even when their telomeres and TPs are of different types (Fig. 1.4). Such clustering of telomeres has also been seen in eukaryotic chromosomes.

Resolution of sister chromosomes and plasmids

The interactions between the TPs at the telomeres of a linear *Streptomyces* replicon pose two potential complications during post-replication segregation: (i) resolution of pseudodimers formed by two daughter linear replicons; and (ii) decatenation of the pseudocircular daughter replicons.

Firstly, if the association between the parental TPs on a linear replicon remains after replication of the DNA, the 5′ ends of the two parental DNAs would remain linked with each other, thus creating a Möbius strip-like situation (Chen, 2007), and the two daughter DNA molecules would retain a pseudodimer structure. To resolve the pseudodimers, either the parental TPs must dissociate from each other and each find a proper new TP partner or DNA recombination must occur to resolve the structures. In the former case, a mechanism must exist that achieves the proper sorting of the old and new TPs. In the latter case, the situation is similar to the resolution of authentic dimers formed by homologous recombination between two circular replicons. For circular chromosomes, such dimer resolution is carried out by XerCD recombinase-catalysed site-specific recombination (reviewed by Barre *et al.*, 2001). In this regard, no XerCD homologues are found in *Streptomyces*. It is not clear whether a different and as yet unknown site-specific resolution system may exist in *Streptomyces*. Moreover, homologous recombination between two pseudocircular linear replicons would also generate a pseudodimer that could be resolved by either of the same two schemes proposed.

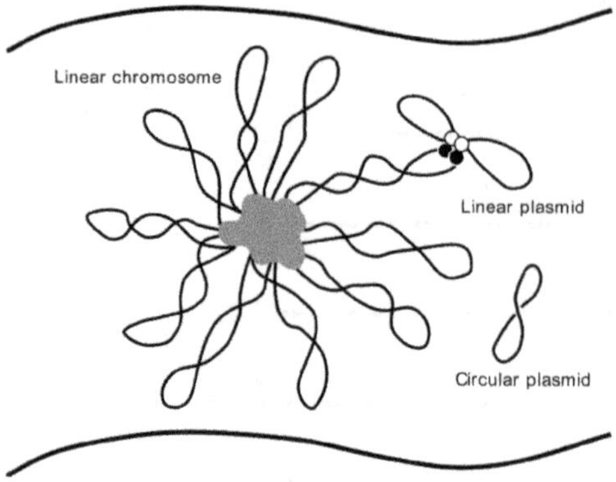

Figure 1.4 Model of association between the linear host chromosome and a linear plasmid in *Streptomyces* mycelium. The linear chromosomal DNA molecule is pictured in a 'rosette' with many superhelical loops, and the TPs (black spheres) on the two ends are associated with each other. The linear plasmid DNA molecule is also presented in superhelical form with the two ends held together by the interacting TPs (white spheres). The linear plasmid is attached to the chromosome through intermolecular interactions between their TPs. The circular plasmid, in contrast, is free.

The second potential complication would be decatenation of the two pseudocircular daughter replicons after replication, if the intramolecular interactions between the parental TPs and between the new TPs persist. This would be similar to that faced by the intertwined sister circular DNA molecules (concatemers), which are decatenated by topoisomerase Topo IV (encoded by *parC* and *parD*) (Adams *et al.*, 1992; Zechiedrich and Cozzarelli, 1995). *Streptomyces* genomes contain homologues of *parC* and *parD*. Attempts to knock out these genes in *S. coelicolor* were successful in strains with a linear chromosome, but not in strains with a circular chromosome (T.-W. Huang, unpublished results). This suggests that the concatemers formed by linear chromosomes may be untangled by a mechanism independent of Topo IV, perhaps by the dissociation and re-association of the TPs on the same replicon. More discussion on this topic may be found in Chen (2007).

Genome dynamics

Structural instability is a hallmark of *Streptomyces* genomes. The most frequently reported instability is spontaneous deletion of the terminal sequences, accompanied by fusion of the two broken arms to form a circular chromosome. The deletions range from hundreds of kb to about 1 Mb (Lin *et al.*, 1993; Redenbach *et al.*, 1993), and the frequencies of such spontaneous events observed in the laboratory are about 10^{-2} to 10^{-3}. These deletion events are detected by the loss of genetic markers (such as *argG*) located in the terminal regions, and therefore the actual deletion events probably occur at a significantly higher frequency.

Unequal crossing over is also often observed among *Streptomyces* genomes, and can result in gross structural changes (reviewed by Chen *et al.*, 2002). Recombination between opposite arms of sister chromosomes would result in recombinant chromosomes, one with two 'left' arms and another with two 'right' arms. This would lead to new sets of TIRs with increased lengths. The first reported case of such an exchange yielded the 480- and 850-kb TIRs of *S. ambofaciens* (Fischer *et al.*, 1998). Later, a similar rearrangement was found in *S. griseus* 2247 (Uchida *et al.*, 2003). Most surprisingly, it was discovered that many of the stock cultures of *S. coelicolor* A3(2) possessed 1-Mb TIR, which might also be generated by exchange between the two arms on sister chromosomes (Weaver *et al.*, 2004). Recently, the longest TIRs to date (over 2.5 Mb) generated by recombination between sister chromosomes were observed in *S. coelicolor* (Widenbrant *et al.*, 2007).

On the other hand, TIR shortening may result from recombination between heterologous chromosomes (with heterologous terminal sequences) during conjugation. Although recombinant chromosomes generally possess the same termini (either the donor or recipient), occasionally recombinant chromosomes may emerge with mixed termini with very short (< 200 bp) TIRs (Chen *et al.*, 2002; Wang *et al.*, 1999). The unusually short TIRs observed in several linear plasmids and chromosomes may reflect such recombination, such as in SLP2 (44-bp imperfect; Chen *et al.*, 1993) and the *S. avermitilis* chromosome (174-bp; Ikeda *et al.*, 2003).

The most dramatic rearrangement is probably recombination between the linear plasmid SCP1 and the host chromosomes in *S. coelicolor* 2106 (a derivative of the A3(2) strain), which produced two separate chimeric chromosomes (Yamasaki and Kinashi, 2004) with no TIR at all. Studies using artificially constructed linear plasmids showed no obvious disadvantage for a linear plasmid that possesses heterologous telomeres as long as the necessary TPs and accessory proteins are readily available (Huang *et al.*, 2007). Despite the fact that TIRs appear to have no obvious biochemical function, the duplication may serve as a template for repair of damage to one terminus (Qin and Cohen, 2002; Uchida *et al.*, 2003). In this regard, two TIRs might provide an evolutionary advantage.

The reason for the high structural instability of *Streptomyces* chromosomes is not clear. One possible cause may be the presence of relatively large numbers of transposable elements (TEs) on *Streptomyces* chromosomes (Lin and Chen, 1997), concentrated in the terminal regions (Chen *et al.*, 2002). Transposition mediated by the TEs may cause the instability. Alternatively, homologous TEs located at different places on the chromosomes may serve as sites for unequal crossing over (Huang *et al.*, 2003; Weaver *et al.*, 2004).

One would expect that circularization of *Streptomyces* chromosomes, which is basically an irreversible event, should drive any *Streptomyces* culture to contain mostly if not exclusively circular chromosomes. However, this seems not to be the case. Linearity of the chromosomes is maintained in laboratory cultures without deliberate selection. One possibility is that mutants containing circular chromosomes grow more slowly than the wild type containing linear chromosomes. While this is true for some mutants, others do not appear to suffer any obvious defect in growth or sporulation. Therefore, if circular chromosomes pose a growth disadvantage, it may be very slight, but sufficient to drive evolution.

Another possibility is that circular chromosomes may be at a disadvantage during mating. Circular chromosomes often lose the terminally located genes encoding the TP (e.g. Tpg) and telomere-associated protein (e.g. Tap), and therefore cannot maintain some linear plasmids, such as SLP2 in *S. lividans*, that depend on one or more of these genes for replication (Lin, 1998). These plasmidless hyphae would become recipients for mating with plasmid-containing hyphae. The products of such conjugation are dominated by progeny containing linear chromosomes (H.-H. Lee and H.-H. Tsai, unpublished results). The reason for this remains unclear.

Genomics of *Streptomyces*

The size of the *Streptomyces* genome

When the complete genome sequence of *S. coelicolor* was published (Bentley *et al.*, 2002), the genome was the largest known, at 8,6677,507 bp carrying 7,912 genes and 7,769 protein coding genes (http://www.ncbi.nlm.nih.gov; October 2009). It still remains one of the largest bacterial chromosomes, and the other full genome sequences from the same genus are also large (Ikeda *et al.*, 2003; Ohnishi *et al.*, 2008). Specifically, *S. avermitilis* has a genome size of 9.025,608 bp with 7,669 genes and 7,580 protein coding genes and *S. griseus* IFO13350 has a genome size of 8,545,929 with 7,224 genes and 7,136 protein coding genes (http://www.ncbi.nlm.nih.gov; October 2009). Both the above are exceeded by the provisional information on *S. scabies*, which has a genome size of 10,148,695 bp and 9,107 provisional genes (http://www.sanger.ac.uk/Projects/S_scabies/). None of the other provisional genome lengths available for other *Streptomyces* species approach that of *S. scabies*, perhaps because *S. scabies* is a plant pathogen and contains within its genome a well-studied large pathogenic island. Nonetheless, the *Streptomyces* genomes, even that of *S. scabies*, are significantly smaller than that of *Sorangium cellulosum*, a myxobacterium and a member of the delta-Proteobacteria, which has a genome size of 13,033,779 bp and a gene complement of 9,704 (Schneiker *et al.*, 2007) (http://www.ncbi.nlm.nih.gov; October 2009). Table 1.1 shows the available information on *Streptomyces* genomes, both complete and provisional, as well as the same information for some of the more closely related actinobacteria (http://www.ncbi.nlm.nih.gov; October 2009). *Streptosporangium roseum* stands out from the non-*Streptomyces* genome sequences as having a larger provisional genome size of 9,430 genes. This indicates that *Streptomyces* species do not have a monopoly on larger genomes

Table 1.1 Details of various actinobacterial genomes

Organism	CDS	Size	%GC	*tpg*	*tap*	*ttrA*
Streptomyces avermitilis	7580	9.03	70.7	+	+	+
Streptomyces coelicolor A3(2)	7769	8.67	72.2	+	+	+
Streptomyces griseus IFO13350	7136	8.55	72.2	?	?	–
Streptomyces scabies[a]	9107	10.15	71.4	+	+	+
Streptomyces albus[a]	6026	6.62	73.1	–	–	–
Streptomyces clavuligerus[a]	6448	6.73	71.1	+	+	+
Streptomyces ghanaensis[a]	7750	8.22	72.2	+	+	+
Streptomyces griseoflavus[a]	7125	7.36	71.7	+	+	–
Streptomyces hygroscopicus[a]	9498	10.47	71.2	+	+	–
Streptomyces lividans[a]	7547	8.19	72.2	+	+	+
Streptomyces pristinaespiralis[a]	5741	6.15	70.4	+	+	–
Streptomyces roseosporus[a]	7056	7.76	71.4	+	+	–
Streptomyces sp. C[a]	7672	7.92	72.6	–	–	–
Streptomyces sp. Mg1[a]	6975	7.11	71.2	+	+	+
Streptomyces sp. SPB74[a]	4933	5.01	71.9	+	+	–
Streptomyces sviceus[a]	8265	8.47	70.1	?	?	–
Streptomyces viridochromogenes[a]	7739	8.55	71.1	+	+	–
Frankia alni ACN14a	6711	7.50	72.8	–	–	–
Frankia sp. CcI3	4499	5.43	70.1	–	–	–
Frankia sp. EAN1pec	7191	8.98	71.2	–	–	–
Micromonospora sp. ATCC 39149[a]	5633	6.51	72.3	–	–	–
Rhodococcus erythropolis PR4	6030	6.52	62.3	–	–	–
Rhodococcus jostii RHA1	7211	7.80	67.0	?	?	–
Rhodococcus opacus B4	7246	7.91	67.9	–	–	–
Saccharomonospora viridis	3838	4.31	67.3	–	–	–
Saccharopolyspora erythraea	7197	8.21	71.0	–	–	–
Salinispora arenicola	4917	5.79	69.5	–	–	–
Salinispora tropica	4536	5.18	69.5	–	–	–
Streptosporangium roseum[a]	9430	10.12	70.8	–	–	–
Thermobifida fusca YX	3184	3.64	67.5	–	–	–
Thermomonospora curvata[a]	5047	5.56	71.6	–	–	–

[a]Draft sequence information; *tpg*: terminal protein gene; *tap*, telomere-associated protein gene; *ttrA*, terminal transfer protein gene; ?: possible evolutionarily divergent homologue.

in the actinobacteria and that other sporulating genera may also have large genomes.

The linearity of *Streptomyces* chromosomes

The first experimental demonstration of linearity of the chromosome in a *Streptomyces* species (*S. lividans*) was based on extensive molecular biological investigations (Lin *et al.*, 1993). The extremely large size of *Streptomyces* chromosomes makes determination of topology very difficult, even with advanced pulsed-field gel electrophoresis (PFGE) techniques. Analysing intact chromosomal DNA with PFGE, Lin *et al.* (1993)

produced evidence that the chromosomes of six *Streptomyces* species other than *S. lividans* were also linear. So far, no exceptions to this general rule of chromosome linearity have been discovered in all the *Streptomyces* chromosomes investigated, including those with sequenced genomes.

Although whole-genome sequencing is currently done with great speed and efficiency, linearity of the chromosomes cannot be confidently established without isolation of the terminal DNA fragments (usually by glass binding of the TP–DNA complex) and confirmation by detection of TP attachment (usually by retarded electrophoretic movement). These criteria were met for *S. coelicolor* (Bentley *et al.*, 2002; Huang *et al.*, 1998) and *S. avermitilis* (Omura *et al.*, 2001).

Identification of the terminal sequences of the *S. griseus* IFO13350 chromosome was more problematic, because its terminal DNA fragments could not be isolated by glass binding, but were isolated by ligation-mediated PCR. The isolated putative telomere sequence was atypical, differing from the archetypal telomere and other atypical telomere sequences.

While the presence of *tpg*, *tap*, and *ttrA* is no guarantee for linearity of the chromosome, most *Streptomyces* chromosomes contain *tpg* and *tap* homologues. This is indicative that most *Streptomyces* chromosomes possess archetypal telomeres. Up to the present, other than the *Streptomyces* genomes, the *Rhodococcus jostii* RHA1 chromosome is the only other actinobacterial chromosome that is identified as linear, and contains a pair of genes that are possibly homologous to *tpg* and *tap*. Thus, although 124 completely sequenced actinobacterial genomes are available at the time of writing, covering a wide range of genera, proven linearity is uncommon. *Streptosporangium roseum*, the actinobacterium with the largest genome other than *S. scabies*, does not seem to contain *tpg*, *tap* or *ttrA* homologues.

Intra-*Streptomyces* genome comparisons

As more complete genome sequences become available, we also need to determine the size of the 'core genome' and 'pan-genome' of *Streptomyces*. Using reciprocal BLAST on eight sequenced genomes, we estimated that the core genome of *Streptomyces*, namely those orthologues that are present in all *Streptomyces* species, to be quite small at around 2000 genes. The exact number depends on the exact criteria used for assigning orthologues. The core genome would seem to be a lot smaller than the figure of 3566 genes given by Ventura *et al.* (2007), which was based on only four sequenced genomes.

On the other hand, the pan-genome of *Streptomyces* is very large and has been increasing significantly as each new genome sequence is added, at a rate of about 2000–3000 genes per new genome sequenced. These two trends support the idea that the *Streptomyces* genome is highly flexible around a core of genes that make a *Streptomyces* genetically a *Streptomyces* and that it is this flexibility that has allowed *Streptomyces* species to become as ubiquitous in nature as they are. This also explains the diversity of secondary metabolic products produced by *Streptomyces* species, but does not provide a reason for why production of these secondary metabolic products, including most of the major antibiotics, is largely confined to *Streptomyces* and a number of closely related genera.

The available *Streptomyces* genome sequences, including a number of draught genomes, were compared using MAUVE (http://gel.ahabs.wisc.edu.mauve) and the results are shown in Fig. 1.5. It is well known that *S. lividans* is very close phylogenetically to *S. coelicolor* A3(2) and was the first *Streptomyces* to be identified as having a linear genome (Chen *et al.*, 1993); the genome similarity is borne out by the results in Fig. 1.5. However, not withstanding this similarity, breakpoints between the left terminal region and the centre of the genome and between the right terminal region and the centre of the genome are visible on the MAUVE plot. This agrees with the hypothesis derived from the actinobacterial comparisons and expounded by Bentley *et al.* (2004) and Ventura *et al.* (2007), whereby the terminal regions act as repositories of novel species-specific genes, while the central core region of the chromosome is conserved. The terminal regions of the other species analysed show only very limited gene conservation, too. The left region between the left terminus and the core genome, termed the 'left *Streptomyces*-specific region' in Fig. 1.5,

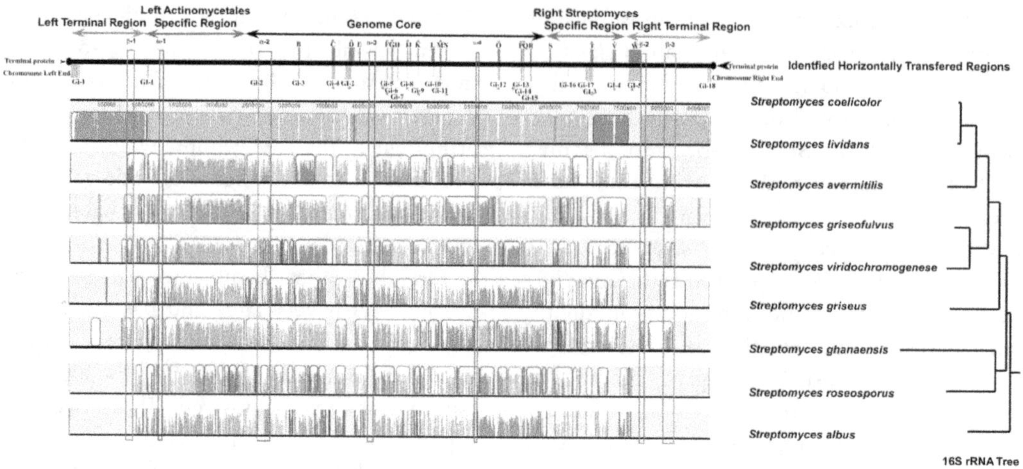

Figure 1.5 Alignment and similarity analysis of nine *Streptomyces* chromosomes. The alignment (left) was constructed with the MAUVE genome alignment program using the *S. coelicolor* chromosome as the standard. The boxes with whiskers represent areas of homology between that species and *S. coelicolor* detected by MAUVE. The 16S rRNA phylogenetic tree (right) was constructed initially using 36 *Streptomyces* chromosomes and 14 other actinobacterial chromosomes, and the minimum tree for the nine *Streptomyces* chromosomes analysed was extracted from it and used to order the genome comparisons.

shows strong conservation and synteny across the *Streptomyces* genomes, which contrasts with the same region across the actinobacterial genomes (see later). This confirms that while some genes in this region may be present in some actinobacteria, they are not found in the majority and they are not in general syntenous. Nevertheless, the genes in this region are highly represented in most *Streptomyces* species. The 'right *Streptomyces* specific region' shows both greater gene divergence and less gene synteny when compared across *Streptomyces,* which supports regionalization of the genome in this genus. This regionalization of the *Streptomyces* chromosome across all the complete and draught genomes available, together with the microarray data from a further 18 *Streptomyces* species (Kirby, unpublished results), expands, consolidates and confirms the ideas on chromosomal topology put forward by Bentley *et al.* (2004) and Ventura *et al.* (2007) There is one exception and that is *Streptomyces rimosus*; in this case the microarray results do not fit the pattern obtained with the other *Streptomyces* species (Kirby *et al.*, 2008). This suggests a novel or highly reorganized topology for this species, which, in terms of phylogeny, is found at the edge of the *Streptomyces* clade.

What is also clear from Fig. 1.5 is that the linear genomes of *Streptomyces* species are not symmetric in terms of genes conservation and synteny. This extends beyond *Streptomyces* to other actinobacteria, as will be described later, with the core genome being conserved in terms of both sequence and synteny across many actinobacteria (see Fig. 1.6). In *Streptomyces,* the left *Streptomyces*-specific region is conserved perhaps to an even higher degree than the core genome, suggesting that it may contain many of the genes that make a *Streptomyces* into a *Streptomyces* rather than another actinobacterium. The right *Streptomyces*-specific region is more divergent than either the core region or the left *Streptomyces*-specific region, but does contain a number of groups of genes that are specifically conserved across the genus. This region also seems to vary in size between *Streptomyces* species more than the core or left *Streptomyces*-specific regions. The left terminal region is highly divergent, with only a few genes and gene groups found in common across the genus. The right terminal region is similar, with perhaps even fewer genes and gene groups in common. Both of these regions also vary in size significantly between species.

However, it could be argued that the boundary for the left terminal region should be moved a little to the left to about 0.8 kb, which would move the conserved gene group β1 into the left *Streptomyces*-specific region. A similar slight move to the right for the right terminal region boundary would allow conserved group β2 to be excluded. However, conserved group β3 remains problematic. This group may represent a gene cluster than is important to many but not all *Streptomyces* species and seems to be available for interspecific exchange in the right terminal region. In this context, *tpg* and *tap* are found within the right terminal region of *S. coelicolor* and similarly near the ends of other *Streptomyces* genomes. As these genes are essential for terminal replication of the linear genome, such a position is not inexplicable. *ttrA*, a gene that seems to be involved in conjugal transfer, lies very close to both ends of *Streptomyces* chromosomes (SCO0002 and SCO7768 in *S. coelicolor*), hinting at the possibility that the transfer of the linear chromosomes (and the linear plasmids) starts from the telomere(s) (Huang *et al.*, 2003).

If the above information is correlated with the fact that, as stated earlier, the core genome of *Streptomyces* is 2000 genes or fewer, this has two implications. Firstly, that the core region, as identified in Fig. 1.5, but excluding any horizontally transferred regions, seems to contain most of the housekeeping genes, with some additional genes probably being present in the left *Streptomyces*-specific region and a few, such as those that encode linear genome function proteins like *tap*, *tpg* and *ttrA*, being elsewhere. More analysis using the new genome sequences is needed to exactly pinpoint these core genes. Secondly, the relatively small core genome relative to the pan genome implies that *Streptomyces* species must obtain the additional genes from somewhere, with the most likely source being horizontal gene transfer. The left and right terminal regions together with the right *Streptomyces*-specific region, which is less well conserved than the left *Streptomyces*-specific region, are the likely homes of most of these genes. As will be seen below, there are specifically identified horizontally transferred regions in all *Streptomyces* species analysed, but these are not large enough in total to contain all of the novel genes in each species. This supports the hypothesis that genetic exchange at the ends of the linear *Streptomyces* chromosome is important evolutionarily and part of the development of new species. The end regions of the *Streptomyces* chromosome generally lack essential core genes and are topographically distinct, which will allow evolutionary flexibility without a great genetic load. Even if the genes for linear genome replication and chromosome transfer are lost by a circularization event during evolution, they can be regained easily for the linear plasmids that abound in the actinobacteria.

Horizontally transferred regions in the *Streptomyces* genomes

Bentley *et al.* (2002) identified a large number of potentially horizontally transferred regions (HTRs) in the *S. coelicolor* A3(2) genome using bioinformatics. DNA/DNA microarray analysis of the *S. lividans* genome compared with the *S. coelicolor* genome (Jayapal *et al.*, 2007) and a number of *Streptomyces* genomes compared with the *S. coelicolor* genome (Hsiao and Kirby, 2008) confirmed that these potential regions were absent from the tested *Streptomyces* genomes, strongly supporting their status as HTRs. The presence of transposases and genes associated with horizontal transfer in these regions as well as a lower percentage of G+C provides further supportive evidence for these new HTRs. This is the same criteria that were used by Bentley *et al.* (2002). Furthermore, these studies were able to pinpoint further potential HTRs within the genome of *S. coelicolor*. An example of this type of analysis is shown in Fig. 1.7. These regions are also identifiable in Fig. 1.5, and the MAUVE genome sequence comparison supports almost all of these regions as HTRs based on their absence from most of the *Streptomyces* genomes analysed. Interestingly, four new potential HTRs can be pinpointed using the MAUVE results, namely α1 to α4. All four of these regions are found in *S. lividans* and thus must have been present before the evolutionary split between *S. coelicolor* and *S. lividans*. Furthermore, one (α2) shows heterogeneity in terms of presence/absence across the genomes analysed by MAUVE. This suggests that some of the *S. coelicolor* HTRs are also involved in core

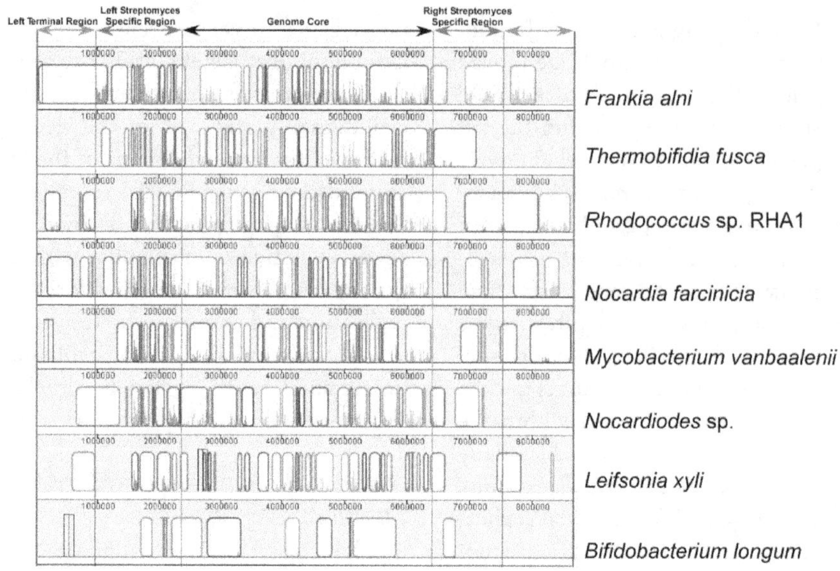

Figure 1.6 Alignment and similarity comparison between the *S. coelicolor* A3(2) chromosome and eight other actinobacterial chromosomes. The alignment was constructed using MAUVE with the *S. coelicolor* chromosome as the standard. The boxes with whiskers represent areas of homology between that species and *S. coelicolor* detected by MAUVE.

genome heterogeneity in *Streptomyces* rather than just novel gene acquisitions by *S. coelicolor*.

The identification of 29 potential HTRs in the *S. coelicolor* genome raises the question of the frequencies of HTRs in the other genomes, specifically the complete published genomes of *S. avermitilis* and *S. griseus* IFO13350. When a similar analysis was carried out for these two species, 42 and 49 potential HTRs were identified. These are also shown in Fig. 1.6. The number of potential HTRs in these two species is somewhat higher than those identified in *S. coelicolor*, which is quite possible, but further in-depth analysis of these regions across all available *Streptomyces* genomes is needed to confirm their status as HTRs.

Gene groups with high representation in *Streptomyces* genomes

Notwithstanding the postulated relatively small core genome of *Streptomyces* as evidenced by recent genome sequencing data, the presence of the highly syntenous core region and left *Streptomyces*-specific region as well as the somewhat less syntenous right *Streptomyces*-specific region implies that there are some groups of genes that may be highly represented in the genus *Streptomyces*. Examples include two-component signal transduction systems, multidrug efflux systems and others.

The two-component signal transduction systems consist of a membrane-bound sensor kinase (SK) and a cytosolic response regulator (RR). In total, (Hutchings *et al.*, 2004) identified 84 SK genes and 88 RR genes in the *S. coelicolor* A3(2) genome; the value for (Hsiao and Kirby, 2009) is slightly higher at 97 SKs and 95 RRs, together with two fused SK/RR genes based on a more inclusive definition of the two genes. This makes up about 2% of the *S. coelicolor* genome, making this group of genes highly significant. A comparative study by (Wei *et al.*, 2007) identified 67 SKs and 68 RRs in the *S. avermitilis* genome. This compares with 35 SKs and 35 RRs in *Bacillus subtilis*. Usually an SK gene and an RR gene form a cognate pair, but there are significant numbers of orphan SK and RR genes in the *S. coelicolor*, *S. avermitilis* and *S. griseus* genomes. In total there are 56 cognate pairs of SK/RR genes next to each other in both *S. coelicolor* and *S. avermitilis* and all of these have a homologue in one or more of *S. griseus, Saccharopolyspora erythrea, Frankia alni, Nocardia farcinica, Thermobifida fusca* and *Salinispora tropica*. In fact only eight of the pairs are absent from *S. griseus*. This supports the idea that there is a representative core of two-component system gene pairs that are evolutionarily conserved in *Streptomyces* and closely related genera. Furthermore, ten orphan SKs and 11 orphan RRs are also present in these other actinobacteria, with only one SK and three RRs without an *S. griseus* equivalent. In this context, it can be postulated that these core genes are needed to respond to the general environmental changes encountered by *Streptomyces* species, while the diverse remainder, which vary from genome to genome, reflect adaption to particular environmental needs

Another group of genes that are found in high number in the genomes of *Streptomyces* are those encoding the so-called 'multidrug resistance' (MDR) efflux pumps. In total, there are 147 in the *S. coelicolor* A3(2) genome, and 64 of these are conserved in one or more of the other available *Streptomyces* genomes (Lee *et al.*, 2007). Interestingly, *Streptomyces* species in general are intrinsically sensitive to any drugs tested, because these efflux pump genes are repressed and not induced by the drugs. This is based on the fact that mutants were isolated in which the repressor genes were inactivated. It was therefore proposed that the physiological role of these efflux systems is not for removal of extrinsic drugs, but rather the export of endogenous waste chemicals (Lee *et al.*, 2007). Under this premise, about half of the efflux pump genes vary from species to species, suggesting that the metabolic diversity of the *Streptomyces* genome (including the secondary metabolites) results in by-products that may be deleterious to the cell. This problem is solved by these export genes, which help to minimize the toxic effects of these chemicals.

There are many other groups of genes that are over-represented in the *Streptomyces* genome. For example, proteins that are exported by the TAT system are much more common in *Streptomyces* genomes (Widdick *et al.*, 2006; see Chapter 4). Oxygenases are also present in high numbers in all *Streptomyces* genomes. The reason for the former is not clear, but may simply represent an evolutionary choice by the *Streptomyces* clade at an early stage in the split from other actinobacteria. The latter may be because, in the soil, *Streptomyces* species need to detoxify many phenolic compounds and these enzymes are the best way to do this. Other highly represented gene classes remain to be explored.

The *Streptomyces* genome compared with other actinobacterial genomes

Given that chromosomal circularity is the norm in the other actinobacteria, suggesting that linearity was a late-developing trait, Fig. 1.6 shows a comparison of the genomes of *S. coelicolor* A3(2) against a series of actinobacteria. Bentley *et al.* (2002), using dot-plot comparison, showed significant syntenous similarity towards the centre of the *S. coelicolor* genome with two divergent actinobacteria, *Mycobacterium tuberculosis* and *Corynebacterium diptheriae*. It was postulated that the core of *S. coelicolor* generally retained a similar gene order and complement, with only a few limited inversion events, compared with the actinobacterium ancestral to all three species. This

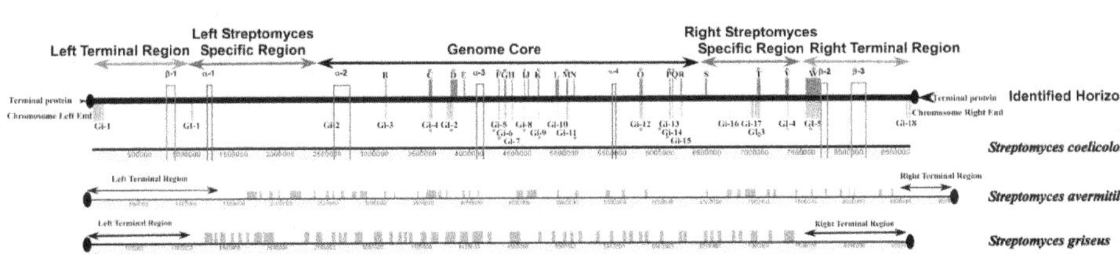

Figure 1.7 Potentially horizontally transferred regions (HTRs) identified in *S. coelicolor* A3(2), *S. avermitilis* and *S. griseus*. The HTRs in *S. avermitilis* and *S. griseus* were identified by multiple genome alignment.

hypothesis would suggest that other actinobacterial genomes should also have retained this core set of genes. Fig. 1.6 shows a different approach to comparing genomes to that utilized by Bentley *et al.* (2002). The program MAUVE was used to compare various actinobacterial genomes with that of *S. coelicolor* and these comparisons were then aligned against each other. This analysis strongly supports the core genome hypothesis with the good gene conservation in all the tested actinobacteria except for *Bifidobacterium*. Even *Leifsonia xyli*, an organism with a minimized genome because of its lifestyle, still contains most of the core genes, albeit considerably reorganized.

From Fig. 1.6, it is also quite clear that the genes in the left and right terminal regions of the *S. coelicolor* genome are not usually present in the genomes of other actinobacteria, with some minor exceptions. To the left of the *S. coelicolor* core region in Fig. 1.6, two regions can be observed. The outermost is the so-called 'left terminal region', where very few genes are conserved between *S. coelicolor* and any of the other actinobacteria. Between this region and the 'genome core' is a region with a significant number of genes in groups that are conserved and in many cases in a syntenous manner. The equivalent region on the right of the *S. coelicolor* genome core is harder to separate into two equivalent regions using the results in Fig. 1.6. Generally, there are fewer conserved genes and the whole of this region resembles the 'left terminal region' in appearance. However, when compared with the similar analysis of *Streptomyces* presented in Fig. 1.5, there is very obviously greater diversity outside the core region, but with the left *Streptomyces*-specific region still retaining a slightly higher level of gene conservation than the right *Streptomyces*-specific region and the two terminal regions.

We can, at this point, reach a number of conclusions. Firstly, the genome core of *S. coelicolor* has a similar gene content to and significant synteny with the same region of other actinobacteria. Secondly, the left terminal region and the complete region to the right of the core contain many genes that are absent from other actinobacteria. Thirdly, the stretch of DNA between the left terminal region and the core genome of *S. coelicolor* contains genes that are found in a limited number of other actinobacteria; however, this conservation is not consistent and not necessarily syntenous across the genomes

Evolution of *Streptomyces* genomes

The linearity of *Streptomyces* chromosomes poses an interesting evolutionary question regarding their origin. That they arise from integration of a linear plasmid into a circular progenitor chromosomes has been proposed by Volff and Altenbuchner (2000) and Chen *et al.* (2002). Several lines of evidence that support this model are derived from the genomic sequence of *S. coelicolor* A3(2) (Chen *et al.*, 2002), such as (i) the uneven lengths, GC skews, G+C contents, and mobile element contents of the two arms; and (ii) the terminal locations of genes specific for end patching (e.g. *tap*, *tpg*) and conjugal transfer (e.g. *ttrA*) of linear replicons. The newly sequenced *Streptomyces* chromosomes also exhibit these features, further strengthening the evidence supporting this hypothesis.

The uneven length of the left and right arms of the sequenced *S. coelicolor* chromosome has been attributed to an 'off-centre' integration of a linear plasmid. Interestingly, the other sequenced *Streptomyces* chromosomes also exhibit a longer right arm and a shorter left arm. This suggests that the hypothetical linearization of the *Streptomyces* chromosome perhaps occurred only once or perhaps only a very few times during evolution.

Linear plasmids may have existed in *Streptomyces* before its separation from the other actinobacteria, because they are also present in many actinobacteria with a circular chromosome. Alternatively, they may have been horizontally transferred into these actinobacteria from *Streptomyces*. In any case, it is noteworthy that the linear plasmids in the other bacteria generally have significantly less conserved and less complex telomere sequences than those in *Streptomyces*.

Conclusions

Overall, the chromosomes of members of the genus *Streptomyces* represent a unique structure in both physical terms, namely linearity and physical organization, and genetic/biochemical terms, namely gene complement, unique replication/

conjugal transfer mechanisms and novel metabolic pathways. The position of the genus as by far the major producer of different types of antibiotics compared with other genera, both prokaryotic and eukaryotic, may reflect these unique physical and genetic aspects of the chromosome structure. These include the relatively large size of the genomes, the highly dynamic nature of the chromosomes, the topological segregation of core genes from both *Streptomyces*-specific and species-specific genes, the high number and diversity of important gene classes and an average gene content of their genome at the high extreme of the prokaryotes. These have probably evolved to allow successful adaptation to diverse terrestrial and other environments. This would involve competition with other soil microbes, including other actinobacteria, and co-existence with others. Their novel linear genomes may be the source of their success.

Acknowledgements

We thank Dr Zhongjun Qin for communication of unpublished results, Dr Hsiu-Hui Tsai for providing Fig. 1.3 and critical reading of the manuscript, and Dr David Hopwood for critical reading of the manuscript.

References

Adams, D.E., Shekhtman, E.M., Zechiedrich, E.L., Schmid, M.B., and Cozzarelli, N.R. (1992). The role of topoisomerase IV in partitioning bacterial replicons and the structure of catenated intermediates in DNA replication. Cell *71*, 277–288.

Aguilar, A., and Hopwood, D.A. (1982). Determination of methylenomycin A synthesis by the pSV1 plasmid from *Streptomyces violaceus-ruber* SANK 95570. J. Gen. Microbiol. *128*, 1893–1901.

Arnberg, A., and Arwert, F. (1976). DNA-protein complex in circular DNA from Bacillus bacteriophage GA-1. J. Virol. *18*, 783–784.

Bao, K., and Cohen, S.N. (2001). Terminal proteins essential for the replication of linear plasmids and chromosomes in *Streptomyces*. Genes Dev. *15*, 1518–1527.

Bao, K., and Cohen, S.N. (2003). Recruitment of terminal protein to the ends of *Streptomyces* linear plasmids and chromosomes by a novel telomere-binding protein essential for linear DNA replication. Genes Dev. *17*, 774–785.

Bao, K., and Cohen, S.N. (2004). Reverse transcriptase activity innate to DNA polymerase I and DNA topoisomerase I proteins of *Streptomyces* telomere complex. Proc. Natl. Acad. Sci. U.S.A. *101*, 14361–14366.

Barre, F.X., Soballe, B., Michel, B., Aroyo, M., Robertson, M., and Sherratt, D. (2001). Circles: the replication-recombination-chromosome segregation connection. Proc. Natl. Acad. Sci. U.S.A. *98*, 8189–8195.

Bentley, S.D., Brown, S., Murphy, L.D., Harris, D.E., Quail, M.A., Parkhill, J., Barrell, B.G., McCormick, J.R., Santamaria, R.I., Losick, R., *et al.* (2004). SCP1, a 356, 023 base pair linear plasmid adapted to the ecology and developmental biology of its host, *Streptomyces coelicolor* A3(2). Mol. Microbiol. *51*, 1615–1628.

Bentley, S.D., Chater, K.F., Cerdeño-Tárraga, A.-M., Challis, G.L., Thomson, N.R., James, K.D., Harris, D.E., Quail, M.A., Kieser, H., Harper, D., *et al.* (2002). Complete genome sequence of the model actinomycete *Streptomyces coelicolor* A3(2). Nature *417*, 141–147.

Bentley, S.D., and Parkhill, J. (2004). Comparative genomic structure of prokaryotes. Annu. Rev. Genet. *38*, 771–792.

Chaconas, G., and Chen, C.W. (2005). Linear chromosomes in bacteria: no longer going around in circles. In The Bacterial Chromosome, Higgins, N.P., ed. (Washington, D.C., American Society for Microbiology), pp. 525–539.

Chang, P.C., and Cohen, S.N. (1994). Bidirectional replication from an internal origin in a linear *Streptomyces* plasmid. Science *265*, 952–954.

Chang, P.C., Kim, E.S., and Cohen, S.N. (1996). *Streptomyces* linear plasmids that contain a phage-like, centrally located, replication origin. Mol. Microbiol. *22*, 789–800.

Chen, C.W. (1996). Complications and implications of linear bacterial chromosomes. Trends Genet. *12*, 192–196.

Chen, C.W. (2007). *Streptomyces* linear plasmids: replication and telomeres. In Microbial Linear Plasmids, Friedhelm, M., and Roland, K., eds (Heidelberg, Springer).

Chen, C.W., Huang, C.-H., Lee, H.-H., Tsai, H.-H., and Kirby, R. (2002). Once the circle has been broken: dynamics and evolution of *Streptomyces* chromosomes. Trends Genet. *18*, 522–529.

Chen, C.W., Yu, T.-W., Lin, Y.S., Kieser, H.M., and Hopwood, D.A. (1993). The conjugative plasmid SLP2 of *Streptomyces lividans* is a 50 kb linear molecule. Mol. Microbiol. 7, 925- 932.

Chen, Y.-T. (2000). Linear chromosomes and linear plasmids in Actinomycetes. In Institute of Genetics (Taipei, National Yang-Ming University).

Chou, S.-H., Zhu, L., and Reid, B.R. (1997). Sheared purine x purine pairing in biology. J. Mol. Biol. *267*, 1055–1067.

Choulet, F., Aigle, B., Gallois, A., Mangenot, S., Gerbaud, C., Truong, C., Francou, F.X., Fourrier, C., Guerineau, M., Decaris, B., *et al.* (2006). Evolution of the terminal regions of the *Streptomyces* linear chromosome. Mol. Biol. Evol. *23*, 2361–2369.

Fischer, G., Wenner, T., Decaris, B., and Leblond, P. (1998). Chromosomal arm replacement generates a high level of intraspecific polymorphism in the terminal inverted repeats of the linear chromosomal DNA of *Streptomyces ambofaciens*. Proc. Natl. Acad. Sci. U.S.A. *95*, 14296–14301.

Flett, F., de Mello Jungmann-Campello, D., Mersinias, V., Koh, S.L., Godden, R., and Smith, C.P. (1999). A 'gram-negative-type' DNA polymerase III is essential for replication of the linear chromosome of *Streptomyces coelicolor* A3(2). Mol. Microbiol. *31*, 949–958.

Flett, F., Platt, J., and Oliver, S.G. (1992). Isolation and characterization of temperature-sensitive mutants of *Streptomyces coelicolor* A3(2) blocked in macromolecular synthesis. J. Gen. Microbiol. *138*, 579–585.

Goshi, K., Uchida, T., Lezhava, A., Yamasaki, M., Hiratsu, K., Shinkawa, H., and Kinashi, H. (2002). Cloning and analysis of the telomere and terminal inverted repeat of the linear chromosome of *Streptomyces griseus*. J. Bacteriol. *184*, 3411–3415.

Hiratsu, K., Mochizuki, S., and Kinashi, H. (2000). Cloning and analysis of the replication origin and the telomeres of the large linear plasmid pSLA2-L in *Streptomyces rochei*. Mol. Gen. Genet. *263*, 1015–1021.

Hirochika, H., Nakamura, K., and Sakaguchi, K. (1985). A linear DNA plasmid from *Streptomyces rochei* with an inverted repetition of 614 base pairs. EMBO J. *3*, 761–766.

Hopwood, D.A. (1967). Genetic analysis and genome structure in *Streptomyces coelicolor*. Bacteriol. Rev. *31*, 373–403.

Hopwood, D.A., and Kieser, T. (1993). Conjugative plasmids of *Streptomyces*. In Bacterial Conjugation, Clewell, D.B., ed. (New York, Plenum Press), pp. 293–311.

Hsiao, N.H., and Kirby, R. (2008). Comparative genomics of *Streptomyces avermitilis*, *Streptomyces cattleya*, *Streptomyces maritimus* and *Kitasatospora aureofaciens* using a *Streptomyces coelicolor* microarray system. Antonie Van Leeuwenhoek *93*, 1–25.

Hsiao, N.H., and Kirby, R. (2009). Two-component signal transduction systems in *Streptomyces* and related organisms studied using DNA comparative microarray analysis. Antonie Van Leeuwenhoek *95*, 189–206.

Huang, C.-H., Lin, Y.-S., Yang, Y.-L., Huang, S.-w., and Chen, C.W. (1998). The telomeres of *Streptomyces* chromosomes contain conserved palindromic sequences with potential to form complex secondary structures. Mol. Microbiol. *28*, 905–926.

Huang, C.-H., Tsai, H.-H., Tsay, Y.-G., Chien, Y.-N., Wang, S.-L., Cheng, M.-Y., Ke, C.-H., and Chen, C.W. (2007). The telomere system of the *Streptomyces* linear plasmid SCP1 represents a novel class. Mol. Microbiol. *63*, 1710–1718.

Huang, C.H., Chen, C.Y., Tsai, H.H., Chen, C., Lin, Y.S., and Chen, C.W. (2003). Linear plasmid SLP2 of *Streptomyces lividans* is a composite replicon. Mol. Microbiol. *47*, 1563–1576.

Huang, T.W., and Chen, C.W. (2008). DNA polymerase I is not required for replication of linear chromosomes in *Streptomyces*. J. Bacteriol. *190*, 755–758.

Hutchings, M.I., Hoskisson, P.A., Chandra, G., and Buttner, M.J. (2004). Sensing and responding to diverse extracellular signals? Analysis of the sensor kinases and response regulators of *Streptomyces coelicolor* A3(2). Microbiology *150*, 2795–2806.

Ikeda, H., Ishikawa, J., Hanamoto, A., Shinose, M., Kikuchi, H., Shiba, T., Sakaki, Y., Hattori, M., and Omura, S. (2003). Complete genome sequence and comparative analysis of the industrial microorganism *Streptomyces avermitilis*. Nat. Biotechnol. *21*, 526–531.

Jayapal, K.P., Lian, W., Glod, F., Sherman, D.H., and Hu, W.S. (2007). Comparative genomic hybridizations reveal absence of large *Streptomyces coelicolor* genomic islands in Streptomyces lividans. BMC Genomics *8*, 229.

Keegstra, W., Van Wielink, P., and Sussenbach, J. (1977). The visualization of a circular DNA-protein complex from adenovirions. Virology *76*, 444–447.

Kinashi, H., Shimaji-Murayama, M., and Hanafusa, T. (1991). Nucleotide sequence analysis of the unusually long terminal inverted repeats of a giant linear plasmid, SCP1. Plasmid *26*, 123–130.

Kirby, R., Gan, T.K., Hunter, I., Herron, P., and Tilley, E. (2008). The genome of *Streptomyces rimosus* subsp. rimosus shows a novel structure compared to other *Streptomyces* using DNA/DNA microarray analysis. Antonie Van Leeuwenhoek *94*, 173–186.

Kirby, R., Wright, L.F., and Hopwood, D.A. (1975). Plasmid-determined antibiotic synthesis and resistance in *Streptomyces coelicolor*. Nature *254*, 265–267.

Lee, L.-F., Chen, Y.-J., Kirby, R., Chen, C., and Chen, C.W. (2007). A multidrug efflux system is involved in colony growth in *Streptomyces lividans*. Microbiology *153*, 924–934.

Lin, Y.-L. (1998). Involvement of the terminal sequence of the *Streptomyces* chromosome in the maintenance of linear plasmids. PhD thesis (Taipei, National Yang-Ming University), p. 69.

Lin, Y.-S., and Chen, C.W. (1997). Instability of artificially circularized chromosomes of *Streptomyces lividans*. Mol. Microbiol. *26*, 709–719.

Lin, Y.-S., Kieser, H.M., Hopwood, D.A., and Chen, C.W. (1993). The chromosomal DNA of *Streptomyces lividans* 66 is linear. Mol. Microbiol. *10*, 923–933.

McLeod, M.P., Warren, R.L., Hsiao, W.W.L., Araki, N., Myhre, M., Fernandes, C., Miyazawa, D., Wong, W., Lillquist, A.L., Wang, D., *et al.* (2006). The complete genome of *Rhodococcus* sp. RHA1 provides insights into a catabolic powerhouse. PNAS *103*, 15582–15587.

Mochizuki, S., Hiratsu, K., Akita, A., and Kinashi, H. (2001). Complete nucleotide sequence of the linear plasmid pSLA2-L revealed unusual gene organizations for secondary metabolism. Paper presented at: Twelfth International Symposium on the Biology of Actinomycetes (Vancouver, Canada).

Ohnishi, Y., Ishikawa, J., Hara, H., Suzuki, H., Ikenoya, M., Ikeda, H., Yamashita, A., Hattori, M., and Horinouchi, S. (2008). Genome sequence of the streptomycin-producing microorganism *Streptomyces griseus* IFO 13350. J. Bacteriol. *190*, 4050–4060.

Omura, S., Ikeda, H., Ishikawa, J., Hanamoto, A., Takahashi, C., Shinose, M., Takahashi, Y., Horikawa, H., Nakazawa, H., Osonoe, T., *et al.* (2001). Genome sequence of an industrial microorganism *Streptomyces avermitilis*: deducing the ability of producing secondary metabolites. Proc. Natl. Acad. Sci. U.S.A. *98*, 12215–12220.

Ortin, J., Vinuela, E., Salas, M., and Vasquez, C. (1971). DNA-protein complex in circular DNA from phage φ29. Nature – New Biology *234*, 275–277.

Possoz, C., Ribard, C., Gagnat, J., Pernodet, J.L., and Guerineau, M. (2001). The integrative element pSAM2 from *Streptomyces*: kinetics and mode of conjugal transfer. Mol. Microbiol. *42*, 159–166.

Qin, Z., and Cohen, S.N. (1998). Replication at the telomeres of the *Streptomyces* linear plasmid pSLA2. Mol. Microbiol. *28*, 893–904.

Qin, Z., and Cohen, S.N. (2002). Survival mechanisms for *Streptomyces* linear replicons after telomere damage. Mol. Microbiol. *45*, 785–794.

Redenbach, M., Bibb, M., Gust, B., Seitz, B., and Spychaj, A. (1999). The linear plasmid SCP1 of *Streptomyces coelicolor* A3(2) possesses a centrally located replication origin and shows significant homology to the transposon Tn*4811*. Plasmid *42*, 174–185.

Redenbach, M., Flett, F., Piendl, W., Glocker, I., Rauland, U., Wafzig, O., Leblond, P., and Cullum, J. (1993). The *Streptomyces lividans* 66 chromosome contains a 1 Mb deletogenic region flanked by two amplifiable regions. Mol. Gen. Genet. *241*, 255–262.

Redenbach, M., Scheel, J., and Schmidt, U. (2000). Chromosome topology and genome size of selected actinomycetes species. Antonie Van Leeuwenhoek *78*, 227–235.

Reeves, A.R., Post, D.A., and Vanden Boom, T.J. (1998). Physical-genetic map of the erythromycin-producing organism *Saccharopolyspora erythraea*. Microbiology *144*, 2151–2159.

Robinson, A.J., Younghusband, H.B., and Bellett, A.J. (1973). A circular DNA-protein complex from adenoviruses. Virology *56*, 54–69.

Sakaguchi, K. (1990). Invertrons, a class of structurally and functionally related genetic elements that include linear plasmids, transposable elements, and genomes of adeno-type viruses. Microbiol. Rev. *54*, 66–74.

Schneiker, S., Perlova, O., Kaiser, O., Gerth, K., Alici, A., Altmeyer, M.O., Bartels, D., Bekel, T., Beyer, S., Bode, E., *et al.* (2007). Complete genome sequence of the myxobacterium Sorangium cellulosum. Nat. Biotechnol. *25*, 1281–1289.

Shiffman, D., and Cohen, S.N. (1992). Reconstruction of a *Streptomyces* linear plasmid replicon from separately cloned DNA fragment: Existence of a cryptic origin of circular replication within the linear plasmid. Proc. Natl. Acad. Sci. U.S.A. *89*, 6129–6133.

Suzuki, H., Marushima, K., Ohnishi, Y., and Horinouchi, S. (2008). A novel pair of terminal protein and telomere-associated protein for replication of the linear chromosome of *Streptomyces griseus* IFO13350. Biosci. Biotechnol. Biochem. *72*, 2973–2980.

Tsai, H.H., Huang, C.H., Lin, A.M., and Chen, C.W. (2008). Terminal proteins of *Streptomyces* chromosome can target DNA into eukaryotic nuclei. Nucleic Acids Res. *36*, e62.

Uchida, T., Miyawaki, M., and Kinashi, H. (2003). Chromosomal arm replacement in *Streptomyces griseus*. J. Bacteriol. *185*, 1120–1124.

Ventura, M., Canchaya, C., Tauch, A., Chandra, G., Fitzgerald, G.F., Chater, K.F., and van Sinderen, D. (2007). Genomics of *Actinobacteria*: tracing the evolutionary history of an ancient phylum. Microbiol. Mol. Biol. Rev. *71*, 495–548.

Volff, J.N., and Altenbuchner, J. (2000). A new beginning with new ends: linearisation of circular chromosomes during bacterial evolution. FEMS Microbiol. Lett. *186*, 143–150.

Wang, S.-J., Chang, H.-M., Lin, Y.-S., Huang, C.-H., and Chen, C.W. (1999). *Streptomyces* genomes: circular genetic maps from the linear chromosomes. Microbiology *145*, 2209–2220.

Weaver, D., Karoonuthaisiri, N., Tsai, H.H., Huang, C.H., Ho, M.L., Gai, S., Patel, K.G., Huang, J., Cohen, S.N., Hopwood, D.A., *et al.* (2004). Genome plasticity in *Streptomyces*: Identification of 1 Mb TIRs in the *S. coelicolor* A3(2) chromosome. Mol. Microbiol. *51*, 1530–1550.

Wei, W., Wang, W., Cao, Z., Yu, H., Wang, X., Zhao, J., Tan, H., Xu, H., Jiang, W., and Li, Y. (2007). Comparative analysis of two-component signal transduction system in two streptomycete genomes. Acta Biochim. Biophys. Sin. (Shanghai) *39*, 317–325.

Widdick, D.A., Dilks, K., Chandra, G., Bottrill, A., Naldrett, M., Pohlschroder, M., and Palmer, T. (2006). The twin-arginine translocation pathway is a major route of protein export in *Streptomyces coelicolor*. Proc. Natl. Acad. Sci. U.S.A. *103*, 17927–17932.

Widenbrant, E.M., Tsai, H.H., Chen, C.W., and Kao, C.M. (2007). *Streptomyces coelicolor* undergoes spontaneous chromosomal end replacement. J. Bacteriol. *189*, 9117–9121.

Wong, M., and Hsu, M. (1989). Linear adenovirus DNA is organized into supercoiled domains in virus particles. Nucleic Acids Res. *17*, 3535–3550.

Xu, M., Zhu, Y., Zhang, R., Shen, M., Jiang, W., Zhao, G., and Qin, Z. (2006). Characterization of the genetic components of *Streptomyces lividans* linear plasmid SLP2 for replication in circular and linear modes. J. Bacteriol. *188*, 6851–6857.

Yamasaki, M., and Kinashi, H. (2001). *Streptomyces coelicolor* 2106 contains two chimeric chromosomes formed by single crossover of SCP1 and the wild-type chromosome. Paper presented at: Twelfth International Symposium on the Biology of Actinomycetes (Vancouver, Canada).

Yamasaki, M., and Kinashi, H. (2004). Two chimeric chromosomes of *Streptomyces coelicolor* A3(2) generated by single crossover of the wild-type chromosome and linear plasmid scp1. J. Bacteriol. *186*, 6553–6559.

Yang, C.-C., Chen, Y.-H., Tsai, H.-H., Huang, C.-H., Huang, T.W., and Chen, C.W. (2006). In vitro deoxynucleotidylation of terminal prortein of *Streptomyces* linear chromosomes. J. Bacteriol. 72, 7959–7961.

Yang, C.-C., Huang, C.-H., Li, C.-Y., Tsay, Y.-G., Lee, S.-C., and Chen, C.W. (2002). The terminal proteins of linear *Streptomyces* chromosomes and plasmids:

A novel class of replication priming proteins. Mol. Microbiol. *43*, 297–305.

Zawilak-Pawlik, A., Kois, A., Majka, J., Jakimowicz, D., Smulczyk-Krawczyszyn, A., Messer, W., and Zakrzewska-Czerwinska, J. (2005). Architecture of bacterial replication initiation complexes: orisomes from four unrelated bacteria. Biochem. J. *389*, 471–481.

Zechiedrich, E.L., and Cozzarelli, N.R. (1995). Roles of topoisomerase IV and DNA gyrase in DNA unlinking during replication in *Escherichia coli*. Genes Dev. *9*, 2859–2869.

Zhang, R., Yang, Y., Fang, P., Jiang, C., Xu, L., Zhu, Y., Shen, M., Xia, H., Zhao, J., Chen, T., *et al.* (2006). Diversity of telomere palindromic sequences and replication genes among *Streptomyces* linear plasmids. Appl. Environ. Microbiol. *72*, 5728–5733.

Zhu, L., Chou, S.-H., Xu, J., and Reid, B.R. (1995). Structure of a single-cytidine hairpin loop formed by the DNA triplet GCA. Nat. Struct. Biol. *2*, 1012–1017.

Streptomyces Conjugative Genetic Elements

2

Jutta Vogelmann, Wolfgang Wohlleben and Günther Muth

Abstract

Antibiotic-producing actinomycetes contain a huge variety of different plasmids, distinguished in size, topology, replication mechanism and copy number. Some are able to integrate into the chromosome by site specific recombination.

With the exception of the huge linear plasmids, *Streptomyces* plasmids encode only functions involved in replication, stable maintenance and conjugative transfer. The *Streptomyces* conjugation system is unique, requiring a single plasmid-encoded protein, TraB. TraB is a hexameric ring ATPase with similarity to the septal DNA translocator proteins FtsK/SpoIIIE which are involved in chromosome segregation during cell division and sporulation. TraB binds non-covalently to 8bp TRS repeats present in the *clt* locus and transfers double stranded plasmid DNA from the donor to the recipient. Presence of *clt*-like sequences in the chromosome of *S. coelicolor* suggests that chromosomal genes are mobilized independently from the plasmid.

Following primary transfer from the donor into the recipient, the plasmid is translocated via septal cross-walls resulting in intramycelial plasmid spreading. Plasmid spreading involves five to seven plasmid-encoded Spd proteins. Protein–protein interaction studies with Spd proteins of the conjugative plasmid pSVH1 suggest formation of a large DNA-translocation apparatus. One component, the integral membrane protein SpdB2 was shown to form pore structures in lipid bilayers.

Introduction

Antibiotic-producing soil bacteria of the genus *Streptomyces* are the natural source of resistance determinants found in pathogenic bacteria (Davies, 1986; D'Costa *et al.*, 2006; Wright, 2007). The resistance genes evolved as part of the biosynthetic gene cluster of the respective antibiotic to protect the producing organism from the toxic action of its own antibiotic. By horizontal gene transfer (HGT) the resistance genes may have found their way into pathogenic bacteria (Benveniste and Davies, 1973) causing major health problems. The main route of HGT is conjugation, the transfer of mobile genetic elements by direct cell-to-cell contact.

A great variety of different plasmids have been identified in mycelium-forming *Streptomyces* and related actinomycetes. Several of these elements have been characterized in great detail and the function of many plasmid-encoded genes was elucidated. After shortly describing the characteristic features of the various plasmid classes the review will focus on the conjugal transfer functions that constitute a unique DNA translocation system, clearly different from the paradigm of bacterial conjugation described in textbooks (Zechner *et al.*, 2000; Llosa *et al.*, 2002).

Plasmid classes and their characteristics

Mobile elements from mycelial actinomycetes fall into four distinct classes:

1. small multi-copy plasmids that replicate via the rolling-circle mode (RCR);

2 low-copy plasmids replicating via a theta mechanism (Schrempf and Goebel, 1977; Haug *et al.*, 2003);
3 Actinomycete integrative and conjugative elements (AICE) (Te Poele *et al.*, 2008a);
4 (huge) linear plasmids (Kinashi and Shimaji, 1987; Qin and Cohen, 1998; Chen *et al.*, 1993).

Plasmids from *Streptomyces* and related actinomycetes are conjugative and mobilize chromosomal markers (CMA) at high frequency (Hopwood and Kieser, 1993; Kieser *et al.*, 1982). In general, they encode only proteins involved in replication, stable maintenance and conjugal transfer. With the exception of the huge linear plasmids that can carry antibiotic production pathways, no plasmid-encoded traits have been described. Sequence analyses of *Streptomyces* plasmids show that many plasmids have a modular architecture with regard to transfer and replication functions (Reuther *et al.*, 2006b; Servín-González *et al.*, 1995).

Small multi-copy plasmids replicating via the rolling-circle mode (RCR)

As in other Gram-positive bacteria, most of the *Streptomyces* plasmids replicate via the rolling-circle-mechanism (RCR). Several plasmids have been characterized by subcloning, linker insertion analysis and nucleotide sequencing (Kendall and Cohen, 1988; Kataoka *et al.*, 1994a; Servín-González *et al.*, 1995). In general, *Streptomyces* RCR plasmids possess only very limited sequence identity to each other and do not seem to be highly interrelated. A detailed comparison of *Streptomyces* plasmids revealed a modular architecture, suggesting that plasmids commonly exchange functional modules, e.g. for transfer or replication (Servín-González *et al.*, 1995; Reuther *et al.*, 2006b). As it is characteristic for RCR plasmids from other bacteria all genes encoded by the *Streptomyces* plasmids are transcribed in the same direction as the *rep* gene. The only exceptions are the regulatory genes *traR (korA)*. For many of these plasmids rolling-circle-type replication has been proved by the detection of ss-DNA that was accumulated by small deletion derivatives (Muth *et al.*, 1995; Kataoka *et al.*, 1994c). All Rep proteins possess the sequence motifs characteristic for RCR plasmids. Highest similarity is found to the replication initiator proteins of the pC194 family (Muth *et al.*, 1995). Using antibodies, the pSG5 Rep protein was localized in the membrane fraction of its host (Muth, unpublished), suggesting that replication of *Streptomyces* RCR plasmids takes place at the cytoplasmic membrane.

As is characteristic for the pC194 family, the double-stranded origin *dso* is located 5′ to the *rep* gene. The *dso* of pIJ101 and pJV1 were mapped by replicon fusion experiments. Fusion of a pIJ101 derivative with a pJV1 derivative resulted in structurally unstable plasmids. The Rep proteins initiated replication at one origin by a nicking reaction but prematurely terminated replication at the second origin, resulting in the deletion of the intermediate sequences. By sequence analysis of the derived deletion derivatives the nicking sites of the origins could exactly be identified (Servín-González, 1993). The *dso* regions of the *Streptomyces* RCR plasmids showed no striking sequence similarity. This probably reflects the rather high diversity of the Rep proteins. Even the nicking site of the RCR plasmids is not absolutely conserved (Muth *et al.*, 1995). However, several *Streptomyces* plasmids contain a consensus sequence upstream of the (putative) nicking site (Fig. 2.1).

Since initiation of second strand synthesis at the single-stranded origin (*sso*) is initiated by host factors (Kramer *et al.*, 1998) the *sso* region is expected to be conserved in all plasmids. The *sso* was mapped for plasmids pIJ101 (Deng *et al.*, 1988), pSN22 (Suzuki *et al.*, 1997) and pSG5 (Muth, unpublished. In many plasmids, *sso* is located downstream of *traR*. *sso* regions are rich in secondary structures and contain imperfect repeat structures. Alignment of the *sso* regions revealed significant sequence similarity with several stretches of absolutely identical bases (Reuther *et al.*, 2006b) (Fig. 2.2).

Low-copy plasmids replicating via a theta mode

Only few low-copy plasmids have been characterized in *Streptomyces*. pSG2 (Wohlleben and Pühler, 1987) or SCP2 (Schrempf and Goebel, 1977), the first circular plasmid that has been

```
pSNA1      CCCGCCGATCCAAAAACGGACAGGCGCCTTG GTA
pSVH1      CGGCGGTCATCAAAAAGGGACGCGG-CCTTG GTA
pSN22      GTCCGCGACCCAAAACGTGTCGCG--CCTTG GGA
pSY10      CGGCCGCCAGCAAATCCGGACA----CCTCG GTA
pJV1       CCGCCCCTGGCAAAAAGGGACG----CCTAG GTA
pIJ101     CCGCCCGAGGCAAAAGCGAACA----CCTTG GGA
pSB24.2    CCGCCCGAGGCAAAAGCGAACA----CCTTG GGA
pTSC2      GCCGCGCCCCCAAAAAGGGACG----CCTTG GGA
```

Figure 2.1 Sequence conservation at the nicking site of several *Streptomyces* RCR plasmids. The (putative) nicking site is underlined. The arrow indicates position of nicking. Conserved nucleotides are written in bold. pSNA1 (AY243257); pSVH1 (AM087403); pSN22 (D14281); pSY10 (AB016241); pJV1 (U23762); pIJ101 (M21778); pSB24.2 (X03756); pTSC2 (GQ984145).

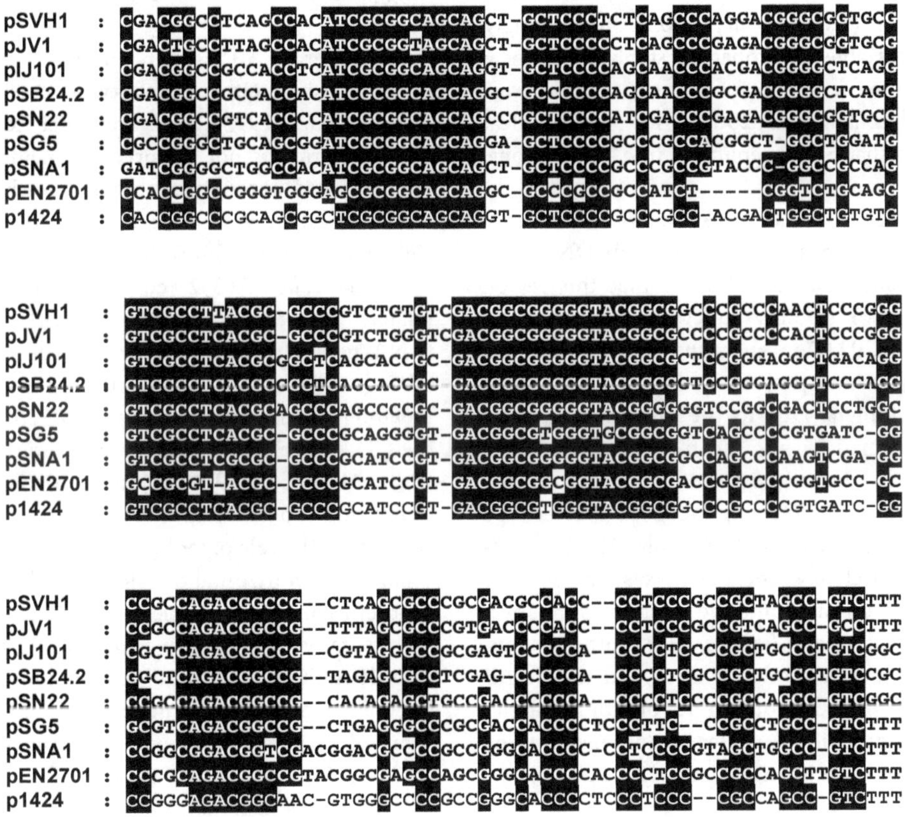

Figure 2.2 Conserved single-stranded origin (sso) region on *Streptomyces* RCR plasmids. Starting at the *sso*, the single-stranded replication intermediate is converted to a double-stranded molecule by host factors. pSVH1 (AM087403); pJV1 (U23762); pIJ101 (M21778); pSB24.2 (X03756); pSN22 (D14281); pSG5 (X80774); pSNA1 (AY243257); pEN2701 (AF533985); p1425 (Muth, unpublished).

isolated from *S. coelicolor* A3(2), belong to this group. SCP2, with a molecular size of 31,317 bp has a copy number of one to four (Bibb *et al.*, 1980) and is stably inherited. Due to their stability and their capability to accept large inserts, as for example the complete biosynthetic gene cluster for actinorhodin biosynthesis (Malpartida and Hopwood, 1984), SCP2 derivatives are valuable cloning vectors (Lydiate *et al.*, 1985).

Information concerning their replication mechanism is limited. The minimal replication region, identified by subcloning analysis was localized to a 5.9 kb fragment. A smaller deletion derivative had an ultra-high-copy number of ~ 1000, suggesting the presence of a negative regulator of copy number control (Larson and Hershberger, 1986). A more detailed characterization of the SCP2 minimal replicon based on the nucleotide sequence identified a 650 bp *ori* region and two short orfs, *repI* and *repII* that showed sequence similarity and were required for stable replication (Haug *et al.*, 2003). Interestingly, SCP2 was able to propagate stably in a linear mode, when telomeres of a linear plasmid were attached (Zhang *et al.*, 2008). This provides clear evidence that SCP2 replicates in a bidirectional mode.

The region necessary for stable maintenance of SCP2 was localized on a ~ 5 kb DNA fragment. Haug *et al.* could show that this region encodes two stability functions (Haug *et al.*, 2003): SCP2.07c, encoding a recombinase, probably ensures proper segregation by resolving plasmid dimers. Due to the proposed function it was named *mrpA*, for multimer resolution protein. SCP2.05c (*parA*) and the translationally coupled SCP2.04.c gene, encode a Walker-type partitioning ATPase and probably a ParB homologue, respectively. The ParAB system of SCP2 might actively distribute the plasmid to the mycelial compartments/spores. For both systems, the multimer resolution system and the ParAB system, their contribution to plasmid stability during sporulation was clearly demonstrated by subcloning experiments (Haug *et al.*, 2003).

Actinomycete integrative and conjugative elements (AICEs)

As revealed by genome sequencing approaches, actinomycetes genomes often contain integrated plasmids. These plasmids are distinguished from 'normal' plasmids in that they encode a phage-like integrase system that allows integration of the plasmid into a chromosomal attachment site. Actinomycete-derived integrative and conjugative elements have been designated AICEs (Te Poele *et al.*, 2008b). AICEs have a molecular size between 11.2 kb (Hosted *et al.*, 2005) and 23.3 kb (Te Poele *et al.*, 2008b). As is typical for an ICE, all integrate into a chromosomal tRNA gene without disruption of its coding frame. The AICE-encoded integrase mediates recombination between the plasmid encode *attP* site and the chromosomal *attB* site overlapping with the 3′ end of a tRNA gene. Different ICEs can integrate into different tRNA genes. Since tRNA genes are quite conserved, the host range of the integration system is broad. For the *Streptomyces ambofaciens* AICE pSAM2 integration into *Mycobacterium smegmatis* has been observed (Martin *et al.*, 1991).

In general, AICEs can propagate autonomously using either theta-type mechanisms (Nordstroem, 1993) or a rolling-circle mode, but replication is not required for maintenance. Excision and autonomous replication however, seems to be necessary for conjugal transfer, as shown for pSAM2 (Smokvina *et al.*, 1991).

S. coelicolor A(3)2 contains six pSAM2-related elements, three integrated into a tRNA gene. Interestingly, two of the pSAM2-related elements are next to secondary metabolite gene clusters. AICESco3250 flanks the CDA (calcium-dependent antibiotic) cluster, while AICESco5349 is adjacent to the WhiB-polyketide spore pigment synthesis cluster, supporting the role of mobile elements in genome shaping and horizontal transfer of antibiotic biosynthetic clusters (Te Poele *et al.*, 2008b).

Integrative vectors derived from SLP1, pSAM2 or pMEA300 proved to be valuable tools for the integration of DNA fragments into the chromosome of various *Streptomyces* and *Amycolatopsis* strains (Hopwood *et al.*, 1987; Vrijbloed *et al.*, 1995b; Smokvina *et al.*, 1990).

Linear plasmids

In contrast to the genomes of most other bacteria, *Streptomyces* chromosomes are usually linear (Lin *et al.*, 1994). Many linear plasmids with a size of 12 kb to 1700 kp (Keen *et al.*, 1988; Kinashi and Shimaji, 1987; Bentley *et al.*, 2004) have been also described. The 5′ ends of the linear replicons are capped with covalently bound telomerase proteins. Tight interactions of the telomerase proteins might cause a circular configuration, as suggested by the circular genetic linkage map of the *S. coelicolor* chromosome (Kieser *et al.*, 1992).

This class of plasmids has a unique replication mechanism with a large variety and diversity in replication loci (Zhang *et al.*, 2006; 2008). Rep-proteins initiate a bidirectional replication from a centrally located origin towards the ends that contain long inverted repeats ranging from 44 bp (Chen *et al.*, 1993) to 180 kb (Pandza *et al.*, 1998). All telomere sequences contain palindromic sequences capable of forming complex secondary structures (Huang *et al.*, 1998). This mode of replication leaves single-stranded gaps of 250–300 nt at the 3′ ends. The telomerase proteins prime the conversion of the single-stranded overhangs into a double strand. When telomeres are deleted, the linear plasmids can also replicate as a circular molecule (Shiffman and Cohen, 1992; Redenbach *et al.*, 1999).

Recently a promoter was identified in the telomere region of the *S. lividans* chromosome, which is highly conserved in linear *Streptomyces* replicons and it has been speculated that this promoter might be involved in priming terminal fragment synthesis (Lin *et al.*, 2009). Most linear plasmids and chromosomes of *Streptomyces* encode a helicase (*ttrA*) at the ends with an unclear function. Inactivation of SLP2 *ttrA* did not interfere with replication but affected efficiency of transfer, suggesting a role in conjugal transfer (Lin *et al.*, 2009).

Linear plasmids seem to contribute significantly to the shaping and evolution of genomes by various interactions, as integration, exchange of chromosomal ends and conjugal transfer of large chromosomal fragments.

Conjugal transfer of *Streptomyces* plasmids

Genetic exchange in *Streptomyces* was first detected more than 50 years ago by the isolation of prototrophic recombinants from mixed cultures of auxotrophic strains (Alikhanian and Mindlin, 1957; Hopwood, 1959). Although conjugal DNA transfer in *Streptomyces* is based on the presence of conjugative plasmids, its mechanism is clearly distinguished from the complex conjugation system of Gram-negative bacteria. The classical conjugation system described in textbooks (Zechner *et al.*, 2000) can be regarded as derived from two cellular processes: a rolling-circle-type DNA replication system and a type IV protein secretion system (T4SS; Christie and Cascales, 2005). Both systems are linked by a coupling protein (Cabezon and de la Cruz, 2006). The key enzyme in this conjugation system is the conjugative relaxase that initiates a rolling-circle-type replication by introducing a nick at the *oriT* region (Willets and Wilkins, 1984). The single-stranded plasmid molecule with the relaxase covalently attached to its 5′ end is then pumped by the coupling protein, a hexameric ring ATPase, through the channel of the T4SS system into the recipient (Schröder and Lanka, 2003; Gomis-Ruth *et al.*, 2002).

Characteristics of the conjugation process

Since small RCR plasmids, like pIJ101 (Kieser *et al.*, 1982) with a molecular size of less than 10 kb are self-transmissible, it was quite clear from the beginning that *Streptomyces* conjugation does not rely on such a complex system involving up to 30 plasmid-encoded proteins. Later on it was revealed that a single plasmid-encoded protein (TraB/KilA) was sufficient to promote efficient DNA transfer (Kieser *et al.*, 1982; Pettis and Cohen, 1994; Servín-González *et al.*, 1995). As was shown by the elegant work of Possoz *et al.*, *Streptomyces* conjugal plasmid transfer most likely involves double-stranded DNA (Possoz *et al.*, 2001). This was demonstrated by the sensitivity of pSAM2 transfer to the presence of the SalI restriction/modification system in the recipient, which only restricts incoming double-stranded DNA but not single-stranded DNA (Possoz *et al.*, 2001). In agreement to this finding, the small RCR plasmids of *Streptomyces* do not code for a protein containing typical relaxase motifs (beside Rep).

Due to the mycelial growth characteristics of *Streptomyces*, conjugation involves two distinct DNA translocation events. First, the plasmid is transferred form the plasmid carrying donor to a plasmid free recipient. This primary transfer is followed by the subsequent translocation of the plasmid within the recipient mycelium, resulting in the spreading of the plasmid within the recipient colony. If matings are performed with an excess of recipients, conjugal transfer and subsequent plasmid spreading is associated with the formation of 'pock'-structures. Pocks are inhibition zones of up to 3 mm in size where development of recipient mycelium is temporarily retarded and indicate the transconjugant area. Although the concept of plasmid spreading (Bibb and Hopwood, 1981; Hopwood and Kieser, 1993) via septal cross-walls in the recipient is still hypothetical, experimental evidence that supports this model is accumulating (Tiffert *et al.*, 2007; see below).

TraB mediated primary DNA transfer

For many different *Streptomyces* plasmids it was demonstrated that a single plasmid-encoded protein was essential for conjugal DNA transfer (Pettis and Cohen, 1994; Servín-González *et al.*, 1995; Kataoka *et al.*, 1994a). Beside TraB only a small non-coding sequence *clt* was required for plasmid transfer, but not for CMA (Pettis and Cohen, 1994; Servín-González, 1996). Interestingly, *traB* is part of a *kil/kor* system. Cloning of the *traB* gene without its upstream region, encoding a GntR-type transcriptional repressor protein (TraR/KorA) is lethal (Reuther *et al.*, 2006b; Kendall and Cohen, 1987; Kataoka *et al.*, 1994b). Toxic effects of *traB* expression have been also observed in *E. coli* (Reuther *et al.*, 2006a).

Comparison of different TraB proteins showed only very low sequence conservation, but identified a FtsK/SpoIIIE Pfam domain for each TraB protein, grouping TraB to the AAA+ ATPases and septal DNA translocator proteins. In addition, all TraB-proteins have a very similar domain organization to that of FtsK with an N-terminal membrane region containing predicted transmembrane helices, the DNA translocase/ATPase domain with Walker A and B boxes and a C-terminal wHTH motif. Whereas sequence similarity among *Streptomyces* TraB-proteins is highest in the ATPase domain the C-terminal wHTH domain shows only low conservation.

During cell division and sporulation FtsK and SpoIIIE were shown to translocate ds-chromosomal DNA from the replication origin towards the terminus (Wu and Errington, 1994; Burton *et al.*, 2007; Massey *et al.*, 2006). Polarity of chromosome translocation is determined by the interaction of the FtsK γ-domain with KOPS, a 8 bp sequence motif whose distribution is highly skewed from the origin to the terminus. The FtsK γ-domain also interacts with the XerCD recombinases to support resolution of chromosome dimers at *dif* sites located within the terminus region.

Electron microscopy and crystallization studies detected hexameric FtsK rings around ds-DNA (Massey *et al.*, 2006). In the current model of chromosome segregation each chromosomal arm is translocated by a separate hexameric/dodecameric FtsK ring involving a rotary inchworm type mechanism (Lowe *et al.*, 2008; Burton *et al.*, 2007).

The similarity of TraB proteins to the septal DNA translocator FtsK indicates that *Streptomyces* use a similar mechanism to translocate DNA, not between two cellular compartments, but from a donor to a recipient mycelium. Whereas FtsK and SpoIIIE are septal DNA translocators, localized at the division and sporulation septa, respectively, a $\text{TraB}_{\text{pSG5}}$–eGFP fusion protein was detected by fluorescence microscopy at the tips of the mycelium, suggesting that *Streptomyces* conjugation takes place at the hyphal tips (Reuther *et al.*, 2006a).

To characterize TraB biochemically, $traB_{pSVH1}$ was expressed with an N-terminal eight amino acids StreptagII-encoding sequence under control of the thiostrepton inducible P_{tipA} promoter in *S. lividans*. Induction of TraB expression was tolerated, but morphology was severely affected. Instead of spherical mycelial pellets, large cylindrical aggregates with long protruding hyphae were produced, suggesting that in the presence of TraB tip growth is favoured and branching of the mycelium retarded (Reuther *et al.*, 2006a). From the membrane fractions of such cultures active $\text{Strep-TraB}_{\text{pSVH1}}$ protein could be recovered, as demonstrated by its ATPase activity.

Oligomeric structure of TraB

Since FtsK was shown to form hexameric and dodecameric rings (Massey *et al.*, 2006), the ability of $TraB_{pSVH1}$ to form oligomeric structures was analysed by chemical cross-linking, bacterial two-hybrid analyses and electron microscopy. Cross-linking of $TraB_{pSVH1}$ with glutaraldehyde resulted in large aggregates that were not able to enter a SDS-PAGE (12.5%). Therefore, a soluble $TraB_{pSVH1}$ derivative ($TraB_{270-778}$) lacking the N-terminal TM-domain was expressed and purified. Cross-linking experiments of soluble $TraB_{270-778}$ detected oligomeric structures corresponding to trimers and hexamers (Vogelmann and Muth, unpublished).

Oligomerization of TraB could be visualized by electron microscopy. Both TraB proteins, the full length protein as well as its N-terminally truncated protein formed ring-shaped structures resembling a hexamer (Fig. 2.3), similar to those structures reported for *P. aeruginosa* FtsK (Massey *et al.*, 2006).

TraB–clt interaction

TraB is the only plasmid-encoded protein essential for conjugative plasmid transfer. Beside TraB only a small non-coding sequence (*clt*) is required for plasmid transfer. So, the most probable role of *clt* might be to represent the binding site of TraB. Indeed, gel retardation experiments with purified $TraB_{pSVH1}$-protein demonstrated that TraB bound to a specific plasmid region (Reuther *et al.*, 2006a). In these gel shift assays, the retarded fragment did not enter the agarose gel indicating that the TraB–DNA complex was too large to enter the agarose gel. This suggested that TraB multimers were bound. The TraB binding site on pSVH1, localized downstream of *traB,* showed several direct and indirect repeats, previously reported to be characteristic for *clt* regions (Franco *et al.*, 2003). Deletion of the TraB binding site from the pSVH1 derivative pEB211, abolished conjugative transfer (Reuther and Muth, unpublished) confirming that the *TraB*-binding site represents the *clt* locus. Sequence analysis of other *Streptomyces* plasmids identified series of imperfect repeats in the majority of *Streptomyces* plasmids immediately downstream of the *traB* gene, suggesting that *Streptomyces* plasmids contain a transfer module consisting of the DNA translocase TraB and its recognition site *clt* closely linked to it.

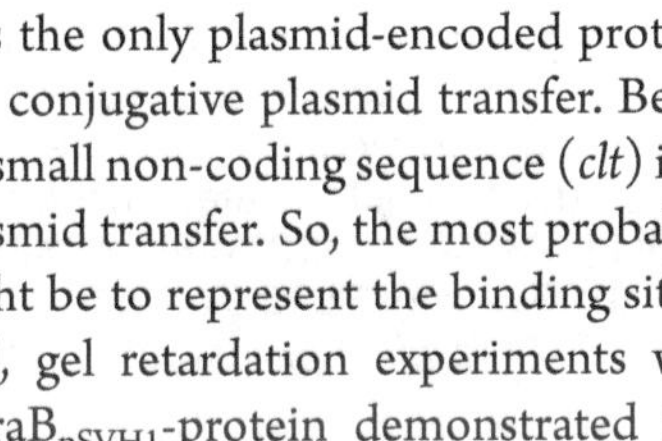

Figure 2.3 EM images of TraB oligomers. Negative stain (uranyl acetate) of purified TraB protein. Full-length $TraB_{pSVH1}$ protein (A) and soluble, N-terminal truncated $TraB_{pSVH1}$ (B) form oligomeric (hexameric) ring shaped structures with a diameter of about 12 and 13 nm (white bar), respectively.

Whereas FtsK proteins recognize KOPS sequences distributed all over the chromosome (Bigot *et al.*, 2005), TraB proteins specifically recognize only the corresponding *clt* locus of the respective plasmid, but do not bind to *clt* loci of unrelated plasmids. To identify the TraB-region responsible for sequence specific DNA binding, chimeric *traB* genes were constructed, each one consisting of distinct parts of two different *traB* genes. All chimeric proteins were expressed as a StreptagII-tagged protein and purified by affinity chromatography. To characterize their DNA binding activity, *clt* loci of different plasmids were amplified and subcloned. The resulting plasmids were used in gel retardation experiments. Thus, the TraB-domain for *clt* recognition could be narrowed down to the very C-terminal end of TraB, predicted to contain the wHTH motif (Vogelmann *et al.*, unpublished. The TraB-wHTH domain corresponds to the γ-domain of FtsK that was recently shown to interact specifically with 8 bp KOPS sequences (Bigot *et al.*, 2005; Lowe *et al.*, 2008). Since the (putative) *clt* regions of *Streptomyces* plasmids also contain imperfect 8 bp repeats (Fig. 2.4) it is tempting to speculate that *Streptomyces* TraB-proteins recognize the TRS repeats (TraB recognition sequence) in the *clt* region in a similar manner as the γ-domain of FtsK interacts with KOPS.

Chromosome mobilizing activity

Streptomyces plasmids mediate transfer of chromosomal markers with a frequency of about 0.1% (Hopwood and Kieser, 1993; Kieser *et al.*, 1982) and a detailed genetic map of the *S. coelicolor* chromosome was constructed by mating experiments (Kieser *et al.*, 1992). However, no

pJV1	p1119	pSN22	SLP1	pSG2	pRL1	SLP2
CCCGCACA	CCCGCACA	CCCGCACA	CGTCACAC	CCTGACAC	CTGTCAGC	ACCGCCAC
CCCGCACA	GCCGCACA	GCCGCACA	CGTCACAC	CCTGACAC	CTGTCAGC	ACCGCCAC
CCCGCACA	GCCGCACA	GCCGCACA	CCTCACAG	CCTGACAG	CTGACACC	ACCGCCAT
GCCGCACA	GCCTCACA	GCCGCACA	CGTCACAC	CCTGACAC	CTGACACC	TCCGCCAT
CCCGCACA	CCCGCACA	GCCGGACC	CGTCACAC	TCTGACAC	CTGACACC	TCCGCCAT
CCCGCACA	CTCGCACA	ACCGCACA	CGTCACAC	CCTGACAC	CTGACACC	GCCGCCTC
GCCGCACA	GCCGCACA	CCCGCACA		CCTGACAC		TCCGCCAC
CCCGCACA	CCCGCACA			TCTGACAC		ACCGCCAT
CCCGCACA						ACCGCCAT
						ACCGCCAC
						TCCGCCAT
						TCCGCCAT
CCCGCACA	CCCGCACA	NCCGCACA	CGTCACAC	CCTGACAC	CTGACACC	WCCGCCAY

pIJ101	p1424	pSVH1	pSG5	pFP11	pMEA300	$FtsK_{coli}$
TGGAAACC	TCAACTCG	GACCCGGC	GCCCGGAC	GACCCGGA	CTCCCTCC	GGGNAGGG
CGCAACCT	GCAACTTC	GACCCGGA	CCCCGGAC	CACCCTGG	CTCCCTCA	
TGCAACTT	TCAACCGG	CGCCCGGA	GCCCGCCC	CGCCCGGT	CTCCATTC	
GTCAACCG	GCAACCGG	GACCCGGA	GCCCGGTC	CCCCCGGA	TTCCCTAC	
TGCAACGG	GCAACCGG	GACCGGGC	TCCCGGAC	GACCTGGA	TTCCCTCG	
CTCAACCG	GCACCCGG	GACCCGGA	CCCCGCAG	GACCTGGA	CTCCCTGC	
GGCAACCC		GACCCGGA	GCCCTGAC	GACCTGCA	TTCCCTAC	
GGAAACCA		GGCCCGGA	GCCCGGAC	GACCCCGA		
		GACCGGGA	GCCCGGAC	GACCTGGA		
		AGCCCGGA		AGCCCGGA		
		GACCCGGA		TGCCCGGC		
		GGCCCGGT		GACCTGGA		
NGCAACCN	GCAACCGG	GACCCGGA	SCCCGGAC	SACCYGGA	YTCCCTNN	

Figure 2.4 Imperfect 8 bp repeats in (putative) *clt* regions that might represent the TraB Recognition Sequence (TRS). *clt* regions of actinomycetes plasmids contain imperfect direct repeat sequences with varying spacing in between. For comparison, 8 bp KOPS sequence recognized by *E. coli* FtsK is also given. Bold characters mark conserved nucleotides. Grey colour indicates sequence motifs located on the opposite strand. In the bottom line the consensus TRS sequence is given. pJV1 (U23762); p1119 (Muth, unpublished); pSN22 (D14281); SLP1 (AL939120); pSG2 (ABYA01000616); pRL1 (DQ322649); SLP2 (AY225511); pIJ101 (M21778); p1425 (Muth, unpublished); pSVH1 (AM087403); pSG5 (X80774); pFP11(AY943952); pMEA300 (L36679).

physical interaction of the conjugative plasmids with the chromosome could be demonstrated. The only exceptions are the integrative plasmids such as pSAM2 which integrate site-specifically into the chromosomal attachment site and the 350 kb linear plasmid SCP1. This indicates that the mechanism of chromosome mobilizing activity (CMA) does not involve a covalent interaction of the plasmid DNA with the chromosome as was initially observed in HFR matings and later on found to be the paradigm in T4SS mediated mating experiments (Zechner *et al.*, 2000). *clt*-independent CMA in *Streptomyces* matings has been also experimentally demonstrated and it was suggested that the chromosome already contains *clt*-like loci (Pettis and Cohen, 1994).

Recognition of a 8 bp sequence motif (TRS) by TraB might explain how chromosomal markers are mobilized during *Streptomyces* conjugation. If TraB recognizes just a 8 bp TRS sequence motif, it is plausible that with lower efficiency it can also assemble at chromosomal locations, if several of the 8 bp TRS motifs are close by. Bioinformatic analysis of the *S. coelicolor* genome for the presence of putative chromosomal clt_{pSVH1}-like sequences predicted 23 putative chromosomal *clt* regions containing four TRS_{pSVH1} sequences within a 100 bp fragment.

Intramycelial plasmid spreading

Generation of pock structures during *Streptomyces* conjugation has been interpreted as the result of intramycelial plasmid spreading following the primary DNA transfer from a donor into the recipient (Hopwood and Kieser, 1993; Grohmann *et al.*, 2003). Whereas tip-localization of the main transfer protein TraB indicated that the primary transfer event occurs at the hyphal tip (Reuther *et al.*, 2006a), intramycelial plasmid spreading is thought to involve DNA translocation via the septal cross-walls. The process of plasmid spreading (and pock formation) involves several plasmid-encoded proteins (Spd) in addition to TraB. Inactivation of a single *spd* gene reduces size of pock structures (Kataoka *et al.*, 1994a; Servín-González *et al.*, 1995; Kieser *et al.*, 1982; Vrijbloed *et al.*, 1995a). Many Spd proteins are small with a size of less than 10 kDa. With few exceptions, the Spd proteins have no significant similarity to functionally characterized proteins hindering prediction of their putative function. The *spd* genes are mostly organized in operons, in part with overlapping stop and start codons, indicating translational coupling. This suggests that the Spd proteins cooperate, probably forming a large protein complex involved in plasmid translocation via septal cross-walls. Evidence for such a multi-protein complex was obtained for plasmid pSVH1 by bacterial two-hybrid analyses (Guezguez and Muth, unpublished). In addition, several protein–protein interactions were revealed by cross-linking and pulldown experiments using purified proteins (Tiffert *et al.*, 2007).

Intrahyphal spreading might also contribute to the stable maintenance of *Streptomyces* plasmids. Hyphal compartments that have lost a plasmid can recover a plasmid from the neighbouring compartment. This recovery most probably involves intramycelial plasmid spreading, since a clear effect of *spd1* inactivation on stable maintenance of the linear plasmid SLP2 was demonstrated (Hsu and Chen, 2009).

Characteristic features of selected Spd proteins
Comparative analysis of actinomycetes plasmids aiming to identify a minimal set of core genes involved in plasmid spreading, does not result in a clear evidence. Only few genes show some conservation, but many *spd* genes are unique to a given plasmid and have no homologues on other plasmids. *spd* genes have been previously characterized by their involvement in the formation of pock structures. However, their precise role in plasmid spreading is waiting to be elucidated and only for few Spd proteins, described below, a molecular function can be postulated.

SpdA Many plasmids (but not all) encode a 120–180 aa protein SpdA. SpdA shows low sequence similarity to putative regulatory proteins and contains a sequence motif (SRDD/E–X_{6-8}–SRDD/E which is conserved in most SpdA proteins. Interestingly, this motif overlaps with a palindromic DNA sequence CGTCACGTGAC-N_{18-24}–CGTCACGTGACG (Reuther *et al.*, 2006b; Servín-González *et al.*, 1995) that is also found upstream of *spdA* in pSVH1 and pEN2701 (Fig. 2.5). Although the role of this repeat structure in plasmid spreading is unknown, one can speculate that SpdA might be a DNA-binding protein with the palindromic sequence representing the binding site within its coding region.

```
pSVH1:   -48   AGCCGTCACTGTGACG---ACTAGCTTCAGTTCGTAAGT-CGTCACTGTGACGGCT
pEN2701: -96r  AGCCGTCACTGTGACA---GCTATCCGGCGCGAGAAGACCCGTCACTGTGACGGCA
pJV1:    +200  TTTCGTCAC-GTGACG---ATCGCCCGACTCCTCCGATTCCGTCAC-GTGACGAAA
p1119:   +185  ATTCGTCAC-GTGTCG---AAAGCCGGTCCACCCCGGTTCCGTCAC-GTGACGAAA
p1424:   +218  TTTCGTCAC-GTGACA---CGAGCGGAGCGGCGGATCGGCCGTCAC-GTGACGAAA
pIJ101:  +185  ATACGTCAC-GTGACG---AAATGCAGCAGCCTTCCATTCCGTCAC-GTGACGAAA
pSNA1:   +305  CAAAGTCAC-GTGACG---ATGCA---GGCGACCAGGCAACGTCAC-GTGACGTTG
pSG2:    +191  CAACGTCAC-GTGACG---ATGAACTGACCCTCCGAGCAACGTCAC-GTGACGATG
pSN22:   +203  GAAAGTCAC-GTGACTTTCCCGGTCGGCAGGCCCCAGCAAAGTCAC-GTGACTTTC
```

Figure 2.5 Conserved palindromic sequence motif within *spdA* genes. Arrows above the alignment mark palindromic sequences. + or – indicates nucleotide position in relation to the *spdA* start codon. r denotes sequence motif on the complementary strand. pSVH1 (AM087403); pEN2701 (AF533985); pJV1(U23762); p1119 (Muth, unpublished); p1425 (Muth, unpublished); pIJ101 (M21778); pSNA1 (AY243257); pSG2 (ABYA01000616); pSN22 (D14281).

SmeA homologues Plasmid pFP11 encodes a small 61 aa protein (pFP11.19c) with significant similarity to the *S. coelicolor* SmeA protein. SmeA was shown to be required to locate the FtsK homologue SffA to the septum of *S. coelicolor* (Ausmees *et al.*, 2007). A probable function of pFP11.19c might be the relocalization of TraB from the hyphal tip to the septal cross-wall to allow intramycelial plasmid spreading. SmeA homologues are also detected on pFP1 (pFP1.27) and on the linear plasmid pSV2 (pSV2.52c) but have not been identified on small RCR plasmids.

SpdB2 SpdB2, an integral membrane protein with four predicted transmembrane helices, is the most interesting Spd protein. Surprisingly, homologues of SpdB2 are also encoded by actinophages (Smith *et al.*, 1999; Van Dessel *et al.*, 2005) but their role during phage infection is not known. Proteins, showing the structural characteristics of SpdB2, are encoded on every *Streptomyces* plasmid implicating a central role of SpdB2 in plasmid spreading. Recently, SpdB2 of the *S. venezuelae* plasmid pSVH1 has been heterologously expressed. The purified protein bound unspecific to ds-DNA and interacted with another small Spd protein (Spd79) which is expressed from the same operon as *spdB2*. By chemical cross-linking and electron microscopy evidence was obtained that SpdB2 formed oligomeric ring shaped structures (Tiffert *et al.*, 2007; Heichlinger *et al.*, unpublished).

Pore formation of SpdB2$_{pSVH1}$ in artificial membranes

Since EM analysis suggested a pore-like structure, the ability of SpdB2$_{pSVH1}$ to form pores in artificial membranes was analysed by single channel conductance measurements. In such an experimental setup a phosphatidylcholine membrane isolates two compartments filled with KCl. Therefore, no current flow is detectable if voltage is applied via Ag-electrodes placed into each compartment. Current flow only occurs if a pore forming protein is applied and a single open pore structure inserts into the membrane.

After adding purified SpdB2$_{pSVH1}$ to one site of the setup, SpdB2 assembled to a pore structure within the membrane resulting in a conductance of 0,14 nS (−30 mV, 1 M KCl, pH 7.0). Fig. 2.6 displays a screenshot showing opening and closure of several SpdB$_{pSVH1}$ pores.

Control of conjugal DNA transfer in *Streptomyces*

Regulation of conjugative plasmid transfer in *Streptomyces* is hardly understood. In all *Streptomyces* plasmids, *traB* is under control of a

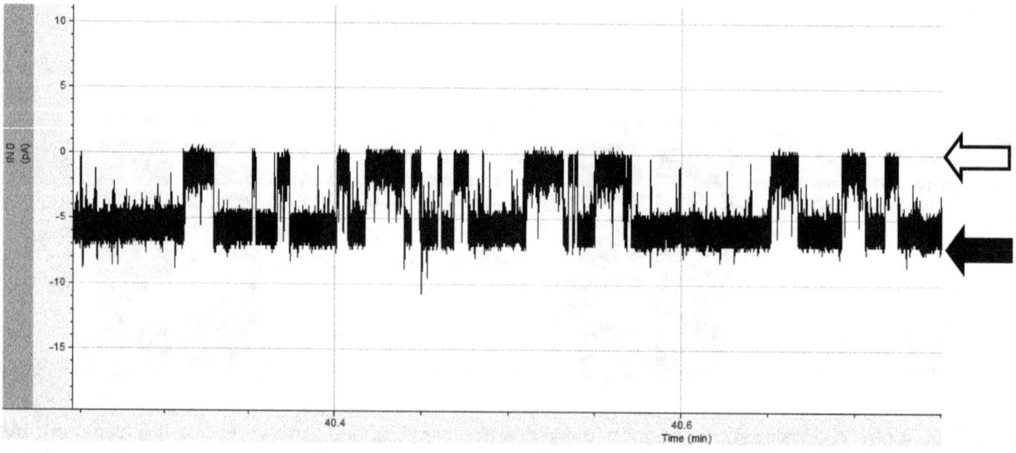

Figure 2.6 Pore formation of SpdB2$_{pSVH1}$ detected by single channel conductance measurements. Formation of an open pore structure is detected by change in conductance. A filled arrow marks open pores, while a conductance of 0, corresponding to no-pore/closed pore is indicated by an open arrow.

GntR-type transcriptional repressor TraR that binds to *tre*-boxes in the promoter region of *traB* (Kataoka *et al.*, 1994b; Kataoka *et al.*, 2008). Despite being repressed, *traB* is able to mediate efficient conjugal transfer; this indicates that TraR senses a signal to allow expression of *traB*. The nature of this signal is unknown, but it could be a low-molecular-weight compound, since repressors of the GntR-type loose their DNA binding activity upon interaction with a signal compound (Gorelik *et al.*, 2006).

A second transcriptional repressor (KorB) was shown to control expression of *kilB* of plasmid pIJ101. *kilB* encodes a toxic protein of unknown molecular function involved in pock formation and plasmid spreading (Pettis *et al.*, 2001). KorB is proteolytically processed (Tai and Cohen, 1993), but the processing signal or the processing protease is unknown. Control of KilB expression is even more complex, since KorB-independent transcriptional read through from the upstream *tra* operon contributes significantly to transcription of *kilB* in a temporal manner (Pettis *et al.*, 2001; Schully and Pettis, 2003).

Entry exclusion, or conjugal immunity, is a characteristic feature making plasmid containing cells poor recipients in additional conjugation rounds. This phenomenon is widely spread and it is considered to be important for plasmid stability (Garcillan-Barcia and de la Cruz, 2008). There is also a report of entry exclusion in *Streptomyces*. The integrative pSAM2 plasmid was shown to encode a gene, *pif*, which mediates immunity to pSAM2 transfer (Possoz *et al.*, 2003). When this gene is already present in the recipient cell, it inhibits excision of pSAM2 from the chromosome of the donor cell, the first step in conjugal pSAM2 transfer. *pif* encodes a Nudix hydrolase, indicating that a di-nucleotide linked to an unidentified moiety is involved in signalling pSAM2 transfer. Homologues of this protein have been detected only on other actinomycetes integrative plasmids (AICE) (Te Poele *et al.*, 2008b), but not in autonomously replicating *Streptomyces* plasmids.

Model of the *Streptomyces* DNA translocation machinery and its relation to T4SS systems

Whereas most bacteria transfer DNA by a combination of a rolling-circle plasmid replication system and a T4SS protein secretion system, *Streptomyces* have evolved a unique conjugation system. The similarity of TraB to the septal DNA-translocators FtsK/SpoIIIE suggests that TraB-mediated plasmid translocation resembles chromosome segregation by FtsK/SpoIIIE. Based on the elucidation of the molecular function of key enzymes involved in transfer and intramycelial plasmid spreading, the following model of the *Streptomyces* DNA-translocation machinery is proposed (see Fig. 2.7).

Streptomyces plasmids do not encode a system for the finding of mating partners. Since non-motile streptomycetes grow as a mycelium, there is no need for special genes for pilus formation or aggregation to mediate cell–cell contact. This could be achieved just by growing together. Contact or an unidentified signal induces expression of the main transfer protein TraB.

During conjugation the DNA has to pass the cytoplasmic membrane and the peptidoglycan layer of the donor and the membrane and the cell wall of the recipient. From sequence prediction, *Streptomyces* plasmids normally do not encode a lytic transglycosylase for the penetration of the murein sacculus. If conjugation takes place at the hyphal tips, as demonstrated by the localization of a TraB–eGFP fusion protein (Reuther *et al.*, 2006a), one can speculate that a plasmid-encoded lytic transglycosylase is not required. *Streptomyces* growth by apical tip extension implies that all enzymes required for peptidoglycan remodelling are already there. Nevertheless, pMR2 and the linear plasmid SLP2 encode a lytic transglycosylase (Xu *et al.*, 2006), but their role in conjugation has not been addressed so far.

The hexameric ring ATPases TraB forms a pore structure involving partial fusion of the membranes of donor and recipient as suggested by the membrane fusion activity reported for SpoIIIE of *B. subtilis* (Liu *et al.*, 2006). TraB

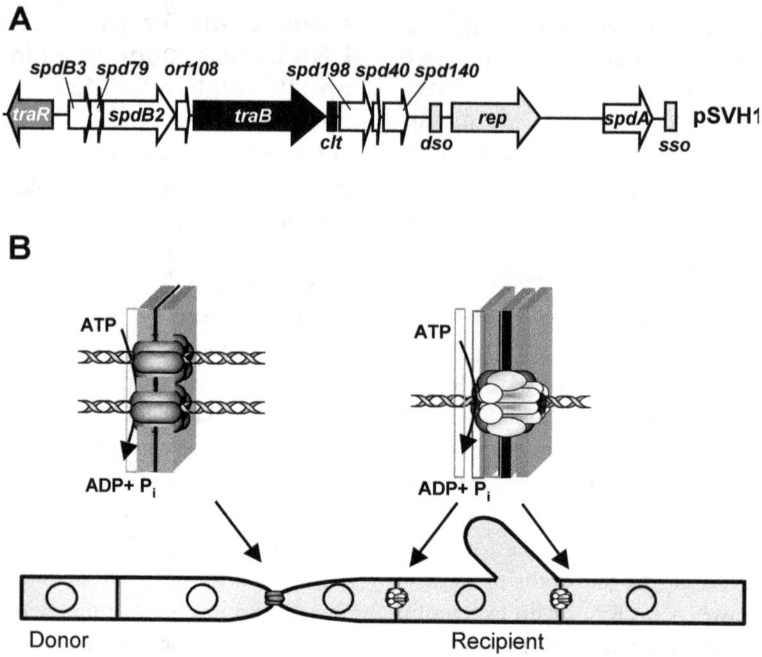

Figure 2.7 Model of DNA translocations during *Streptomyces* conjugation. Panel A shows a genetic map of plasmid pSVH1. The regulatory gene *traR* is drawn in grey, *traB* and *clt* in black, genes involved in pock formation and plasmid spreading are given in white, while genes and regions involved in replication are depicted in light grey. *Streptomyces* conjugation involves two putative DNA translocation complexes (B). The primary plasmid transfer from a donor into a recipient is mediated by hexameric TraB (left inset). Subsequent plasmid spreading via septal cross-walls involves a large multi-protein complex (right inset) consisting of several Spd proteins and TraB (for explanations see text).

interacts with the 8bp TRS sequence motifs of the *clt* region by its C-terminal wHTH motif and pumps double-stranded DNA probably by a rotary inchworm type mechanism (Massey *et al.*, 2006) to the recipient. Most probably, more than one TraB complex is required for plasmid transfer. Although no plasmid processing enzyme has been identified yet, one has to speculate that topoisomerases might be involved in the translocation process. ATPase activity of purified TraB suggests that DNA-transfer is energized by ATP hydrolysis (Reuther *et al.*, 2006a). Since chromosomes also contain *clt*-like sequences, TraB might also assemble at those sites and pump chromosomal DNA to the recipient. This is in agreement with the fact that CMA in *Streptomyces* does not involve physical interaction of the plasmid with the chromosome.

In the recipient, the *spd* genes are induced and assemble to a septum traversing complex with the integral membrane protein SpdB2 as the major pore forming protein. By protein–protein interaction the DNA translocase TraB is redirected to the septal cross-wall, interacts with the Spd complex and pumps the DNA to the neighbouring mycelial compartments. Since unregulated expression of *traB* and *spd* genes, as *kilB* (Kieser *et al.*, 1982) or *spdB2* (Tiffert *et al.*, 2007) is toxic, conjugation and plasmid spreading results in the formation of inhibition zones, pock structures.

As depicted above, *Streptomyces* conjugation relies on an FtsK-homologous hexameric ring ATPase TraB. Specificity for the transfer of a given plasmid by TraB is determined by a C-terminal wHTH-motif that recognizes specific TRS repeats within the *clt* locus.

The classical T4SS conjugation system also involves a membrane localized hexameric ring ATPase (TraG/TrwB) with similarity to FtsK. In the T4SS-conjugation system, the relaxase confers sequence specific plasmid nicking producing a single-stranded DNA molecule. By protein–protein interaction the relaxase contacts the TrwB/TraG-ATPase which pumps the ss-DNA molecule with the relaxase covalently attached. From bioinformatic analyses, the main difference between FtsK/TraB and the coupling proteins TraG/TrwB is the presence/absence of the C-terminal wHTH-motif, conferring sequence specific DNA binding.

Interestingly, sequence analysis of *Streptomyces* plasmids identified conjugative relaxases (TraA) on pFP11 and pFP1 in addition to TraB. Furthermore, relaxases are also encoded by the *Amycolatopsis* plasmid pA387 (Malhotra *et al.*, 2008) and the *Nocardiopsis* plasmid pSQ10 (ABD48725). Although involvement of TraA of pFP11 or FP1 in conjugal DNA transfer has not been demonstrated (Zhang *et al.*, 2008), the presence of relaxases raises the provocative question whether a second conjugation mechanism relying on the transfer of a single-stranded plasmid molecule might exist in *Streptomyces*.

Acknowledgements

We thank D. Linke and T. Arnold for help with the single channel conductance measurements and Y. Stierhof for performing electron microscopy. Part of the project was funded by Deutsche Forschungsgemeinschaft (SFB766).

References

Alikhanian, S.I., and Mindlin, S.Z. (1957). Recombinations in *Streptomyces rimosus*. Nature *180*, 1208–1209.

Ausmees, N., Wahlstedt, H., Bagchi, S., Elliot, M.A., Buttner, M.J., and Flardh, K. (2007). SmeA, a small membrane protein with multiple functions in *Streptomyces* sporulation including targeting of a SpoIIIE/FtsK-like protein to cell division septa. Mol. Microbiol. *65*, 1458–1473.

Bentley, S.D., Brown, S., Murphy, L.D., Harris, D.E., Quail, M.A., Parkhill, J., Barrell, B.G., McCormick, J.R., Santamaria, R.I., Losick, R., *et al.* (2004). SCP1, a 356, 023 bp linear plasmid adapted to the ecology and developmental biology of its host, *Streptomyces coelicolor* A3(2). Mol. Microbiol. *51*, 1615–1628.

Benveniste, R., and Davies, J. (1973). Aminoglycoside antibiotic-inactivating enzymes in actinomycetes similar to those present in clinical isolates of antibiotic-resistant bacteria. Proc. Natl. Acad. Sci. U.S.A. *70*, 2276–2280.

Bibb, M.J., and Hopwood, D.A. (1981). Genetic studies of the fertility plasmid SCP2 and its SCP2* variants in *Streptomyces coelicolor* A3(2). J. Gen. Microbiol. *126*, 427–442.

Bibb, M.J., Schottel, J.L., and Cohen, S.N. (1980). A DNA cloning system for interspecies gene transfer in antibiotic-producing *Streptomyces*. Nature *284*, 526–531.

Bigot, S., Saleh, O.A., Lesterlin, C., Pages, C., El Karoui, M., Dennis, C., Grigoriev, M., Allemand, J.F., Barre, F.X., and Cornet, F. (2005). KOPS: DNA motifs that control *E. coli* chromosome segregation by orienting the FtsK translocase. EMBO J. *24*, 3770–3780.

Burton, B.M., Marquis, K.A., Sullivan, N.L., Rapoport, T.A., and Rudner, D.Z. (2007). The ATPase SpoIIIE transports DNA across fused septal membranes during sporulation in *Bacillus subtilis*. Cell *131*, 1301–1312.

Cabezon, E., and de la Cruz, F. (2006). TrwB: an F(1)-ATPase-like molecular motor involved in DNA transport during bacterial conjugation. Res. Microbiol. *157*, 299–305.

Chen, C.W., Yu, T.W., Lin, Y.S., Kieser, H.M., and Hopwood, D.A. (1993). The conjugative plasmid SLP2 of *Streptomyces lividans* is a 50 kb linear molecule. Mol. Microbiol. 7, 925–932.

Christie, P.J., and Cascales, E. (2005). Structural and dynamic properties of bacterial type IV secretion systems. Mol. Membr. Biol. *22*, 51–61.

D'Costa, V.M., McGrann, K.M., Hughes, D.W., and Wright, G.D. (2006). Sampling the antibiotic resistome. Science *311*, 374–377.

Davies, J. (1986). A new look at antibiotic resistance. FEMS Microbiol. Rev. *39*, 363–371.

Deng, Z.X., Kieser, T., and Hopwood, D.A. (1988). 'Strong incompatibility' between derivatives of the *Streptomyces* multi-copy plasmid pIJ101. Mol. Gen. Genet. *214*, 286–294.

Franco, B., Gonzalez-Ceron, G., and Servin-Gonzalez, L. (2003). Direct repeat sequences are essential for function of the cis-acting locus of transfer (*clt*) of *Streptomyces phaeochromogenes* plasmid pJV1. Plasmid *50*, 242–247.

Garcillan-Barcia, M.P., and de la Cruz, F. (2008). Why is entry exclusion an essential feature of conjugative plasmids? Plasmid *60*, 1–18.

Gomis-Ruth, F.X., de la Cruz, F., and Coll, M. (2002). Structure and role of coupling proteins in conjugal DNA transfer. Res. Microbiol. *153*, 199–204.

Gorelik, M., Lunin, V.V., Skarina, T., and Savchenko, A. (2006). Structural characterization of GntR/HutC family signaling domain. Protein Sci. *15*, 1506–1511.

Grohmann, E., Muth, G., and Espinosa, M. (2003). Conjugative plasmid transfer in gram-positive bacteria. Microbiol. Mol. Biol. Rev. *67*, 277–301, table.

Haug, I., Weissenborn, A., Brolle, D., Bentley, S., Kieser, T., and Altenbuchner, J. (2003). *Streptomyces coelicolor* A3(2) plasmid SCP2*: deductions from the complete sequence. Microbiology *149*, 505–513.

Hopwood, D.A. (1959). Linkage and the mechanism of recombination in *Streptomyces coelicolor*. Ann. N.Y. Acad. Sci. *81*, 887–898.

Hopwood, D.A., Bibb, M.J., Chater, K.F., and Kieser, T. (1987). Plasmid and phage vectors for gene cloning and analysis in *Streptomyces*. Methods Enzymol. *153*, 116–165.

Hopwood, D.A., and Kieser, T. (1993). Conjugative Plasmids of *Streptomyces*. In Bacterial Conjugation, Clewell, D.B., ed. (New York, Plenum Press), pp. 293–311.

Hosted, Jr., T.J., Wang, T., and Horan, A.C. (2005). Characterization of the *Micromonospora rosaria* pMR2 plasmid and development of a high G+C codon optimized integrase for site-specific integration. Plasmid *54*, 249–258.

Hsu, C.C., and Chen, C.W. (2010). Linear plasmid SLP2 is maintained by partitioning, intrahyphal spread, and conjugal transfer in *Streptomyces*. J. Bacteriol. *192*, 307–315.

Huang, C.H., Lin, Y.S., Yang, Y.L., Huang, S.W., and Chen, C.W. (1998). The telomeres of *Streptomyces* chromosomes contain conserved palindromic sequences with potential to form complex secondary structures. Mol. Microbiol. *28*, 905–916.

Kataoka, M., Kiyose, Y.M., Michisuji, Y., Horiguchi, T., Seki, T., and Yoshida, T. (1994a). Complete nucleotide sequence of the *Streptomyces nigrifaciens* plasmid, pSN22: Genetic organization and correlation with genetic properties. Plasmid *32*, 55–69.

Kataoka, M., Kosono, S., Seki, T., and Yoshida, T. (1994b). Regulation of the transfer genes of *Streptomyces* plasmid pSN22: In vivo and in vitro study of the interaction of TraR with promoter regions. J. Bacteriol. *176*, 7291–7298.

Kataoka, M., Kuno, N., Horiguchi, T., Seki, T., and Yoshida, T. (1994c). Replication of the *Streptomyces* plasmid pSN22 through single- stranded intermediates. Mol. Gen. Genet. *242*, 130–136.

Kataoka, M., Tanaka, T., Kohno, T., and Kajiyama, Y. (2008). The carboxyl-terminal domain of TraR, a *Streptomyces* HutC family repressor, functions in oligomerization. J. Bacteriol. *190*, 7164–7169.

Keen, C.L., Mendelovitz, S., Cohen, G., Aharonowitz, Y., and Roy, K.L. (1988). Isolation and characterization of a linear DNA plasmid from Streptomyces clavuligerus. Mol. Gen. Genet. *212*, 172–176.

Kendall, K.J., and Cohen, S.N. (1987). Plasmid transfer in *Streptomyces lividans*: Identification of a kil-kor system associated with the transfer region of pIJ101. J. Bacteriol. *169*, 4177–4183.

Kendall, K.J., and Cohen, S.N. (1988). Complete nucleotide sequence of the *Streptomyces lividans* plasmid pIJ101 and correlation of the sequence with genetic properties. J. Bacteriol. *170*, 4634–4651.

Kieser, H.M., Kieser, T., and Hopwood, D.A. (1992). A combined genetic and physical map of the *Streptomyces coelicolor* A3(2) chromosome. J. Bacteriol. *174*, 5496–5507.

Kieser, T., Hopwood, D.A., Wright, H.M., and Thompson, C.J. (1982). pIJ101, a multi-copy broad host-range *Streptomyces* plasmid: functional analysis and development of DNA cloning vectors. Mol. Gen. Genet. *185*, 223–238.

Kinashi, H., and Shimaji, M. (1987). Detection of giant linear plasmids in antibiotic producing strains of *Streptomyces* by the OFAGE technique. Antibiot. J. (Tokyo) *40*, 913–915.

Kramer, M.G., Espinosa, M., Misra, T.K., and Khan, S.A. (1998). Lagging strand replication of rolling-circle plasmids: specific recognition of the *ssoA*-type origins in different gram-positive bacteria. Proc. Natl. Acad. Sci. U.S.A. *95*, 10505–10510.

Larson, J.L., and Hershberger, C.L. (1986). The minimal replicon of a streptomycete plasmid produces an ultrahigh level of plasmid DNA. Plasmid *15*, 199–209.

Lin, Y.R., Hahn, M.Y., Roe, J.H., Huang, T.W., Tsai, H.H., Lin, Y.F., Su, T.S., Chan, Y.J., and Chen, C.W. (2009). *Streptomyces* telomeres contain a promoter. J. Bacteriol. *191*, 773–781.

Lin, Y.S., Kieser, H.M., Hopwood, D.A., and Chen, C.W. (1994). The chromosomal DNA of *Streptomyces lividans* 66 is linear. Mol. Microbiol. *14*, 1103.

Liu, N.J., Dutton, R.J., and Pogliano, K. (2006). Evidence that the SpoIIIE DNA translocase participates in membrane fusion during cytokinesis and engulfment. Mol. Microbiol. *59*, 1097–1113.

Llosa, M., Gomis-Ruth, F.X., Coll, M., and de la Cruz, F. (2002). Bacterial conjugation: a two-step mechanism for DNA transport. Mol. Microbiol. *45*, 1–8.

Lowe, J., Ellonen, A., Allen, M.D., Atkinson, C., Sherratt, D.J., and Grainge, I. (2008). Molecular mechanism of sequence-directed DNA loading and translocation by FtsK. Mol. Cell *31*, 498–509.

Lydiate, D.J., Malpartida, F., and Hopwood, D.A. (1985). The *Streptomyces* plasmid SCP2*: its functional analysis and development into useful cloning vectors. Gene *35*, 223–235.

Malhotra, S., Majumdar, S., Kumar, M., Bhasin, V.K., Gartemann, K.H., and Lal, R. (2008). Nucleotide sequence of plasmid pA387 of *Amycolatopsis benzoatilytica* and construction of a conjugative shuttle vector. J. Basic Microbiol. *48*, 177–185.

Malpartida, F., and Hopwood, D.A. (1984). Molecular cloning of the whole biosynthetic pathway of a *Streptomyces* antibiotic and its expression in a heterologous host. Nature *309*, 462–464.

Martin, C., Mazodier, P., Mediola, M.V., Gicquel, B., Smokvina, T., Thompson, C.J., and Davies, J. (1991). Site specific integration of the *Streptomyces* plasmid pSAM2 in *Mycobacterium smegmatis*. Mol. Microbiol. *5*, 2499–2502.

Massey, T.H., Mercogliano, C.P., Yates, J., Sherratt, D.J., and Lowe, J. (2006). Double-stranded DNA translocation: structure and mechanism of hexameric FtsK. Mol. Cell *23*, 457–469.

Muth, G., Farr, M., Hartmann, V., and Wohlleben, W. (1995). *Streptomyces ghanaensis* plasmid pSG5: nucleotide sequence analysis of the self-transmissible minimal replicon and characterization of the replication mode. Plasmid *33*, 113–126.

Nordstroem, K. (1993). Plasmid replication and maintenance. In Plasmids, A Practical Approach, Hardy, K.G., ed. (Oxford, IRL Press), pp. 1–38.

Pandza, S., Biukovic, G., Paravic, A., Dadbin, A., Cullum, J., and Hranueli, D. (1998). Recombination between the linear plasmid pPZG101 and the linear chromosome of *Streptomyces rimosus* can lead to exchange of ends. Mol. Microbiol. *28*, 1165–1176.

Pettis, G.S., and Cohen, S.N. (1994). Transfer of the pIJ101 plasmid in *Streptomyces lividans* requires a *cis*-acting function dispensable for chromosomal gene transfer. Mol. Microbiol. *13*, 955–964.

Pettis, G.S., Ward, N., and Schully, K.L. (2001). Expression characteristics of the transfer-related *kilB* gene product of *Streptomyces* plasmid pIJ101: implications for the plasmid spread function. J. Bacteriol. *183*, 1339–1345.

Possoz, C., Gagnat, J., Sezonov, G., Guerineau, M., and Pernodet, J.L. (2003). Conjugal immunity of *Streptomyces* strains carrying the integrative element pSAM2 is due to the pif gene (pSAM2 immunity factor). Mol. Microbiol. *47*, 1385–1393.

Possoz, C., Ribard, C., Gagnat, J., Pernodet, J.L., and Guerineau, M. (2001). The integrative element pSAM2 from *Streptomyces*: kinetics and mode of conjugal transfer. Mol. Microbiol. *42*, 159–166.

Qin, Z.J., and Cohen, S.N. (1998). Replication at the telomeres of the *Streptomyces* linear plasmid pSLA2. Mol. Microbiol. *28*, 893–903.

Redenbach, M., Bibb, M., Gust, B., Seitz, B., and Spychaj, A. (1999). The linear plasmid SCP1 of *Streptomyces coelicolor* A3(2) possesses a centrally located replication origin and shows significant homology to the transposon Tn4811. Plasmid *42*, 174–185.

Reuther, J., Gekeler, C., Tiffert, Y., Wohlleben, W., and Muth, G. (2006a). Unique conjugation mechanism in mycelial streptomycetes: a DNA-binding ATPase translocates unprocessed plasmid DNA at the hyphal tip. Mol. Microbiol. *61*, 436–446.

Reuther, J., Wohlleben, W., and Muth, G. (2006b). Modular architecture of the conjugative plasmid pSVH1 from *Streptomyces venezuelae*. Plasmid *55*, 201–209.

Schrempf, H., and Goebel, W. (1977). Characterization of a plasmid from *Streptomyces coelicolor* A3(2). J. Bacteriol. *131*, 251–258.

Schröder, G., and Lanka, E. (2003). TraG-like proteins of type IV secretion systems: functional dissection of the multiple activities of TraG (RP4) and TrwB (R388). J. Bacteriol. *185*, 4371–4381.

Schully, K.L., and Pettis, G.S. (2003). Separate and coordinate transcriptional control mechanisms link expression of the potentially lethal KilB spread locus to the upstream transmission operon on *Streptomyces* plasmid pIJ101. J. Mol. Biol. *334*, 875–884.

Servín-González, L. (1993). Relationship between the replication functions of *Streptomyces* plasmids pJV1 and pIJ101. Plasmid *30*, 131–140.

Servín-González, L. (1996). Identification and properties of a novel *clt* locus in the *Streptomyces phaeochromogenes* plasmid pJV1. J. Bacteriol. *178*, 4323–4326.

Servín-González, L., Sampieri, A., Cabello, J., Galván, L., Juárez, V., and Castro, C. (1995). Sequence and functional analysis of the *Streptomyces phaeochromogenes* plasmid pJV1 reveals a modular organization of *Streptomyces* plasmids that replicate by rolling circle. Microbiology *141*, 2499–2510.

Shiffman, D., and Cohen, S.N. (1992). Reconstruction of a *Streptomyces* linear replicon from separately cloned DNA fragments: existence of a cryptic origin of circular replication within the linear plasmid. Proc. Natl. Acad. Sci. U.S.A. *89*, 6129–6133.

Smith, M.C., Burns, R.N., Wilson, S.E., and Gregory, M.A. (1999). The complete genome sequence of the *Streptomyces* temperate phage straight phiC31: evolutionary relationships to other viruses. Nucleic Acids Res. *27*, 2145–2155.

Smokvina, T., Boccard, F., Pernodet, J.-L., Friedmann, A., and Guerineau, M. (1991). Functional analysis of the *Streptomyces ambofaciens element* pSAM2. Plasmid *25*, 40–52.

Smokvina, T., Mazodier, P., Boccard, F., Thompson, C.J., and Guerineau, M. (1990). Construction of a series of pSAM2-based integrative vectors for use in actinomycetes. Gene *94*, 53–59.

Suzuki, I., Kataoka, M., Seki, T., and Yoshida, T. (1997). Three single-strand origins located on both strands of the *Streptomyces* rolling circle plasmid pSN22. Plasmid *37*, 51–64.

Tai, J.T.N., and Cohen, S.N. (1993). The active form of the KorB protein encoded by the *Streptomyces* plasmid pIJ101 is a processed product that binds differentially to the two promoters it regulates. J. Bacteriol. *175*, 6996–7005.

Te Poele, E.M., Bolhuis, H., and Dijkhuizen, L. (2008a). Actinomycete integrative and conjugative elements. Antonie Van Leeuwenhoek *94*, 127–143.

Te Poele, E.M., Samborskyy, M., Oliynyk, M., Leadlay, P.F., Bolhuis, H., and Dijkhuizen, L. (2008b). Actinomycete integrative and conjugative pMEA-like elements of *Amycolatopsis* and *Saccharopolyspora* decoded. Plasmid *59*, 202–216.

Tiffert, Y., Gotz, B., Reuther, J., Wohlleben, W., and Muth, G. (2007). Conjugative DNA transfer in *Streptomyces*: SpdB2 involved in the intramycelial spreading of plasmid pSVH1 is an oligomeric integral membrane protein that binds to dsDNA. Microbiology *153*, 2976–2983.

Van Dessel, W., Van Mellaert, L., Liesegang, H., Raasch, C., De Keersmaeker, S., Geukens, N., Lammertyn, E., Streit, W., and Anne, J. (2005). Complete genomic nucleotide sequence and analysis of the temperate bacteriophage VWB. Virology *331*, 325–337.

Vrijbloed, J.W., Van der Put, N.M.J., and Dijkhuizen, L. (1995a). Identification and functional analysis of the transfer region of plasmid pMEA300 of the methylotrophic actinomycete *Amycolatopsis methanolica*. J. Bacteriol. *177*, 6499–6505.

Vrijbloed, J.W., Vlieg, J.V., Van der Put, N.M.J., Hessels, P.G.I., and Dijkhuizen, L. (1995b). Molecular cloning with a pMEA300-derived shuttle vector and characterization of the *Amycolatopsis methanolica* prephenate dehydratase gene. J. Bacteriol. *177*, 6666–6669.

Willets, N., and Wilkins, B. (1984). Processing of plasmid DNA during bacterial conjugation. Microbiol. Rev. *48*, 24–41.

Wohlleben, W., and Pühler, A. (1987). The *Streptomyces ghanaensis* low copy plasmid pSG2 and its use for vector construction. Arch. Microbiol. *148*, 298–304.

Wright, G.D. (2007). The antibiotic resistome: the nexus of chemical and genetic diversity. Nat. Rev. Microbiol. *5*, 175–186.

Wu, L.J., and Errington, J. (1994). *Bacillus subtilis* SpoIIIE protein required for DNA segregation during asymmetric cell division. Science *264*, 572–575.

Xu, M., Zhu, Y., Zhang, R., Shen, M., Jiang, W., Zhao, G., and Qin, Z. (2006). Characterization of the genetic components of *Streptomyces lividans* linear plasmid SLP2 for replication in circular and linear modes. J. Bacteriol. *188*, 6851–6857.

Zechner, E.L., de la Cruz, F., Eisenbrandt, R., Grahn, A.M., Koraimann, G., Lanka, E., Muth, G., Pansegrau, W., Thomas, C.M., Wilks, M., *et al.* (2000). Conjugative-DNA transfer processes. In The Horizontal Gene Pool – Bacteriol Plasmids and Gene Spread, Thomas, C.M., ed. (Amsterdam, Harwood Academic Publishers), pp. 87–175.

Zhang, R., Yang, Y., Fang, P., Jiang, C., Xu, L., Zhu, Y., Shen, M., Xia, H., Zhao, J., Chen, T., and Qin, Z. (2006). Diversity of telomere palindromic sequences and replication genes among *Streptomyces* linear plasmids. Appl. Environ. Microbiol. *72*, 5728–5733.

Zhang, R., Zeng, A., Fang, P., and Qin, Z. (2008). Characterization of replication and conjugation of *Streptomyces* circular plasmids pFP1 and pFP11 and their ability to propagate in linear mode with artificially attached telomeres. Appl. Environ. Microbiol. *74*, 3368–3376.

Differentiation in *Streptomyces*: the Properties and Programming of Diverse Cell Types

3

Keith F. Chater

Abstract

Streptomyces colonies are complex differentiated organisms, generated from a single ovoid spore by filamentous growth and branching. Eventually, much of this biomass is converted to large numbers of spores in long chains on specialized aerial hyphae. During colony development, different cellular compartments have different physiology and metabolism, and exoskeletal and cytoskeletal elements bring about different morphological changes. These cellular differentiating processes are underpinned by a large number of regulatory genes, often operating in cascades. During the transition from biomass accumulation to reproductive development, antibiotics are made, sometimes under the control of developmental regulators.

Introduction

Hundreds of millions of years ago, streptomycetes assumed their key place in soil ecology, because of their ability to solubilize chitin, cellulose and other recalcitrant macromolecular components of soil by producing numerous hydrolytic exoenzymes (Chater *et al.*, 2010). Streptomycetes accumulate biomass as branching substrate mycelium. They characteristically achieve dispersal by forming a specialized reproductive aerial mycelium, which grows largely parasitically on the substrate mycelial biomass. Much of the aerial mycelium is eventually turned into long chains of round spores (Fig. 3.1).

In the course of colony differentiation, some resources are also given over to the production of antibiotics, a physiological differentiation that involves some regulatory elements in common with morphological differentiation. This has led to streptomycetes becoming major industrial organisms, supplying many of the antibiotics and related molecules that underpin treatments of infectious diseases, as well as agents valuable for other medical, veterinary and agricultural purposes. The principal driving force for the virtuosity of *Streptomyces* in antibiotic production may well be the defence of nutrients released both from the solubilization of soil materials, and from the *Streptomyces* biomass during its conversion into spores (Chater and Merrick, 1979). To understand fully the mechanisms that influence antibiotic production, the broader developmental biology of streptomycetes needs to be taken into account. The aim of this chapter is to provide something approaching a comprehensive overview of the growth and development of *Streptomyces*.

Under laboratory conditions, a single *Streptomyces* spore can give rise to millions more in the course of a few days. This simple statement masks substantial developmental complexity, and raises many questions. How are the transitions between round and filamentous cell morphology brought about? What machinery ensures a suitable frequency and positioning of hyphal branching? What are the structural and physiological requirements for growth into the air, away from nutrient supplies? How is it all controlled? This chapter summarizes, and attempts to integrate, current understanding of the ways in which cytoskeletal and exoskeletal elements bring about these morphological transitions, under

Figure 3.1 Round and filamentous cell types in a colony of *Streptomyces coelicolor* A3(2). Chains of ovoid spores held together by sheath material cover the colony surface. Below them can be seen cylindrical hyphae, some of which are intact (thick arrow), while others have undergone lysis to provide nutrients for sporulation (thin arrow). Scanning electron micrograph provided by Kim Findlay.

the control of a suite of regulatory elements, and supported by changes in the ways nutrients are supplied and utilized.

The overall topic of *Streptomyces* development has been reviewed quite frequently and from various perspectives. A recent wide-ranging account is by Elliot *et al.* (2008), and selected other reviews have emphasized physiology (Hodgson, 1992), integration with secondary metabolism (Champness and Chater, 1994), genetic analysis (Chater, 1998), interspecific comparisons and the involvement of autoregulatory molecules such as A-factor (Chater and Horinouchi, 2003), extracellular aspects (Chater, 2006; Chater *et al.*, 2010), evolution (Chater and Chandra, 2006) and cell biology (Flardh and Buttner, 2009).

Part 1. From quiescence to biomass accumulation

Spores: quiescent near-spherical cells

Structure of the outer layers of spores

In soil samples, the vast majority of *Streptomyces* genomes are usually present in uninucleoidal spores (Ensign, 1978), ovoid, thick-walled cells with low metabolic activity, about 1 µm in diameter. Some details of spore physiology were summarized by Kalakoutski and Agre (1976), Hirsch and Ensign (1978), Ensign (1978) and Garcia Diaz *et al.* (1983). The ca. 50-nm-thick peptidoglycan spore wall and a thin outer layer together confer resistance to rough physical treatment (such as sonication), chemicals (such as 0.01 N HCl), or degradative enzymes. The moderate heat resistance of spores is mildly impaired in the thin-walled spores that result from mutations in certain genes for peptidoglycan hydrolases (Haiser *et al.*, 2009). It is more severely compromised by mutations in *whiD*, a sporulation regulatory gene, or *mreB*, a cytoskeletal sporulation protein, which give rise to irregularly sized spores with relatively thin walls, possibly with other pleiotropic deficiencies (Molle *et al.*, 2000; Mazza *et al.*, 2006).

In most streptomycetes, aromatic pigments derived through the agency of type II or type III polyketide synthases are tightly bound to spore walls (Davis and Chater, 1990; Metsä-Ketelä *et al.*, 2002; Funa *et al.*, 1999, 2005). These probably serve to lignify and thereby further toughen the walls, increasing the ability of spores to survive passage through the guts of earthworms or other animals that might frequently consume *Streptomyces*-containing material, as suggested for fungal spore pigments (Bloomfield and Alexander, 1967). Pigmentation of *Bacillus* endospores, in contrast, is a property of one of the coat proteins (Driks and Setlow, 2000).

The outermost thin sculptured sheath layer of *Streptomyces* spores comprises rodlins and chaplins, amphipathic proteins that confer considerable hydrophobicity, with rodlins organizing chaplin fibres into paired rodlets that give a basketwork-like appearance to the surface of *S. coelicolor* spores (Wildermuth, 1970b; Wildermuth *et al.*, 1971; Claessen *et al.*, 2002, 2004; Fig. 3.2A). An absence of paired rodlets from the smooth spores of *S. avermitilis* is consistent with the absence of rodlin genes from its genome. In other streptomycetes, spiny, warty or hairy projections from spores also appear to be composed of paired rodlets, probably made of chaplins and rodlins (Fig. 3.2B and C) (Wildermuth, 1970b, 1972a,b; Williams *et al.*, 1972, and references therein). Perhaps the ornamentations are involved in attachment of spores to different surfaces.

Rodlins and chaplins can be extracted only by treatment with such severe solvents as trifluoroacetic acid (Claessen *et al.*, 2002, 2004; Elliot *et al.*, 2003). Other proteins ('Saps', for spore-associated proteins) have been extracted from the surface of spores by relatively mild detergent treatment (SapA, -B, -C, -D, -E: Guijarro *et al.*, 1988; Willey *et al.*, 1991; Bentley *et al.*, 2004). It is not known how Saps contribute to the properties of spores, though SapB is important in the formation of aerial hyphae under some conditions (see below). Cytological evidence has shown that the spore wall also harbours NepA protein, which has been implicated in maintaining spore dormancy; but NepA could not be extracted under the conditions tested (de Jong *et al.*, 2009a).

Resistance properties involving intracellular functions

Although the spore DNA is not packaged by proteins equivalent to the SASPs of *Bacillus* spores (Driks and Setlow, 2000), some kind of DNA condensation is visible by appropriate microscopy during the formation of spores in aerial hyphae (e.g. Flärdh *et al.*, 1999). One component of the machinery for holding the chromosomes in this condensed state is a histone-like protein, HupS,

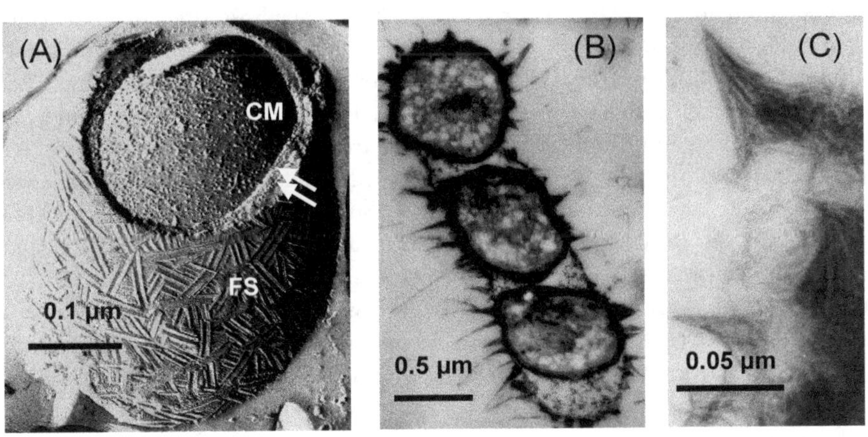

Figure 3.2 Layers on the spore surface. (A) Freeze-fractured spore of *S. coelicolor* showing the basketwork-like mosaic of the outermost fibrous sheath (FS) overlying a double-layered spore wall (twin arrows) and the cytoplasmic membrane (CM). (B). Hair-like extensions from the outer sheath of a developing spore chain of *Streptomyces glaucescens* (thin section, transmission electron microscopy). (C). Spiny protrusions from the sheath of *Streptomyces viridochromogenes*, showing their paired rodlet appearance (negative staining, transmission electron microscopy). The photographs were taken by Hansreudi Wildermuth (Wildermuth *et al.*, 1971; Wildermuth, 1972a,b).

that is maximally abundant in spores (Salerno *et al.*, 2009). HupS-like proteins are found in many actinomycetes, and consist of a ca. 90-aa N-terminal domain similar to the HU proteins present in most bacteria, and an actinobacteria-specific C-terminal domain of ca. 190 aa, rich in basic residues, that resembles histone H1 of eukaryotes. Spores of a *hupS* null mutant have chromosomes that are not fully condensed, and such spores are markedly reduced in heat resistance (Salerno *et al.*, 2009). *Streptomyces coelicolor* also has three orthologues of Dps proteins, which function in other bacteria to protect DNA against various stresses. The three proteins appear to interact to control chromosome condensation during sporulation, and this has effects on spore heat resistance: in particular, mutations in *dpsB* show a loss of heat resistance (Facey *et al.*, 2009). However, some other *dps* single and double mutants were unaffected in heat resistance, despite having changes in DNA condensation (Facey *et al.*, 2009).

Wild-type spores are very resistant to desiccation. To some extent, this resistance property may be attributable to a high content of trehalose (α-1,2-linked diglucose), which is associated with protection against various stresses in bacteria and fungi (Elbein *et al.*, 2003; Martín *et al.*, 1986), and can make up as much as 18% of a spore's dry weight (Hey-Ferguson *et al.*, 1973; Braña *et al.*, 1986; Martín *et al.*, 1986; McBride and Ensign, 1987a,b; Rueda *et al.*, 2001). The spores of *Streptomyces griseus* contain a trehalase enzyme, held inactive until spore germination (McBride and Ensign, 1987b, 1990).

Spore germination and outgrowth: the transition from a rounded to a filamentous form, and from metabolism of endogenous resources to utilization of exogenous nutrients

Activation of metabolism

Strictly, spore germination is seen as swelling of the spores and a loss of refractility, which accompany an increased rate of respiration and an increase in ATP content (Hirsch and Ensign, 1978). Germination is dependent on the presence of Ca^{2+} and CO_2 in defined minimal medium. Initiation is usually very asynchronous in spore populations, but various procedures can increase the extent of synchrony (see Kieser *et al.*, 2000). During germination, the excretion of germination inhibitory substances such as germicidin and hypnosin may serve to limit germination to just part of a population of locally abundant spores, for example preventing mass germination *in situ* in spore chains (Grund and Ensign, 1985; Peterson *et al.*, 1993; Challis, 2008; Aoki *et al.*, 2007).

Germination and subsequent germ tube emergence involve the use of endogenous reserves, with up to 10% of endogenous carbon being released as CO_2 in the first 90 min after activation (Hirsch and Ensign, 1978). A major source of this CO_2 is stored trehalose, which is hydrolysed by activation of the spore-associated trehalase mentioned above. This interesting enzyme has not been identified experimentally with any gene product predicted from the genome sequence, though several *Streptomyces* genes encode products with highly significant similarity (about 28–41% amino acid identity) to a biochemically defined trehalase of mycobacteria (Carroll *et al.*, 2007; G. Chandra and K.F. Chater, unpublished). A remarkable feature of the mycobacterial trehalase is its inhibition by polyphosphate, and dependence on free phosphate (Carroll *et al.*, 2007). Polyphosphate can be detected by electron microscopy in *Streptomyces* aerial mycelium prior to sporulation (Plaskitt and Chater, 1995), so it is possible that it could be implicated in holding trehalase in an inactive state: but it is not clear whether polyphosphate persists as spores mature.

Streptomycetes also have *rpf* genes for so-called resuscitation factors. These are secreted proteins with cell wall hydrolytic activity that are widespread among actinobacteria and are important in allowing cells to emerge from quasi-dormant states (Mukamolova *et al.*, 2006). One of the seven in *S. coelicolor* has been disrupted (*rpfA*), and the mutant was found to show delayed germination (Haiser *et al.*, 2009). It will be interesting to learn whether the activity of cell wall hydrolases such as resuscitation factors is controlled by germination inhibitors such as NepA and germicidin (see above).

The onset of macromolecular synthesis

Stable mRNA is present in spores, and is expressed upon germination (Mikulík *et al.*, 1984; Quiros *et al.*, 1985). The proteins made from this stable mRNA have yet to be fully characterized, but at least 15 different spots were detected by 2-D gel electrophoresis of proteins extracted from spores of *Streptomyces granaticolor* induced to germinate synchronously by mechanical disruption of the surface sheath, (Stastná, 1977; Mikulík *et al.*, 2002) (for unknown reasons, this activation procedure does not seem to be widely effective for other streptomycetes).

Within minutes of the activation of germination, de novo transcription and translation begin. The ribosomes present in spores are inactive, so their activation is among the earliest events (Mikulík *et al.*, 1984; Bobek *et al.*, 2004). An important part of this is the increased association of ribosomes with chaperones that are also present in the spore, notably DnaK and peptidyl prolyl *cis-trans* isomerase. This association probably follows from two changes: the increased hydration of the spore cytoplasm, which enhances the mobility of proteins; and the phosphorylation of these and other chaperones (notably GroEL) by the action of undefined protein kinase(s). Likewise, RNA polymerase subunits become freed from aggregated forms. The hundreds of proteins resulting from de novo transcription/translation, before DNA replication starts about an hour later, include components of the translational apparatus, as well as many of the enzymes of central carbon metabolism and amino acid and nucleotide biosynthesis (Bobek *et al.*, 2004).

The ubiquitous signalling molecule cyclic AMP is important for efficient germination, but its role at this or any other part of the developmental cycle in *Streptomyces* remains shadowy (Süsstrunk *et al.*, 1998). In *S. coelicolor*, mutations eliminating adenyl cyclase or CRP, a cAMP-binding protein present in many bacteria, result in very low germination rates (Süsstrunk *et al.*, 1998; Derouaux *et al.*, 2004a,b) [unlike the situation in *E. coli*, CRP has no role in catabolite repression in *S. coelicolor* (Derouaux *et al.*, 2004a), and neither does cAMP (Hodgson, 2000) – see Chapter 5]. No clear-cut DNA targets for CRP have been defined, though changes in abundance of the products of some 45 different genes were observed by proteomic analysis during germination of a *crp* mutant (Piette *et al.*, 2005). A *crp* mutant had reduced levels of several general stress response proteins, consistent with the importance of macromolecular renaturation during germination. There were also deficiencies in ppGppp synthetase, which might result in lower levels of the signal molecules ppGpp/ppGppp, and of a peptidoglycan hydrolase (SCO5466) that may be important in degrading the spore wall during germination (along with other such hydrolases: Haiser *et al.*, 2009).

DNA synthesis starts about one hour after the start of RNA and protein synthesis (Mikulík *et al.*, 1977, 2002; Bobek *et al.*, 2004). The initiation of bacterial chromosome replication, even in the case of the linear chromosomes of *Streptomyces*, takes place at a specific site on the chromosome, the origin of replication or *oriC*. This site is typically located fairly centrally in the chromosome, next to the *dnaA* gene, whose product (DnaA) is the key determinant of replication initiation (see Zawilak-Pawlik *et al.*, 2005 for a review). The *Streptomyces oriC* has an exceptionally high number of binding sites for DnaA (19, compared with 5 in *E. coli*), though most of these sites depart slightly from the strong consensus sequence. Several further clusters of DnaA boxes are located away from *oriC*, and the high number of strong DnaA boxes in these latter clusters affects the availability of DnaA protein, and hence the frequency of chromosome replication initiation and the rate of growth (Smulczyk-Krawczyszyn *et al.*, 2006). It is likely that HupS and other molecules that hold the chromosome in its condensed state in spores must be removed for DnaA and other replication proteins to gain access to *oriC*. The mechanism for this removal, which presumably ends with the proteolytic degradation of the packaging proteins, is completely unstudied. Probably, DnaA protein accumulates during the lag preceding replication, until it reaches a level sufficient to bring about the necessary opening of *oriC*: the components of the replication machinery ('replisome') are presumably also being synthesized and assembled during this period.

Polar growth

Outgrowth takes place about an hour after the initiation of replication, with the emergence of one or two germ tubes (not necessarily synchronously). The positions of germ tube emergence coincide with foci of SsgA, one of a family of paralogous proteins peculiar to complex actinomycetes (Noens *et al.*, 2007; Traag and van Wezel, 2008). These foci are laid down during spore formation (see later section). The walls of the germ tube are co-extensive with the innermost spore wall layer (Hardisson *et al.*, 1978 and references therein; Fig. 3.2A). Thus, the almost spherical spore acquires the ability to grow as a cylindrical hypha.

A critical determinant of the hyphal growth mode is a cytoplasmic coiled coil protein, DivIVA, orthologues of which are present in many bacteria. In bacilli, in which DivIVA was first characterized, its role is to permit centrally located cell division to take place: a propensity of DivIVA to adhere to concave membrane surfaces causes it to collect at cell poles, where it binds proteins that inhibit cell division, holding them away from the cell's midpoint (Lenarcic *et al.*, 2009: Ramamurthi and Losick, 2009). In streptomycetes and other actinobacteria DivIVA is located at hyphal tips (Fig 3.3), perhaps because of its affinity for concave membrane surfaces. It is associated with the cell wall biosynthetic apparatus to determine polar growth (Flärdh, 2003a,b). DivIVA foci are present even in the shortest germ tubes, so they are probably assembled close to SsgA foci in germinated spores just before germ-tube emergence. It is therefore likely that the *divIVA* gene is strongly expressed early in germination (Flärdh, 2003a,b).

According to one report, germ tube extension accelerates periodically as chromosomes go through sequential rounds of replication (Kretschmer, 1988) to a maximum final constant elongation rate of about 20–23 µm/h (Kretschmer, 1988; Jyothikumar *et al.*, 2008). The intervals between rounds of replication at this stage average 47 min, even though the replication of an entire chromosome takes twice as long (Lee *et al.*, 2002). There may therefore be multiple replication forks and a relatively high representation of *oriC*-proximal genes at this stage. The rates of protein synthesis during outgrowth observed by Bobek *et al.* (2004) accelerated with a comparable periodicity. Since ribosomal proteins made up a large part of this synthesis, the acceleration may correspond to the replication of the relevant gene sets.

Early programmed cell death?

Mainly on the basis of staining with dyes that differentially stain live and dead cells, it has been suggested that germ tubes exhibit alternating live and dead segments – a kind of programmed cell death (Manteca *et al.*, 2005a,b, 2006a,b).

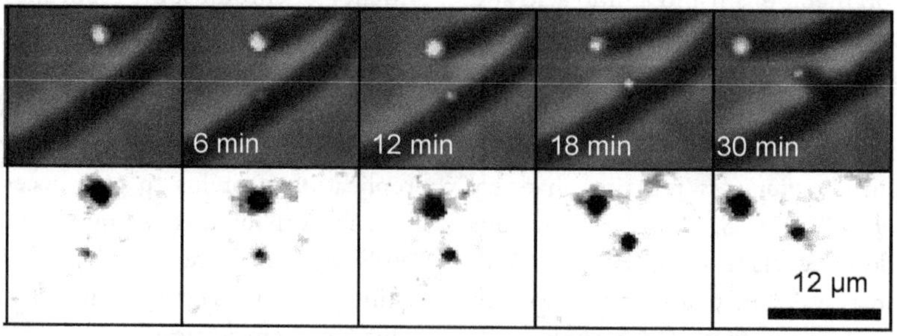

Figure 3.3 DivIVA marks sites of hyphal branching. DivIVA fused to EGFP (enhanced green fluorescent protein) is visualized by fluorescence microscopy. The fused gene was introduced into a strain that has an intact copy of *divIVA*, and images were captured at the times indicated (these images, taken by Antje Hempel, are from a longer time-lapse sequence: Hempel *et al.*, 2008). In the upper set of panels, the fluorescent images are superimposed on phase contrast images, and in the lower set the fluorescent images are presented with the grey scale inverted to give increased sensitivity, clearly showing that DivIVA accumulates at a future branch-point.

Electron microscopic and fluorescence images provided some support for multiple septa in germ tubes (Manteca *et al.*, 2005b), but these images were not unambiguous; and in my opinion it remains to be established more directly that this is a correct interpretation, since the differentially staining segments occur before there is any evidence of the formation of the Z-rings that are expected to precede septation (see, for example, Jyothikumar *et al.*, 2008). If such compartments do exist, they will make it difficult to explain the accelerating growth rate of the tips of germlings, which has been considered to reflect the increasing number of chromosome copies in the germ-tube cytoplasm. More work is needed to clarify the relationships between the observations of different laboratories using different technical approaches to study the early stages of growth.

The establishment of a branching mycelium with flexible metabolism

Branch emergence

Hyphal extension rates at tips are likely to be limited by the catalytic properties of the cell wall biosynthetic machinery, the physical constraints of the tip dimensions, and the supplies of energy and precursors. Thus, an exponentially increasing growth rate keeping pace with successive rounds of chromosome doubling cannot be expected to be maintained for long by a germ tube. To solve this potential growth limitation, new growth points must emerge. In the first place, another germ tube may emerge from the spore; but branches eventually arise in the germ tubes, and a protomycelium is formed. Branches can emerge even from quite short germ tubes under some conditions (Jyothikumar *et al.*, 2008), while in other conditions the branches are up to 42 μm behind the growing germling tip (Kretschmer, 1991). In older, more well-developed, mycelium, branch points are usually well behind the growing tip (for example, about 10 μm in the work of Jyothikumar *et al.*, 2008, at least 16 μm in that of Allan and Prosser, 1987, and up to 44 μm in that of Kretschmer, 1992). Positions of imminent branch emergence are marked by an accumulation of DivIVA, which subsequently remains at the advancing tip of the branch after its emergence (Fig. 3.3; Hempel *et al.*, 2008). It is not known what determines the position of the DivIVA deposits that precede branch emergence, but branches are frequently associated with the outside of bends in hyphae (Hempel *et al.*, 2008). This may reflect the increased concavity of the inside of the membrane, which might make it a preferential target for DivIVA (see above). SsgA protein is also associated with vegetative hyphal tips (Noens *et al.*, 2007), but it is not clear at what point during branch emergence the SsgA foci form.

There is some evidence for an involvement of specific peptidoglycan hydrolases in branch growth: when four of the 60 such hydrolases encoded by the *S. coelicolor* genome were separately eliminated by targeted mutagenesis, two of the mutants (*swlB, swlC*) had relatively short side branches (Haiser *et al.*, 2009).

The formation of cross-walls

As the mycelium develops, cross-walls (vegetative septa) form at some distance behind the growing tips, delineating tip compartments up to 41 μm long (and correspondingly containing tens of copies of the genome), and shorter subapical compartments (Kretschmer, 1989). Although cross-walls are often close to branch points, there is no obligatory association (see Flärdh, 2003b). It has been suggested that, at least at fast growth rates, the position of septum formation may first be determined by the midpoint of the apical compartment, but that the tip will have moved on by the time that the cross-wall has become visible (Kretschmer, 1989). This model begs the question of how the mid-point might be determined in cells that, unlike rod-shaped bacteria, are asymmetrical.

In streptomycetes, many of the proteins that make up the cell division apparatus (the 'divisome') are encoded by a large gene cluster – the *dcw* cluster, for 'division and cell wall'- that is present in all but wall-less bacteria. The most well-known product of the cluster is the tubulin-like GTPase FtsZ, a cytoskeletal protein that can assemble into rings round the circumference of the cytoplasm, acting as scaffolding for the assembly of the divisome. Curiously, streptomycetes and other actinomycetes lack FtsA and other divisome components that anchor Z-rings to the cell membrane in many other bacteria.

FtsZ is essential in most bacteria. It was therefore a major surprise to find that mycelial growth and branching can still take place, albeit relatively feebly, when septation is eliminated by mutations in *ftsZ* (McCormick *et al.*, 1994). Mutants in *ftsW, ftsI* and *ftsQ* form vegetative cross-walls quite normally, but are defective in the specialized septation involved in sporulation (McCormick and Losick, 1996; Mistry *et al.*, 2008; see also below).

Possibly, a benefit of septation is that separation of the mycelium into compartments makes it more resilient against damage such as infection by phages or mechanical injury (McCormick, 2009). It also permits physiological differentiation between compartments. For reasons that are still unexplored, *ftsZ* mutants of *S. coelicolor* have hyperpigmented colonies (McCormick *et al.*, 1994).

Chromosome replication in multigenomic compartments

Although growth is confined to the tip, DNA is distributed throughout the apical compartment, raising the questions of whether all the chromosomes replicate, and what determines their position. A comparison of the number of replisomes with the number of chromosomes indicated that not all chromosomes are replicating at any one time, and that some compartments have no organized complexes (Ruban-Ośmiałowska *et al.*, 2006). Apical compartments show the highest frequency of replisomes. The replisomes are dispersed irregularly through compartments. The chromosome nearest to the tip is nearly always complexed with a large focus of the chromosome partitioning protein ParB, about 1.4 μm from the tip, though a minority of tips do not show such a focus. Foci result from binding of ParB protein to a large number (ca. 20) of specific target sequences (*parS* sites) clustered around the replication origin, nucleating the formation of a large ParB–DNA complex (Jakimowicz *et al.*, 2002, 2005a; Ruban-Ośmiałowska *et al.*, 2006). These complexes presumably interfere with binding of DnaA protein to the cluster of DnaA box sequences at *oriC* (Smulczyk-Krawczyszyn *et al.*, 2006), preventing the chromosome nearest the tip from replicating unless or until there is some (transient?) release of the chromosome from ParB (possibly represented by the tips without foci). The efficient formation of tip-located ParB chromosome complexes involves interaction of ParB with ParA, an ATPase with cytoskeletal attributes (see below) whose genetic determinant forms an operon with the ParB determinant (the *parAB* operon, located very close to *oriC*). The absence of ParA or ParB does not have any obvious effects on hyphal growth and branching, though DNA partitioning into spores is irregular (Kim *et al.*, 2000; Jakimowicz *et al.*, 2007; see below). Presumably, some other system also contributes to the distribution of chromosomes along the vegetative mycelium.

Attachment to surfaces

Vegetative hyphae can attach strongly to hydrophobic surfaces. Atomic force microscopy showed that an extracellular matrix is implicated in this attachment (Del Sol *et al.*, 2007). Possibly this matrix corresponds to the attachment-mediating fimbriae described by de Jong *et al.* (2009b), which consist of a core of cellulose, on which are assembled amyloid complexes of strongly surface-active amphipathic chaplin proteins. In addition to cellulose, another β-1,4-linked glucan may be able to nucleate fimbriae, since inactivation of the gene for cellulose synthase does not completely eliminate either fimbriation or the ability of calcofluor white (a β-1,4-glucan-specific stain) to stain fimbriae. There are genes for eight chaplins in the *S. coelicolor* genome; five small, and three longer (Elliot *et al.*, 2003a; Claessen *et al.*, 2003). Only small chaplins are important in adhesion (de Jong *et al.*, 2009b). The large ones possess a domain needed for direct linkage to the cell wall, and have an important role in aerial mycelium formation (see below).

Metabolic flexibility of the vegetative mycelium

Once the endogenous spore reserves have been exhausted, the protomycelium can reasonably be assumed to have expressed the additional enzymes and transport systems needed to utilize soluble nutrients such as amino acids, sugars and amino sugars that may be readily available in the environment. If levels of these nutrients become growth-limiting, this information is perceived by global regulatory systems. For example, the presence of N-acetyl glucosamine maintains a state of carbon catabolite repression through the DasR global regulator (Rigali *et al.*, 2006, 2008 – see Chapter 5); and regulation in response to nitrogen availability is largely channelled through the OmpR-like protein GlnR (recently reviewed by Amon *et al.*, 2010 – see Chapter 6). When such regulatory systems permit, the promoters of genes needed for the utilization of less readily accessible alternative nutrient sources can become activated if the relevant substrate is available – a situation perceived by pathway-specific regulators of gene expression. Thus, extracellular chitinases may be produced that enable the mycelium to use fungal cell walls (sometimes on living fungi) as a source of carbon and nitrogen (reviewed by Chater *et al.*, 2010). If these are produced by a mycelium that already has significant biomass, the production of a sufficient concentration of enzyme to be effective in solubilizing chitin requires a relatively small contribution from each hyphal compartment.

Hyphal growth direction

The rigidity of hyphae needed to permit more or less smooth growth in one direction is not only a function of turgor pressure – it also involves filaments of a cytoskeletal protein, FilP, that extend along hyphal compartments (Bagchi *et al.*, 2008). Hyphae of a *filP* mutant change direction very frequently, and atomic force microscopy shows that they are unusually soft. FilP, which is structurally related to the intermediate filament proteins of eukaryotes, also forms foci at hyphal tips, possibly through interaction with DivIVA or tip-located cellulose synthase (Bagchi *et al.*, 2008).

Early in mycelium formation, growing hyphal tips generally seem to avoid contact with other hyphae (Hopwood, 1960); but as the mycelium becomes more dense, this avoidance mechanism is somehow overridden. Hyphal avoidance may be due to tropism towards environmental nutrients or away from 'staling compounds' produced during hyphal growth (Allan and Prosser, 1987). Later, nutrients released by dying parts of the existing mycelium might serve as the target for growth: thus, no specific avoidance mechanism would be needed. On rich medium, hyphal avoidance is not observed, and hyphae can grow as rope-like complexes (Hopwood, 1960).

Part 2. The complex transition from growth to development

Complex metabolic transitions at the end of rapid growth, and their relationship to developmental changes

Global metabolic regulatory systems may also repress the adoption of physiological states and developmental stages associated with cessation of biomass accumulation. These states most notably include antibiotic production and reproductive aerial growth. Accelerated or ectopic sporulation have been seen in carbon catabolite derepressed mutants resulting from deletion of the glucose kinase gene and flanking DNA (Kelemen *et al.*, 1995); *relA* mutants unable to make the signalling molecule ppGpp, which coordinates information about growth rate with the expression of genes important for rapid growth, show delayed and reduced morphological differentiation (Chakraburtty and Bibb, 1997; Hesketh *et al.*, 2007); and nutritional downshift can lead to sporulation of some streptomycetes in submerged culture (reviewed by Hodgson, 1992). A number of careful and detailed analyses, reviewed in the following subsections, have also pointed to pleiotropic physiological changes during growth, secondary metabolism and development, but the precise regulatory connections between nutrient limitation and these physiological changes are still little understood.

A transition phase precedes a final round of growth

Radiolabelling and proteomic and transcriptomic studies of *Streptomyces* growth during surface culture on cellophane discs overlaid on agar medium (Granozzi *et al.*, 1990), or in liquid culture (Vohradsky *et al.*, 1997, 2000; Hesketh *et al.*, 2007; Nieselt *et al.*, 2010), have all revealed a sequence of major switches in metabolism after the first phase of biomass accumulation. The details of these observations depend on medium composition and culture conditions. In broad outline, a slow-down in bulk macromolecular biosynthesis takes place at the first limitation for a preferred nutrient source – for example, glutamate was limiting in the experiments of Novotna *et al.* (2003), and phosphate was limiting in the work of Nieselt *et al.* (2010). This is accompanied by a downshift in expression of genes for ribosomal components and the TCA cycle, along with several rapid transitions in global gene expression (Blanco *et al.*, 1994; Vohradsky *et al.*, 2000; Novotna *et al.*, 2003; Nguyen *et al.*, 2005; Nieselt *et al.*, 2010). After this transition phase (which is detectable as a short pause in growth in certain conditions), a changed pattern of gene expression is associated with a further short round of growth, in which somewhat less favoured nutrients may be used (e.g. maltose for carbon, and ammonia for nitrogen, in Novotna *et al.*, 2003). On exhaustion of one of these nutrients, stationary phase ensues. In surface culture, a transition phase precedes aerial growth, a second growth phase corresponds to aerial growth, and sporulation takes place in stationary phase (Granozzi *et al.*, 1990). However, there are certainly some differences in gene expression patterns between liquid-grown and surface-grown cultures.

Storage metabolism and development

Among the changes in metabolism associated with transition phase, some affect the storage of carbon. Braña *et al.* (1982, 1986) first showed that glycogen accumulates, to up to 5% of total mycelial dry weight, as rapid growth comes to an end (transition phase). With suitable staining, this accumulation is visualized as small granules around the cell periphery of hyphae in the central

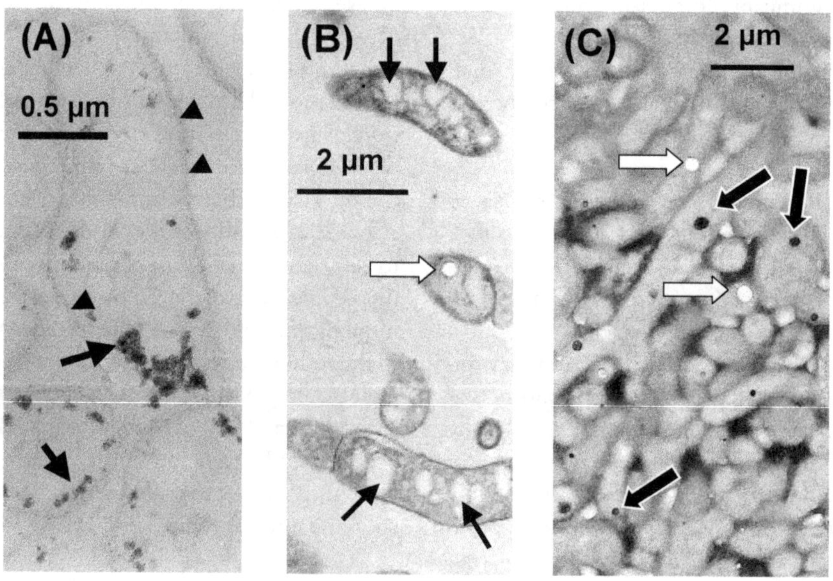

Figure 3.4 Storage deposits in hyphae near the upper surface of *Streptomyces coelicolor* colonies (transmission electron micrographs taken by Kitty Plaskitt; Plaskitt and Chater, 1995). (A) Glycogen (large arrows). The walls of the hyphae (small arrows) are not optimally visible in material that has been silver-stained in this way. (B). Triacyl glycerol. Deposits of triglycerides, seen as bubble-like clusters, are particularly abundant in some morphological mutants, such as this *bldA* mutant. (C). Polyphosphate. In this unstained section, which is much thicker than usual sections analysed by transmission electron microscopy, circular black bodies (black arrows) are believed to be deposits of polyphosphate, and circular clear zones (white arrows) are areas in which such polyphosphate granules have been destroyed by the electron beam.

parts of the colony (Fig. 3.4A). Emerging aerial hyphae do not contain glycogen, but granules reappear at an early stage of sporulation, declining as spores mature (Braña *et al.*, 1986; Plaskitt and Chater, 1995: see below).

Direct interplay between glycogen deposits and the trehalose pool underpins these observations. In *Streptomyces brasiliensis*, which undergoes extensive sporulation in liquid culture, submerged sporulation is preceded by massive glycogen accumulation (ca. 20% dry weight), and as spores mature nearly all the glycogen is converted into trehalose (Rueda *et al.*, 2001). This carbon cycling is achieved by the products of operons containing four genes: *glgE*, predicted from studies of its mycobacterial orthologue to encode a maltosyl transferase that adds maltosyl units to α-1,4 glucan chains (Kalscheuer *et al.*, 2010); *treS*, encoding a trehalose synthase able to interconvert maltose and trehalose; *pep2*, encoding a maltose kinase; and *glgB*, encoding a glycogen branching enzyme (Schneider *et al.*, 2000; Jarling *et al.*, 2004; Bruton *et al.*, 1995). These enzymes allow interconversion of glycogen, trehalose and maltose. In most streptomycetes there are two closely similar copies of these gene sets: operon I functions in the centre of the colony (phase I), and operon II in the sporulating parts of the aerial mycelium (phase II) (Bruton *et al.*, 1995; Schneider *et al.*, 2000; Yeo and Chater, 2005; see below). Gene pairs for the conventional synthesis of trehalose and glycogen from glucose-1-phosphate, and for the conversion of glycogen into trehalose, are present elsewhere in the genome. This conveys a picture in which phase I glycogen is converted into soluble, mobile maltose and/or trehalose, which can contribute carbon, energy and turgor for aerial growth. Some of the mobile carbon is eventually reconverted into glycogen deposits present in developing spore chains, under the direction of operon II, and then into the trehalose found abundantly in spores. It has been speculated that turgor changes mediated by the interplay of macromolecular glycogen with mobile trehalose may help to bring about aerial growth cessation and the rounding up of spores (Plaskitt and Chater, 1995).

Transmission electron microscopy often reveals electron-transparent vacuole-like bodies in the central part of mature colonies, approximately corresponding to the zone undergoing transition phase physiology (Fig. 3.4B; Plaskitt and Chater, 1995). These appear to correspond to the deposits of triacylglycerol (TAG) first detected biochemically by Olukoshi and Packter (1994), who found that this material accumulates during stationary phase, sometimes up to 5% of the dry weight. Spores contain lower, but still substantial, amounts. Although there has been some further analysis of TAG biochemistry and metabolism (Arabolaza *et al.*, 2008) its synthesis and degradation have not been clearly related to developmental progression.

Occasional perfectly circular, completely electron transparent areas seen in hyphae in the same area of the colony appear to mark the sites of polyphosphate deposits. Polyphosphate can be seen in thick sections as electron-dense bodies, which are quickly destroyed in the electron beam (Fig. 3.4C; Plaskitt and Chater, 1995). The role of polyphosphate, which is an ancient means of energy storage (Brown and Kornberg, 2004), has not been extensively studied in streptomycetes (but see Ghorbel *et al.*, 2006).

Stress-response sigma factors and their interplay with development

A conspicuous feature of *Streptomyces* genomes is the very large number of genes for RNA polymerase sigma factors (Bentley *et al.*, 2002). Among the 60 or more sigma factors of *S. coelicolor*, nine belong to the Gram-positive-specific group 3 family represented in *Bacillus subtilis* by the extensively studied σ^B (involved in stress responses) and its paralogues σ^F and σ^G (both involved in sporulation).

Cellular responses to externally applied stresses change as hyphae go through different growth and developmental stages (Puglia *et al.*, 1995; Novotna *et al.*, 2003). Moreover, proteins associated with certain stress responses (heat, cold and salt) are synthesized in the transition phase in the absence of applied stresses, while proteins normally associated with treatment with ethanol or an antibiotic (pristinamycin) are not upregulated at this stage (Vohradsky *et al.*, 2000). It thus seems that some stress responses are intimately bound up with the early stages of development, a conclusion borne out by the finding that developmental phenotypes are associated with mutations affecting the activity of at least seven of the nine group 3 sigma factors of *S. coelicolor* (Table 3.1). In other bacteria, Group 3 sigmas are characteristically regulated by anti-sigma factors and anti-anti-sigma factors, in a manner that depends on the phosphorylation state of the anti-anti-sigma factors (Price, 2000). The phosphorylation levels are determined by the balance between the protein kinase activity of the anti-sigma factors and the counter-activity of signal-sensitive phosphatases. Some of the effects on *Streptomyces* development are apparent only though the phenotypes of mutants lacking anti- or anti-anti-sigmas.

In *S. coelicolor*, an important player in these developmental effects is σ^B, which responds to osmotic stress, and is involved in the increased expression of σ^L, σ^H and σ^M (Lee *et al.*, 2005); while σ^K appears to have a negative effect on both development and antibiotic production (Mao *et al.*, 2009a). Both σ^F and σ^N are developmentally regulated, and play more strictly developmental roles (Potuckova *et al.*, 1995; Dalton *et al.*, 2007). It is difficult to interpret these mutant phenotypes, for several reasons. First, there appears to be considerable overlap in the promoter specificity of the different sigmas of this group (Lee *et al.*, 2005). Secondly, several of the sigmas exhibit positive autoregulation and/or contribute to the transcription of each other's determinants (Lee *et al.*, 2005; Viollier *et al.*, 2003b). Thirdly, some of them are subject to post-transcriptional proteolytic processing, the significance of which is difficult to assess (Viollier *et al.*, 2003a; Bibb and Buttner, 2003). Fourthly, the interactions with anti-sigmas, which are critical for understanding many aspects of group 3 sigma function, have largely resisted in-depth investigation. This fourth point is likely to be highly significant, because there appear to be many more anti-sigmas than group 3 sigmas sigmas in streptomycetes (Mittenhuber, 2002; Gaskell *et al.*, 2007), suggesting that each sigma may interact with more than one anti-sigma, while the anti-sigmas may also interact with more than one anti-anti-sigma (Kim *et al.*, 2008b) and anti-anti sigmas with more than one anti-sigma (Parashar *et al.*, 2009). Viewed from the perspective of the partner-switching concept, in which the anti-sigmas are sequestered by one or the other of their alternative interacting partners (sigma or anti-anti-sigma) (Price, 2000), a change in the abundance of any one of these proteins may bring about wide indirect effects on the relative activities of several – perhaps many – of this group of sigmas. What is more, these systems are likely to be responsive to (still unknown) signals, whose transduction classically operates through the anti-anti-sigmas. There are comparatively few 'obvious' genes for anti-anti-sigmas in *S. coelicolor* (13 according to Kim *et al.*, 2008b). However, about 40% of the candidate anti-sigmas of *S. coelicolor* are multidomain proteins that also have sensory domains of the PAS and/or GAF families. These large proteins also have phosphatase domains related to that of the archetypal SpoIIE, which counteracts the consequences of the protein kinase activity of a sporulation-specific anti-sigma of *B. subtilis* (Dalton *et al.*, 2007; Kim *et al.*, 2008b; Mittenhuber, 2002; Aínsa *et al.*, 2010).

All in all, these observations imply that the progression of development in streptomycetes is sensitive to diverse signal inputs, interpreted through what amounts to a 'neural network' of interacting signal transduction systems: a concept first articulated in a paper delivered by Charles Thompson to the 12th International Symposium on Actinomycete Biology in Vancouver in 2001, and by the application of the term 'checkpoint multiplex' to *Streptomyces* development (Chater, 2001). It may be that the unravelling of the network will depend on the kind of large-scale *in vitro* approach exemplified by Skerker *et al.* (2005) in the study of two-component regulatory systems in *Caulobacter crescentus*.

The switch to the production of antibiotics and other secondary metabolites

Spatial determination of secondary metabolism

Secondary metabolism and its regulation are extensively dealt with in other chapters of this book. Here it is examined specifically in relation to colony development, since it is evident from vertical sections of colonies that the distribution

Table 3.1 Group 3 sigma factors whose activity affects development in *S. coelicolor*

Sigma factor	SCO number	Anti-sigma?	Anti-anti-sigma?	Induction	Mutant morphological phenotype	References
σ^B	0600	SCO0599 (*rsbA*)	SCO0598 (*rsbB*)	Osmotic	Bald	Cho *et al.* (2001), Lee *et al.* (2004a,b, 2005)
σ^L	7278	SCO7277		Osmotic σ^B-regulated	White, low actinorhodin	Lee *et al.* (2005)
σ^I	3068	SCO3066 (*prsI*)	SCO3067 (*arsI*)	Osmotic	Wild-type	Lee *et al.* (2005)
σ^N	4034	SCO3548 (*apgA*)?	SCO3549 (*bldG*)?	Developmental	Conditionally bald; *bldG* mutant bald.	Dalton *et al.* (2007), Champness (1988), Bignell *et al.* (2000), Parashar *et al.* (2009)
σ^F	4035	SCO4677 (*rsfA*)	SCO0781 SCO0869	Developmental	Thin-walled σ^F mutant spores	Potúková *et al.* (1995), Kelemen *et al.* (1996), Kim *et al.* (2008b)
σ^H	5243	SCO5244 (*prsH* or *ushX*)		Osmotic, σ^B-regulated	Wild-type or mild sporulation defect; *sigh-prsH* double mutant bald	Sevcíková *et al.* (2001), Sevcíková and Kormanec (2002). Viollier *et al.* (2003a,b), Kormanec *et al.* (2000)
σ^K	6520				Accelerated development and antibiotic production	Mao *et al.* (2009a)
σ^M	7314	SCO7313 (*rsmA*)		Osmotic, σ^B-regulated	Reduced spore numbers	Lee *et al.* (2005)
σ^G	7341					Sevcíková *et al.* (2005)

of pigmented antibiotics is limited to a narrow zone between the vegetative and aerial mycelium (Chater, 1998; Fig. 3.5). Several general observations help to explain this spatial specificity.

First, biosynthesis of many secondary metabolites is starvation-responsive, being subject to C, N and P catabolite repression, and to control by the stringent response (as signalled by an increase in cellular levels of the alarmone ppGpp that results from ribosome pausing when an uncharged tRNA enters the ribosomal A-site) (Bibb, 2005). The most nutritionally stressed part of the colony is the dense mycelium in the centre of the surface, which is probably physiologically equivalent to the transition phase described above.

Secondly, the biosynthesis of many (but far from all) antibiotics is subject to a form of quorum sensing, in which the extracellular concentration of a small secreted signalling molecule increases most rapidly in the most densely packed part of the mycelium, to reach a level high enough for its perception by a receptor protein. Most often, the molecules used by streptomycetes for this purpose are γ-butyrolactones (see the chapter by Takano, this volume). These molecules have lipophilic side chains that, at least below a certain length, permit them to diffuse freely through membranes. The cytoplasmic location of the receptor protein is therefore not an impediment to its sensing role. On binding to the γ-butyrolactone, the properties of the receptor protein are changed. Most typically, this reduces the ability of the protein to bind to specific target sequences in the promoters of certain genes critical for antibiotic production, which therefore become free of repression, leading to the onset of production. In some cases, the repressible targets are specific activator genes for particular antibiotic pathways, while in the case of A-factor the target promoter controls the expression of *adpA*, which encodes a universal regulator of some aspects of development and antibiotic production among streptomycetes (Ohnishi and Horinouchi, 2004; Takano, 2006; Horinouchi, 2007).

Sensitivity of antibiotic production to diverse signals and checkpoints

Even if all the above regulatory devices are permissive for antibiotic production, production does not necessarily follow. This is because, quite apart from the stress-responsive sigma factor effects already summarized, a substantial number of other regulatory systems also impinge on expression of the pathway-specific regulatory genes (Bibb, 2005). Thus, many pathway-specific activator genes contain the very rare codon TTA (Chandra and Chater, 2008). The tRNA for the corresponding UUA codon (one of six leucine codons) is encoded by the *bldA* gene, and, unusually, is more abundant after, rather than during,

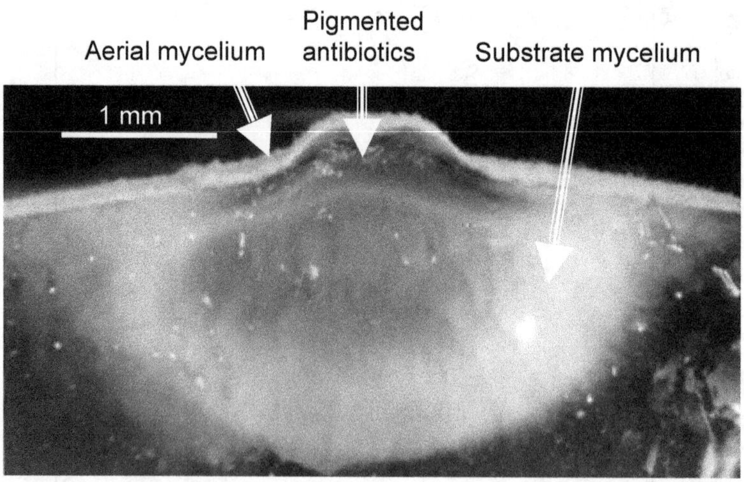

Figure 3.5 Vertical section through a colony of *Streptomyces coelicolor*, showing morphologically and physiologically differentiated parts of the colony (Chater, 1998). Photograph taken by Jamie Ryding.

rapidly primary growth (Trepanier *et al.*, 1997). Other pleiotropically acting regulatory proteins interact with the promoters of the pathway-specific regulatory genes (Uguru *et al.*, 2005; Park *et al.*, 2009; Rigali *et al.*, 2008; Pan *et al.*, 2009). It is very likely that these and the pathway-specific regulatory proteins respond to the presence of specific ligands, though very few such ligands have been identified.

In view of the fact that most streptomycetes have the genetic potential to produce numerous different secondary metabolites, it is an unexplored possibility that the production profile is not the same in all compartments of the mycelium that are undergoing secondary metabolism. Speculation on the possible heterogeneity of cellular responses is made plausible by the discovery that physiologically distinct subpopulations of cells can coexist in clonal cultures of *Bacillus subtilis* in what appear to be homogeneous environments, their origins being explained by bimodal (or 'bistable') switching underpinned by positive feedback regulatory circuits (reviewed by López and Kolter, 2010). Basically, if the suite of genes regulated by a global activator includes its own gene, there is massive signal amplification in those cells in which the threshold level of activator is passed. Some of the Group 3 sigmas are autoinducing, and therefore fit with this scheme (see above). Likewise, many autoregulators such as γ-butyrolactones induce their own further synthesis (Takano, 2006). If one result of autoinduction is the production of something that prevents uninduced neighbouring cells from following the same course, physiological heterogeneity is stabilized. Furthermore, there may be hierarchies of bimodal switches if the initial activator switches on regulators that are themselves subject to positive feedback loops (López and Kolter, 2010). It is interesting, therefore, that positive feedback by end-products or late intermediates in antibiotic biosynthesis have recently been identified in streptomycetes (Tahlan *et al.*, 2007; Wang *et al.*, 2009), and that the activators specified by regulatory genes associated with biosynthetic clusters are not always straightforwardly pathway specific, but may also interact with the products of other pathways to mediate cross-repression between pathways (Huang *et al.*, 2005; de la Fuente *et al.*, 2002; Xu *et al.*, 2010b), or have an impact on morphological differentiation (Liu *et al.*, 2005). Such effects may well reinforce bimodal differences between cells.

The emergence of aerial hyphae: growth away from nutrients

The growth of aerial branches is dependent on the formation of the specialized outer sheath that eventually becomes the outer rodlet layer of spores (Glauert and Hopwood, 1961; Wildermuth *et al.*, 1971; García, 1995; Elliot *et al.*, 2003a; Elliot and Talbot, 2004; Claessen *et al.*, 2002, 2003, 2004). The sheath has a hydrophilic face in contact with the hydrated hyphal wall, and a hydrophobic face towards the air. The major sheath components are chaplins, including those described above as components of the spore surface and agents of hyphal attachment to surfaces. The number of different chaplins varies among species, but *S. coelicolor* has eight. Among these, the 'large' chaplins A, B, and C have – as well as an additional chaplin domain – C-terminal extensions that contain recognition sites for sortase enzymes, via which they are covalently attached to the cell wall, providing a scaffolding to which the 'small' chaplins can attach (though even in the absence of the large chaplins the small chaplins can assemble and promote aerial growth and sporulation) (Elliot *et al.*, 2003a; Claessen *et al.*, 2003). Because of the extensive functional redundancy among chaplins, multiple gene knockouts are necessary in order to demonstrate their importance for aerial growth (Elliot *et al.*, 2003a; Claessen *et al.*, 2003, 2004; DiBerardo *et al.*, 2008). A strain carrying just the three conserved genes *chpC*, *E* and *H* – which happen to be the only *chp* genes expressed during vegetative growth (see also Nieselt *et al.*, 2010) – shows almost wild-type aerial mycelium development. It is thought that ChpH assembles on cell wall-anchored ChpC, while ChpE, the only chaplin lacking a set of otherwise conserved cysteines, plays a supporting role in the assembly (DiBerardo *et al.*, 2008). Rodlins, which contribute to sheath structure but are not essential for its major functions (Claessen *et al.*, 2002), are encoded by a pair of adjacent genes that are also close to some of the chaplin genes. The rodlins are needed for the sheath to develop into paired rodlet structures (Claessen *et al.*, 2002).

SapB-like peptides

On certain rich media, aerial growth also requires SapB, a highly modified amphipathic peptide generated by processing of the 42-aa product of the *ramS* gene to a complex 21-aa residue oligopeptide (Kodani *et al.*, 2004; Willey *et al.*, 2006; Capstick *et al.*, 2007). Its structural features include lanthionine cross-links (crosslinks involving modified cysteine residues that define lantibiotics, a class of ribosomally synthesized peptide antibiotics). The modification steps generating pro-SapB are carried out by the product of *ramC*, which is next to *ramS*, and a specific transporter (the *ramAB* product) is thought to be the main exporter of SapB. It is not known whether the 21-aa SapB leader sequence is removed before, during or after export. The SapB-like peptides encoded by other *Streptomyces* genomes, such as AmfS of *S. griseus* (Ueda *et al.*, 2002, 2005), are particularly highly conserved in the residues involved in modification, implying general structural similarity. The expression of the *ram* and *amf* gene sets depends on the 'orphan' response regulator specified by the respective adjacent genes, *ramR* and *amfR* (Keijser *et al.*, 2002; Nguyen *et al.*, 2002; Ueda *et al.*, 2005).

The role of SapB in aerial growth and sporulation of *S. coelicolor* can be substituted by a somewhat diverged lanthionine-containing amphipathic peptide (SapT) from *Streptomyces tendae* (Kodani *et al.*, 2005). Even the completely unrelated (but functionally equivalent) hydrophobin protein SC3 of the fungus *Schizophyllum commune* can restore aerial mycelium formation to aerial mycelium-deficient (*bld*) mutants of various classes, which are deficient in both SapB and chaplins (Tillotson *et al.*, 1998), and in the case of *ram* mutants, which specifically lack SapB, SC3 can even restore sporulation (Kodani *et al.*, 2005). The previously mentioned ability of streptomycetes to grow epiphytically on living fungi suggests that this responsiveness to fungal hydrophobins may not be a mere laboratory artefact.

Proposed physicochemical roles of the surfactant components of the sheath

The very high surface activity of SapB, chaplins and rodlins has led to the idea that their main role is to allow the aerial hyphae to breach the surface tension of the aqueous environment surrounding growing vegetative hyphae (reviewed by Wösten and Willey, 2000). This would be consistent with the secretion of some of them into the surrounding medium. However, the proteins continue to be secreted by aerial hyphae after their emergence into the air, to generate the outer sheath. The cylindrical compartment between the sheath and the cell membrane (which is largely occupied by the hydrated cell wall, to which the inner face of the sheath presumably adheres by water contacts as well as by covalent attachment of large chaplins) may therefore be a channel through which nutrients and other metabolites can diffuse from the substrate to the growing tips of the aerial hyphae (Fig. 3.6; Chater and Chandra, 2006; Chater *et al.*, 2010). This would circumvent a long-standing, but seldom-addressed, problem of how nutrient supplies could be transported across the cross-walls that are usually formed in the lower parts of aerial hyphae: in the absence of the extracellular channel, there would be a need to export nutrients from basal compartments and then to re-import them into the compartment above. The analogous hydrophobins of some fungi may also provide such an extracellular channel (Chater *et al.*, 2010).

As sporulation takes place, the shedding of a surface layer reveals the mosaic of paired rodlets characteristic of mature spores (Del Sol *et al.*, 2007). It was reported previously that the appearance of the mature mosaic spreads back from the hyphal tip (Szabó *et al.*, 1977). Thus, it is possible that the interactions between the various sheath components undergo some kind of change at this time.

In view of the role of cellulose in the assembly of small chaplins into the fimbriae involved in attachment of vegetative hyphae to surfaces, it is interesting that a mutation in the gene for cellulose synthase interferes with aerial mycelium formation (a phenotype partially rescued by

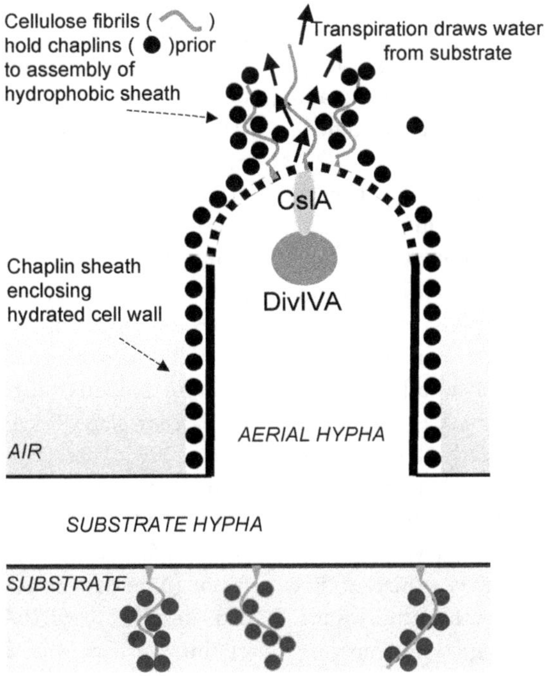

Figure 3.6 Roles of cellulose and chaplins at different stages of colony growth and development. Cellulose fibrils under substrate hyphae nucleate the assembly of amyloid complexes of chaplins to attach hyphae to the substrate (De Jong *et al.*, 2009b). Fibrils are also extruded at aerial hyphal tips, where cellulose synthase (Cs1A) is complexed with the polarity determinant DivIVA (Li *et al.*, 2007). The tip-located cellulose appears to help chaplins to form the hydrophobic sheath, whose hydrophilic inner face is predicted to hold a water column (in the form of a water-saturated cell wall), maintaining hydration of the aerial hypha. On this model, the sheath is open at the growing tip, with the resulting evaporation of water bringing about transpiration-mediated movement of nutrients from the base to the upper parts of the aerial hypha. The diagram is based on that of Chater *et al.* (2010). The surfactant properties of chaplins are also thought to aid the penetration of aerial hyphae through the water surface of hydrated substrates (Elliot and Talbot, 2004).

adding extracts rich in chaplins and rodlins) (Xu *et al.*, 2008a). Cellulose synthase is localized to aerial hyphal tips (possibly by interacting with DivIVA), where it gives rise to cellulose deposits. This suggests a speculative model in which cellulose aids the organization of chaplins and rodlins into the extending sheath at the growing hyphal tip. Evaporation of water from the partially exposed tip may well function to draw water and associated solutes up from lower parts of the colony (Xu *et al.*, 2008a; Chater *et al.*, 2010; Fig. 3.6). It has recently emerged that mutation of *cebR*, a key regulatory gene for cellulose catabolism, results in very poor aerial growth (Marushima *et al.*, 2009). This phenotype is similar to that of a cellulose synthase mutant, and may imply a broad link between components of cellulose metabolism and aerial growth.

The apparent involvement of several *bld* genes in checkpoint responses, leading to the expression of sheath components

The formation of aerial mycelium has been studied extensively in *S. coelicolor* through the use of aerial mycelium-deficient *bld* mutants, which do not form a fuzzy aerial mycelium (at least under some conditions). Early studies focused on six *bld* genes (*bldA, -B, -C, -D*: Merrick, 1976; *bldG, -H*: Champness, 1988), providing a broad framework that later studies have filled in and extended considerably: now, more than 20 genes whose inactivation gives a Bld phenotype have been recognized (listed by Elliot *et al.*, 2008). They include some TCA cycle genes, whose mutant phenotypes are partially due to medium acidification (Viollier *et al.*, 2001a,b).

Two lines of evidence indicated that five of the six canonical *bld* genes defined in early studies are components of a checkpoint cascade, one of whose endpoints is the increased expression of the structural genes for chaplins, rodlins and SapB. First, *bldB, -C, -D, -G and -H* encode regulatory proteins (reviewed by Chater and Chandra, 2006) (as mentioned earlier, *bldA* is the determinant of a tRNA). Because nearly all regulatory genes apart from those of phages respond to signals, this immediately suggests that the *bld* gene products transmit information relevant to the decision to differentiate. Secondly, growth of different *bld* mutants close to each other often results in cross-stimulation of aerial growth, implying that extracellular signals are transmitted (Willey *et al.*, 1993). Most of the mutants could be placed on a single linear cascade of extracellular signals, in the order [*bldA, bldH*], *bldG, bldC, bldD* (a *bldB* mutant did not fit into this cascade) (Willey *et al.*, 1993). The cascade ends with regulators that are closely implicated in the expression of the genes for chaplins, rodlins and SapB (see Chater and Chandra, 2006). The cascade also leads to the expression of *sigN*, the determinant of a group 3 sigma factor implicated in early developmental events in the emerging aerial hyphae (Table 3.1 and next section). Unfortunately, the signals in this cascade have so far resisted characterization, except for the first, which is an oligopeptide of unknown origin that is taken up by an ATP-dependent oligopeptide transporter (named BldK) (Nodwell *et al.*, 1996; Nodwell and Losick, 1998). For those members of the cascade that have been investigated sufficiently, the picture emerging is that each of them has targets that include other developmentally significant genes.

AdpA, a key regulator of development encoded by bldH, *and its connection with* bldA

The AdpA protein was first discovered in *S. griseus*, where its production is directly dependent on the presence of A-factor (see Horinouchi, 2007). In most other streptomycetes AdpA production does not depend on a gamma-butyrolactone. AdpA, a member of the AraC/XylS class of bacterial transcriptional regulators, binds to several tens of promoters in any one *Streptomyces* genome (e.g. Ohnishi *et al.*, 2005; Akanuma *et al.*, 2009). Most of these promoters depend on AdpA for expression, though some are AdpA-repressible. In *S. griseus,* the genes controlled by AdpA include both pathway-specific activator genes for antibiotic production and genes important for morphological development (Fig. 3.7), so *adpA* mutants (also known as *bldH* mutants in *S. coelicolor*: Takano *et al.*, 2003; Nguyen *et al.*, 2003) are impaired in aerial mycelium formation as well as in production of some antibiotics. AdpA-dependent developmental regulatory genes of *S. griseus* include *adsA,* encoding a sigma factor (Yamazaki *et al.*, 2000), and *amfR,* which activates the *amfTSBA* operon for production of the AmfS peptide needed for aerial growth (Yamazaki *et al.*, 2004). Other direct targets of AdpA also include components of a developmentally significant extracellular protease cascade (see below; Akanuma *et al.*, 2009). Many, but not all, of the biological consequences of AdpA action are extracellular (Fig. 3.7).

In every free-living streptomycete for which there is sequence information, *adpA* has a *bldA*-dependent TTA codon. In *S. coelicolor,* this accounts for the finding that *adpA* and *bldA* mutants behave similarly in their extracellular interactions with other *bld* mutants (Willey *et al.*, 1993; Takano *et al.*, 2003; Nguyen *et al.*, 2003). The *bldA-adpA* link was evidently present in the ancient progenitor of all these strains, and must be assumed to confer a significant adaptive benefit (Chater and Chandra, 2008). Since AdpA protein is autorepressing at the transcriptional level (Kato *et al.*, 2005c), translational control mediated by the availability of charged *bldA* tRNA may determine the level of activation of the AdpA regulon under some conditions. As mentioned earlier, the pattern of abundance of the *bldA* tRNA differs from that of most tRNAs: instead of being maximally abundant during rapid growth, its abundance is highest in stationary phase (Trepanier *et al.*, 1997).

At least in *S. coelicolor,* but possibly more widely, AdpA levels are also controlled at the level

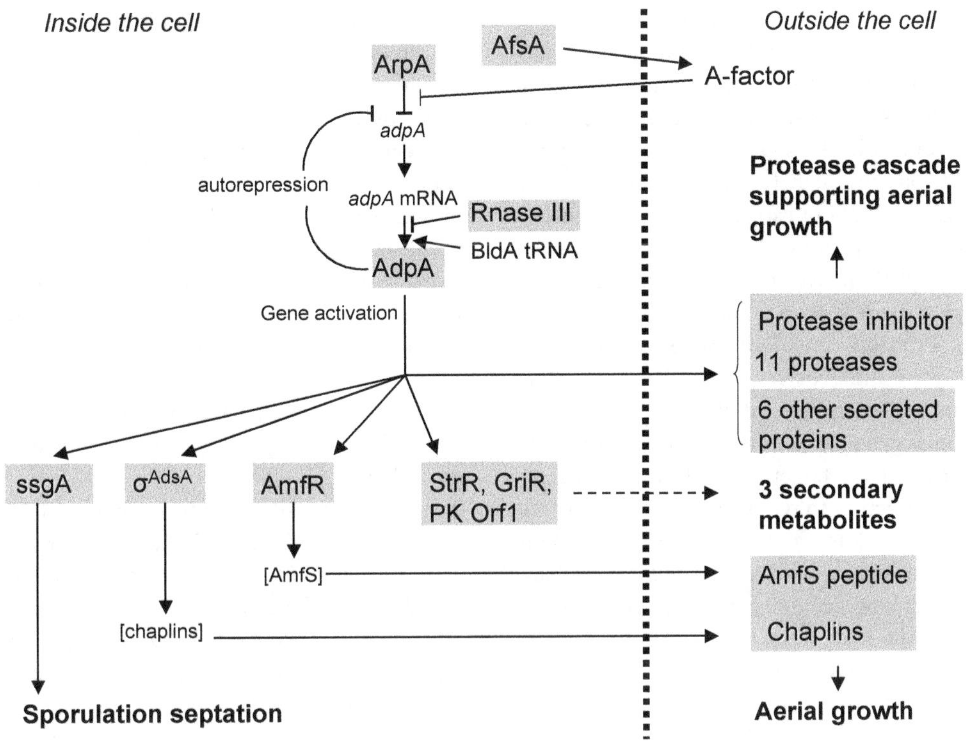

Figure 3.7 The regulation and targets of AdpA in *S. griseus*. The involvement of Rnase III in *adpA* mRNA processing is based on work in *S. coelicolor* (Xu *et al.*, 2010a). Otherwise, the figure is mainly based on Akanuma *et al.* (2009).

of *adpA* mRNA processing: RNase III cleaves *adpA* mRNA, and a mutant lacking Rnase III overexpresses many AdpA target genes, including intracellular proteases that, it is speculated, may inactivate RNase III in a feedback loop (Xu *et al.*, 2010a) (the RNase III gene, *absB*, was originally identified in a search for antibiotic-negative mutants that were morphologically normal: Price *et al.*, 1999 and references therein; Chang *et al.*, 2005). The feedback loop is further complicated by the sensitivity of the *absB* mRNA itself to cleavage by RNase III (Xu *et al.*, 2008b).

The spectrum of genes directly dependent on bldA

TTA-containing *bldA* target genes make up about 2–3% of the content of typical *Streptomyces* genomes (Li *et al.*, 2007; Chater and Chandra, 2008). Most of them are not widely conserved, and may have been laterally acquired, probably conferring strain-specific adaptations. Of the few that are widely conserved, only *bldH/adpA* has been attributed function (see above). Deletion of *bldA* greatly reduces the efficiency with which most UUA codons can be translated, but does not interfere with the ability to grow vegetatively (Leskiw *et al.*, 1991); so TTA-containing genes are dispensable for normal growth, at least under laboratory conditions. However, the ability of *bldA* mutant colonies of diverse streptomycetes to make antibiotics or form a sporulating aerial mycelium is severely compromised under most conditions. This phenotypic effect can be accounted for by the defective translation of *adpA* mRNA and of UUA-containing mRNAs for, on one hand, pathway-associated regulators of antibiotic pathway genes (Chandra and Chater, 2008), and on the other, for components of a developmentally important extracellular protease cascade described below.

BldG, a likely anti-anti-sigma factor

The *bldG* gene product is a paralogue of anti-anti-sigma factors, and *bldG* lies next to a gene for an anti-sigma paralogue, ApgA (Bignell *et al.*, 2000; Table 3.1). The two gene products interact, but although (as with other such anti-anti sigmas) BldG phosphorylation is important for its function, ApgA is not responsible for this phosphorylation, lacking a functional kinase domain (Bignell *et al.*, 2003; Parashar *et al.*, 2009). Probably, therefore, BldG interacts with at least one other partner, and the interactions are sensitive to checkpoint signals. Although the *bldG, apgA* gene pair is not located close to a gene encoding a likely target sigma factor, it is likely that they function to regulate one or more of the σ^B-like sigmas known to influence development (see above; Table 3.1). Dalton *et al.* (2007) suggested that one such target might be σ^N, while pointing out that, if so, the antibiotic-negative phenotype of *bldG* mutants would require that there should be a second target sigma. Since inactivation of *apgA* is lethal, it is likely that the uncontrolled activity of an ApgA-regulated sigma factor prevents growth (Parashar *et al.*, 2009).

The bldD regulon

BldD, an autoregulatory DNA-binding protein of the XRE family (Elliot *et al.*, 1998, 2003b; Elliot and Leskiw, 1999; Kim *et al.*, 2006), represses several developmental regulatory genes, including *whiG*, *bldN* and *sigH* in *S. coelicolor* (Elliot *et al.*, 2001; Kelemen *et al.*, 2001). It also represses the *amfTSBA* operon for the AmfS peptide in *S. griseus* (Ueda *et al.*, 2005). These observations suggest that a checkpoint signal may be necessary for BldD to detach from these promoters. Since a constructed complete deletion of *bldD* had the same phenotype as the original point mutant (Elliot *et al.*, 2003b), there must either be a BldD target gene that is itself a repressor of development, or BldD must activate a gene needed for development to proceed (Chater, 2001).

BldC, an exceptionally small regulatory protein

The 68-aa *bldC* gene product is related to the DNA binding domains of the MerR family of transcriptional regulators (Hunt *et al.*, 2005). Its direct targets, and the reason(s) for the morphological deficiency of null mutants, remain to be determined. BldC is required, directly or indirectly, to maximize transcription of regulatory genes for actinorhodin and undecylprodigiosin biosynthesis in *S. coelicolor* (Hunt *et al.*, 2005). Its small size suggests that it is not directly responsive to a signal. Perhaps its abundance or function will turn out to be influenced by interaction with one or more other proteins.

The bldN-specified sigma factor is directly required for transcription of another regulatory gene, bldM

Among the later-discovered *bld* genes of *S. coelicolor*, *bldN* encodes an ECF sigma factor (Bibb *et al.*, 2000), and *bldM* an aberrant response regulator (Molle and Buttner, 2000). Both belong to classes of regulatory elements whose members typically respond indirectly to extracellular signals. Both *in vivo* and *in vitro*, transcription of *bldM* is dependent on σ^{BldN} (Bibb *et al.*, 2000). In addition, the expression of all the chaplin and rodlin genes is dependent on σ^{BldN} and BldM (Elliot *et al.*, 2003a).

BldB – a small protein thought to transduce a signal by protein–protein interactions

BldB is a member of a family of some 19 paralogues encoded in the *S. coelicolor* genome, all in the 63- to 96-aa size range (Harasym *et al.*, 1990; Pope *et al.*, 1998; Gehring *et al.*, 2000; Eccleston *et al.*, 2002; Aínsa *et al.*, 2010). Most *bldB*-like genes are located next to genes encoding members of a family of predicted DNA-binding proteins (Gehring *et al.*, 2000; Aínsa *et al.*, 2010). This genomic intimation of an interaction between the two kinds of gene product has been substantiated by genetic evidence in the case of the *whiJ* locus, and it has been suggested that the ability of the WhiJ DNA-binding protein to repress development and antibiotic production is counteracted by the BldB-like protein in response to an unknown checkpoint signal (Aínsa *et al.*, 2010). Although *bldB* itself is not adjacent to any such conserved genes, BldB may also engage in interactions with one or more unidentified partner proteins, as well as being able to form an asymmetric homodimer (Eccleston *et al.*, 2006). These conserved gene pairs are also frequently associated with a member of a third family of paralogous genes, encoding putative anti-sigmas (see above).

Part 3. Growth of sporogenic aerial hyphae and their metamorphosis into spore chains

An early differentiation event in the formation of sporulating aerial hyphae

In a targeted analysis of one of the nine σ^B-like sigma factors, σ^N, it emerged that *sigN* expression depends on the *bld* gene-mediated extracellular cascade, and is involved in the establishment of a specialized aerial hyphal compartment, the subapical stem compartment (Dalton *et al.*, 2007; Fig. 3.8). The establishment of sufficient σ^N levels is brought about in part by a dual expression system for *sigN*. Pre-developmental expression from the *sigN*P1 promoter establishes a basal level of SigN, and in response to the completion of the *bld* gene-mediated extracellular cascade it appears that this SigN becomes active, and directs transcription of *sigN* from a second promoter, P2, as well as increasing P1 activity. Dalton *et al.* (2007) speculated that the BldG anti-anti-sigma and its partner anti-sigma may be involved in the activation of SigN and hence in the autoactivating circuitry. Strong expression of *sigN* appears to be confined to the subapical stem compartment.

A single cross-wall separates this compartment from the apical compartment of aerial hyphae, which is normally destined to differentiate into a chain of spores. Differentiation of the subapical stem compartment should therefore be delayed until the sheath-coated hypha has emerged, and become long enough for this basal cross-wall to form. The formation of a specialized basal cross-wall, involving a cell-wall biosynthetic enzyme

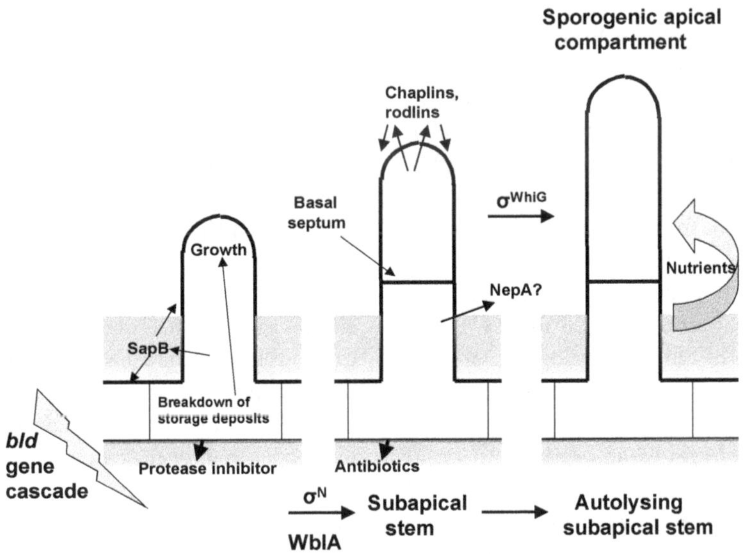

Figure 3.8 Speculative model for the emergence of a sporogenic aerial hypha. In this simplistic model, the *bld* gene cascade activates the production of SapB surfactant and the degradation of storage metabolites, permitting the emergence of an aerial hyphal initial. The action of σ^N and WblA results in the formation of a specialized basal septum, generating a subapical stem compartment (the possible site of production of NepA protein and of antibiotics) and an extending apical compartment that produces the sheath components (rodlins and chaplins). The apical compartment then becomes committed to sporulation under the action of σ^{WhiG}, while the subapical stem compartment undergoes autolysis to provide nutrients for aerial growth. The autolysis involves the extracellular protease cascade illustrated in Fig. 3.9.

with a specific sensitivity to antibiotics, was suggested particularly by studies of *S. griseus* mycelium induced to sporulate in submerged culture by nutritional downshift (Kwak *et al.*, 2001a). It is possible that the emergent aerial hypha in which basal septation has yet to take place may correspond physiologically to the transition phase described earlier. After septation, the subapical compartment may be fated to undergo eventual autolysis and provide nutrition for the further extension and sporulation of the apical compartment (see Fig. 3.9).

A marker gene for the subapical stem compartment, *nepA*, was discovered by microarray analysis and fluorescence microscopy (Dalton *et al.*, 2007). NepA is a small secreted protein of unknown function, which is required for efficient differentiation under some conditions. Transcription of *nepA* probably involves the action of an upstream activator sequence responsive to some unknown signal, and is σ^N-dependent (directly or indirectly). Perhaps this is part of the 'Sky pathway' proposed by Claessen *et al.* (2006) to activate processes specific to the aerial mycelium in response to the emergence of sheathed nascent aerial hyphae into the air. However, this picture of NepA is controversial: another paper found that *nepA* expression was specific to sporulating compartments rather than to subapical compartments (De Jong *et al.*, 2009a). The reasons for this disparity have yet to be clarified. De Jong *et al.* (2009a) also found that NepA is a tightly bound component of the spore surface, and is an agent of spore dormancy.

Recent studies of another developmental gene, *wblA*, suggest that it may be important in the formation of the subapical stem compartment. Elimination of *wblA* results in overproduction of pigmented antibiotics and failure of most aerial hyphae to sporulate (K. Fowler *et al.*, in preparation). This phenotype is consistent with *wblA* mutant aerial hyphae being fixed in a recapitulating physiological state (perhaps close to that of the subapical stem compartment) that normally precedes commitment to sporogenic growth. Overexpression of *wblA* has been found to cause a general repression of antibiotic production (Kang *et al.*, 2007).

An extracellular protease cascade that aids developmentally regulated autolysis of the non-sporulating parts of the colony

Various cytological, genetic, physiological and biochemical approaches, using several different streptomycetes, have indicated that a kind of programmed cell death of non-sporulating parts of the mycelium provides nutrients to support aerial growth (Wildermuth, 1970a; Méndez *et al.*, 1985; Kim *et al.*, 1995; Kang and Lee, 1997; Lee, 1998: Miguélez *et al.*, 1999; Nicieza *et al.*, 1999; Fernández and Sánchez, 2002; reviewed by Chater *et al.*, 2010). In the preceding section it was suggested that the subapical stem compartment might be a central part of this. Such autolysis has turned out to be at least partially regulated by an extracellular protease cascade, which is described in this section (Fig. 3.9).

Involvement of protease inhibitors in the control of the protease cascade

Developmental autolysis, which involves several extracellular proteases, is held in check by protease inhibitors, usually but not always themselves proteins known as SSIs (*Streptomyces* subtilisin inhibitors) (Taguchi *et al.*, 1993; Kim and Lee, 1995a,b, 1996; Kim *et al.*, 1998; Kuramoto *et al.*, 1996; Van Mellaert *et al.*, 1998). SSIs are conserved dimers of identical subunits of about 100 amino acids (Takeuchi *et al.*, 1991, 1992; Mitsui *et al.*, 1977). They interact with various types of target proteases (see Chater *et al.*, 2010, for references), and, significantly, one SSI has been shown to be able to interact simultaneously with two different types of proteases (Hiraga *et al.*, 2000). Since SSIs are fairly well conserved, this dual interaction ability may well be a general feature. Probably, one of the two interaction faces binds to and inhibits certain generic types of protease, while the other mediates a very specific interaction with a specialized regulatory domain – the P-domain – present in just a small number of proteases (Kim *et al.*, 2008a). Thus, in *S. coelicolor*, the SSI encoded by SCO0762 binds to the P-domains of two proteases, encoded by SCO5447 and SCO1355 (Kim *et al.*, 2008a), while in *S. griseus* the SSI encoded by SGR551 (SgiA) interacts

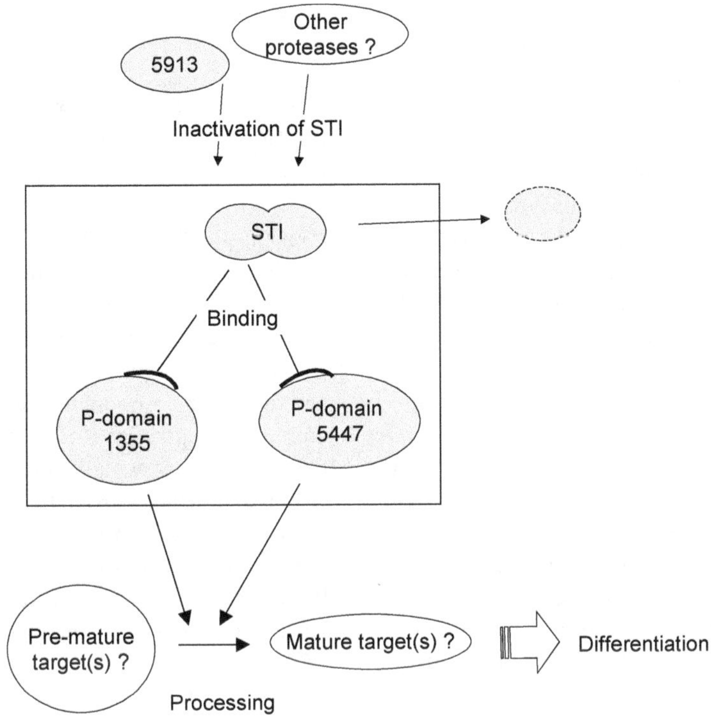

Figure 3.9 An extracellular protease cascade that supports aerial growth in *Streptomyces coelicolor.* The boxed part of the diagram shows the situation before the action of the proteases is needed: key proteases specified by SCO1355 and SCO5447 are held in check by a protease inhibitor (STI). In appropriate circumstances, proteases (including the SCO5913 product) capable of inactivating STI accumulate, eliminating STI and releasing SCO1355 and SCO5447 proteases, whose role appears to be the activation of proenzymes that participate in autolysis (Kim *et al.*, 2008a). The production of several components of the cascade is dependent on AdpA (see Fig. 3.7).

with the SGR2095 protease orthologous with that of SCO5447 (Hirano *et al.*, 2006).

In order for the P-domain-containing proteases to be released from inhibition, the SSI must be inactivated. This is achieved by SSI-specific proteases that accumulate in older cultures. In the case of *S. coelicolor*, a major role in SSI inactivation is played by the protease encoded by SCO5913, though it appears that at least one other protease also acts on the SSI (Kim *et al.*, 2008a). Since SCO5913 is not an orthologue of any gene present in other characterized *Streptomyces* genomes, the interaction between SSIs and their inactivating partners may be species-specific, limiting cross-talk between the cascades of different streptomycetes sharing the same niche.

Regulation of genes involved in the extracellular protease cascade

The protease cascade in *S. coelicolor* is strongly dependent on AdpA, and therefore on the *bldA*-specified tRNA needed to translate the UUA codon in *adpA* mRNA (as demonstrated for the SSI by Kim *et al.*, 2005); and, in *S. griseus*, the cascade also depends on the accumulation of A-factor. Indeed, in *S. griseus*, genes for at least 11 proteases and peptidases, as well as the SSI gene, are transcriptionally activated by AdpA (Kato *et al.*, 2002, 2005a,b; Tomono *et al.*, 2005; Akanuma *et al.*, 2009; Fig. 3.7). The SSI genes of several other streptomycetes are also AdpA-dependent (Hirano *et al.*, 2006; Chater *et al.*, 2010). In *S. coelicolor*, it seems likely that production of the SSI is a function of the earliest stage of aerial growth, since a *wblA* mutant unable to progress from stem compartment physiology (see above) overexpresses SCO0762 (K. Fowler *et al.*, in preparation).

Possible targets of the protease cascade

The P-domain is homologous with a domain present on certain eukaryotic proteases that fulfil specialized roles in the cell cycle or developmental pathways, particularly by activating proenzymes. This may be the case also for the P-domain-containing proteases of streptomycetes: in *Streptomyces* (formally *Streptoverticillium*) *mobaraensis*, a protease called TAMEP activates the proprotein form of an extracellular transglutaminase (Pasternack *et al.*, 1998), and is regulated by an SSI-type protease inhibitor (Zotzel *et al.*, 2003a,b). The experimentally determined N-terminal sequence of TAMEP corresponds to that of the P-domain-containing SCO5447 and SGR2095 proteases that interact with SSI in *S. coelicolor* and *S. griseus*. Similar findings have recently been reported for a *Streptomyces hygroscopicus* strain (Ren *et al.*, 2008; Zhang *et al.*, 2008a,b).

Extracellular transglutaminases, currently the only known targets of the protease cascade, are industrially interesting enzymes that catalyse the cross-linking of proteins, generating high molecular weight aggregates. Most sequenced *Streptomyces* genomes do not encode homologues of known transglutaminases, but nevertheless it has been suggested that cell wall-associated transglutaminase may cross-link surface proteins of aerial hyphae. Intriguingly, the *S. mobaraensis* protease inhibitor is itself a substrate for the transglutaminase (Schmidt *et al.*, 2008). This may well amount to a positive feedback circuit. Activation of other kinds of proenzymes has been described in the context of *Streptomyces* growth and development, though it is not clear whether the protease cascade is implicated in any of these. For example, in *S. antibioticus*, one of the nucleases involved in sporulation is activated through processing by a serine protease (Nicieza *et al.*, 1999; Fernández and Sánchez, 2002), while there are several examples of proteolytic separation of the substrate-binding and catalytic domains of extracellular cellulases and xylanases (e.g. Fernández-Abalos *et al.*, 2003).

The apical compartment of aerial hyphae and its metamorphosis into a chain of rounded spores

Growth of aerial hyphae

Fuelled by nutrients obtained from the lower parts of the mycelium, the aerial hyphal apical compartments grow rapidly upwards. Among the genes upregulated during this elongation are those for components of a branched chain amino-acid uptake system, which, on the model described earlier, would contribute to the assimilation of nutrients from the hydrated cell wall encased by the sheath; while the *amtB* gene for the ammonium transporter, which is unlikely to be necessary for the reuse of previously assimilated nitrogen, is downregulated in aerial hyphae (Tian *et al.*, 2007).

Using a GFP-tagged DnaN protein (DNA polymerase III β-subunit) to track the positions of replisomes, and therefore of ongoing chromosome replication, it was found that replication takes place at several foci within both vegetative and aerial hyphal compartments, and that there is no overall juxtaposition of tip-located cell wall growth and chromosome replication (Ruban-Ośmiałowska *et al.*, 2006; Fig. 3.10). Apical aerial compartments always had several bright foci, whereas subapical compartments often had none. This is consistent with the quite distinct roles of these compartments. However, the number of foci in apical aerial compartments was generally smaller than expected, and their spacing greater. This suggests either that some chromosomes are not replicating or that more than one chromosome is replicating at each focus. It is likely that after the initiation of the final round of replication before sporulation, each newborn *oriC* moves rapidly to its final position as a partitioning complex while the rest of its chromosome is still being replicated (see below for details of partitioning). By the time sporulation septation is complete, replisome foci have disappeared, leaving transient low levels of fluorescence.

To reach spore chain lengths of the order of 100, c. 50 additional chromosomes would be

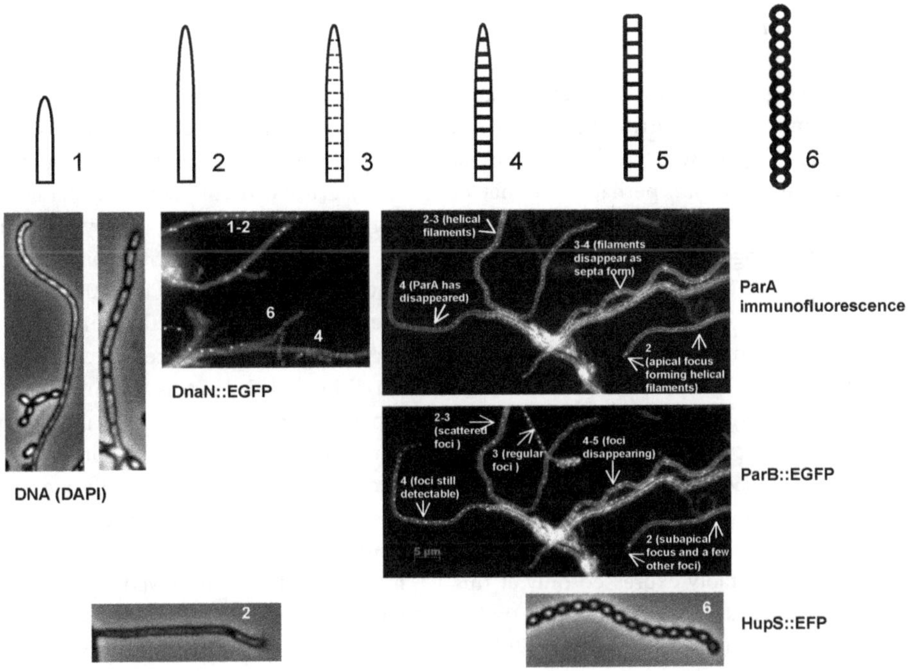

Figure 3.10 Localization of DNA and associated proteins during sporulation of aerial hyphae. Numbers on figure panels refer to the developmental stages indicated in the diagrams [for details, see Ruban-Ośmiałowska *et al.* (2006), Jakimowicz *et al.* (2007) and Dedrick *et al.* (2009)]. Sources of images: DNA and HupS::EGFP, Klas Flärdh; DnaN::EGFP, ParA immunofluorescence, ParB::EGFP, Dagmara Jakimowicz.

generated in the final two-hour round of replication, requiring that the compartment would increase in length by that number of 'unit cell lengths'. On the basis of average intervals of about 1.3 μm between sporulation septa, this would imply an increase in length of up to 65 μm in two hours. This is nearly twice the maximum rates reported for vegetative hyphae (see above), leading Chater (2000) to speculate that there may also be significant synthesis along the lateral walls (the distended spore phenotype of CrgA mutants has also been interpreted as indicating lateral wall extension: Del Sol *et al.*, 2006, see below).

It may be worth considering an alternative hypothesis. The extension rate of aerial hyphae may be lower than the above model predicts, if we take into account the greater width often observed for growing aerial hyphae compared with most vegetative hyphae (e.g. Hopwood, 1960). Applying the concept of a unit cell *volume*, a small increase in hyphal diameter would permit the required rate of length extension to be similar to rates known for vegetative growth. For example, increasing the diameter by a factor of $\sqrt{2}$ (i.e. 1.4) would halve the required rate. After the final round of chromosome replication in such wide aerial hyphae, some remodelling of the lateral wall might then lead to the eventual c. 1.3 μm spacing between sporulation septa. Such remodelling might be responsible for moderate staining of the lateral wall of apical compartments with a treatment that recognizes incompletely cross-linked peptidoglycan (Jakimowicz *et al.*, 2005). Furthermore, the increased surface area of a wider hyphal tip would permit increased occupation by DivIVA, and thus enhanced local cell wall biosynthetic activity. It is interesting that the overexpression of *ssgA* causes a doubling of the diameter of vegetative hyphae (van Wezel *et al.*, 2000), so it is possible that the increased *ssgA* expression normally observed in aerial hyphae (Yamazaki *et al.*, 2003; Traag *et al.*, 2004) may bring about just such an increase in diameter in the aerial hyphae.

Growth cessation and the cell biology of sporulation

When extension growth stops (perhaps because of the stress imposed by nutritional limitations on the accelerating growth rate), there must be a change in metabolism, perhaps like that of vegetative cultures entering stationary phase. For example, at this stage cytologically observable granules of glycogen are formed, partially under the direction of one copy of a duplicated gene set involved in the interconversion of trehalose, maltose and glycogen (Fig. 3.11). (As discussed earlier, the other copy of this gene set functions exclusively in older parts of the substrate mycelium: Bruton *et al.*, 1995.)

The genes for several proteins involved in condensation of nucleoids during sporulation (see below) are detectably expressed only at this stage, but in no case is it known how this spatial specificity is achieved (Dedrick *et al.*, 2009; Kois *et al.*, 2009; Facey *et al.*, 2009).

In broad outline, the first cytologically observable events specific to sporulation involve the ParA protein, which (together with ParB) guides the positioning of chromosomes (Jakimovicz *et al.*, 2007; Fig. 3.10), and the SsgA protein, which, together with the SsgA paralogues SsgB, C and G, appears to guide the positioning of sporulation septa: SsgA is the archetypal representative of the SsgA-like 'SALP' family of paralogues, which play significant, but still poorly understood, roles at different stages of *Streptomyces* growth and development (Noens *et al.*, 2005; Traag and van Wezel, 2008; Xu *et al.*, 2009; Fig. 3.12). ParA and SsgA are initially concentrated at the growing aerial hyphal tip, but as the final round of DNA replication nears completion, and the apical compartment approaches its maximum length, ParA becomes rearranged as a helix along the length of the apical compartment, and SsgA complexes mark off the compartment at pre-spore-sized intervals. This is followed by the assembly of complexes of ParB with chromosomal replication origins at regular intervals along the ParA helix, in between the SsgA markers (Jakimowicz *et al.*, 2005a). At the same time, the key cell division protein FtsZ polymerizes along the length of the compartment (Grantcharova *et al.*, 2005), the

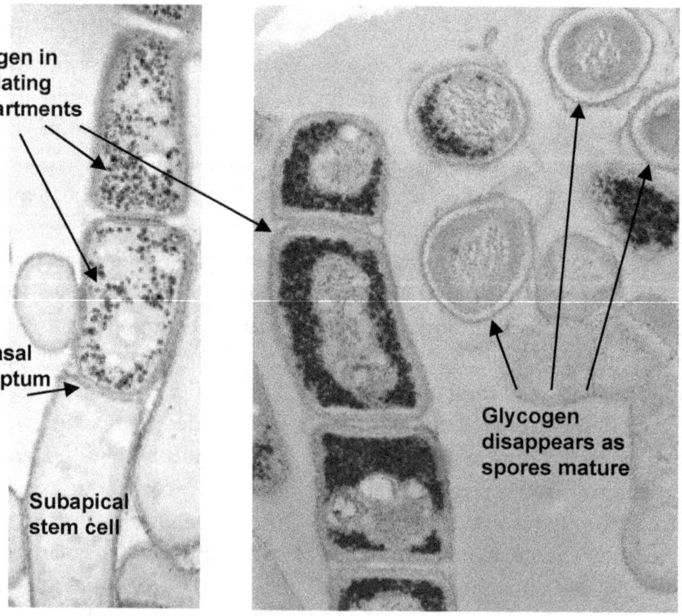

Figure 3.11 Glycogen accumulation in sporulating compartments, and its disappearance as spores mature (Plaskitt and Chater, 1995).

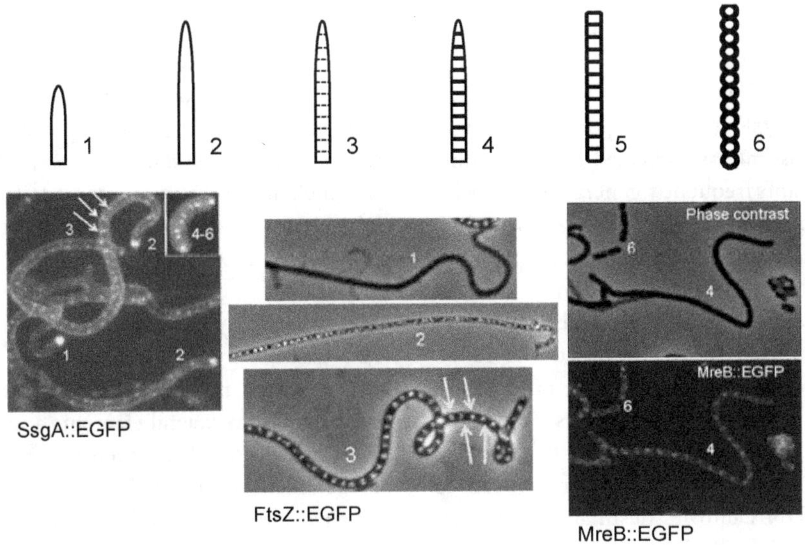

Figure 3.12 Localization of division-related proteins during sporulation of aerial hyphae. Numbers on figure panels refer to the developmental stages indicated in the diagrams. Arrows show regularly spaced foci of SsgA::EGFP and FtsZ::EGFP. Sources of images: SsgA::EGFP, Gilles van Wezel; FtsZ::EGFP, MreB::EGFP, Klas Flärdh. For further details, see Noens *et al.* (2005), Grantcharova *et al.* (2005) and Mazza *et al.* (2006).

polymer then fragmentating before reassembly into the ladder of Z-rings that will guide sporulation septation (Schwedock *et al.*, 1997; Willemse and van Wezel, 2009; Fig. 3.12). Probably, the positioning of the Z-rings is determined by the SsgA markers (G. van Wezel, personal communication), while the timing of Z-ring appearance may be determined by a small membrane protein, CrgA (Del Sol *et al.*, 2003). Thus, Z-rings and ParB/oriC complexes alternate along the length of the compartment, and sporulation septation is initiated.

In many rod-shaped bacteria, Z-ring formation is prevented wherever nucleoids are present (nucleoid occlusion: Bramkamp and van Baarle, 2009), but it has been considered that nucleoid occlusion does not affect Z-ring positioning in streptomycetes because DNA can often be seen traversing the sites of ingrowing sporulation septa (Flardh, 2003b). However, recent work of Facey *et al.* (2009) suggests that nucleoid occlusion may interfere with sporulation septation in *dps* mutants that have abnormalities in their chromosome condensation.

The ingrowing septa have an organization perceptibly different from that of vegetative septa, being led by a double edge that seems to have evolved to permit cell separation after septation (Williams and Sharples, 1970; Wildermuth and Hopwood, 1970; McVittie, 1974; Hardisson and Manzanal, 1976; Manzanal and Hardisson, 1978; Strunk, 1978). This difference may in part depend on a sporulation-specific interaction of FtsZ with some other component(s), since certain mutant forms of FtsZ retain the ability to form vegetative cross-walls while losing the ability to generate sporulation septa (Grantcharova *et al.*, 2002). Unlike vegetative septation, sporulation septation requires FtsI and FtsW, without which FtsZ spirals cannot form rings, and FtsQ, which is needed for the closure of ingrowing sporulation septa (Mistry *et al.*, 2008).

In order to avoid guillotining of segments of chromosomes that traverse the cell division plane, closure of sporulation septa is associated with pumping of the residual parts of chromosomes into their destination compartments, through a pore in the septum consisting of the motor protein FtsK. It is probable that FtsK of streptomycetes, like that of other bacteria, recognizes asymmetric sequence motifs ('KOPS') present predominantly in one strand of the chromosome to achieve the correct partitioning of the misplaced DNA (Bigot *et al.*, 2007; Hendrickson and Lawrence, 2006; Flardh and Buttner, 2009). The importance of FtsK in ensuring that spores have entire chromosomes is indicated by the finding that FtsK mutants frequently generate spores with incomplete chromosomes (Wang *et al.*, 2007; Ausmees *et al.*, 2007; Dedrick *et al.*, 2009). FtsK is not the only agent by which chromosomes are packaged into individual spores. The nucleoids become condensed under the action of HupS proteins (Salerno *et al.*, 2009) and other DNA packaging proteins such as Smc (Kois *et al.*, 2009; Dedrick *et al.*, 2009) and DpsA, DpsB and DpsC (Facey *et al.*, 2009), and this process appears to help to pull DNA into its destination compartment. A second FtsK-like protein, SffA, is also present in ingrowing sporulation septa, its insertion depending on a membrane protein, SmeA. Curiously, while the elimination of SmeA causes aberrations in sporulation, the loss of SffA has no perceptible effect (Ausmees *et al.*, 2007).

When septation is complete, deposits of the actin-like MreB protein against the septa may guide the synthesis and controlled degradation of cell wall material leading to the proper separation of the prespore compartments (Mazza *et al.*, 2006; Fig. 3.12). MreB is then redistributed over the entire inner face of the prespore membrane, probably controlling the shape change from short cylinders to ovoid spores, and the deposition of the thick spore wall. MreB is thought to interact with the membrane proteins MreC and MreD, encoded in the *mreBCD* operon, to form a complex that is associated with Pbp83 (high-molecular-weight penicillin-binding protein) and Sfr (member of the SpoVE/FtsW/RodA family), which are also encoded by adjacent genes (Burger *et al.*, 2000). Spores of an *mreB* mutant are swollen, somewhat thin-walled, and sensitive to heat and SDS, but the mutant is unimpaired in other aspects of its growth (Mazza *et al.*, 2006).

The morphological stages of sporulation are completed by the production of spore pigment and its likely cross-linking to the spore wall. In many species, the spore pigment is a type II aromatic polyketide whose synthesis is catalysed by the products of the *whiE* cluster (Davis and Chater, 1990; Metsä-Ketelä *et al.*, 2002), the transcription of which, from a divergent promoter region, depends on all the well-characterized *whi* sporulation regulatory genes (see below: Kelemen *et al.*, 1998). In *S. griseus,* the pigment is a melanin-like substance made via a type III (chalcone synthase-like) polyketide synthase (Funa *et al.*, 1999, 2005).

The regulatory underpinning of sporulation

As indicated in the previous section, several important cytoskeletal elements and other major determinants of cellular events are more abundant in the apical compartment than in other parts of the mycelium. This involves increased expression of the relevant genes. Thus, in *S. griseus* AdpA is needed for the up-regulation of *ssgA* (Yamazaki *et al.*, 2003), while in *S. coelicolor* the *ssgR* regulatory gene upstream of *ssgA* is essential for the increased *ssgA* expression (Traag *et al.*, 2004). Increases in ParA, ParB and FtsZ are brought about by the activation of extra, developmentally specific promoters for the relevant genes. These promoters depend, directly or indirectly, on sporulation regulatory genes (Jakimowicz *et al.*, 2006; Flardh *et al.*, 2000; Kwak *et al.*, 2001b) that were initially discovered through the study of mutants that retain the chalky white appearance associated with immature aerial mycelium (*whi* mutants: Hopwood *et al.*, 1970; Chater, 1972; Ryding *et al.*, 1999). Just as with the *bld* genes involved in the formation of aerial mycelium, the *whi* genes nearly all encode regulatory proteins; but as time has gone on, reverse genetics has identified further genes that might have been called *whi* if they had been identified in the original mutant hunt, and some of these do not seem to be regulatory (e.g. *ssgA*). It is not clear why such genes were not encountered in primary mutant hunts, but it is worth recalling that the early mutants were isolated on minimal medium containing glucose, whereas in most of the more recent work mannitol has been the carbon source: the white phenotype of *ssgA* mutants was only seen on mannitol, not on glucose (van Wezel *et al.*, 2000; Jiang and Kendrick, 2000).

The whiG, -H, -I branch of the regulatory cascade leading to spore formation

The regulators encoded by the *whi* genes originally described by Chater (1972) are implicated in a converging cascade (Fig. 3.13). One tributary to the cascade is initiated by the activity of a sigma factor, σ^{WhiG}, that closely resembles sigma factors required for the formation of flagella in other bacteria (Chater *et al.*, 1989). Only eight promoters are known that are recognized by σ^{WhiG} RNA polymerase: the promoters of *whiH* and *whiI* (Ryding *et al.*, 1998; Aínsa *et al*,, 1999; Kormanec *et al.*, 1999); and several for genes whose developmental significance is unknown (two in *S. coelicolor* and four in *S. aureofaciens*; Tan and Chater, 1993; Tan *et al.*, 1998; Nováková *et al.*, 1998; Kormanec *et al.*, 2001). The function of the phase II sporulation-associated glycogen/trehalose cluster (see above) is also rather specifically dependent on *whiG,* but it is not clear whether the cluster is transcribed by σ^{WhiG} RNA polymerase (Yeo and Chater, 2005). The accumulation of σ^{WhiG} RNA polymerase may partially depend on release of BldD protein from the *whiG* promoter (Elliot *et al.*, 2001).

WhiH and WhiI appear to be involved in both self-regulation and cross-regulation, but it is not known whether this is direct (Ryding *et al.*, 1998; Aínsa *et al.*, 1999; Fig. 3.13). The *whiH* promoter is activated during the transition phase in submerged culture (Nieselt *et al.*, 2010). This suggests that σ^{WhiG} is active at this stage. In *S. aureofaciens,* two proteins were found to bind the *whiH* promoter: one (RwhB, binding around the transcription start site) was suggested to be WhiH itself, and the other (RwhA, binding much further upstream) was postulated to repress *whiH* expression in substrate mycelium (Homerová and

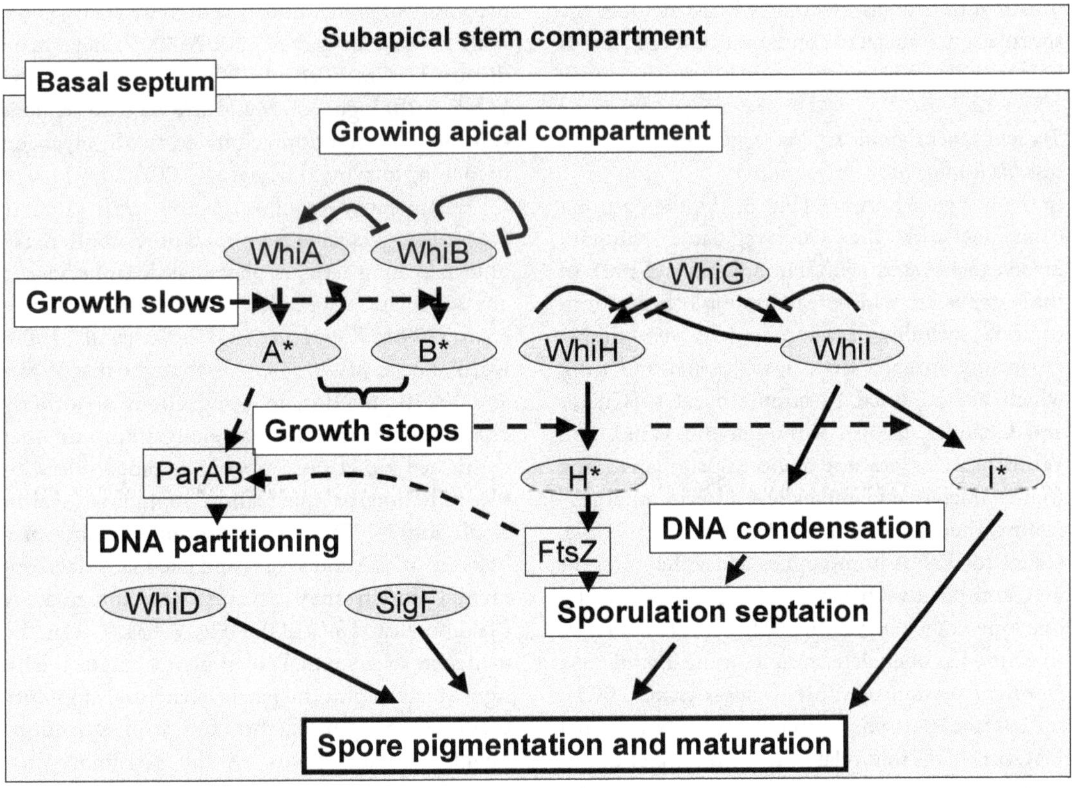

Figure 3.13 A regulatory cascade leading to sporulation in the aerial hyphal apical compartment (redrawn from Chater and Chandra, 2006, with additional information).

Kormanec, 2001). The resemblance of WhiH to the GntR family of metabolite-sensitive regulators suggests that WhiH may be responsive to a small molecule whose concentrations change in the developing apical compartment (Ryding *et al.*, 1998). The phenotype of *whiH* mutants indicates that WhiH may influence the level of FtsZ and/or other cell division proteins (Ryding *et al.*, 1998; Flardh *et al.*, 1999; Flardh *et al.*, 2000).

WhiI is an aberrant response regulator that possibly responds to an unknown ligand, rather than being controlled by phosphorylation (Aínsa *et al.*, 1999; Tian *et al.*, 2007). Microarray analysis of *whiI* site-directed mutants has indicated that a substantial number of genes are affected directly or indirectly by the action of WhiI, some by the supposed ligand-free form and some by the supposed ligand-bound form (Tian *et al.*, 2007). One of the WhiI-influenced genes, SCO3900, is itself regulatory, appearing to repress an operon of which it is the first gene (Zhang *et al.*, 2009). The broad range of areas affected by WhiI includes the sporulation-associated condensation of nucleoids (Aínsa *et al.*, 1999).

The whiA, whiB branch of the regulatory cascade leading to spore formation

In the *whi* gene network (Fig. 3.13), the part that converges with the *whiG*–regulated tributary involves *whiA* and *whiB*. Homologues of both of these genes are widespread among other actinobacteria, including developmentally very simple organisms, in contrast to *whiG, whiH* and *whiI*, which are confined to streptomycetes (Chater and Chandra, 2006). Although the WhiA and WhiB proteins are not obviously similar to any well-characterized families of transcriptional regulators, there is growing evidence for such a role. Genes for WhiA homologues are widely present in Gram-positive bacteria, and there is usually only one copy per genome (Aínsa *et al.*, 2000). A structure has been determined for the *Thermatoga maritima* version of WhiA (Kaiser *et al.*, 2009). It comprises two domains: an N-terminal LAGLIDADG fold related to a homing endonuclease domain, and a C-terminal helix-turn-helix domain whose closest functionally characterized representative is the C-terminal domain of sigma factors (the part that recognizes the –35 sequences of promoters). The LAGLIDADG domain is likely to interact with DNA but not to cleave it, in view of the absence of residues critical for the endonuclease function. Thus, WhiA may have two DNA binding domains.

WhiB is the archetypal member of a substantial family of actinobacteria-specific small redox-sensitive iron–sulphur proteins (Wbl proteins, including WblA, see above) (Davis and Chater, 1992; Soliveri *et al.*, 2000; Gao *et al.*, 2006; Jakimowicz *et al.*, 2005b; Singh *et al.*, 2007; Alam *et al.*, 2007, 2009; den Hengst and Buttner, 2008; Crack *et al.*, 2009). Experimental (Steyn *et al.*, 2002) and genomic evidence (Chater and Chandra, 2006) suggests that Wbl proteins interact with sigma factors – there is even one case of a protein with a Wbl-like N-terminus and a sigma-like C-terminus (WblP; Bentley *et al.*, 2004). The *in vitro* study of Wbl proteins has been frustratingly difficult, and claims that they have significant protein disulphide reductase activity (Garg *et al.*, 2007, 2009; Alam *et al.*, 2007, 2009) have been disputed (Crack *et al.*, 2009). Curiously, the WblE orthologue of *Mycobacterium tuberculosis* can be co-immunoprecipitated with glycogen branching enzyme (Garg *et al.*, 2007, 2009).

The phenotypes of *whiA* and *whiB* mutants are indistinguishable from each other: both make unusually long aerial hyphae that do not undergo any sporulation septation (Chater, 1972; Flardh *et al.*, 1999; Xie *et al.*, 2007; Xu *et al.*, 2003; Kormanec *et al.*, 1998). It is thought that WhiA and WhiB function to bring about an orderly cessation of aerial hyphal elongation, and that continued growth is probably incompatible with the initiation of sporulation septation (Flardh *et al.*, 1999). There is transcriptional interplay between WhiA and WhiB, and like many developmental proteins they appear to be autoregulatory (Jakimowicz *et al.*, 2006; Fig. 3.13). Given the evidence that some Wbl proteins interact with sigma factors, and the partial structural similarity of WhiA to sigma factors, future investigations might profitably focus on the possibility that a WhiB–WhiA complex has sigma factor-like activity, and that the formation or structure of the complex is responsive to redox changes.

Other genes needed for the production of mature spores

Another Wbl protein, WhiD, is also important during sporulation (Chater, 1972; Molle *et al.*, 2000). In the absence of WhiD, sporulation levels are somewhat reduced, but where sporulation septation occurs, it appears to be uncontrolled, often generating minicompartments with longitudinally oriented septa that indicate a second round of septation; and such spores as are formed are thin-walled and temperature-sensitive (McVittie, 1974; Molle *et al.*, 2000).

The deficiencies in spore pigment production in most *whi* mutants of *S. coelicolor* are because of the dependence of transcription of the *whiE* spore pigment gene cluster both on the *whi* gene regulatory cascade and on the resultant morphological processes leading to sporulation (Kelemen *et al.*, 1998; Grantcharova *et al.*, 2002). The sigma factor σ^F, which has a similar pattern of *whi* gene dependence to that of *whiE* and is important for spore maturation, may well be implicated in transcription from one of the *whiE* promoters (Kelemen *et al.*, 1998), but otherwise there is little information to clarify the genetic regulation of events at these later stages, such as the presumptive upregulation of *mreB* and of the genes for proteins involved in DNA condensation, or the metabolic transitions that cause glycogen to disappear (Fig. 3.11), trehalose to accumulate, and inactive trehalase to be deposited.

Concluding remarks: some inconvenient fellow-travellers, and future research directions

This chapter is incomplete in many ways in its coverage of the literature. In part, this reflects difficulties in assimilating some aspects into the narrative framework. Here I list references that can be used to follow themes that I have not been able to cover. They include such matters as whether calcium ions have important roles in development (Wang *et al.*, 2008); the apparently pivotal involvement of the membrane-bound GTP-binding protein Obg in the onset of differentiation (Okamoto and Ochi, 1998); the roles of several ECF sigma factors in development (see Mao *et al.*, 2009b, for further references); the importance of intracellular protein turnover for development, as indicated by the Bld phenotype of Clp protease mutants (de Crécy-Lagard *et al.*, 1999); the often species-specific developmental effects of some secondary metabolites, such as borrelidin (Schüz and Zähner, 1993) and pamamycin (McCann and Pogell., 1979; Natsume *et al.*, 1995; Hashimoto *et al.*, 2003); the ability of the extracellular protein Factor C to restore sporulation to non-sporulating mutants of *S. griseus*, including A-factor-defective mutants (Chater *et al.*, 2010); the role of ADP-ribosylation of proteins in development (Penyige *et al.*, 2009); the stimulation of development by copper (Endo *et al.*, 2002); the requirement for a particular iron siderophore, desferrioxamine E, for aerial growth (Yamanaka *et al.*, 2005); the influence of medium osmolarity on development (Fernández Martínez *et al.*, 2009); and sporulation in submerged culture of some streptomycetes, notably *S. griseus* (Kwak *et al.*, 2001a,b) and *S. venezuelae* (Flärdh and Buttner, 2009). The combination of genome-based analytical tools, metabolomics and advanced cytological procedures that are now available will underpin much future research into all of these topics, as well as into the many areas of ignorance revealed in this article. The results should provide new insights into the increasingly complex, but also increasingly beautiful, picture of development that is emerging for these remarkable organisms.

Acknowledgements

I am grateful to Klas Flärdh, Dagmara Jakimowicz and Gilles van Wezel for providing photomicrographs and comments; to Govind Chandra for help with comparative genome analysis; and the John Innes Foundation for an Emeritus Fellowship.

References

Aínsa, J.A., Parry, H.D., and Chater, K.F. (1999). A response regulator-like protein that functions at an intermediate stage of sporulation in *Streptomyces coelicolor* A3(2). Mol. Microbiol. *34*, 607–619.

Aínsa, J.A., Ryding, N.J., Hartley, N., Findlay, K.C., Bruton, C.J., and Chater, K.F. (2000). WhiA, a protein of unknown function conserved among gram-positive bacteria, is essential for sporulation in *Streptomyces coelicolor* A3(2). J. Bacteriol. *182*, 5470–5478.

Aínsa, J.A., Bird, N., Ryding, N.J., Findlay, K.C., and Chater, K.F. (2010). The complex *whiJ* locus mediates environmentally sensitive repression of development of *Streptomyces coelicolor* A3(2). Antonie Van Leeuwenhoek *98*, 225–236.

Akanuma, G., Hara, H., Ohnishi, Y., and Horinouchi, S. (2009). Dynamic changes in the extracellular proteome caused by absence of a pleiotropic regulator AdpA in *Streptomyces griseus*. Mol. Microbiol. *73*, 898–912.

Alam, M.S., Garg, S.K., and Agrawal, P. (2007). Molecular function of WhiB4/Rv3681c of *Mycobacterium tuberculosis* H37Rv: a [4Fe-4S] cluster co-ordinating protein disulphide reductase. Mol. Microbiol. *63*, 1414–1431.

Alam, M.S., Garg, S.K., and Agrawal, P. (2009). Studies on structural and functional divergence among seven WhiB proteins of *Mycobacterium tuberculosis* H37Rv. FEBS J. *276*, 76–93.

Allan, E.J., and Prosser, J.I. (1987). Colony growth of *Streptomyces coelicolor* A3(2) under conditions of continuous nutrient supply. FEMS Microbiol. Lett. *43*, 139–142.

Amon, J., Titgemeyer, F., and Burkovski, A. (2010). Common patterns – unique features: Nitrogen metabolism and regulation in Gram-positive bacteria. FEMS Microbiol. Rev. *34*, 588–605.

Aoki, Y., Yoshida, Y., Yoshida, M., Kawaide, H., Abe, H., and Natsume, M. (2007). Isolation and characterization of a spore germination inhibitor from *Streptomyces* sp. CB-1-1, a phytopathogen causing root tumor of melon. Biosci. Biotech. Biochem. *71*, 986–992.

Arabolaza, A., Rodriguez, E., Altabe, S., Alvarez, H., and Gramajo, H. (2008). Multiple pathways for triacylglycerol biosynthesis in *Streptomyces coelicolor*. Appl. Environ. Microbiol. *74*, 2573–2782.

Ausmees, N., Wahlstedt, H., Bagchi, S., Elliot, M.A., Buttner, M.J., and Flärdh, K. (2007). SmeA, a small membrane protein with multiple functions in *Streptomyces* sporulation including targeting of a SpoIIIE/FtsK-like protein to cell division septa. Mol. Microbiol. *65*, 1458–1473.

Bagchi, S., Tomenius, H., Belova, L.M., and Ausmees, N. (2008). Intermediate filament-like proteins in bacteria and a cytoskeletal function in *Streptomyces*. Mol. Microbiol. *70*, 1037–1050.

Bentley, S.D., Chater, K.F., Cerdeno-Tarraga, A.M., Challis, G.L., Thomson, N.R., James, K.D., Harris, D.E., Quail, M.A., Kieser, H., Harper, D., *et al.* (2002). Complete genome sequence of the model actinomycete *Streptomyces coelicolor* A3(2). Nature *417*, 141–147.

Bentley, S.D., Brown, S., Murphy, L.D., Harris, D.E., Quail, M.A., Parkhill, J., Barrell, B.G., McCormick, J.R., Santamaría, R.I., Losick, R., *et al.* (2004). SCP1, a 356, 023 bp linear plasmid adapted to the ecology and developmental biology of its host, *Streptomyces coelicolor* A3(2). Mol. Microbiol. *51*, 1615–1628.

Bibb, M.J. (2005). Regulation of secondary metabolism in streptomycetes. Curr. Opin. Microbiol. *8*, 208–215.

Bibb, M.J., and Buttner, M.J. (2003). The *Streptomyces coelicolor* developmental transcription factor sigma-BldN is synthesized as a proprotein. J. Bacteriol. *185*, 2338–2345.

Bibb, M.J., Molle, V., and Buttner, M.J. (2000). Sigma(BldN), an extracytoplasmic function RNA polymerase sigma factor required for aerial mycelium formation in *Streptomyces coelicolor* A3(2). J. Bacteriol. *182*, 4606–4616.

Bignell, D.R., Warawa, J.L., Strap, J.L., Chater, K.F., and Leskiw, B.K. (2000). Study of the *bldG* locus suggests that an anti-anti-sigma factor and an anti-sigma factor may be involved in *Streptomyces coelicolor* antibiotic production and sporulation. Microbiology *146*, 2161–2173.

Bignell, D.R., Lau, L.H., Colvin, K.R., and Leskiw, B.K. (2003). The putative anti-anti-sigma factor BldG is post-translationally modified by phosphorylation in *Streptomyces coelicolor*. FEMS Microbiol. Lett. *225*, 93–99.

Bigot, .S., Sivanathan, V., Possoz, C., Barre, F.X., and Cornet, F. (2007). FtsK, a literate chromosome segregation machine. Mol. Microbiol. *64*, 1434–1441.

Blanco, G., Rodicio, M.R., Puglia, A.M., Méndez, C., Thompson, C.J., and Salas, J.A. (1994). Synthesis of ribosomal proteins during growth of *Streptomyces coelicolor*. Mol. Microbiol. *12*, 375–385.

Bloomfield, B.J., and Alexander, M. (1967). Melanins and resistance of fungi to lysis. J. Bacteriol. *93*, 1276–1280.

Bobek, J., Halada, P., Angelis, J., Vohradsky, J., and Mikulík, K. (2004). Activation and expression of proteins during synchronous germination of aerial spores of *Streptomyces granaticolor*. Proteomics *4*, 3864–3880.

Braña, A.F., Manzanal, M.B., and Hardisson, C. (1982). Characterization of intracellular polysaccharides of *Streptomyces*. Can. J. Microbiol. *28*, 1320–1323.

Braña, A.F., Méndez, C., Diaz, L.A., Manzanal, M.B., and Hardisson, C. (1986). Glycogen and trehalose accumulation during colony development in *Streptomyces antibioticus*. J. Gen. Microbiol. *132*, 1319–1326.

Bramkamp, M., and van Baarle, S. (2009). Division site selection in rod-shaped bacteria. Curr. Opin. Microbiol. *12*, 683–688.

Brown, M.R., and Kornberg, A. (2004). Inorganic polyphosphate in the origin and survival of species. Proc. Natl. Acad. Sci. U.S.A, *101*, 16085–16087.

Bruton, C.J., Plaskitt, K.A., and Chater, K.F. (1995). Tissue-specific glycogen branching isoenzymes in a multicellular prokaryote, *Streptomyces coelicolor* A3(2). Mol. Microbiol. *18*, 89-99.

Burger, A., Sichler, K., Kelemen, G., Buttner, M., and Wohlleben, W. (2000). Identification and characterization of the *mre* gene region of *Streptomyces coelicolor* A3(2). Mol. Gen. Genet. *263*, 1053–1060.

Capstick, D.S., Willey, J.M., Buttner, M.J., and Elliot, M.A. (2007). SapB and the chaplins: connections between morphogenetic proteins in *Streptomyces coelicolor*. Mol. Microbiol. *64*, 602–613.

Carroll, J.D., Pastuszak, I., Edavana, V.K., Pan, Y.T., and Elbein, A.D. (2007). A novel trehalase from

Mycobacterium smegmatis – purification, properties, requirements. FEBS J. *274*, 1701–1714.

Chakraburtty, R., and Bibb, M. (1997). The ppGpp synthetase gene (*relA*) of *Streptomyces coelicolor* A3(2) plays a conditional role in antibiotic production and morphological differentiation. J. Bacteriol. *179*, 5854–5861.

Challis G.L. (2008). Mining microbial genomes for new natural products and biosynthetic pathways. Microbiology *154*, 1555–1569.

Champness, W.C. (1988). New loci required for *Streptomyces coelicolor* morphological and physiological differentiation. J. Bacteriol. *170*, 1168–1174.

Champness, W.C and Chater, K.F. (1994). The regulation and interaction of antibiotic production and morphological differentiation in *Streptomyces* spp. In Regulation of Bacterial Differentiation, Piggot, P., Moran, C., and Youngman, P., eds (Washington, American Society for Microbiology), pp. 61–93.

Chandra, G., and Chater, K.F. (2008). Evolutionary flux of potentially *bldA*-dependent *Streptomyces* genes containing the rare leucine codon TTA. Antonie Van Leeuwenhoek *94*, 111–126.

Chang, S.A., Bralley, P., and Jones, G.H. (2005). The *absB* gene encodes a double strand-specific endoribonuclease that cleaves the read-through transcript of the *rpsO-pnp* operon in *Streptomyces coelicolor*. J. Biol. Chem. *280*, 14555–14561.

Chater, K.F. (1972). A morphological and genetic mapping study of white colony mutants of *Streptomyces coelicolor*. J. Gen. Microbiol. *72*, 9–28.

Chater, K.F. (1998). Taking a genetic scalpel to the *Streptomyces* colony. Microbiology *144*, 1465–1478.

Chater, K.F. (2000). Developmental decisions during sporulation in the aerial mycelium in *Streptomyces*. In Prokaryotic Development, Brun, Y.V., and Shimkets, L.J., eds (Washington, American Society for Microbiology), pp. 33–48.

Chater, K.F. (2001). Regulation of sporulation in *Streptomyces coelicolor* A3(2): a checkpoint multiplex? Curr. Opin. Microbiol. *4*, 667–673.

Chater, K.F. (2006). *Streptomyces* inside out: a new perspective on the bacteria that provide us with antibiotics. Phil. Trans. Roy. Soc. B. *361*, 761–768.

Chater, K.F., and Chandra, G. (2006). The evolution of development in *Streptomyces* analysed by genome comparisons. FEMS Microbiol. Rev. *30*, 651–672.

Chater, K.F., and Chandra, G. (2008). The use of the rare UUA codon to define 'expression space' for genes involved in secondary metabolism, development and environmental adaptation in *Streptomyces*. J. Microbiol. *46*, 1–11.

Chater, K.F., and Horinouchi, S. (2003). Signalling early developmental events in two highly diverged *Streptomyces* species. Mol. Microbiol. *48*, 9–15.

Chater K.F and Merrick M.J. (1979). Streptomycetes. In Developmental Biology of Prokaryotes, Parish, J.H., ed. (Oxford, Blackwell), pp. 93–114.

Chater, K F., Bruton, C.J., Plaskitt, K.A., Buttner, M.J., Méndez, C., and Helmann, J.D. (1989). The developmental fate of *S. coelicolor* hyphae depends upon a gene product homologous with the motility σ factor of *B. subtilis*. Cell *59*, 133–143.

Chater, K.F., Biró, S., Lee, K.J., Palmer, T., and Schrempf, H. (2010). The evolution of *Streptomyces* has involved an amazingly complex extracellular biology. FEMS Microbiol. Rev. *24*, 171–198.

Cho, Y.H., Lee, E.J., Ahn, B.E., and Roe, J.H. (2001). SigB, an RNA polymerase sigma factor required for osmoprotection and proper differentiation of *Streptomyces coelicolor*. Mol. Microbiol. *42*, 205–214.

Claessen, D., Wosten, H.A.B., van Keulen, G., Faber, O.G., Alves, A.M.C.R., Meijer, W.G., and Dijkhuizen, L. (2002). Two novel homologous proteins of *Streptomyces coelicolor* and *Streptomyces lividans* are involved in the formation of the rodlet layer and mediate attachment to a hydrophobic surface. Mol. Microbiol. *44*, 1483–1492.

Claessen, D., Rink, R., de Jong, W., Siebring, J., de Vreugd, P., Boersma, F.G., Dijkhuizen, L., and Wösten, H.A. (2003). A novel class of secreted hydrophobic proteins is involved in aerial hyphae formation in *Streptomyces coelicolor* by forming amyloid-like fibrils. Genes Dev. *17*, 1714–1726.

Claessen, D., Stokroos, I., Deelstra, H.J., Penninga, N.A., Bormann, C., Salas, J.A., Dijkhuizen, L., and Wösten, H.A. (2004). The formation of the rodlet layer of streptomycetes is the result of the interplay between rodlins and chaplins. Mol. Microbiol. *53*, 433–443.

Claessen, D., de Jong, W., Dijkhuizen, L., and Wösten, H.A. (2006). Regulation of *Streptomyces* development: reach for the sky! Trends Microbiol. *14*, 313–319.

Crack, J.C., den Hengst, C.D., Jakimowicz, P., Subramanian, S., Johnson, M.K., Buttner, M.J., Thomson, A.J., and Le Brun, N.E. (2009). Characterization of [4Fe-4S]-containing and cluster-free forms of *Streptomyces* WhiD. Biochemistry *48*, 12252–12264.

Dalton, K.A., Thibessard, A., Hunter, J.I., and Kelemen, G.H. (2007). A novel compartment, the 'subapical stem' of the aerial hyphae, is the location of a sigN-dependent, developmentally distinct transcription in *Streptomyces coelicolor*. Mol. Microbiol. *64*, 719–737.

Davis, N.K., and Chater, K.F. (1990). Spore colour in *Streptomyces coelicolor* A3(2) involves the developmentally regulated synthesis of a compound biosynthetically related to polyketide antibiotics. Mol. Microbiol. *4*, 1679–1691.

Davis, N.K., and Chater, K.F. (1992). The *Streptomyces coelicolor whiB* gene encodes a small transcription factor-like protein dispensable for growth but essential for sporulation. Mol. Gen. Genet. *232*, 351–358.

de Crécy-Lagard, V., Servant-Moisson, P., Viala, J., Grandvalet, C., and Mazodier, P. (1999). Alteration of the synthesis of the Clp ATP-dependent protease affects morphological and physiological differentiation in *Streptomyces*. Mol. Microbiol. *32*, 505–517.

De Jong, W., Manteca, A., Sanchez, J., Bucca, G., Smith, C.P., Dijkhuizen, L., Claessen, D., and Wösten, H.A. (2009a). NepA is a structural cell wall protein involved in maintenance of spore dormancy in *Streptomyces coelicolor.* Mol. Microbiol. *71,* 1591–1603.

De Jong, W., Wösten, H.A.B., Dijkhuisen, L., and Claessen, D. (2009b). Attachment of *Streptomyces coelicolor* is mediated by amyloidal fimbriae that are anchored to the cell surface via cellulose. Mol. Microbiol. 73, 1128–1140.

De la Fuente, A., Lorenzana, L.M., Martin, J.F., and Liras, P. (2002). Mutants of *Streptomyces clavuligerus* with disruptions in different genes for clavulanic acid biosynthesis produce large amounts of holomycin: possible cross-regulation of two unrelated secondary metabolic pathways. J. Bacteriol. *184,* 6559–6565.

Dedrick, R.M., Wildschutte, H., and McCormick, J.R. (2009). Genetic interactions of *smc, ftsK,* and *parB* genes in *Streptomyces coelicolor* and their developmental genome segregation phenotypes. J. Bacteriol. *191,* 320–332.

Del Sol, R., Pitman, A., Herron, P., and Dyson, P. (2003). The product of a developmental gene, *crgA,* that coordinates reproductive growth in *Streptomyces* belongs to a novel family of small actinomycete-specific proteins. J. Bacteriol. *185,* 6678–6685.

Del Sol, R., Mullins, J.G., Grantcharova, N., Flärdh, K., and Dyson, P. (2006). Influence of CrgA on assembly of the cell division protein FtsZ during development of *Streptomyces coelicolor.* J. Bacteriol. *188,* 1540–1550.

Del Sol, R., Armstrong, I., Wright, C., and Dyson, P. (2007). Characterization of changes to the cell surface during the life cycle of *Streptomyces coelicolor*: atomic force microscopy of living cells. J. Bacteriol. *189,* 2219–2225.

den Hengst, C.D., and Buttner, M.J. (2008). Redox control in actinobacteria. Biochim. Biophys. Acta *1780,* 1201–1216.

Derouaux, A., Halici, S., Nothaft, H., Neutelings, T., Moutzourelis, G., Dusart, J., Titgmeyer, F., and Rigali, S. (2004a). Deletion of a cyclic AMP receptor protein homologue diminishes germination and affects morphological development of *Streptomyces coelicolor.* J. Bacteriol. *186,* 1893–1897.

Derouaux, A., Dehareng, D., Lecocq, E., Halici, S., Nothaft, H., Giannotta, F., Moutzourelis, G., Dusart, J., Devreese, B., Titgmeyer, F., *et al.* (2004b). Crp of *Streptomyces coelicolor* is the third transcription factor of the large CRP-FNR superfamily able to bind cAMP. Biochem. Biophys. Res. Comm. *325,* 983–990.

DiBerardo, C., Capstick, D.S., Bibb, M.J., Findlay, K., Buttner, M.J., and Elliott, M.A. (2008). Function and redundancy of the chaplin cell-surface proteins in aerial hyphae formation, rodlet assembly, and viability in *Streptomyces coelicolor.* J. Bacteriol. *190,* 5879–5889.

Driks, A., and Setlow, P. (2000). Morphogenesis and properties of the bacterial spore. In Prokaryotic Development, Brun, Y.V., and Shimkets, L.J., eds (Washington, D.C., American Society for Microbiology), pp. 191–218.

Eccleston, M., Ali, R., Seyler, R., Westpheling, J., and Nodwell, J. (2002). Structural and genetic analysis of the BldB protein of *Streptomyces coelicolor.* J. Bacteriol. *184,* 4270–4276.

Eccleston, M., Willems, A., Beveridge, A., and Nodwell, J.R. (2006). Critical residues and novel effects of overexpression of the *Streptomyces coelicolor* developmental protein BldB: evidence for a critical interacting partner. J. Bacteriol. *188,* 8189–8195.

Elbein, A.D., Pan, Y.T., Patuszak, I., and Carroll, D. (2003). New insights on trehalose: a multifunctional molecule. Glycobiology *13,* 17R-27R.

Elliot, M.A., and Leskiw, B.K. (1999). The BldD protein from *Streptomyces coelicolor* is a DNA-binding protein. J. Bacteriol. *181,* 6832–6835.

Elliot, M., and Talbot, N.J. (2004). Building filaments in the air: aerial morphogenesis in bacteria and fungi. Curr. Opin. Microbiol. *7,* 594–601.

Elliot, M.A., Damji, F., Passantino, R., Chater, K., and Leskiw, B. (1998). The *bldD* gene of *Streptomyces coelicolor* A3(2) : a regulatory gene involved in morphogenesis and antibiotic production. J. Bacteriol. *180,* 1549–1555.

Elliot, M.A., Bibb, M.J., Buttner, M.J., and Leskiw, B.K. (2001). BldD is a direct regulator of key developmental genes in *Streptomyces coelicolor* A3(2). Mol. Microbiol. *40,* 257–269.

Elliot, M.A., Karoonuthaisiri, N., Huang, J., Bibb, M.J., Cohen, S.N., Kao, C.M., and Buttner, M.J. (2003a). The chaplins: a family of hydrophobic cell-surface proteins involved in aerial mycelium formation in *Streptomyces coelicolor.* Genes Dev. *17,* 1727–1740.

Elliot, M.A., Locke, T.R., Galibois, C.M., and Leskiw, B.K. (2003b). BldD from *Streptomyces coelicolor* is a non-essential global regulator that binds its own promoter as a dimer. FEMS Microbiol. Lett. *225,* 35–40.

Elliot, M.A., Buttner, M.J., and Nodwell, J.R. (2008). Multicellular development in *Streptomyces.* In Myxobacteria: Multicellularity and Differentiation, Whitworth, D.E., ed. (Washington, D.C., American Society for Microbiology), pp. 419–438.

Endo, K., Hosono, K., Beppu, T., and Ueda, K. (2002). A novel extracytoplasmic phenol oxidase of *Streptomyces*: its possible involvement in the onset of morphogenesis. Microbiology *148,* 1767–1776.

Ensign, J.C. (1978). Formation, properties and germination of actinomycete spores. Annu. Rev. Microbiol. *35,* 182–219.

Facey, P.D., Hitchings, M.D., Saavedra-Garcia, P., Fernandez-Martinez, L., Dyson, P.J., and Del Sol, R. (2009). *Streptomyces coelicolor* Dps-like proteins: differential dual roles in response to stress during vegetative growth and in nucleoid condensation during reproductive cell division. Mol. Microbiol. *73,* 1186–1202.

Fernández, M., and Sánchez, J. (2002). Nuclease activities and cell death processes associated with the development of surface cultures of *Streptomyces antibioticus.* Microbiology *148,* 405–412.

Fernández-Abalos, J.M., Reviejo, V., Díaz, M., Rodríguez, S., Leal, F., and Santamaría, R.I. (2003). Posttranslational processing of the xylanase Xys1L from *Streptomyces halstedii* JM8 is carried out by secreted serine proteases. Microbiology *149,* 1623–1632.

Fernández Martínez, L., Bishop, A., Parkes, L., Del Sol, R., Salerno, P., Sevcíková, B., Mazuráková, V., Kormanec, J., and Dyson, P. (2009). Osmoregulation in *Streptomyces coelicolor*: modulation of SigB activity by OsaC. Mol. Microbiol. *71*, 1250–1262.

Flärdh, K. (2003a). Essential role of DivIVA in polar growth and morphogenesis in *Streptomyces coelicolor* A3(2). Mol. Microbiol. *49*, 1523–1536.

Flärdh, K. (2003b). Growth polarity and cell division in *Streptomyces*. Curr. Opin. Microbiol. *6*, 564–571.

Flärdh, K., and Buttner, M.J. (2009). *Streptomyces* morphogenetics: dissecting differentiation in a filamentous bacterium. Nature Rev. Microbiol. *7*, 36–49.

Flärdh, K., Findlay, K.C., and Chater, K.F. (1999). Association of early sporulation genes with suggested developmental decision points in *Streptomyces coelicolor* A3(2). Microbiology *145*, 2229–2243.

Flärdh, K., Leibowitz, E., Buttner, M.J., and Chater, K.F. (2000). Generation of a non-sporulating strain of *Streptomyces coelicolor* A3(2) by the manipulation of a developmentally controlled *ftsZ* promoter. Mol. Microbiol. *38*, 737–749.

Funa, N., Ohnishi, Y., Fujii, I., Shibuya, M., Ebizuka, Y., and Horinouchi, S. (1999). A new pathway for polyketide synthesis in microorganisms. Nature *400*, 897–899.

Funa, N., Funabashi, M., Ohnishi, Y., and Horinouchi, S. (2005). Biosynthesis of hexahydroxyperylenequinone melanin via oxidative aryl coupling by cytochrome P-450 in *Streptomyces griseus*. J. Bacteriol. *187*, 8149–8155.

Gao, B., Paramanathan, R., and Gupta, R.S. (2006). Signature proteins that are distinctive characteristics of Actinobacteria and their subgroups. Antonie Van Leeuwenhoek *90*, 69–91.

García, M. (1995). A membrane-like structure envelops substrate mycelium during colony development in *Streptomyces*. FEMS Microbiol. Lett. *131*, 107–111.

Garcia Diaz, L.A., Salas, J.A., and Hardisson, C. (1983). Intracellular pool of *Streptomyces* spores: amino acids, adenine nucleotide levels and energy charge. FEMS Microbiol. Lett. *19*, 215–219.

Garg, S.K., Alam, M.S., Kishan, K.V., and Agrawal, P. (2007). Expression and characterization of alpha-(1, 4)-glucan branching enzyme Rv1326c of *Mycobacterium tuberculosis* H37Rv. Protein Expr. Purif. *51*, 198–208.

Garg, S., Alam, M.S., Bajpai, R., Kishan, K.R., and Agrawal, P. (2009). Redox biology of *Mycobacterium tuberculosis* H37Rv: protein–protein interaction between GlgB and WhiB1 involves exchange of thiol-disulfide. BMC Biochem. *10*, 1.

Gaskell, A.A., Crack, J.C., Kelemen, G.H., Hutchings, M.I., and Le Brun, N.E. (2007). RsmA is an anti-sigma factor that modulates its activity through a [2Fe-2S] cluster cofactor. J. Biol. Chem. *282*, 31812–31820.

Gehring, A., Nodwell, J., Beverley, S., and Losick, R. (2000). Genomewide insertional mutagenesis in *Streptomyces coelicolor* reveals additional genes involved in morphological differentiation. Proc. Natl. Acad. Sci. U.S.A. *97*, 9642–9647.

Ghorbel, S., Smirnov, A., Chouayekh, H., Sperandio, B., Esnault, C., Kormanec, J., and Virolle, M.J. (2006). Regulation of *ppk* expression and *in vivo* function of Ppk in *Streptomyces lividans* TK24. J. Bacteriol. *188*, 6269–6276.

Glauert, A.M., and Hopwood, D.A. (1961). The fine structure of *Streptomyces violaceoruber* (*S. coelicolor*). J. Biophys. Biochem. Cytol. *10*, 505–516.

Granozzi, C., Billetta, R., Passantino, R., Sollazzo, M., and Puglia, A.M. (1990). A breakdown in macromolecular synthesis preceding differentiation in *Streptomyces coelicolor* A3(2). J. Gen. Microbiol. *136*, 713–716.

Grantcharova, N., Lustig, U., and Flärdh, K. (2005). Dynamics of FtsZ assembly during sporulation in *Streptomyces coelicolor* A3(2). J. Bacteriol. *187*, 3227–3237.

Grantcharova, N., Ubhayasekera, W., Mowbray, S.L., McCormick, J.R., and Flärdh, K. (2002). A missense mutation in *ftsZ* differentially affects vegetative and developmentally controlled cell division in *Streptomyces coelicolor* A3(2). Mol. Microbiol. *47*, 645–656.

Grund, A.D., and Ensign, J.C. (1985). Properties of the germination inhibitor of *Streptomyces viridochromogenes* spores. J. Gen. Microbiol. *131*, 833–847.

Guijarro, J., Santamaría, R., Schauer, A., and Losick, R. (1988). Promoter determining the timing and spatial localization of a cloned *Streptomyces coelicolor* gene encoding a spore-associated polypeptide. J. Bacteriol. *170*, 1895–1901.

Haiser, H.J., Yousef, M.R., and Elliot, M.A. (2009). Cell wall hydrolases affect germination, vegetative growth and sporulation in *Streptomyces coelicolor*. J. Bacteriol. *191*, 6501–6512.

Harasym, M., Zhang, L.H., Chater, K., and Piret, J. (1990). The *Streptomyces coelicolor* A3(2) *bldB* region contains at least two genes involved in morphological development. J. Gen. Microbiol. *136*, 1543–1550.

Hardisson, C., and Manzanal, M.B. (1976). Ultrastructural studies of sporulation in *Streptomyces*. J. Bacteriol. *127*, 1443–1454.

Hardisson, C., Manzanal, M.B., Salas, J.A., and Suárez, J.E. (1978). Fine structure, physiology and biochemistry of arthrospore germination in *Streptomyces antibioticus*. J. Gen. Microbiol. *105*, 203–214.

Hashimoto, M., Kondo, T., Kozone, I., Kawaide, H., Abe, H., and Natsume, M. (2003). Relationship between response to and production of the aerial mycelium-inducing substances pamamycin-607 and A-factor. Biosci. Biotech. Biochem. *67*, 803–808.

Hempel, A.M., Wang, S.B., Letek, M., Gil, J.A., and Flärdh, K. (2008). Assemblies of DivIVA mark sites for hyphal branching and can establish new zones of cell wall growth in *Streptomyces coelicolor*. J. Bacteriol. *190*, 7579–7583.

Hendrickson, H., and Lawrence, J.G. (2006). Selection for chromosome architecture in bacteria. J. Mol. Evol. *62*, 615–629.

Hesketh, A., Bucca, G., Laing, E., Flett, F., Hotchkiss, G., Smith, C.P., and Chater, K.F. (2007). New pleiotropic effects of eliminating a rare tRNA from *Streptomyces coelicolor*, revealed by combined proteomic and transcriptomic analysis of liquid cultures. BMC Genomics *8*, 261.

Hesketh, A., Chen, W.J., Ryding, J., Chang, S., and Bibb, M. (2007). The global role of ppGpp synthesis in morphological differentiation and antibiotic production in *Streptomyces coelicolor* A3(2). Genome Biol. *8*, R161.

Hey-Ferguson, A., Mitchell, M., and Elbein, A.D. (1973). Trehalose metabolism in germinating spores of *Streptomyces hygroscopicus*. J. Bacteriol. *116*, 1084–1085.

Hiraga, K., Suzuli, T., and Oda, K. (2000). A novel double headed proteinaceous inhibitor for metalloproteinase and serine proteinase. J. Biol. Chem. *275*, 25173–25179.

Hirano, S., Kato, J.Y., Ohnishi, Y., and Horinouchi, S. (2006). Control of the *Streptomyces* subtilisin inhibitor gene by AdpA in the A-factor regulatory cascade in *Streptomyces griseus*. J. Bacteriol. *188*, 6207–6216.

Hirsch, C.F., and Ensign, J.C. (1978). Some properties of *Streptomyces viridochromogenes* spores. J. Bacteriol. *134*, 1056–1063.

Hodgson, D.A. (1992). Differentiation in actinomycetes. In Prokaryotic Structure and Function: A New Perspective. Society for General Microbiology Symposium Volume 47. Mohan, S., Dow, C., and Cole, J., eds (Cambridge, UK, Cambridge University Press), pp. 407–440.

Hodgson, D.A. (2000). Primary metabolism and its control in streptomycetes: a most unusual group of bacteria. Adv. Microb. Physiol. *42*, 47–238.

Homerová, D., and Kormanec, J. (2001).Identification of DNA-binding proteins involved in regulation of expression of the *Streptomyces aureofaciens whiH* gene encoding a sporulation transcription factor. Folia Microbiol. (Praha) *46*, 527–534.

Hopwood, D.A. (1960). Phase-contrast observations on *Streptomyces coelicolor*. J. Gen. Microbiol. *22*, 295–302.

Hopwood, D.A., Wildermuth, H., and Palmer, H.M. (1970). Mutants of *Streptomyces coelicolor* defective in sporulation. J. Gen. Microbiol. *61*, 397–408.

Horinouchi, S. (2007). Mining and polishing of the treasure trove in the bacterial genus *Streptomyces*. Biosci. Biotechnol. Biochem. *71*, 283–299.

Huang, J., Shi, J., Molle, V., Sohlberg, B., Weaver, D., Bibb, M.J., Karoonuthaisiri, N., Li, C.J., Kao, C.M., Buttner, M.J., *et al.* (2005). Cross-regulation among disparate antibiotic biosynthetic pathways of *Streptomyces coelicolor*. Mol. Microbiol. *58*, 1276–1287.

Hunt, A.C., Servín-González, L., Kelemen, G.H., and Buttner, M.J. (2005). The *bldC* developmental locus of *Streptomyces coelicolor* encodes a member of a family of small DNA-binding proteins related to the DNA-binding domains of the MerR family. J. Bacteriol. *187*, 716–728.

Jakimowicz, D., Chater, K.F., and Zakrzewska-Czerwinska, J. (2002). The ParB protein of *Streptomyces coelicolor* A3(2) recognizes a cluster of *parS* sequences within the origin-proximal region of the linear chromosome. Mol. Microbiol. *45*, 1365–1377.

Jakimowicz, D., Gust, B., Zakrzewska-Czerwinska, J., and Chater, K.F. (2005a). Developmental stage-specific assembly of ParB complexes in *Streptomyces coelicolor* hyphae. J. Bacteriol. *187*, 3572–3580.

Jakimowicz, P., Cheesman, M.R., Bishai, W.R., Chater, K.F., Thomson, A.J., and Buttner, M.J. (2005b). Evidence that the *Streptomyces* developmental protein WhiD, a member of the WhiB family, binds a [4Fe-4S] cluster. J. Biol. Chem. *280*, 8309–8315.

Jakimowicz, D., Mouz, S., Zakrzewska-Czerwinska, Z., and Chater, K.F. (2006). Regulation of developmentally controlled *parAB* promoter leading to formation of ParB complexes in *Streptomyces coelicolor* aerial hyphae. J. Bacteriol. *188*, 1710–1720.

Jakimowicz, D., Zydek, P., Kois, A., Zakrzewska-Czerwińska, J., and Chater, K.F. (2007). Alignment of multiple chromosomes along helical ParA scaffolding in sporulating *Streptomyces* hyphae. Mol. Microbiol. *65*, 625–641.

Jarling, M., Cauvet, T., Grundmeier, M., Kuhnert, K., and Pape, H. (2004). Isolation of *mak1* from *Actinoplanes missouriensis* and evidence that Pep2 from *Streptomyces coelicolor* is a maltokinase. J. Basic Microbiol. *44*, 360–373.

Jiang, H., and Kendrick, K.E. (2000). Characterization of *ssfR* and *ssgA*, two genes involved in sporulation of *Streptomyces griseus*. J. Bacteriol. *182*, 5521–5529.

Jyothikumar, V., Tilley, E.J., Wali, R., and Herron, P.R. (2008). Time lapse photography of *Streptomyces coelicolor* growth and sporulation. Appl. Environ. Microbiol. *74*, 6774–6781.

Kaiser, B.K., Clifton, C., Shen, B.W., and Stoddard, B.L. (2009). The structure of a bacterial DUF199/WhiA protein: domestication of an invasive endonuclease. Structure *17*, 1368–1376.

Kalakoutski, L.V., and Agre, N.S. (1976). Comparative aspects of development and differentiation in actinomycetes. Bacteriol. Rev. *40*, 469–524.

Kalscheuer, R., Syson, K., Veeraraghavan, U., Weinrick, B., Biermann, K.E., Lui, Z., Sacchettini, J.C., Besra, G., Bornemann, S., and Jacobs, W.R. (2010). Self-poisoning of *Mycobacterium tuberculosis* by targeting GlgE in an α-glucan pathway. Nat. Chem. Biol. *6*, 376–384.

Kang, S.G., and Lee, K.J. (1997). Kinetic analysis of morphological differentiation and protease production in *Streptomyces albidoflavus* SMF301. Microbiology *143*, 2709–2714.

Kang, S.H., Huang, J., Lee, H.N., Hur, Y.A., Cohen, S.N., and Kim, E.S. (2007). Interspecies DNA microarray analysis identifies *wblA* as a pleiotropic downregulator of antibiotic biosynthesis in *Streptomyces*. J. Bacteriol. *189*, 4315–4319.

Kato, J., Suzuki, A., Yamazaki, H., Ohnishi, Y., and Horinouchi, S. (2002). Control by A-factor of a metalloendopeptidase gene involved in aerial mycelium formation in *Streptomyces griseus*. J. Bacteriol. *184*, 6016–6025.

Kato, J., Chi, W.J., Ohnishi, Y., Hong, S.K., and Horinouchi, S. (2005a). Transcriptional control by

A-factor of two trypsin genes in *Streptomyces griseus*. J. Bacteriol. *187*, 286–295.

Kato, J., Hirano, S., Ohnishi, Y., and Horinouchi, S. (2005b). The *Streptomyces* subtilisin inhibitor (SSI) gene in *Streptomyces* coelicolor A(3)2. Biosci. Biotechnol. Biochem. *69*, 1624–1629.

Kato, J.Y., Ohnishi, Y., and Horinouchi, S. (2005c). Autorepression of AdpA of the AraC/XylS family, a key transcriptional activator in the A-factor regulatory cascade in *Streptomyces griseus*. J. Mol. Biol. *350*, 12–26.

Keijser, B.J., van Wezel, G.P., Canters, G.W., and Vijgenboom, E. (2002). Developmental regulation of the *Streptomyces lividans ram* genes: involvement of RamR in regulation of the *ramCSAB* operon. J. Bacteriol. *184*, 4420–4429.

Kelemen, G.H., Plaskitt, K.A., Lewis, C.G., Findlay, K.C., and Buttner, M.J. (1995). Deletion of DNA lying close to the *glkA* locus induces ectopic sporulation in *Streptomyces coelicolor* A3(2). Mol. Microbiol. *17*, 221–230.

Kelemen, G.H., Brown, G.L., Kormanec, J., Potúcková, L., Chater, K.F., and Buttner, M.J. (1996). The positions of the sigma-factor genes, *whiG* and *sigF*, in the hierarchy controlling the development of spore chains in the aerial hyphae of *Streptomyces coelicolor* A3(2). Mol. Microbiol. *21*, 593–603.

Kelemen, G.H., Brian, P., Flärdh, K., Chamberlin, L., Chater, K.F., and Buttner, M.J. (1998). Developmental regulation of transcription of whiE, a locus specifying the polyketide spore pigment in *Streptomyces coelicolor* A3 (2). J. Bacteriol. *180*, 2515–2521.

Kelemen, G.H., Viollier, P.H., Tenor, J.L., Marri, L., Buttner, M.J., and Thompson, C.J. (2001). A connection between stress and development in the multicellular prokaryote *Streptomyces coelicolor* A3(2). Mol. Microbiol. *40*, 804–814.

Kieser, T., Bibb, M.J., Buttner, M.J., Chater, K.F., and Hopwood, D.A. (2000). Practical *Streptomyces* Genetics (Norwich, UK, The John Innes Foundation).

Kim, I.S., and Lee, K.J. (1995a). Physiological roles of leupeptin and extracellular proteases in mycelium development of *Streptomyces exfoliatus* SMF13. Microbiology *141*, 1017–1025.

Kim, I.S., and Lee, K.J. (1995b). Kinetic study on the production and degradation of leupeptin in *Streptomyces exfoliatus* SMF13. J. Biotechnol. *42*, 35–44.

Kim, I.S., and Lee, K.J. (1996). Trypsin like protease in *Streptomyces exfoliatus* SMF13, as a potential agent for mycelium differentiation. Microbiology *142*, 1797–1806.

Kim, I.S., Kang, S.G., and Lee, K.J. (1995). Physiological importance of trypsin like protease during morphological differentiation of *Streptomyces* spp. J. Microbiol. *33*, 315–321.

Kim, I.S., Kim, Y.B., and Lee, K.J. (1998). Characterization of the leupeptin- inactivating enzyme from *Streptomyces exfoliatus* SMF13 which produces leupeptin. Biochem. J. *331*, 539–545.

Kim, H.J., Calcutt, M.J., Scmidt, F.J., and Chater, K.F. (2000). Partitioning of the linear chromosome during sporulation of *Streptomyces coelicolor* A3(2) involves an *oriC*-linked *parAB* locus. J. Bacteriol. *182*, 1313–1320.

Kim, D.W., Chater, K.F., Lee, K.J., and Hesketh, A.J. (2005). Changes in the extracellular proteome caused by the absence of the *bldA* gene product, a developmentally significant tRNA, reveal a new target for the pleiotropic regulator AdpA in *Streptomyces coelicolor*. J. Bacteriol. *187*, 2957–2966.

Kim, I.K., Lee, C.J., Kim, M.K., Kim, J.M., Kim, J.H., Yim, H.S., Cha, S.S., and Kang, S.O. (2006). Crystal structure of the DNA-binding domain of BldD, a central regulator of aerial mycelium formation in *Streptomyces coelicolor* A3(2). Mol. Microbiol. *60*, 1179–1193.

Kim, D.W., Hesketh, A., Kim, E.S., Song, J.Y., Lee, D.H., Kim, I.S., Chater, K.F., and Lee, K.J. (2008a). Complex extracellular interactions of proteases and a protease inhibitor influence multicellular development of *Streptomyces coelicolor*. Mol. Microbiol. *70*, 1180–1193.

Kim, E.S., Song, J.Y., Kim, D.W., Ko, E.J., Chater, K.F., and Lee, K.J. (2008b). A possible extended family of regulators of sigma factor activity in *Streptomyces coelicolor*. J. Bacteriol. *190*, 7559–7566

Kelemen, G.H., Plaskitt, K.A., Lewis, C.G., Findlay, K.C., and Buttner, M.J. (1995). Deletion of DNA lying close to the *glkA* locus induces ectopic sporulation in *Streptomyces coelicolor* A3(2). Mol. Microbiol. *17*, 221–230.

Kodani, S., Hudson, E., Durrant, M.C., Buttner, M.J., Nodwell, J.R., and Willey, J.M. (2004). The SapB morphogen is a lantibiotic-like peptide derived from the product of the developmental gene *ramS* in *Streptomyces coelicolor*. Proc. Natl. Acad. Sci. U.S.A. *101*, 11448–11453.

Kodani, S., Lodato, M.A., Durrant, M.C., Picart, F., and Willey, J.M. (2005). SapT, a lanthionine-containing peptide involved in aerial hyphae formation in the streptomycetes. Mol. Microbiol. *58*, 1368–1380.

Kois, A., Swiatek, M., Jakimowicz, D., and Zakrzewska-Czerwinska, J. (2009). SMC protein-dependent condensation during aerial hyphal development in *Streptomyces*. J. Bacteriol. *191*, 310–319.

Kormanec, J., Sevcíková, B., Sprusanský, O., Benada, O., Kofronová, O., Nováková, R., Rezuchová, B., Potúcková, L., and Homérová, D. (1998). The *Streptomyces aureofaciens* homologue of the *whiB* gene is essential for sporulation; its expression correlates with the developmental stage. Folia Microbiol. (Praha). *43*, 605–612.

Kormanec, J., Nováková, R., Homerová, D., and Sevcíková, B. (1999). The *Streptomyces aureofaciens* homologue of the sporulation gene *whiH* is dependent on *rpoZ*-encoded sigma factor. Biochim. Biophys. Acta *1444*, 80–84.

Kormanec, J., Sevcíková, B., Halgasová, N., Knirschová, R., and Resuchová, B. (2000). Identification and transcriptional characterization of the gene encoding the stress-response sigma factor sigma (H) in *Streptomyces coelicolor* A3 (2). FEMS Microbiol. Lett. *189*, 31–38.

Kormanec, J., Nováková, R., Homerová, D., and Rezuchová, B. (2001). *Streptomyces aureofaciens* sporulation-specific sigma factor sigma (RpoZ) directs expression of a gene encoding protein similar to hydrolases involved in degradation of the lignin-related biphenyl compounds. Res. Microbiol. *152*, 883–888.

Kretschmer, S. (1988). Stepwise increase in elongation rate in individual hyphae of *Streptomyces granaticolor* during outgrowth. J. Basic Microbiol. *28*, 35–43.

Kretschmer, S. (1989). Septation behaviour of the apical cell in *Streptomyces granaticolor* mycelia. J. Basic Microbiol. *29*, 587–595.

Kretschmer, S. (1991). Correlation between branching and elongation in germ tubes of *Streptomyces granaticolor*. J. Basic Microbiol. *31*, 259–264.

Kretschmer, S. (1992). Location of branches within the apical hyphal region of *Streptomyces granaticolor* mycelia. J. Basic Microbiol. *32*, 35–42.

Kwak, J., McCue, L.A., Trczianka, K., and Kendrick, K.E. (2001a). Identification and characterization of a developmentally regulated protein, EshA, required for sporogenic hyphal branches in *Streptomyces griseus*. J. Bacteriol. *183*, 3004–3015.

Kwak, J., Dharmatilake, A.J., Jiang, H., and Kendrick, K.E. (2001b) Differential regulation of *ftsZ* transcription during septation of *Streptomyces griseus*. J. Bacteriol. *183*, 5092–5101.

Kuramoto A., Lezhava, A., Taguchi, S., Momose, H., and Kinashi, H. (1996). The location and deletion of the genes which code for SSI-like protease inhibitors in *Streptomyces* species. FEMS Microbiol. Lett. *139*, 37–42.

Lee, K.J. (1998). Dynamics of morphological and physiological differentiation in actinomycetes group and quantitative analysis of the differentiation. J. Microbiol. Biotech. *8*, 1–7.

Lee, E.J., Cho, Y.H., Kim, H.S., Ahn, B.E., and Roe, J.H. (2004a). Regulation of σ^B by an anti- and anti-anti-sigma factor in *Streptomyces coelicolor* in response to osmotic stress. J. Bacteriol. *186*, 8490–8498.

Lee, E.J., Cho, Y.H., Kim, H.S., and Roe, J.H. (2004b). Identification of σ^B-dependent promoters using consensus-directed search of *Streptomyces coelicolor* genome. J. Microbiol. *42*, 147–151.

Lee, L.F., Yeh, S.H., and Chen, C.W.. (2002). Construction and synchronization of *dnaA* temperature-sensitive mutants *of Streptomyces*. J. Bacteriol. *184*, 1214–1218.

Lee, E., Karoonuthaisiri, N., Kim, H., Park, J., Cha, C., Kao, C., and Roe, J. (2005). A master regulator sigma governs osmotic and oxidative response as well as differentiation via a network of sigma factors in *Streptomyces coelicolor*. Mol. Microbiol. *57*, 1252–1264.

Lenarcic, R., Halbedel, S., Visser, S., Shaw, M., Wu, L.J., Errington, J., Marenduzzo, D., and Hamoen, L.W. (2009). Localisation of DivIVA by targeting to negatively curved membranes. EMBO J. *28*, 2272–2282.

Leskiw, B.K., Lawlor, E.J., Fernández-Abalos, J.M., and Chater, K.F. (1991). TTA codons in some genes prevent their expression in a class of developmental, antibiotic-negative, *Streptomyces* mutants. Proc. Natl. Acad. Sci. U.S.A. *88*, 2461–2465.

Li, W., Wu, J., Tao, W., Zhao, C., Wang, Y., He, X., Chandra, G., Xhou, X., Deng, Z., Chater, K.F., and Tao, M. (2007). A genetic and bioinformatic analysis of *Streptomyces coelicolor* genes containing TTA codons, possible targets for regulation by a developmentally significant tRNA. FEMS Microbiol. Lett. *266*, 20–28.

Liu, G., Tian, Y., Yang, H., and Tan, H. (2005). A pathway-specific transcriptional regulatory gene for nikkomycin biosynthesis in *Streptomyces ansochromogenes* that also influences colony development. Mol. Microbiol. *55*, 1855–1866.

López, D., and Kolter, R. (2010). Extracellular signals that define distinct and coexisting states in *Bacillus subtilis*. FEMS Microbiol. Rev. *34*, 134–149.

McBride, M.J., and Ensign, J.C. (1987a). Effects of intracellular trehalose content on *Streptomyces* spores. J. Bacteriol. *169*, 4995–5001.

McBride, M.J., and Ensign, J.C. (1987b). Metabolism of endogenous trehalose by *Streptomyces griseus* spores and by spores or cells of other actinomycetes. J. Bacteriol. *169*, 5002–5007.

McBride, M.J., and Ensign, J.C. (1987a). Effects of intracellular trehalose content on *Streptomyces* spores. J. Bacteriol. *169*, 4995–5001.

McBride, M.J., and Ensign, J.C. (1990). Regulation of trehalose metabolism by *Streptomyces griseus* spores. J. Bacteriol. *172*, 3637–3643.

McCann, P.A., and Pogell, B.M. (1979). Pamamycin: a new antibiotic and stimulator of aerial mycelia formation. J. Antibiot. *32*, 673–678.

McCormick, J.R. (2009). Cell division is dispensable but not irrelevant in *Streptomyces*. Curr. Opin. Microbiol. *12*, 689–698.

McCormick, J.R., and Losick, R. (1996). Cell division gene *ftsQ* is required for efficient sporulation but not growth and viability in *Streptomyces coelicolor* A3(2). J. Bacteriol. *178*, 5295–5301.

McCormick, J.R., Su, E.P., Driks, A., and Losick, R. (1994). Growth and viability of *Streptomyces coelicolor* mutant for the cell division gene *ftsZ*. Mol. Microbiol. *14*, 243–254.

McVittie, A. (1974). Ultrastructural studies on sporulation in wild-type and white colony mutants of *Streptomyces coelicolor*. J. Gen. Microbiol. *81*, 291–302.

Manteca, A., Fernández, M., and Sánchez, J. (2005a). A death round affecting a young compartmentalized mycelium precedes aerial mycelium dismantling in confluent surface cultures of *Streptomyces antibioticus*. Microbiology *151*, 3689–3697.

Manteca, A., Fernández, M., and Sánchez, J. (2005b). Mycelium development in *Streptomyces antibioticus* ATCC11891 occurs in an orderly pattern which determines multiphase growth curves. BMC Microbiol. *5*, 51.

Manteca, A., Fernández, M., and Sánchez, J. (2006a). Cytological and biochemical evidence for an early cell dismantling event in surface cultures of *Streptomyces antibioticus*. Res. Microbiol. *157*, 143–152.

Manteca, A., Mader, U., Connolly, B.A., and Sánchez, J. (2006b). A proteomic analysis of *Streptomyces coelicolor* programmed cell death. Proteomics *6*, 6008–6022.

Manzanal, M.B., and Hardisson, C. (1978). Early stages of arthrospore maturation in *Streptomyces*. J. Bacteriol. *133*, 293–297.

Mao, X.M., Zhou, Z., Hou, X.P., Guan, J.P., and Li, Y.Q. (2009a). The reciprocal regulation between SigK and differentiation programs in *Streptomyces coelicolor*. J. Bacteriol. *191*, 6473–6481.

Mao, X.M., Zhou, Z., Cheng, L.Y., Hou, X.P., Guan, J.P., and Li, Y.Q. (2009b). Involvement of SigT and RstA in the differentiation of *Streptomyces coelicolor*. FEBS Lett. *583*, 3145–3150.

Martín, M.C., Díaz, L.A., Manzanal, M.B., and Hardisson, C. (1986). Role of trehalose in the spores of *Streptomyces*. FEMS Microbiol. Lett. *35*, 49–54.

Marushima, K., Ohnishi, Y., and Horinouchi, S. (2009). CebR as a master regulator for cellulose/cellooligosaccharide catabolism affects morphological development in *Streptomyces griseus*. J. Bacteriol. *191*, 5930–5940.

Mazza, P., Noens, E.E., Schirner, K., Grantcharova, N., Mommaas, A.M., Koerten, H.K., Muth, G., Flärdh, K., van Wezel, G.P., and Wohlleben, W. (2006). MreB of *Streptomyces coelicolor* is not essential for vegetative growth but is required for the integrity of aerial hyphae and spores. Mol. Microbiol. *60*, 838–852.

Méndez, C., Braña, A.F., Manzanal, M.B., and Hardisson, C. (1985). Role of substrate mycelium in colony development in *Streptomyces*. Can. J. Microbiol. *31*, 446–450.

Merrick, M. (1976). A morphological and genetic mapping study of bald colony mutants of *Streptomyces coelicolor*. J. Gen. Microbiol. *96*, 299–315.

Metsä-Ketelä, M., Halo, L., Munukka, E., Mäntsälä, P., and Ylihonko, K. (2002). Molecular evolution of aromatic polyketides and comparative sequence analysis of polyketide ketosynthase and 16S ribosomal DNA genes from various *Streptomyces* species. Appl. Env. Microbiol. *68*, 4472–4479.

Miguélez, E.M., Hardisson, C., and Manzanal, M.B. (1999). Hyphal death during colony development in *Streptomyces antibioticus*: morphological evidence for the existence of a process of cell deletion in a multicellular prokaryote. J. Cell Biol. *145*, 515–525.

Mikulík, K., Janda, I., Mašková, H., Štastná, J., and Jiráňová, A. (1977). Macromolecular synthesis accompanying the transition from spores to vegetative forms of *Streptomyces granaticolor*. Fol. Microbiol. *22*, 254–261.

Mikulík, K., Janda, I., Weiser, J., Štastná, J., and Jiráňová, A. (1984). RNA and ribosomal protein patterns during aerial spore germination in *Streptomyces granaticolor*. Eur. J. Biochem. *145*, 381–388.

Mikulík, K., Bobek, J., Bezousková, S., Benada, O., and Kofronová, O. (2002). Expression of proteins and protein kinase activity during germination of aerial spores of *Streptomyces granaticolor*. Biochem. Biophys. Res. Commun. *299*, 335–342.

Mistry, B.V., Del Sol, R., Wright, C., Findlay, K., and Dyson, P. (2008). FtsW is a dispensable cell division protein required for Z-ring stabilization during sporulation septation in *Streptomyces coelicolor*. J. Bacteriol. *190*, 5555–5566.

Mitsui, Y., Satow, Y., Sakamaki, T., and Iitaka, Y. (1977). Crystal structure of a protein proteinase inhibitor. *Streptomyces* subtilisin inhibitor, at 2.3 angstrom resolution. J. Biochem. *82*, 295–298.

Mittenhuber G. (2002). A phylogenomic study of the general stress response sigma factor σ^B of *Bacillus subtilis* and its regulatory proteins. J. Mol. Microbiol. Biotechnol. *4*, 427–452.

Molle, V., and Buttner, M.J. (2000). Different alleles of the response regulator gene *bldM* arrest *Streptomyces coelicolor* development at distinct stages. Mol. Microbiol. *36*, 1265–1278.

Molle, V., Palframan, W., Findlay, K.C., and Buttner, M.J. (2000). WhiD and WhiB, homologous proteins required for different stages of sporulation in *Streptomyces coelicolor* A3(2). J. Bacteriol. *182*, 1286–1295.

Mukamolova, G.V., Murzin, A.G., Salina, E.G., Demina, G.R., Kell, D.B., Kaprelyants, A.S., and Young, M. (2006). Muralytic activity of *Micrococcus luteus* Rpf and its relationship to physiological activity in promoting bacterial growth and resuscitation. Mol. Microbiol. *59*, 84–98.

Natsume, M., Tazawa, J., Abe, H., Kudo, Y., Kondo, S., and Marumo, S. (1995). Transient increase in intracellular calcium in *Streptomyces alboniger* produced by pamamycin-607, an aerial mycelium-inducing substance. Biosci. Biotechnol. Biochem. *59*, 152–154.

Nguyen, K.T., Willey, J.M., Nguyen, L.D., Nguyen, L.T., Viollier, P.H., and Thompson, C.J. (2002). A central regulator of morphological differentiation in the multicellular bacterium *Streptomyces coelicolor*. Mol. Microbiol. *46*, 1223–1238.

Nguyen, K.T., Tenor, J., Stettler, H., Nguyen, L.T., Nguyen, L.D., and Thompson, C.J. (2003). Colonial differentiation in *Streptomyces coelicolor* depends on translation of a specific codon within the *adpA* gene. J. Bacteriol. *185*, 7291–7296.

Nguyen, L.D., Kalachová, L., Novotná, J., Holub, M., Kofronová, O., Benada, O., Thompson, C.J., and Weiser, J. (2005). Cultivation system using glass beads immersed in liquid medium facilitates studies of *Streptomyces* differentiation. Appl. Environ. Microbiol. *71*, 2848–2852.

Nicieza, R.G., Huergo, J., Connolly, B.A., and Sánchez, J. (1999). Purification, characterization, and role of nucleases and serine proteases in *Streptomyces* differentiation. J. Biol. Chem. *274*, 20366–20375.

Nieselt, K., Battke, F., Herbig, A., Bruheim, P., Wentzel, A., Jakobsen, O.M., Sletta, H., Alam, M.T., Merlo, M.E., Moore, J., *et al.* (2010). The dynamic architecture of the metabolic switch in *Streptomyces coelicolor*. BMC Genomics *11*, 10.

Nodwell, J.R., McGovern, K., and Losick, R. (1996). An oligopeptide permease responsible for the import of an extracellular signal governing aerial mycelium formation in *Streptomyces coelicolor*. Mol. Microbiol. *22*, 881–893.

Nodwell, J.R., and Losick, R. (1998). Purification of an extracellular signaling molecule involved in production of aerial mycelium by *Streptomyces coelicolor*. J. Bacteriol. *180*, 1334–1337.

Noens, E.E.E., Mersinias, V., Traag, B.A., Smith, C.P., Koerten. H.K., and van Wezel, G.P. (2005). SsgA-like proteins determine the fate of peptidoglycan during sporulation of *Streptomyces coelicolor*. Mol. Microbiol. *58*, 929–944.

Noens, E.E., Mersinias, V., Willemse, J., Traag, B.A., Laing, E., Chater, K.F., Smith, C.P., Koerten, H.K., and van Wezel, G.P. (2007). Loss of the controlled localization of growth stage-specific cell-wall synthesis pleiotropically affects developmental gene expression in an *ssgA* mutant of *Streptomyces coelicolor*. Mol. Microbiol. *64*, 1244–1259.

Nováková R., Sevcíková, B., and Kormanec, J. (1998). A method for the identification of promoters recognized by RNA polymerase containing a particular sigma factor: cloning of a developmentally regulated promoter and corresponding gene directed by the *Streptomyces aureofaciens* sigma factor RpoZ. Gene *208*, 43–50.

Novotna, J., Vohradský, J., Berndt, P., Gramajo, H., Langen, H., Li, X.M., Minas, W., Orsaria, L., Roeder, D., and Thompson, C.J. (2003). Proteomic studies of diauxic lag in the differentiating prokaryote *Streptomyces coelicolor* reveal a regulatory network of stress-induced proteins and central metabolic enzymes. Mol. Microbiol. *48*, 1289–1303.

Ohnishi, Y., and Horinouchi, S. (2004). The A-factor regulatory cascade that leads to morphological development and secondary metabolism in *Streptomyces*. Biofilms *1*, 319–328.

Ohnishi, Y., Yamazaki, H., Kato, J., Tomono, A., and Horinouchi, S. (2005). AdpA, a central transcriptional regulator in the A-factor regulatory cascade that leads to morphological development and secondary metabolism in *Streptomyces griseus*. Biosci. Biotechnol. Biochem. *69*, 431–439.

Okamoto, S., and Ochi, K. (1998). An essential GTP-binding protein functions as a regulator for differentiation in *Streptomyces coelicolor*. Mol. Microbiol. *30*, 107–119.

Olukoshi, E.R., and Packter, N.M. (1994). Importance of stored triacylglycerols in *Streptomyces*: possible carbon source for antibiotics. Microbiology *4*, 931–943.

Pan, Y., Liu, G., Yang, H., Tian, Y., and Tan, H. (2009). The pleiotropic regulator AdpA-L directly controls the pathway-specific activator of nikkomycin biosynthesis in *Streptomyces ansochromogenes*. Mol. Microbiol. *72*, 710–723.

Parashar, A., Colvin, K.R., Bignell, D.R., and Leskiw, B.K. (2009). BldG and SCO3548 interact antagonistically to control key developmental processes in *Streptomyces coelicolor*. J. Bacteriol. *191*, 2541–2450.

Park, S.S., Yang, Y.H., Song, E., Kim, E.J., Kim, W.S., Sohng, J.K., Lee, H.C., Liou, K.K., and Kim, B.G. (2009). Mass spectrometric screening of transcriptional regulators involved in antibiotic biosynthesis in *Streptomyces coelicolor* A3(2). J. Ind. Microbiol. Biotechnol. *36*, 1073–1083.

Pasternack, R., Dorsch, S., Otterbach, J.T., Robenek, I.R., Wolf, S., and Fuchsbauer, H.L. (1998). Bacterial pro-transglutaminase from *Streptoverticillium mobaraense*. Purification, chracterisation and sequence of the zymogen. Eur. J. Biochem. *257*, 570–576.

Peterson, F., Zahner, H., Metzger, J.W., Freund, S., and Hummel, R.P. (1993). Germicidin, an autoregulative germination inhibitor of *Streptomyces viridochromogenes* NRRL B-1551. J. Antibiot. *46*, 1126–1138.

Penyige, A., Keseru, J., Fazakas, F., Schmelczer, I., Szirák, K., Barabás, G., and Biró, S. (2009). Analysis and identification of ADP-ribosylated proteins of *Streptomyces coelicolor* M145. J. Microbiol. *47*, 549–556.

Piette, A., Derouaux, A., Gerkens, P., Noens, E.E.E., Mazzucchelli, G., Vion, S., Koerten, H.K., Titgemeyer, F., De Pauw, E., Leprince, P., *et al.* (2005). From dormant to germinating spores of *Streptomyces coelicolor* A3(2): new perspectives from the *crp* null mutant. J. Proteome Res. *4*, 1699–1708.

Plaskitt, K.A., and Chater, K.F. (1995). Influences of development genes on localized glycogen deposition in colonies of a mycelial prokaryote, *Streptomyces coelicolor* A3(2): a possible interface between metabolism and morphogenesis. Phil. Trans. Roy. Soc. B *347*, 105–121.

Pope, M.K., Green, B., and Westpheling, J. (1998). The *bldB* gene encodes a small protein required for morphogenesis, antibiotic production, and catabolite control in *Streptomyces coelicolor*. J. Bacteriol. *180*, 1556–1562.

Potúcková, L., Kelemen, G.H., Findlay, K.C., Lonetto, M.A., Buttner, M.J., and Kormanec, J. (1995). A new RNA polymerase sigma factor, sigma(F), is required for the late stages of morphological differentiation in *Streptomyces spp*. Mol. Microbiol. *17*, 37–48.

Price, C.W. (2000). Protective function and regulation of the general stress response in *Bacillus subtilis* and related bacteria. In Bacterial Stress Responses. Stortz, G., and Hengge-Aronis, R., eds (Washington, D.C., American Society for Microbiology), pp. 179–197.

Price, B., Adamidis, T., Kong, R., and Champness, W. (1999). A *Streptomyces coelicolor* antibiotic regulatory gene, *absB*, encodes an RNase III homolog. J. Bacteriol. *181*, 6142–6151.

Puglia, A.M., Vohradský, J., and Thompson, C.J. (1995). Developmental control of the heat-shock stress regulon in *Streptomyces coelicolor*. Mol. Microbiol. *17*, 737–746.

Quiros, L.M., Hardisson, C., and Salas, J.A. (1985). Stable mRNA is a common feature of some *Streptomyces* spores. FEMS Microbiol. Lett. *26*, 323–327.

Ramamurthi, K.S., and Losick, R. (2009). Negative membrane curvature as a cue for subcellular localization of a membrane protein. Proc. Natl. Acad. Sci. U.S.A. *106*, 13541–13545.

Ren, Z., Zhang, D., Yu, M., Zhao, Q., Du, G., Chen, W., and Wu, J. (2008). [Identification of the gene encoding transglutaminase zymogen from *Streptomyces*

hygroscopicus and its expression in *Escherichia coli*] [Article in Chinese] Wei Sheng Wu Xue Bao *48*, 480–485.

Rigali, S., Nothaft, H., Noens, E.E., Schlicht, M., Colson, S., Müller, M., Joris, B., Koerten, H.K., Hopwood, D.A., Titgemeyer, F., *et al.* (2006). The sugar phosphotransferase system of *Streptomyces coelicolor* is regulated by the GntR-family regulator DasR and links N-acetyl glucosamine metabolism to the control of development. Mol. Microbiol. *61*, 1237–1251.

Rigali, S., Titgemeyer, F., Barends, S., Mulder, S., Thomae, A.W., Hopwood, D.A., and van Wezel, G.P. (2008). Feast or famine: the global regulator DasR links nutrient stress to antibiotic production by *Streptomyces*. EMBO Rep. *9*, 670–675.

Ruban-Ośmiałowska, B., Jakimowicz, D., Smulczyk-Krawczyszyn, A., Chater, K.F., and Zakrzewska-Czerwinska, J. (2006). Replisome localization in vegetative and aerial hyphae of *Streptomyces coelicolor*. J. Bacteriol. *188*, 7311–7316.

Rueda, B., Miguélez, E.M., Hardisson, C., and Manzanal, M. (2001). Changes in glycogen and trehalose content of *Streptomyces brasiliensis* hyphae during growth in liquid cultures under sporulating and non-sporulating conditions. FEMS Microbiol. Lett. *194*, 181–185.

Ryding, N.J., Kelemen, G.H., Whatling, C.A., Flärdh, K., Buttner, M.J., and Chater, K.F. (1998). A developmentally regulated gene encoding a repressor-like protein is essential for sporulation in *Streptomyces coelicolor* A3(2). Mol. Microbiol. *29*, 343–357.

Ryding, N.J., Bibb, M.J., Molle, V., Findlay, K.C., Chater, K.F., and Buttner, M.J. (1999). New sporulation loci in *Streptomyces coelicolor* A3(2). J. Bacteriol. *181*, 5419–5425.

Salerno, P., Larsson, J., Bucca, G., Laing, E., Smith, C.P., and Flärdh, K. (2009). One of the two genes encoding nucleoid-associated HU proteins in *Streptomyces coelicolor* is developmentally regulated and specifically involved in spore maturation. J. Bacteriol. *191*, 6489–6500.

Schmidt S., Adolf F and Fuchsbauer H.L. (2008). The transglutaminase activating metalloprotease inhibitor from *Streptomyces mobaraensis* is a glutamine and lysine donor substrate of the intrinsic transglutaminase. FEBS Lett. *582*, 3132–3138.

Schneider, D., Bruton, C.J., and Chater, K.F. (2000). Duplicated gene clusters suggest an interplay of glycogen and trehalose metabolism during sequential stages of aerial mycelium development in *Streptomyces coelicolor* A3(2). Mol. Gen. Genet. *263*, 543–553.

Schüz, T.C., and Zähner, H. (1993). Induction of aerial mycelium formation in a *bld* mutant of *Streptomyces tendae* by Borrelidin – a macrolide antibiotic. FEMS Microbiol. Lett. *111*, 41–46.

Schwedock, J., McCormick, J.R., Angert, E.R., Nodwell, J.R., and Losick, R. (1997). Assembly of the cell division protein FtsZ into ladder-like structures in the aerial hyphae of *Streptomyces coelicolor*. Mol. Microbiol. *25*, 847–858.

Sevcíková, B., Benada, O., Kofronova, O., and Kormanec, J. (2001). Stress response sigma factor sigma(H) is essential for morphological differentiation of *Streptomyces coelicolor* A3(2). Arch. Microbiol. *177*, 98–106.

Sevcíková, B., and Kormanec, J. (2002). Activity of the *Streptomyces coelicolor* stress-response sigma factor sigmaH is regulated by an anti-sigma factor. FEMS Microbiol. Lett. *209*, 229–235.

Sevcíková, B., Mazuráková, V., and Kormanec, J. (2005). Characterization of the alternative sigma factor sigmaG in *Streptomyces coelicolor* A3(2). Folia Microbiol. (Praha) *50*, 47–58.

Singh, A., Guidry, L., Narasimhulu, K.V., Mai, D., Trombley, J., Redding, K.E., Giles, G.I., Lancaster, J.R. Jr, and Steyn, A.J. (2007). *Mycobacterium tuberculosis* WhiB3 responds to O_2 and nitric oxide via its [4Fe-4S] cluster and is essential for nutrient starvation survival. Proc. Natl. Acad. Sci. U.S.A. *104*, 11562–11567.

Skerker, J.M., Prasol, M.S., Perchuk, B.S., Biondi, E.G., and Laub, M.T. (2005). Two-component signal transduction pathways regulating growth and cell cycle progression in a bacterium: a system-level analysis. PLOS Biol. *10*, 1771–1788.

Smulczyk-Krawczyszyn, A., Jakimowicz, D., Ruban-Ośmiałowska, B., Zawilak-Pawlik, A., Majka, J., Chater, K., and Zakrzewska-Czerwinska, J. (2006). A cluster of DnaA boxes involved in the regulation of *Streptomyces* chromosome replication: from *in silico* to *in vivo* studies. J. Bacteriol. *188*, 6184–6194.

Soliveri, J.A., Gomez, J., Bishai, W.R., and Chater, K.F. (2000). Multiple paralogous genes related to the *Streptomyces coelicolor* developmental regulatory gene *whiB* are present in *Streptomyces* and other actinomycetes. Microbiology *146*, 333–343.

Stastná, J. (1977). A method of rapid wetting and synchronous germination of streptomycete spores. Folia Microbiol. (Praha) 22, 137–138.

Steyn, A.J.C., Collins, D.M., Hondalus, M.K., Jacobs Jr., W.R., Kawakami, R.P., and Bloom, B.R. (2002). *Mycobacterium tuberculosis* WhiB3 interacts with RpoV to affect host survival but is dispensable for *in vivo* growth. Proc. Natl. Acad. Sci. U.S.A. *99*, 3149–3152.

Strunk, C. (1978). Sporogenesis in *Streptomyces melanochromogenes*. Arch. Microbiol. *118*, 309–316.

Süsstrunk, U., Pidoux, J., Taubert, S., Ullmann, A., and Thompson, C.J. (1998). Pleiotropic effects of cAMP on germination, antibiotic biosynthesis and morphological development in *Streptomyces coelicolor*. Mol. Microbiol. *30*, 33–46.

Szabó, I.M., Kondics, L., Marton, M., and Buti, I. (1977). Changes in the cell surface layer (sheath) of the cell wall of streptomycetes during sporulation. Acta Microbiologica (Acad. Sci. Hung.) *24*, 237–246.

Taguchi S., Kikuchi, H., Suzuki, M., Kojima, S., Terabe, M., Miura, K., Nakase T and Momose H. (1993). *Streptomyces* subtilisin inhibitor-like proteins are distributed widely in streptomycetes. Appl. Environ. Microbiol. *59*, 4338–4341.

Tahlan, K., Ahn, S.K., Sing, A., Bodnaruk, T.D., Willems, A.R., Davidson, A.R., and Nodwell, J.R. (2007). Initiation of actinorhodin export in *Streptomyces coelicolor*. Mol. Microbiol. *63*, 951–961.

Takano, E. (2006). Gamma-butyrolactones: *Streptomyces* signalling molecules regulating antibiotic production and differentiation. Curr. Opin. Microbiol. *9*, 287–294.

Takano, E., Tao, M., Long, F., Bibb, M.J., Wang, L., Li, W., Buttner, M.J., Bibb, M.J., Deng, Z.X., and Chater, K.F. (2003). A rare leucine codon in *adpA* is implicated in the morphological defect of *bldA* mutants of *Streptomyces coelicolor*. Mol. Microbiol. *50*, 475–486.

Takeuchi Y., Satow, Y., Nakamura, K.T and Mitsui Y. (1991). Refined crystal structure of the complex of subtilisin BPN and *Streptomyces* subtilisin inhibitor at 1.8 Å resolution. J. Mol. Biol. *221*, 309–325.

Takeuchi Y., Nonaka, T., Nakamura, K.T., Kojima S and Miura K.I. (1992). Crystal structure of an engineered subtilisin inhibitor complexed with bovine trypsin. Proc. Natl. Acad. Sci. U.S.A. *89*, 4407–4411.

Tan, H., and Chater, K.F. (1993). Two developmentally controlled promoters of *Streptomyces coelicolor* A3(2) that resemble the major class of motility-related promoters in other bacteria. J. Bacteriol. *175*, 933–940.

Tan, H.R., Yang, H.H., Tian, Y.Q., Wu, W., Whatling, C.A., Chamberlin, L.C., Buttner, M.J., Nodwell, J., and Chater, K.F. (1998). The *Streptomyces coelicolor* sporulation-specific sigma(WhiG) form of RNA polymerase transcribes a gene encoding a ProX-like protein that is dispensable for sporulation. Gene *212*, 137–146.

Tian, Y., Fowler, K., Findlay, K., Tan, H., and Chater, K.F. (2007). An unusual response regulator influences sporulation at early and late stages in *Streptomyces coelicolor*. J. Bacteriol. *189*, 2873–2885.

Tillotson, R.D., Wösten, H.A., Richter, M., and Willey, J.M. (1998). A surface active protein involved in aerial hyphae formation in the filamentous fungus *Schizophillum commune* restores the capacity of a bald mutant of the filamentous bacterium *Streptomyces coelicolor* to erect aerial structures. Mol. Microbiol. *30*, 595–602.

Tomono, A., Tsai, Y., Ohnishi, Y., and Horinouchi, S. (2005). Three chymotrypsin genes are members of the AdpA regulon in the A-factor regulatory cascade in *Streptomyces griseus*. J. Bacteriol. *187*, 6341–6353.

Traag, B.A., and van Wezel, G.P. (2008). The SsgA-like proteins in actinomycetes: small proteins up to a big task. Antonie Van Leeuwenhoek *94*, 85–97.

Traag, B.A., Kelemen, G.H., and van Wezel, G.P. (2004). Transcription of the sporulation gene *ssgA* is activated by the IclR-type regulator SsgR in a *whi*-independent manner in *Streptomyces coelicolor* A3(2). Mol. Microbiol. *53*, 985–1000.

Trepanier, N.K., Nguyen, G.D., Leedell. P.J., and Leskiw, B.K. (1997). Use of polymerase chain reaction to identify a leucyl tRNA in *Streptomyces coelicolor*. Gene *193*, 59–63.

Ueda, K., Oinuma, K., Ikeda, G., Hosono, K., Ohnishi, Y., Horinouchi, S., and Beppu, T. (2002). AmfS, an extracellular peptidic morphogen in *Streptomyces griseus*. J. Bacteriol. *184*, 1488–1492.

Ueda, K., Takano, H., Nishimoto, M., Inaba, H., and Beppu, T. (2005). Dual transcriptional control of *amfTSBA*, which regulates the onset of cellular differentiation in *Streptomyces griseus*. J. Bacteriol. *187*, 135–142.

Uguru, G.C., Stephens, K.E., Stead, J.A., Towle, J.E., Baumberg, S., and McDowall, K.J. (2005). Transcriptional activation of the pathway-specific regulator of the actinorhodin biosynthetic genes in *Streptomyces coelicolor*. Mol. Microbiol. *58*, 131–150.

Van Mellaert L., Lammertyn, E., Schacht, S., Proost, P., Van Damme J., Wroblowski, B., Anné, J., Scarcez, T., Sablon, E., Raeymaeckers, J., *et al.* (1998). Molecular characterization of a novel subtilisin inhibitor protein produced by *Streptomyces venezuelae* CBS762.70. DNA Seq. *9*, 19–30.

Van Wezel, G.P., van der Meulen, J., Kawamoto, S., Luiten, R.G.M., Koerten, H.K., and Kraal, B. (2000). *ssgA* is essential for sporulation of *Streptomyces coelicolor* A3(2) and affects hyphal development by stimulating septum formation. J. Bacteriol. *182*, 5653–5662.

Viollier, P.H., Minas, W., Dale, G.E., Folcher, M., and Thompson, C.J. (2001a). Role of acid metabolism in *Streptomyces coelicolor* morphological differentiation and antibiotic biosynthesis. J. Bacteriol. *183*, 3184–3192.

Viollier, P.H., Nguyen, K.T., Minas, W., Folcher, M., Dale, G.E., and Thompson, C.J. (2001b). Roles of aconitase in growth, metabolism, and morphological differentiation of *Streptomyces coelicolor*. J. Bacteriol. *183*, 3193–3203.

Viollier, P.H., Weihofen, A., Folcher, M., and Thompson, C.J. (2003a). Post-transcriptional regulation of the *Streptomyces coelicolor* stress responsive sigma factor, SigH, involves translational control, proteolytic processing, and an anti-sigma factor homolog. J. Mol. Biol. *325*, 637–649.

Viollier, P.H., Kelemen, G.H., Dale, G.E., Nguyen, K.T., Buttner, M.J., and Thompson, C.J. (2003b).Specialized osmotic stress response systems involve multiple SigB-like sigma factors in *Streptomyces coelicolor*. Mol. Microbiol. *47*, 699–714.

Vohradský, J., Li, X.M., and Thompson, C.J. (1997). Identification of procaryotic developmental stages by statistical analyses of two-dimensional gel patterns. Electrophoresis *18*, 1418–1428.

Vohradsky, J., Li, X.M., Dale, G., Folcher, M., Nguyen, L., Viollier, P.H., and Thompson, C.J. (2000). Developmental control of stress stimulons in *Streptomyces coelicolor* revealed by statistical analyses of global gene expression patterns. J. Bacteriol. *182*, 4979–4986.

Wang, L., Yu, Y., He, X., Zhou, X., Deng, Z., Chater, K.F., and Tao, M. (2007). Role of an FtsK-like protein in genetic stability in *Streptomyces coelicolor* A3(2). J. Bacteriol. *189*, 2310–2318.

Wang, S.L., Fan, K.Q., Yang, X., Lin, Z.X., Xu, X.P., and Yang, K.Q. (2008). CabC, an EF-hand calcium-binding protein, is involved in Ca2+-mediated regulation of spore germination and aerial hypha formation in *Streptomyces coelicolor*. J. Bacteriol. *190*, 4061–4068.

Wang, L., Tian, X., Wang, J., Yang, H., Fan, K., Xu, G., Yang, K., and Tan, H. (2009). Autoregulation of antibiotic biosynthesis by binding of the end product to an atypical response regulator. Proc. Natl. Acad. Sci. U.S.A. *106*, 8617–8622.

Wildermuth, H. (1970a). Development and organization of the aerial mycelium in *Streptomyces coelicolor.* J. Gen. Microbiol. *60*, 43–50.

Wildermuth, H. (1970b). Surface structure of *Streptomyces* spores as revealed by negative staining and freeze-etching. J. Bacteriol. *101*, 318–322.

Wildermuth, H. (1972a). The surface structure of spores and aerial hyphae in *Streptomyces viridochromogenes.* Archiv. Fur Mikrobiol. *81*, 309–320.

Wildermuth, H. (1972b). Morphological surface characteristics of *Streptomyces glaucescens* and *S. acrimycini*, two streptomycetes with 'hairy' spores. Archiv. Fur Mikrobiol. *81*, 321–332.

Wildermuth, H., and Hopwood, D.A. (1970). Septation during sporulation in *Streptomyces coelicolor.* J. Gen. Microbiol. *60*, 51–59.

Wildermuth, W., Wehrli, E., and Horne, R.W. (1971). The surface structure of spores and aerial mycelium in *Streptomyces coelicolor.* J. Ultrastruct. Res. *35*, 168–180.

Willemse, J., and van Wezel, G.P. (2009). Imaging of *Streptomyces coelicolor* A3(2) with reduced autofluorescence reveals a novel stage of FtsZ localization. PLoS ONE *4*, e4242.

Willey, J., Santamaría, R., Guijarro, J., Geistlich, M., and Losick, R. (1991). Extracellular complementation of a developmental mutation implicates a small sporulation protein in aerial mycelium formation by *Streptomyces coelicolor.* Cell *65*, 641–650.

Willey, J., Schwedock, J., and Losick, R. (1993). Multiple extracellular signals govern the production of a morphogenetic protein involved in aerial mycelium formation by *Streptomyces coelicolor.* Genes Dev. *7*, 895–903.

Willey, J., Willems, A., Kodani, S., and Nodwell, J.R. (2006). Morphogenetic surfactants and their role in the formation of aerial hyphae in *Streptomyces coelicolor.* Mol. Microbiol. *59*, 731–742.

Williams, S.T., and Sharples, G.P. (1970). A comparative study of spore formation in two *Streptomyces* species. Microbios *5*, 17–26.

Williams, S.T., Bradshaw, R.M., Costerton, J.W., and Forge, A. (1972). Fine structure of the spore sheath of some *Streptomyces* species. J. Gen. Microbiol. 72, 249–258.

Wösten, H.A., and Willey, J.M. (2000). Surface-active proteins enable microbial aerial hyphae to grow into the air. Microbiology *146*, 767–773.

Xie, Z., Li, W., Tian, Y., Liu, G., and Tan. H. (2007). Identification and characterization of *sawC*, a *whiA*-like gene, essential for sporulation in *Streptomyces ansochromogenes.* Arch. Microbiol. *188*, 575–582.

Xu, J., Liu, G., and Tan, H. (2003). [Identification and functional characterization of *sawE*–a gene involved in differentiation of *Streptomyces ansochromogenes* 7100] [Article in Chinese]. Wei Sheng Wu Xue Bao *43*, 48–55.

Xu, H., Chater, K.F., Deng, Z., and Tao, M. (2008a). A cellulose synthase-like protein involved in hyphal tip growth and morphological differentiation in *Streptomyces.* J. Bacteriol. *190*, 4971–4978.

Xu, W., Huang, J., and Cohen, S.N. (2008b). Autoregulation of AbsB (RNase III) expression in *Streptomyces coelicolor* by endoribonucleolytic cleavage of *absB* operon transcripts. J. Bacteriol. *190*, 5526–5530.

Xu, W., Huang, J., Lin, R., Shi, J., and Cohen, S.N. (2010a). Regulation of morphological differentiation in *S. coelicolor* by RNase III (AbsB) cleavage of mRNA encoding the AdpA transcription factor. Mol. Microbiol. *75*, 781–791.

Xu, X., Wang, J., Wang, L., Tian, X., Yang, H., Fan, K., Yang, K., and Tan, H. (2010b). "Pseudo" gamma butyrolactone receptors respond to antibiotic signals to coordinate antibiotic biosynthesis. J. Biol. Chem. *285*, 27440–27448.

Xu, Q., Traag, B.A., Willemse, J., McMullan, D., Miller, M.D., Elsliger, M.A., Abdubek, P., Astakhova, T., Axelrod, H.L., Bakolitsa, C., *et al.* (2009). Structural and functional characterizations of SsgB, a conserved activator of developmental cell division in morphologically complex actinomycetes. J. Biol. Chem. *284*, 25268–25279.

Yamanaka, K., Oikawa, H., Ogawa, H., Hosono, K., Shinmachi, F., Takano, H., Sakuda, S., Beppu, T., and Ueda, K. (2005). Desferrioxamine E produced by *Streptomyces griseus* stimulates growth and development of *Streptomyces tanashiensis.* Microbiology *151*, 2899–2905.

Yamazaki, H., Ohnishi, Y., and Horinouchi, S. (2000). An A-factor-dependent extracytoplasmic sigma factor (σ^{AdsA}) that is essential for morphological development in *Streptomyces griseus.* J. Bacteriol. *182*, 4596–4605.

Yamazaki, H., Ohnishi, Y., and Horinouchi, S. (2003). Transcriptional switch-on of *ssgA* by A-factor, which is essential for spore septum formation in *Streptomyces griseus.* J. Bacteriol. *185*, 1273–1283.

Yamazaki, H., Tomono, A., Ohnishi, Y., and Horinouchi, S. (2004). DNA binding specificity of AdpA, a transcriptional activator in the A-factor regulatory cascade in *Streptomyces griseus.* Mol. Microbiol. *53*, 555–572.

Yeo, M., and Chater, K. (2005). The interplay of glycogen metabolism and differentiation provides an insight into the developmental biology of *Streptomyces coelicolor.* Microbiology *151*, 855–861.

Zawilak-Pawlik, A., Kois, A., Majka, J., Jakimowicz, D., Smulczyk-Krawczyszyn, A., Messer, W., and Zakrzewska-Czerwinska, J. (2005). Architecture of bacterial replication initiation complexes: orisomes from four unrelated bacteria. Biochem. J. *389*, 471–481.

Zhang, D., Wang, M., Du, G., Zhao, Q., Wu, J., and Chen, J. (2008a). Surfactant protein of the *Streptomyces* subtilisin inhibitor family inhibits transglutaminase activation in *Streptomyces hygroscopicus.* J. Agric. Food Chem. *56*, 3403–3408.

Zhang, D., Wang, M., Du, G., Zhao, Q., Wu, J., and Chen, J. (2008b). Two different proteases from *Streptomyces hygroscopicus* are involved in transglutaminase activation. J. Agric. Food. Chem. *56*, 10261–10264.

Zhang, G., Tian, Y., Hu, K., Feng, C., and Tan, H. (2009). SCO3900, co-transcripted with three downstream genes, is involved in the differentiation of *Streptomyces coelicolor*. Curr. Microbiol. *60*, 268–273.

Zotzel, J., Keller, P., and Fuchsbauer, H.L. (2003a). Transglutaminase from *Streptomyces mobaraensis* is activated by an endogenous metalloprotease. Eur. J. Biochem. *270*, 3214–3222.

Zotzel, J., Pasternack, R., Pelzer, C., Ziegert, D., Mainusch, M., and Fuchsbauer, H.L. (2003b). Activated transglutaminase from *Streptomyces mobaraensis* is processed by tripeptidyl aminopeptidase in the final step. Eur. J. Biochem. *270*, 4149–4155.

Protein Secretion in *Streptomyces*

4

Tracy Palmer and Matthew I. Hutchings

Abstract

The saprophytic lifestyle of *Streptomyces* requires them to secrete prolific numbers of proteins. For example, inspection of the genome sequence of *Streptomyces coelicolor* indicates it encodes some 819 proteins with predicted signal peptides. This represents more than 10% of the protein coding genes and is most likely an underestimate. Many secreted proteins are required for nutrient capture, and there is an abundance of secreted hydrolases for the breakdown of complex carbohydrates (including cellulose and chitin), peptides and phospho-compounds. In addition to proteins that are secreted into the milieu, many proteins are covalently anchored to the cell surface by means of either a lipid anchor to the membrane or by covalent attachment to the cell wall through the sortase system. Here we summarize what is known about the different protein secretion systems utilized by *Streptomyces*, and the mechanisms by which proteins are anchored to the extracellular surface.

Protein secretion pathways

Analysis of the available genome sequences for various streptomycetes indicates that they encode at least three different protein secretion systems – the general secretory (Sec) pathway, the twin arginine translocation (Tat) pathway and the ESX-1/type VII secretion pathway. Each of these systems will be discussed in more detail below. In addition, they encode numerous species-specific ABC transporters that are linked to the export of lantibiotics – modified peptides that often show bioactivity. These will not be described further and for more details about lantibiotic biosynthesis and export the reader is referred to a recent review by Lubelski *et al.* (2008).

The Sec pathway

The Sec pathway is a ubiquitous system present in all organisms and is essential for life. In prokaryotes it resides in the cytoplasmic membrane, whilst in eukaryotes it transports proteins across the endoplasmic reticulum (ER) membrane from the cytoplasm to the ER lumen. It is also found in the energy transducing membranes of some organelles, for example the thylakoid membrane of plant chloroplasts (see Driessen and Nouwen, 2008, for a recent review).

The Sec machinery comprises a passive channel with a narrow pore through which unfolded polypeptides are threaded (Van den Berg *et al.*, 2004). In bacteria the channel is comprised of three proteins. Two of these, SecY and SecE, are conserved and essential, and in bacteria they associate with the non essential SecG subunit of the translocon. Recently it was shown that SecG also plays an auxiliary role in Sec-dependent secretion in *Streptomyces lividans* where deletion of the encoding gene was not lethal but resulted in reduced levels of protein secretion and a sporulation defect (Palomino and Mellado, 2008). In the best studied bacterial system, that of *Escherichia coli*, three additional proteins, SecD, SecF and YajC form a second heterotrimeric complex that can interact with SecYEG and that when inactivated shows a Sec-specific protein transport defect (e.g. Nouwen and Driessen, 2005). The genome sequences of streptomycetes reveal that all of the standard Sec components, including the essential proteins SecY and SecE, the auxiliary protein SecG, along with SecD and SecF, which are encoded next to each other on the *S. coelicolor* chromosome, are present.

Secretory proteins are targeted to the Sec pathway by virtue of the fact that they contain a signal peptide at their N-terminus. This peptide is removed at a late stage of protein export by a membrane-bound type I signal peptidase which in *S. lividans* can be any of SipW, X, Y or Z, with SipY playing the major role (Parro *et al.*, 1999; Escutia *et al.*, 2006). Sec-targeting signal peptides in Gram-positive bacteria have a mean length of 29–31 amino acids (von Heijne and Abrahmsen, 1989) and have a tripartite structure comprising n- h- and c-regions as shown in Fig. 4.1A. Overall they lack any notable amino acid sequence conservation with the exception of the signal peptidase cleavage site, which is most commonly Ala-xaa-Ala (Nielsen *et al.*, 1997).

The driving force for protein secretion by the Sec pathway is generally provided through ATP hydrolysis by the motor protein, SecA. The SecA protein from *S. lividans* has been characterized biochemically and shown to be required for the secretion of the Sec-dependent α-amylase *in vitro* (Blanco *et al.*, 1998). In addition to being the

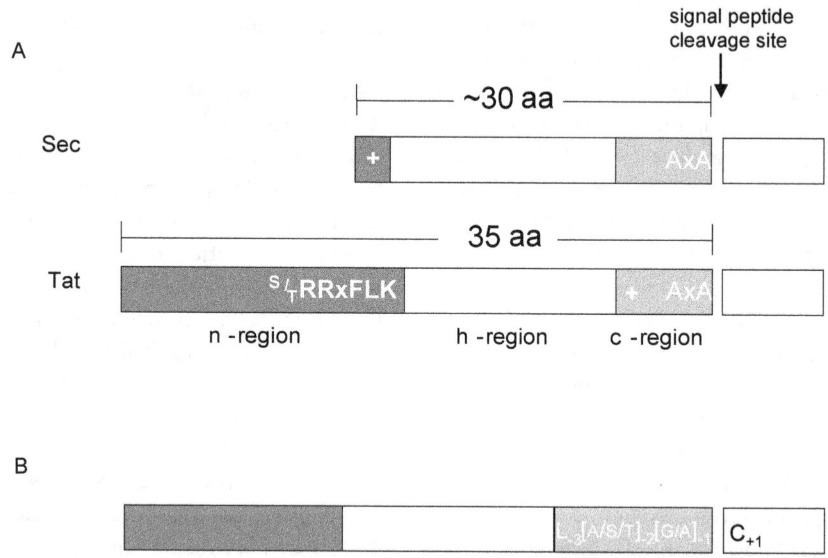

Figure 4.1 Signal peptides in bacteria. (A) Schematic representation of Sec and Tat signal peptides. Both Sec and Tat signal peptides show a recognisable tripartite structure with a polar (basic) n-region, a hydrophobic h-region and a polar c-region that contains the recognition site for a type I signal peptidase (shown as AxA in the figure, although generally residues at the –3 and -1 positions relative to the cleavage site are any amino acid with a small, neutral side chain – Perlman and Halvorson, 1983). Sec signal peptides in Gram-positive bacteria have an average length of 29–31 amino acids (von Heijne and Abrahmsen, 1989). Tat signal peptides have a conserved motif S/T-R-R-x-F-L-K that is found at the n-region/h-region boundary and are generally noticeably longer than Sec signal peptides, and can exceed 60 amino acids in length (the average length shown in the figure is for Tat signal peptides in *S. coelicolor* and represents the average length for the signal peptides of the 80 *S. coelicolor* proteins that are predicted to be Tat dependent by both TATFIND 1.4 and TatP). Other differences between Sec and Tat signal peptides include the fact that the h-regions of Tat signal peptides are less hydrophobic than Sec signals, and that they often contain one or more basic residues in the c-region that act as a Sec-avoidance signal (Cristobal *et al.*, 1999; Bogsch *et al.*, 1997). (B) Lipoproteins can be routed for export by either the Sec or Tat pathways. Their signal peptides contain a conserved motif in the c-region that is typically L_{-3}-[A/S/T]$_{-2}$-[G/A]$_{-1}$ along with an absolutely conserved cysteine at the +1 position of the mature protein. These signal peptides are cleaved by a type II signal peptidase (usually known as Lsp or lipoprotein signal peptidase) that usually recognizes the cysteine after it has been lipidated by the diacylglycerol lipoprotein transferase (Lgt) enzyme.

major route for the transport of most extracellular proteins, the Sec pathway also plays a critical role in the assembly of almost all membrane proteins. Membrane proteins are normally inserted into the Sec channel in a co-translational manner, whilst they are still being synthesized. In this case the ribosome provides the pushing force for threading the polypeptide through the Sec channel, although large extracellular loops also require the action of SecA (Sääf *et al.*, 1995).

The targeting of nascent chains of membrane proteins to the Sec translocon requires the signal recognition particle (SRP) which is a complex of a GTPase (Ffh) and a small stable RNA species (4.5S RNA). SRP interacts with the hydrophobic segments of membrane proteins as they emerge from the ribosome and target the ribosome-nascent chain complex to the SRP receptor, a peripheral membrane protein which binds to anionic phospholipids and to SecYEG (de Leeuw *et al.*, 2000; Angelini *et al.*, 2005). This interaction of the SRP receptor with the Sec channel allows it to hand over the ribosome-nascent chain complex to initiate transport. In the Gram-negative model organism *E. coli* it has been demonstrated that SRP is essential for the targeting of almost all inner membrane proteins (Luirink *et al.*, 2005) and as such the *ffh* gene that encodes the protein component of SRP, and *ftsY* that encodes the SRP receptor are essential genes (Seluanov and Bibi, 1997).

Streptomyces species also encode the RNA and protein components of SRP (Palacin *et al.*, 2003). However, it has been reported that *ftsY* can be deleted from *S. coelicolor*, and although the mutation results in a pleiotropic effect on sporulation and antibiotic production, the strain remains viable (Shen *et al.*, 2008). It is rather surprising that the SRP pathway, which is essential for viability of bacteria, Archaea and eukaryotes, should be dispensable in *Streptomyces*. One possibility is that spontaneous suppressor mutations arose during *ftsY* deletion allowing the strain to survive, much as deletion of the essential *chpE* gene resulted in suppressor mutations that inactivated the Tat pathway (Di Berardo *et al.*, 2008). Indeed an essential role for SRP in the *Streptomyces* is supported by studies with the polypeptide antibiotic goadsporin. This compound is synthesized by *Streptomyces* sp. TP-A0584 and inhibits the growth of other *Streptomyces* strains at concentrations in excess of 1 μM. Immunity to goadsporin is conferred by *godI*, which encodes a protein with spectacular identity to Ffh, strongly implicating SRP as the cellular target for goadsporin action (Onaka *et al.*, 2005).

The Tat pathway

The Tat pathway is the second general protein transport system found in the cytoplasmic membranes of bacteria. However, unlike the ubiquitous Sec pathway, components of the Tat machinery are only encoded by a proportion of the sequenced prokaryotic genomes, although these include Gram-negative and Gram-positive bacteria and Archaea. It is also largely absent in eukaryotes, with the exception of the chloroplasts and mitochondria of plants and a few protists. The Tat pathway is generally not essential for viability and has been inactivated in several bacteria, although it appears to be essential in some halophilic Archaea (Dilks *et al.*, 2005; Thomas and Bolhuis, 2006).

The Tat system differs from Sec in that it transports pre-folded proteins across the cytoplasmic membrane. Proteins are targeted to the Tat pathway by N-terminal signal peptides that harbour a conserved twin arginine amino acid motif (Berks, 1996; discussed in more detail below). It was originally assumed that the Tat pathway was a rather specialized export system, dedicated primarily to the export of proteins that bind complex redox-active cofactors which by necessity are inserted into the pre-protein in the cytoplasmic compartment (Berks *et al.*, 2000). However, some Archaea and bacteria, and in particular *Streptomyces* species, appear to export large numbers of non-cofactor-containing proteins by this pathway indicating that it can be considered a general protein export system analogous to the Sec pathway (Rose *et al.*, 2002; Dilks *et al.*, 2003; Widdick *et al.*, 2006).

Components of the Tat machinery

The Tat machinery in the Gram-negative bacteria and in Gram-positive Actinobacteria comprises three membrane-bound components, TatA, TatB and TatC, and generally all three of them are indispensable for function of the pathway (e.g. Sargent *et al.*, 1998, 1999; Weiner *et al.*, 1998; Bogsch *et al.*, 1998; Schaerlaekens *et al.*, 2004a; De Keersmaeker *et al.*, 2005a; Widdick *et al.*, 2006). Frequently, a second copy of the *tatA* gene is also encoded, and in *E. coli* this is designated *tatE*. The *tatE* gene has an identical function to *tatA* but is produced at much lower levels and is thought to be largely cryptic (Jack *et al.*, 2001). A probable *tatE* gene is also encoded by *S. coelicolor* (see Fig. 4.2) and is present in genomes of other sequenced streptomycetes.

The Tat system has been best characterized in *E. coli*. The TatB and TatC proteins are known to form a complex of high molecular mass, where each of the two proteins is present in a 1:1 ratio (Bolhuis *et al.*, 2001). This complex acts as the receptor for substrates, which bind to TatBC through their signal peptides, with the twin arginine motif being recognized by TatC (e.g. Alami *et al.*, 2003; Tarry *et al.*, 2009; see Fig. 4.3). The TatA protein forms the transport channel, which is composed of many copies of this small protein. TatA has a strong propensity to homo-oligomerize and can form rings of variable diameter that contain different numbers of TatA subunits (deLeeuw *et al.*, 2001; Gohlke *et al.*, 2005). These rings of TatA surround cavities that are large enough to accommodate folded proteins, and the variable diameters suggests that the TatA complexes adopt a range of sizes to facilitate the transport of a wide array of folded proteins of differing dimensions (Gohlke *et al.*, 2005). Under resting conditions TatA forms tetramers in the membrane, and these units of TatA only appear to assemble onto the TatBC complex to form

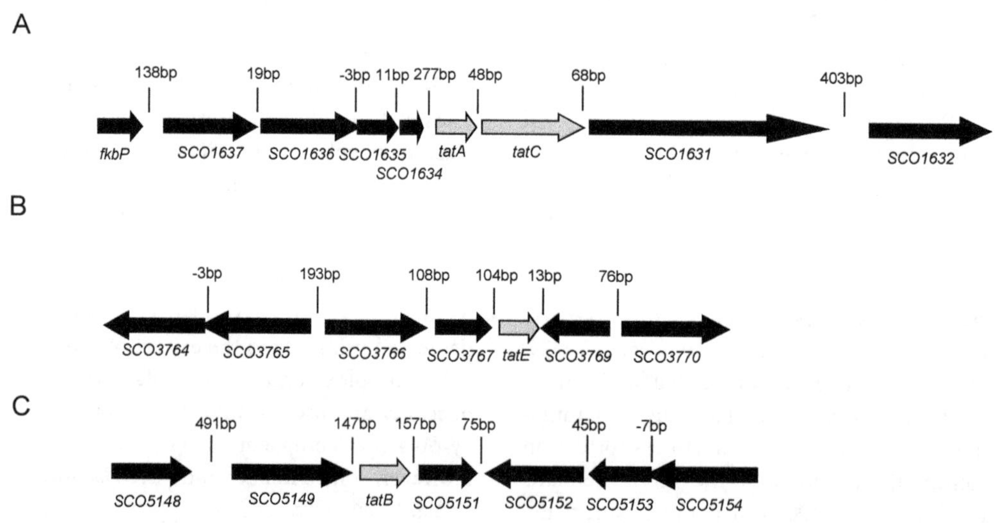

Figure 4.2 Genomic organization of genes encoding Tat components in *S. coelicolor*. In each case probable *tat* genes are shown in grey, ORFs are roughly to scale. (A). Genomic organization around the *tatAC* genes: *SCO1631* codes for a predicted DEAD box helicase; *SCO1630*: a sensor kinase protein component of a *Streptomyces* conservon; *SCO1634*: small membrane protein found only in streptomycetes; *SCO1635*: a possible zinc finger protein; *SCO1636* and *SCO1637* two COG2378 family proteins – probable transcriptional regulators; *fkbP*: a peptidyl-prolyl cis-trans isomerase; (B) Genomic organization around the putative *tatE* gene: *SCO3764* and *SCO3765* encode sequence related proteins that are possible Mg/Co transporters; *SCO3766*: SpoOM homologue; *SCO3767*: possible tellurite resistance protein B; *SCO3769*: TetR family regulator; *SCO3770*: cytochrome P450 oxidoreductase; (C) Genomic organization around the *tatB* gene: *SCO5147* encodes a probable ECF sigma factor; *SCO5148*: a membrane-bound anti-sigma factor; *SCO5149*: a possible serine protease; *SCO5151*: a hypothetical protein only found in streptomycetes; *SCO5152*: ParA-like protein; *SCO5153*: conserved hypothetical membrane protein; *SCO5154*: putative Mg transporter.

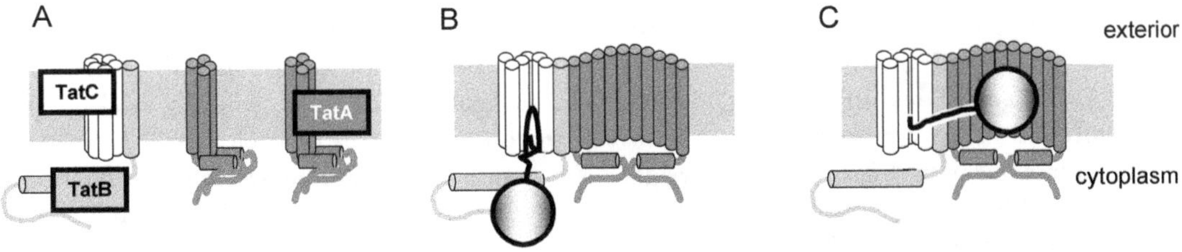

Figure 4.3 Probable mechanism for operation of the Tat protein export pathway. (A) Under resting conditions the TatBC complex and tetrameric units of TatA exist as separate assemblies in the membrane. (B) A folded substrate protein docks at the TatBC complex, binding by virtue of its twin arginine signal peptide. (C) TatA tetramers assemble onto the substrate-bound TatBC complex to form the transport channel. The substrate is transported across the membrane through a channel formed by TatA. After transport the translocated substrate is released, its signal peptide is cleaved and the TatABC complex dissociates to give the TatBC complex and separate multimers of TatA (as depicted in part A).

channels when TatBC is in the substrate-bound state (Mori and Cline, 2002; Leake *et al.*, 2008). Once the substrate has been transported across the membrane TatA dissociates from TatBC to allow another round of protein transport to be initiated. The Tat protein transport cycle is shown schematically in Fig. 4.3.

Interestingly, TatA and TatB are homologous proteins, and arose from a common ancestor. However, they have clearly diversified in function and are not able to cross-complement each other, even when overproduced (Sargent *et al.*, 1998; 1999; De Keersmaeker *et al.*, 2005a). Phylogenetic analysis has indicated that the duplication that gave rise to separate TatA and TatB proteins has occurred more than once during evolution (Yen *et al.*, 2002). Most Gram-positive bacteria (with the exception of the Actinobacteria), and Archaea lack TatB proteins, and their Tat systems comprise only TatA and TatC. This probably represents the ancestral Tat machinery, after which a very early gene duplication event giving rise to TatB occurred. Sequence analysis shows that the TatB proteins of streptomycetes cluster with TatBs from Gram-negative bacteria, and rather surprisingly, the streptomycete TatA proteins also branch from Gram-negative TatA proteins rather than the low G+C Gram-positive TatA proteins (Yen *et al.*, 2002). The relative similarity of the streptomycete Tat proteins with those of Gram-negative bacteria has been supported by an experimental study where it was demonstrated that the TatA, TatB and TatC proteins of *S. coelicolor* were, respectively, able to strongly complement individual *E. coli* mutations in each of these components (Hicks *et al.*, 2006). The genetic organization of the *tat* genes in *S. coelicolor* is shown in Fig. 4.2.

Compared with the number of publications reporting the analysis of Tat components in *E. coli*, there is rather less information about the arrangement of the Tat machinery in *Streptomyces*. A complex of TatBC has been partially purified from membranes of *S. lividans*, and has a similarly large mass to that of the *E. coli* TatBC complex (De Keersmaeker *et al.*, 2005b; 2007). Surprisingly, however, it was noted that upon fractionation of cells, a small proportion of TatB and a significant amount of TatA were found in the soluble fraction. It was further shown that of the TatA and TatB protein that was membrane-associated, some of it could be washed off by carbonate or particularly urea treatment (De Keersmaeker *et al.*, 2005b). This is in contrast to the *E. coli* TatA and TatB proteins which are not extracted from the membrane by urea washing (de Leeuw *et al.*, 2001). However, it should be noted that single transmembrane-spanning proteins from plant thylakoids can also be partially membrane extracted using urea (Thompson *et al.*, 1999) so these observations do not necessarily mean that there is a difference between the organization of TatA and TatB in *Streptomyces* membranes compared with *E. coli*, it may merely reflect differences in membrane lipid composition.

The presence of cytoplasmic forms of the TatA protein is not unique to *S. lividans*, but has also been reported for the TatA proteins of *Bacillus subtilis*. *B. subtilis* has two separate Tat pathways, designate Tat_d and Tat_y. It was observed that the $TatA_d$ protein of this organism, in addition to being found in the membrane, also exists as a cytoplasmic form of high molecular weight (Pop *et al.*, 2003; Westermann *et al.*, 2006). Furthermore, when *B. subtilis* $TatA_y$ was produced in *E. coli* some of the protein was also observed in the cytoplasmic fraction where it formed a large aggregate (of approximately 5MDa; Barnett *et al.*, 2009). Thus the relationship between soluble forms of TatA and the operation of the Tat system is unclear. However, given the fact that *S. coelicolor* Tat proteins can function heterologously with *E. coli* Tat components strongly suggests that the mechanism of operation of the two systems is the same and is likely to follow the scheme outlined in Fig. 4.3.

Tat signal peptides

Proteins are targeted to the Tat pathway by N-terminal signal peptides that superficially resemble Sec signal peptides and, in *E. coli* at least, are cleaved by the same signal peptidase that cleaves Sec signal peptides (Yahr and Wickner, 2001; Lüke *et al.*, 2009). However, the most striking difference from Sec signals is that Tat signal peptides contain two adjacent and invariant arginine residues (Berks, 1996) (see Fig. 4.1A). The twin arginines are part of an extended motif which can be defined as S/T-R-R-x-F-L-K and where, apart from the normally invariant arginines, the other motif residues usually occur with a frequency > 50% and x is usually any polar amino acid (Berks, 1996; Stanley *et al.*, 2000). This motif is always found at the boundary between the signal peptide n-region and the hydrophobic h-region. A second key feature of Tat-targeting signals is that their hydrophobic regions are significantly less hydrophobic than those of Sec signal peptides. This is an important discriminating factor and it has been demonstrated that simply raising the h-region hydrophobicity of a Tat signal peptide can re-route a passenger protein to the Sec pathway (Cristobal *et al.*, 1999).

Despite the features that distinguish Tat targeting sequences from Sec signals, in reality it is not always straight-forward to differentiate between signal peptides that target the two different pathways. To assist in identifying candidate Tat signal peptides, two generally available prediction programs are available. TATFIND was the first such program to be developed, and it uses a rule-based classification of allowed amino acids at certain positions of the consensus motif, and the presence of an uncharged stretch of at least 13 amino acids (Rose *et al.*, 2002; Dilks *et al.*, 2003). More recent versions of this program, the latest of which is TATFIND version 1.4 has improved prediction capacity by increasing the number of allowed residues at the consensus positions, and can be found at http://signalfind.org/tatfind.html.

A second prediction program, TatP, has also been developed to predicted twin-arginine signal peptides. This method uses a similar approach as TATFIND to classify allowed amino acids at the consensus positions, coupled with two artificial neural networks, one for recognition of the cleavage site and the second to determine whether a given amino acid belongs to a Tat signal peptide or not (Bendtsen *et al.*, 2005). TatP can be found at http://www.cbs.dtu.dk/services/TatP/.

Both programs are reasonably good at predicting Tat signal peptides, and usually generate partially, but not fully, overlapping lists of substrates when used to search genome sequences. Based on observations with the known Tat substrates of *E. coli*, those signal peptides that are recognized as Tat signals by *both* programs are almost certainly correct (Dilks *et al.*, 2003; Bendtsen *et al.*, 2005), but this will without doubt give a significant underestimate of Tat substrate number. In addition it should be noted that there may be species-specific bias for certain amino acids at the consensus positions, for example it appears that *Streptomyces* show a preponderance of alanine or glycine at the consensus lysine position (Dilks *et al.*, 2003; Schaerlaekens *et al.*, 2004a; Widdick *et al.*, 2006).

When such prediction programmes are applied to the genome sequence of *S. coelicolor*, substrate numbers of 141 (TATFIND 1.4) and 189 (TatP), respectively, are predicted, and a total of 80 of these pass both of these prediction programmes (see http://www.lifesci.dundee.ac.uk/groups/tracy_palmer/links.html for lists of these proteins). However, the fact that many of the predicted substrates do not apparently contain cofactors, coupled with the observation that the high G+C content of the *S. coelicolor* genome means that it is common to find arginine codons in place of lysine in signal peptides, suggested that these predictions might be an overestimate.

Tat substrates in streptomycetes

A number of approaches have therefore been taken to confirm the *in silico* predictions of candidate *Streptomyces* Tat substrates. These include assaying enzymatic activity of and/or using antibodies raised against candidate Tat substrates of interest to investigate whether their presence is lost in the extracellular fraction of *Streptomyces tat* mutant strains (e.g. Schaerlakens *et al.*, 2001, 2004; Faury *et al.*, 2004; Widdick *et al.*, 2006; Yikmis *et al.*, 2008); mutagenesis of the consensus twin arginine motif to determine whether export is lost (Lucana *et al.*, 2005) and development of Tat-dependent reporter proteins such as xylanase C or agarase which can be C-terminally fused to candidate twin arginine signal peptides (e.g. Li *et al.*, 2005; Widdick *et al.*, 2006, 2008).

A more global way to assess the extent of the Tat secretome is to use proteomic approaches to compare the extracellular proteome of a *tat* mutant with the cognate wild type strain. This approach is not without difficulties, particularly since cell lysis, which is an obligate part of the *Streptomyces* lifecycle, results in severe contamination of the extracellular proteome with cytoplasmic proteins. Widdick *et al.* (2006) got round this issue by growing strains (in this case the *S. coelicolor* wild type and an isogenic Δ*tatC* mutant) on solid medium and carrying out proteomic analysis of proteins released after washing of cell walls. Many differences were noted between the wild type and the Δ*tatC* extracellular proteome, including the presence of 43 secreted proteins that were only found in the wild type cell extract. Whilst intuitively these should all be expected to be substrates of the Tat pathway, in practice it was not straight-forward because seven of these had signal peptides that did not contain a twin arginine motif. Moreover, a further 19 proteins, although they had consecutive arginine residues somewhere within their signal peptides, were not predicted to be Tat substrates by either of the two Tat signal peptide prediction programmes.

Thus it was clear that a second criterion needed to be applied to these candidate Tat substrates to confirm their Tat dependence. To this end, Widdick *et al.* (2006) constructed fusions of the signal peptides of each of these proteins to the mature portion of agarase, a *Streptomyces* Tat dependent reporter protein. In total 25 signal peptides were able to mediate Tat-dependent agarase export whilst the remainder could not and were therefore deemed to be Sec signal sequences. A full list of all of the confirmed Tat substrate proteins in *Streptomyces* is given in Table 4.1. It was concluded that the pleiotropy of the Δ*tatC* mutation resulted in an inability of the organism to induce production of many of these secreted proteins rather than the more obvious conclusion that they were Tat substrates. These observations indicate that the results of proteomic studies should be interpreted with caution and should always be backed up with other approaches to confirm Tat dependence wherever possible. Interestingly, of the 25 confirmed Tat substrates from this study, only two of them were not predicted to be Tat-dependent by either of TATFIND or TatP (and in both cases due to their extremely long signal peptide n-regions) (Widdick *et al.*, 2006). This study therefore helped to validate the utility of the *in silico* analysis and allows us to be reasonably confident that the Tat pathway is indeed a general route of protein export in the streptomycetes.

It is abundantly clear from bioinformatic predictions coupled with experimental verification of Tat substrates that the Tat system is expected to make a major contribution to the physiology of *Streptomyces*. This is supported by phenotypic analysis of *tat* mutant strains, where mutations in *tat* genes of *S. coelicolor*, *S. lividans* and *S. scabies* have all been shown to result in highly pleiotropic phenotypes (Schaerlaekens *et al.*, 2004a, Widdick *et al.*, 2006; Joshi *et al.*, 2010). Δ*tat* strains of *S. lividans* and *S. coelicolor* were observed to grow in a dispersed manner in liquid culture and failed to sporulate on solid media containing sucrose. The *tatC* strain of *S. coelicolor* appeared fragile and prone to lysis, indicating a probable defect with the cell wall. A similar *tatC* mutant of *S. scabies*

Table 4.1 Confirmed Tat substrates from *Streptomyces*

SCO0105[a]	Xylanase C	SCO3471[c]	Agarase
SCO0624[b]	Putative xylosidase	SCO3484[c,d]	Sugar-binding protein
SCO0538[b]	Probable sugar transporter	SCO3790[c]	Protein of PhoX type
SCO0677[b]	Chitosanase	SCO4672[c]	Conserved hypothetical protein
SCO0736[c,d]	Putative peptidoglycan-binding protein	SCO6052[c]	Conserved hypothetical protein
SCO1172[c]	Putative amidase	SCO5145[f]	PHP domain protein
SCO1196[c]	Putative glycerophosphoryl diester phosphodiesterase	SCO6198[c]	Conserved hypothetical protein
SCO1356[c,d,e]	Putative iron–sulphur protein	SCO6272[b,c,e]	Putative FAD-binding protein
SCO1432[c]	Putative peptidase	SCO6281[c,e]	Putative FAD-binding protein
SCO1565[c]	Putative glycerophosphoryl diester phosphodiesterase	SCO6457[c]	Putative beta-galactosidase
SCO1590[c]	Putative transglycosidase	SCO6580[c]	Putative xylose isomerase
SCO1955[b,d,e]	Putative iron sulphur protein	SCO6594[c]	Conserved hypothetical protein
SCO1639[2]	Putative peptidyl-prolyl *cis-trans* isomerase	SCO6691[c]	Putative phospholipase C
SCO1906[c]	PhoX phosphatase	SCO7631[c]	Putative phosphohydrolase
SCO$_2$068[c]	Alkaline phosphatase	SCO7677[c]	Solute-binding protein
SCO$_2$286[b,c]	Alkaline phosphatase	HbpS[g]	Haem-binding protein
SCO$_2$758[c]	Beta-*N*-acetylglucosaminidase	Lcp[h]	Latex clearing protein
SCO$_2$780[c,d]	Fe^{3+}-siderophore binding domain FhuD	MelC1[i]	Tyrosinase transactivator protein
SCO$_2$786[c]	Beta-*N*-acetylhexosaminidase	MelC2[i]	Tyrosinase

[a]Shown to be a Tat substrate in *S. lividans* by Faury *et al.* (2004).
[b]Signal peptides of these proteins shown to mediate the Tat-dependent export of xylanase/chitosanase reporter proteins (Li *et al.*, 2005).
[c]Proteins differentially absent from cell wall washes of *S. coelicolor tatC* strain, and signal peptides of these proteins shown to mediate the Tat-dependent export of the agarase reporter protein (Widdick *et al.*, 2006).
[d]potential lipoprotein.
[e]predicted to be cofactor-containing.
[f]Present only in the cell wall proteome of the wild type and not the *tatC* strain in Widdick *et al.* (2006) but was not tested in the agarase test. However, the signal peptide of the *S. scabies* homologue (SCAB15581) can mediate Tat-dependent export when tested in the agarase reporter assay (Joshi *et al.*, 2010).
[g]HbpS of *S. reticuli* shown to be Tat-dependent by substitution of the twin arginine motif (Lucana *et al.*, 2005).
[h]Lcp from *Streptomyces* species Strain K30 shown to be Tat-dependent when produced in *S. lividans* (Yikmis *et al.*, 2008).
[i]These two proteins are exported together as a complex mediated by a signal peptide only on MelC1 (Chen *et al.*, 1992) shown to be Tat dependent by Schaerlaekens *et al.* (2001, 2004a).

is defective for the production of melanin (Joshi *et al.*, 2010), which is consistent with a failure to secrete the enzyme tyrosinase, a known Tat substrate (Schaerlaekens *et al.*, 2001).

The secretion of tyrosinase is of particular interest because the enzyme, encoded by the *melC2* gene, lacks any N-terminal signal peptide. Instead it is exported as a complex with its chaperone, MelC1 that has a typical Tat-targeting signal (Chen *et al.*, 1992). After transport through the Tat system MelC1 apparently dissociates from tyrosinase as it is not found in the mature enzyme. The dissociation of MelC1 is dependent upon the presence of copper in the medium, and it is likely that in addition to directing its export, MelC1 also plays a role in the insertion of copper into the tyrosinase active site (Chen *et al.*, 1992; Tsai and Lee, 1998). This export of heterodimeric proteins by a signal peptide present on only one of the partner subunits has been termed 'hitch-hiking' and may be relatively common, with at least seven of the 27 Tat substrates in *E. coli* being predicted to have one or more hitch-hiking partners (Rodrigue *et al.*, 1999; Berks *et al.*, 2005). The identification of hitch-hiking partner proteins is notoriously difficult because they lack any kind of N-terminal signal sequence and it is therefore likely that additional hitch-hiking partner proteins may also exist in *Streptomyces*.

Analysis of the known or predicted Tat substrates in *S. coelicolor* indicates that the Tat pathway secretes a diverse array of hydrolytic enzymes and therefore most likely plays a major role in nutrient acquisition, probably during hyphal growth. For example the Tat system plays a key role in the acquisition of phosphate, with several phosphatase enzymes, including alkaline phosphatases of the PhoD and PhoX families as well as phosphodiesterases and phospholipases being secreted Tat-dependently (Widdick *et al.*, 2006; see Table 4.1). In addition, at least one of the three characterized xylanases of *S. lividans* is a Tat substrate (Faury *et al.*, 2004; Schaerlaekens *et al.*, 2004a), and many other enzymes involved in extracellular carbohydrate metabolism are known or predicted to be secreted by the Tat pathway (Widdick *et al.*, 2006). Other Tat substrates include a suite of substrate-binding protein components of ABC transporters, which in Gram-positive bacteria are usually anchored to the outer face of the membrane by lipid modification (Hutchings *et al.*, 2009). Indeed, inspection of many of their signal peptides indicates that they harbour apparent 'lipoboxes' with invariant cysteine residues at the +1 position of the mature protein sequence (shown schematically in Fig. 4.1B – section below for a detailed discussion of lipoproteins and their biogenesis).

Perhaps one of the most interesting and possibly unexpected Tat substrates of *S. coelicolor* is agarase itself. The production of an extracellular agarase enzyme by *S. coelicolor* is rather unusual as most other characterized streptomycetes lack this activity. The regulation and secretion of agarase activity had been well studied in *S. coelicolor* and its signal peptide had been used for heterologous protein production long before the Tat system had been discovered (e.g. Buttner *et al.*, 1987, 1988; Isiegas *et al.*, 1999). Agarase has a canonical Tat signal peptide and is clearly exported by the Tat pathway (Widdick *et al.*, 2006, 2008). The origin of *S. coelicolor* agarase is unclear since the closest enzymes in terms of amino acid sequence similarity are all from Gram-negative marine bacteria and are also Sec-dependent. This would imply that the *S. coelicolor* agarase was acquired through horizontal gene transfer, although it is not apparent why the enzyme should be Tat-dependent in *Streptomyces*. What is clear, however, is that *S. coelicolor* agarase is a very good reporter enzyme for the Tat pathway and has been used to experimentally verify Tat signal peptides from all three domains (Widdick *et al.*, 2008). Application of this reporter system to predicted twin arginine signal peptides from *S. scabies* indicates that a number of candidate virulence factors, as well as the secreted regulatory protein Factor C (an extracellular signalling protein involved in cellular differentiation of certain *Streptomyces* strains – Birkó *et al.*, 1999) are substrates of the Tat pathway (Joshi *et al.*, 2010), suggesting that many other interesting Tat substrates remain to be uncovered.

Biotechnological aspects of protein secretion in the streptomycetes

Bacteria are frequently used in academic and industrial settings for the production of proteins. The two most important organisms for recombinant protein production are *E. coli*, and *Bacillus* species. However, there are inherent problems working with these organisms. For example *E. coli* secretes very few proteins into the growth medium, and therefore proteins of commercial interest need to be purified away from host cell proteins. *Bacillus*, on the other hand, has a high propensity for protein secretion, and is used as a vector for the production of about 60% of the commercially available enzymes, for example those used as biological washing agents (e.g. Westers *et al.*, 2004). However, *Bacillus* species produce large numbers of extracellular proteases that serve to remove misfolded proteins and this is a major bottleneck for the production of heterologous proteins (Westers *et al.*, 2004).

Streptomyces secrete large numbers of proteins and appropriate expression systems are available for high levels of constitutive protein production. Generally *S. lividans* has been the streptomycete of choice as an expression host. This is mainly because it lacks a restriction modification system that in *S. coelicolor* is a major problem for the propagation of vectors that have been derived from *E. coli* (González-Cerón *et al.*, 2009). *S. lividans* also has the advantage that it naturally produces only low levels of extracellular protease activity (e.g. Strickler *et al.*, 1992). Interestingly, it appears that *S. lividans* is also capable of glycosylating some secreted proteins, which also represents a major advantage over the other prokaryotic expression systems (Lara *et al.*, 2004).

Although the Sec pathway has been the major (only?) route by which heterologous proteins have been produced industrially, studies with *Streptomyces* have strengthened the idea that the Tat pathway may also support the high level secretion of recombinant proteins. Indeed, much early work on protein secretion in *Streptomyces* took advantage of the signal peptides of agarase, the MelC1 partner protein of tyrosinase and of *Streptomyces* β-galactosidase, all three of which are Tat targeting sequences (Widdick *et al.*, 2006; Schaerlaekens *et al.*, 2001, 2004a). The presence of the *dagA* gene in multicopy in *S. lividans* resulted in a 500 times greater production of extracellular agarase, accounting for almost 50% of the total extracellular protein (Kendall and Cullum, 1984). Furthermore, high level secretion of *Streptomyces* β-galactosidase could also be achieved when cloned in multicopy, indicating that the Tat pathway of *S. lividans* has an enormous capacity for protein secretion (Eckhardt *et al.*, 1987).

The secretion of two pharmaceutically important proteins, glucagon and interleukin 1β has been achieved in *S. lividans* by employing Tat-targeting signals (Qi *et al.*, 2008; Lichenstein *et al.*, 1988). The secretion of human interleukin 10 and tumour necrosis factor α has also been achieved by the *S. lividans* Tat pathway, although in this case it was demonstrated that both proteins could be secreted to higher levels using a Sec-targeting signal peptide (Schaerlaekens *et al.*, 2004b). Thus, although the utility of *S. lividans* as a heterologous host for secreted protein production, particularly via the Tat pathway shows great promise, further work is required to realize the full potential of this system.

The ESX-1 pathway

The third type of protein secretion system found in streptomycetes is the ESX-1 or type VII secretion system. This is a recently described protein secretion system that appears to be unique to Gram-positive bacteria. It was first described in *Mycobacterium tuberculosis* where it is required for the secretion of two small extracellular proteins, ESAT-6 and CFP10 (Hsu *et al.*, 2003; Pym *et al.*, 2003; Stanley *et al.*, 2003). It was later shown that a related system is found in *Staphylococcus aureus* where it also secretes ESAT-6-like proteins (Burts *et al.*, 2005). Although in both of these organisms the ESX-1 pathway is intimately related with virulence, the delivery of virulence factors cannot be the only function of this transport machinery since it is also encoded in the genomes of many streptomycetes and non-pathogenic mycobacteria (Pallen, 2002).

The genomic signature of the ESX-1 secretion system is the presence of a gene encoding an FtsK/SpoIIIE-like ATPase, in the vicinity of one or more genes encoding small proteins with a conserved W-x-G amino acid motif (Pallen, 2002). The genome of *S. scabies* encodes a single ESX secretion system (see Fig. 4.4A), whereas the other sequenced *Streptomyces* species – *S. griseus, S. avermitilis* and *S. coelicolor* encode two apparently distinct systems (shown for *S. coelicolor* in Fig. 4.4B). In *S. coelicolor* the ATPases are encoded by *sco4508* and *sco5734*, respectively and in *S. scabies* by *ssc58621*, which is most closely related to *sco5734*. The function of any of these systems in *Streptomyces* is currently unknown, although a recent publication implicated a role in sporulation (Akpe San Roman *et al.*, 2010).

Very little is known about how protein secretion is mediated by the ESX-1 machinery.

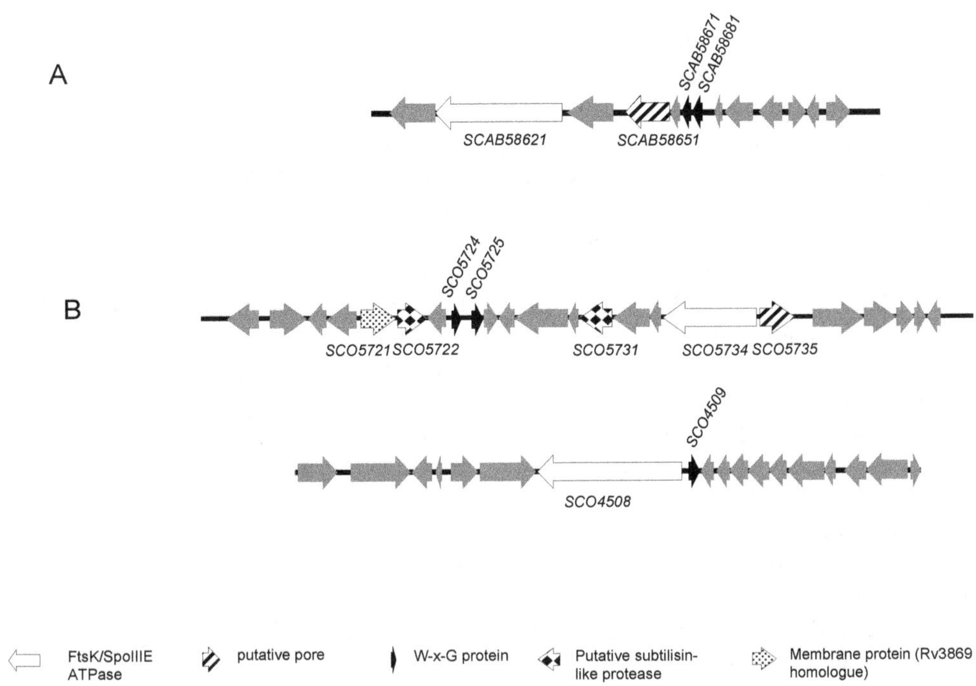

Figure 4.4 Organization of genes encoding putative ESX-1 secretion pathway components in (A) *S. scabies* and (B) *S. coelicolor*. Only those genes encoding probable components of the ESX-1 system are shown and are colour-coded according to probable function. ORFs and intergenic regions not to scale.

Mutagenesis studies, primarily in mycobacteria, have indicated that the FtsK/SpoIIIE ATPase (which is encoded as two separate polypeptides in mycobacteria rather than the single polypeptide in streptomycetes) is essential for secretion of the W-x-G motif proteins and that it recognizes a non-cleavable signal sequence comprising a short stretch of amino acids at the C-terminus of one of these proteins (*M. tuberculosis* CFP-10) (Stanley *et al.*, 2003; Champion *et al.*, 2006). Interestingly, a second W-x-G protein, ESAT-6, forms a complex with CFP-10 and is exported by virtue of the CFP-10 signal sequence in a manner perhaps akin to the hitch-hiking mechanism for transport of some Tat substrates (Stanley *et al.*, 2003; Renshaw *et al.*, 2002; Brodin *et al.*, 2005; Champion *et al.*, 2006). It is likely that secretion is energized through ATP hydrolysis by the ATPase and it has been proposed that the highly hydrophobic protein Rv3877 (encoded by *sco5735* in *S. coelicolor*) may form the translocation pore (Stanley *et al.*, 2003; Brodin *et al.*, 2006).

It should be noted that in addition to the secretion of small proteins with a W-x-G motif, this machinery, at least in *Mycobacterium*, also secretes other proteins that lack any apparent signal sequence (Abdallah *et al.*, 2006; Raghavan *et al.*, 2008). This means that there could be many substrates of this export system waiting to be uncovered in *Streptomyces*.

Lipoprotein biogenesis in *Streptomyces*

Until recently, it was generally held that all lipoproteins were exported as unfolded proteins through the Sec pathway and then targeted to the lipoprotein biogenesis machinery. A lipobox motif in the c-region of their signal peptides, typically L_{-3}–$[A/S/T]_{-2}$–$[G/A]_{-1}$–C_{+1}, where the cysteine is absolutely conserved (see Fig. 4.1B), targets the prolipoprotein first to the enzyme diacylglycerol lipoprotein transferase (Lgt) and then to a specific lipoprotein signal peptidase (Lsp). Lgt covalently attaches a lipid molecule to the conserved cysteine residue in the lipobox motif via a thioether linkage to its sulphydryl group and then Lsp cleaves the signal peptide such that the lipidated cysteine becomes the first residue in the mature lipoprotein. These steps are conserved between Gram-positive and Gram-negative bacteria and in Gram-positive bacteria the pathway usually ends here, with the lipoprotein tethered by its lipid to the outer face of the cytoplasmic membrane (Hutchings *et al.*, 2009). In Gram-negative bacteria a third enzyme, lipoprotein N acyl transferase (Lnt) further lipidates the +1 cysteine on its amino group and the lipoprotein is then either retained in the cytoplasmic membrane or, in the majority of cases, transported to the outer membrane by the lipoprotein localization (Lol) pathway where they usually face into the periplasm. Lnt has been implicated in targeting lipoproteins to the outer membrane because depletion of *lnt* in *E. coli* results in accumulation of lipoproteins in the cytoplasmic membrane and cell death (Robichon *et al.*, 2005). Lnt is not found in Firmicutes (the low GC branch of Gram-positive bacteria) but is present in some Actinobacteria, has been shown to be functional in mycobacteria (Tschumi *et al.*, 2009), which also have an outer membrane (Hoffmann *et al.*, 2008; Zuber *et al.*, 2008). However, speculation that N-acylation may also be involved in targeting lipoproteins to the mycolic acid layer of mycobacteria should be tempered by the fact that streptomycetes, which also encode Lnt enzymes, do not have an outer membrane.

Bioinformatic analysis suggests that *S. coelicolor* encodes at least 219 putative lipoproteins, around 40% of which are substrate-binding proteins for ATP-binding cassette (ABC) transporters while the majority of the remaining 60% are function unknown proteins (Thompson, *et al.*, 2010). *S. coelicolor* encodes a single Lsp enzyme, two Lgt enzymes, which complement each other (Thompson *et al.*, 2010) and two Lnt homologues. It is unclear why *S. coelicolor* has two copies of *lgt* since most streptomycetes encode only a single copy. However, it is striking that both the Lsp and Lgt enzymes of streptomycetes and some other actinomycetes, have cytoplasmic extensions at their N and C termini, respectively, and which are not found in other bacteria. The role of these extensions is currently unclear but may be involved in co-localization of different components of the machinery. Another unusual feature of lipoprotein biogenesis in streptomycetes is that around 20% of their lipoproteins are putative Tat substrates, with a total of 50 putative Tat dependent lipoproteins in *S. coelicolor* (Thompson *et al.*, 2010). As for most Tat substrates in *S. coelicolor*, it is not clear why many of these lipoproteins should utilize Tat as the many of them are substrate-binding proteins, which generally go out through Sec.

Protein glycosylation in *Streptomyces*

Protein glycosylation is a relatively recent discovery in bacteria (Szymanski *et al.*, 1999) but appears to be widespread. Actinomycetes have a general O-glycosylation system, in which the glycan is attached via an O-glycosidic linkage to the hydroxy oxygen of hydroxy amino acids (Lommel and Strahl, 2009). In the mycobacteria many secreted antigens are both lipidated and O-glycosylated, including the phosphate-binding protein PstS (also known as the major 38 kDa antigen) and the superoxide dismutase SodC (Sartain and Belisle, 2008), suggesting some potential overlap between these two pathways (discussed below). Protein O-glycosylation in streptomycetes was first demonstrated in *S. lividans* (Kluepfel *et al.*, 1990) and subsequent studies demonstrated that heterologous proteins, including the *M. tuberculosis* 45/47 kDa glycoprotein, could also be produced in *S. lividans* (Macleod *et al.*, 1992; Ong *et al.*, 1994; Lara *et al.*, 2004).

Glycosylated proteins were discovered in the closely related *S. coelicolor* during a study of phage-resistant receptor mutants. Some, but not all, of these mutants could be complemented by the *sco3514* gene, encoding a putative protein O-mannosyl transferase (Pmt), that is required for glycosylation of the phage ΦC31 receptor (Cowlishaw and Smith, 2001). The remainder of the mutants were complemented by the *sco1423* gene, encoding a putative polyprenol phosphate mannose synthase (Ppm1), thus confirming a role for protein O-glycosylation in phage ΦC31 receptor activity (Cowlishaw and Smith, 2002). Homologues of *S. coelicolor* Pmt and Ppm1 were subsequently demonstrated to have roles in protein glycosylation and lipid mannosylation, respectively, in mycobacteria and corynebacteria (Gurcha *et al.*, 2002; Baulard *et al.*, 2003; Gibson *et al.*, 2003) and, more recently, both proteins have been shown to be required for glycosylation of the phosphate-binding lipoprotein PstS, in *S. coelicolor* (Wehmeier *et al.*, 2008).

The discovery that mycobacterial and streptomycete lipoproteins can be glycosylated suggests there may be some, as yet unidentified, link between the two pathways. This hypothesis was recently strengthened by the identification of a lipoprotein N acyl transferase in mycobacteria (Tschumi *et al.*, 2009). This bifunctional enzyme is annotated as a polyprenol mannose transferase in both *M. tuberculosis* (Mt-Ppm1) and *M. smegmatis* (Ms-Ppm2) and contains a functional Ppm domain at its N-terminus and a functional Lnt domain in its C terminus (Tschumi *et al.*, 2009). In fact, Ppm1 is required for synthesis of the lipoglycans lipoarabinomannan, lipomannan and phosphatidylinositol mannosides in mycobacteria but involvement in protein glycosylation has yet to be demonstrated (Gurcha *et al.*, 2002; Baulard *et al.*, 2003; Gibson *et al.*, 2003). *S. coelicolor* encodes two Lnt homologues, SCO1014, which is homologous to the Mt-Ppm1 and Ms-Ppm2 and contains both Ppm and Lnt domains, and SCO1336 which is the closest homologue to *E. coli* Lnt. The functions of these proteins have yet to be confirmed in streptomycetes but both have proven very difficult to disrupt in *S. coelicolor* (unpublished). Hence, there is clearly overlap between lipid modification and glycosylation of proteins in actinomycetes with some secreted proteins being both lipidated and glycosylated following export from the cell. The order in which these events occur has yet to be determined and, although tantalising, the activity of the Ppm/Lnt hybrids in protein glycosylation and lipidation have yet to be experimentally verified.

Cell wall attachment of proteins

The best characterized pathway for covalent attachment of secreted proteins to the cell wall is catalysed by sortase enzymes that, in the case of *Staphylococcus aureus* SrtA, recognize an LPXTG motif within the cell wall sorting signal. The sortase cleaves the precursor protein between the threonine and glycine residues and covalently links the terminal threonine to the peptidoglycan (Ton That *et al.*, 2004). Sortases are found in almost all Gram-positive bacteria including most actinomycetes, although notably they are absent from the genomes of mycobacteria (Pallen *et al.*, 2001), probably because their outer membranes encapsulate their cell walls (Hoffmann *et al.*, 2008; Zuber *et al.*, 2008). The role of sortases in actinomycetes has been best characterized in *Corynebacterium diptheriae*, where they play an essential role in the assembly and anchoring of pili on the cell surface. In *C. diptheriae* the sortase substrates are translocated through Sec and it has been suggested that the sortase components are built around the SecYEG channel on the extracytoplasmic face of the membrane (Mandlik *et al.*, 2008). This raises interesting questions about the interaction of post secretion machineries for lipidation, glycosylation or cell wall sorting with the Sec and Tat translocons.

S. coelicolor is predicted to encode seven sortase enzymes (Pallen *et al.*, 2001) which, although not characterized, have been implicated in the attachment of chaplins to the surface of aerial hyphae in *S. coelicolor* (Elliot *et al.*, 2003). *S. coelicolor* produces three long chaplins, ChpA, -B and -C, which each contain two chaplin domains and a C terminal cell wall LAXTG sorting signal, and five short chaplins, ChpD, -E, -F, -G and -H which presumably polymerize by binding to the longer, cell wall attached chaplins. All of these chaplins are predicted to go out through Sec. Deletion of the *chpABC* genes delays aerial hyphae formation and blocks sporulation and a proposed model suggests that the hydrophobic chaplin proteins polymerize to coat the aerial hyphae and allow them to break surface tension and emerge into the air. Curiously, the *chpE* gene is essential to the viability of *S. coelicolor* and can only be deleted in a strain lacking the other chaplins, the rodlins, which form another hydrophobic layer (Willey *et al.*, 2006), or the Tat translocon (Di Berardo *et al.*, 2008). This link between Tat and ChpE has not yet been explained, and is confusing since both the chaplins and rodlins are exported via Sec.

Another pathway for the covalent attachment of proteins to cell walls is catalysed by a set of recently discovered L-D transpeptidase (Ldt) enzymes, which are also involved in cross-linking of the mature cell wall. These enzymes all contain the ErfK-YbiS-YhnG (PF03734) domain. Peptidoglycan precursors contain pentapeptide stems, UDP-MurNAc-Ala^1-Glu^2-DAP^3-Ala^4-Ala^5 in *S. coelicolor* (Hong *et al.*, 2005), in which the fourth position alanine (Ala^4) is cross-linked to the third position diaminopimelic acid (DAP^3) by penicillin-binding proteins, using energy released through cleavage of the Ala^4–Ala^5 bond. Ldt enzymes catalyse cross-bridge formation between the position three amino acids in peptidoglycan precursors and may facilitate recycling of mature peptidoglycan (Mainardi *et al.*, 2005). *E. coli* encodes four Ldt enzymes which are involved in both 3–3 cross-linking of the cell wall and cross-linking of the major outer membrane lipoprotein Lpp to the cell wall to stabilize the cell envelope (Magnet *et al.*, 2007). *S. coelicolor* encodes six ErfK domain proteins, all of which are lipoproteins and which likely cross-link proteins or newly synthesized peptidoglycan precursors to the mature cell wall. Of these six, two are putative Tat substrates (SCO3194 and SCO4934) while four (SCO_2153, SCO4868, SCO5457 and SCO5458) are predicted to have Sec signal sequences. SCO4868 also contains a LysM peptidoglycan cell wall binding domain that may link it to the cell wall. The fact that the *S. coelicolor* Ldt homologues are all lipoproteins and, as in the case of *E. coli* Lpp, may link other lipoproteins to the peptidoglycan to stabilize the cell envelope, once again highlights the interactions between different post-translational modification pathways which take place after secretion of the substrates from the bacterial cell.

Acknowledgements

The authors acknowledge funding from the MRC, BBSRC, Wellcome Trust, European Union (through the 6th Framework TatMachine programme) and the Research Councils UK for a Fellowship to MIH. We also thank David Widdick and Govind Chandra for helpful discussion, Joanna Fyans for assistance with preparing Fig. 4.4, and Iain Sutcliffe, Stefan Mann, David Widdick, Benjamin Thompson and Matthew Hicks for unpublished results.

References

Abdallah, A.M., Verboom, T., Hannes, F., Safi, M., Strong, M., Eisenberg, D., Musters, R.J., Vandenbroucke-Grauls, C.M., Appelmelk, B.J., Luirink, J., *et al.* (2006). A specific secretion system mediates PPE41 transport in pathogenic mycobacteria. Mol. Microbiol. *62*, 667–679.

Akpe San Roman, S., Facey, P.D., Fernandez-Martinez, L., Rodriguez, C., Vallin, C., Del Sol, R., and Dyson, P. (2010). A heterodimer of EsxA and EsxB is involved in sporulation and is secreted by a type VII secretion system in *Streptomyces coelicolor*. Microbiology *156*, 1719–1729.

Alami, M., Luke, I., Deitermann, S., Eisner, G., Koch, H.G., Brunner, J., and Muller, M. (2003). Differential interactions between a twin-arginine signal peptide and its translocase in *Escherichia coli*. Mol. Cell *12*, 937–946.

Angelini, S., Deitermann, S., and Koch, H.G. (2005). FtsY, the bacterial signal-recognition particle receptor, interacts functionally and physically with the SecYEG translocon. EMBO Rep. *6*, 476–481.

Barnett, J.P., van der Ploeg, R., Eijlander, R.T., Nenninger, A., Mendel, S., Rozeboom, R., Kuipers, O.P., van Dijl, J.M., and Robinson, C. (2009). The twin-arginine translocation (Tat) systems from *Bacillus subtilis* display a conserved mode of complex organization and similar substrate recognition requirements. FEBS J. *276*, 232–243.

Baulard, A.R., Gurcha, S.S., Engohang-Ndng, J., Gouffi, K., Locht, C., and Besra, G.S. (2003). *In vivo* interaction between the polyprenol phosphate mannose synthase Ppm1 and the integral membrane protein Ppm2 from *Mycobacterium smegmatis* revealed by a bacterial two-component system. J. Biol. Chem. *278*, 2242–2248.

Bendtsen, J.D., Nielsen, H., Widdick, D., Palmer, T., and Brunak, S. (2005). Prediction of twin-arginine signal peptides. B. M. C. Bioinformatics *6*, 167.

Berks, B.C. (1996). A common export pathway for proteins binding complex redox cofactors? Mol. Microbiol. *22*, 393–404.

Berks, B.C., Sargent, F., and Palmer, T. (2000). The Tat protein export pathway. Mol. Microbiol. *35*, 260–274.

Birkó, Z., Sümegi, A., Vinnai, A., van Wezel, G., Szeszák, F., Vitális, S., Szabó, P.T., Kele, Z., Janáky, T., and Biró, S. (1999). Characterization of the gene for factor C, an extracellular signal protein involved in morphological differentiation of *Streptomyces griseus*. Microbiology *145*, 2245–2253.

Blanco, J., Driessen, A.J., Coque, J.J., and Martin, J.F. (1998). Biochemical characterization of the SecA protein of *Streptomyces lividans* – interaction with nucleotides, binding to membrane vesicles and in vitro translocation of proAmy protein. Eur. J. Biochem. *257*, 472–478.

Bogsch, E., Brink, S., and Robinson, C. (1997). Pathway specificity for a delta pH-dependent precursor thylakoid lumen protein is governed by a 'Sec-avoidance' motif in the transfer peptide and a 'Sec-incompatible' mature protein. EMBO J. *16*, 3851–3859.

Bogsch, E., Sargent, F., Stanley, N.R., Berks, B.C., Robinson, C., and Palmer, T. (1998). An essential component of a novel bacterial protein export system with homologues in plastids and mitochondria. J. Biol. Chem. *273*, 18003–18006.

Bolhuis, A., Mathers, J.E., Thomas, J.D., Barrett, C.M., and Robinson, C. (2001). TatB and TatC form a functional and structural unit of the twin-arginine translocase from *Escherichia coli*. J. Biol. Chem. *276*, 20213–20219.

Brodin, P., de Jonge, M.I., Majlessi, L., Leclerc, C., Nilges, M., Cole, S.T., and Brosch, R. (2005). Functional analysis of early secreted antigenic target-6, the dominant T-cell antigen of *Mycobacterium tuberculosis*, reveals key residues involved in secretion, complex formation, virulence, and immunogenicity. J. Biol. Chem. *280*, 33953–33959.

Brodin, P., Majlessi, L., Marsollier, L., de Jonge, M.I., Bottai, D., Demangel, C., Hinds, J., Neyrolle, S.O., Butcher, P.D., Leclerc, C., *et al.* (2006). Dissection of ESAT-6 system 1 of *Mycobacterium tuberculosis* and impact on immunogenicity and virulence. Infect. Immun. *74*, 88–98.

Burts, M.L., Williams, W.A., DeBord, K., and Missiakas, D.M. (2005). EsxA and EsxB are secreted by an ESAT-6-like system that is required for the pathogenesis of *Staphylococcus aureus* infections. Proc. Natl. Acad. Sci. U.S.A. *102*, 1169–1174.

Buttner, M.J., Fearnley, I.M., and Bibb, M.J. (1987). The agarase gene (*dagA*) of *Streptomyces coelicolor* A3(2): nucleotide sequence and transcriptional analysis. Mol. Gen. Genet. *209*, 101–109.

Buttner, M.J., Smith, A.M., and Bibb, M.J. (1988). At least three different RNA polymerase holoenzymes direct transcription of the agarase gene (*dagA*) of *Streptomyces coelicolor* A3(2). Cell *52*, 599–607.

Champion, P.A., Stanley, S.A., Champion, M.M., Brown, E.J., and Cox, J.S. (2006). C-terminal signal sequence promotes virulence factor secretion in *Mycobacterium tuberculosis*. Science *313*, 1632–1636.

Chen, L.Y., Leu, W.M., Wang, K.T., and Lee, Y.H. (1992). Copper transfer and activation of the *Streptomyces* apotyrosinase are mediated through a complex formation between apotyrosinase and its trans-activator MelC1. J. Biol. Chem. *267*, 20100–20107.

Cowlishaw, D.A., and Smith, M.C.M. (2001). Glycosylation of a *Streptomyces coelicolor* A3(2) cell envelope protein is required for infection by bacteriophage phi C31. Mol. Microbiol. *41*, 601–610.

Cowlishaw, D.A., and Smith, M.C.M. (2002). A gene encoding a homologue of dolichol phosphate-beta-D-mannose synthase is required for infection of *Streptomyces coelicolor* A3(2) by phage (phi)C31. J. Bacteriol. *184*, 6081–6083.

Cristobal, S., de Gier, J.W., Nielsen, H., and von Heijne, G. (1999). Competition between Sec- and TAT-dependent protein translocation in *Escherichia coli*. EMBO J. *18*, 2982–2990.

De Keersmaeker, S., Van Mellaert, L., Lammertyn, E., Vrancken, K., Anné, J., and Geukens, N. (2005a). Functional analysis of TatA and TatB in *Streptomyces lividans*. Biochem. Biophys. Res. Commun. *335*, 973–982.

De Keersmaeker, S., Van Mellaert, L., Schaerlaekens, K., Van Dessel, W., Vrancken, K., Lammertyn, E., Anné, J., and Geukens, N. (2005b). Structural organization of the twin-arginine translocation system in *Streptomyces lividans*. FEBS Lett. *579*, 797–802.

De Keersmaeker, S., Vrancken, K., Van Mellaert, L., Anné, J., and Geukens, N. (2007). The Tat pathway in *Streptomyces lividans*: interaction of Tat subunits and their role in translocation. Microbiology *153*, 1087–1094.

de Leeuw, E., te Kaat, K., Moser, C., Menestrina, G., Demel, R., de Kruijff, B., Oudega, B., Luirink, J., and Sinning, I. (2000). Anionic phospholipids are involved in membrane association of FtsY and stimulate its GTPase activity. EMBO J. *19*, 531–541.

De Leeuw, E., Porcelli, I., Sargent, F., Palmer, T., and Berks, B.C. (2001). Membrane interactions and self-association of the TatA and TatB components of the twin-arginine translocation pathway. FEBS Lett. *506*, 143–148.

Di Berardo, C., Capstick, D.S., Bibb, M.J., Findlay, K.C., Buttner, M.J., and Elliot, M.A. (2008). Function and redundancy of the chaplin cell surface proteins in aerial hypha formation, rodlet assembly, and viability in *Streptomyces coelicolor*. J. Bacteriol. *190*, 5879–5889.

Dilks, K., Rose, R.W., Hartmann, E., and Pohlschroder, M. (2003). Prokaryotic utilization of the twin-arginine translocation pathway: a genomic survey. J. Bacteriol. *185*, 1478–1483.

Dilks, K., Giménez, M.I., and Pohlschröder, M. (2005). Genetic and biochemical analysis of the twin-arginine translocation pathway in halophilic archaea. J. Bacteriol. *187*, 8104–8113.

Driessen, A.J., and Nouwen, N. (2008). Protein translocation across the bacterial cytoplasmic membrane. Annu. Rev. Biochem. *77*, 643–667.

Eckhardt, T., Strickler, J., Gorniak, L., Burnett, W.V., and Fare, L.R. (1987). Characterization of the promoter, signal sequence, and amino terminus of a secreted beta-galactosidase from '*Streptomyces lividans*'. J. Bacteriol. *169*, 4249–4256.

Elliot, M.A., Karoonuthaisiri, N., Huang, J., Bibb, M.J., Cohen, S.N., Kao, C.M., and Buttner, M.J. (2003). The chaplins: a family of hydrophobic cell-surface proteins involved in aerial mycelium formation in *Streptomyces coelicolor*. Genes Dev. *17*, 1727–1740.

Escutia, M.R., Val, G., Palacín, A., Geukens, N., Anné, J., and Mellado, R.P. (2006). Compensatory effect of the minor *Streptomyces lividans* type I signal peptidases on the SipY major signal peptidase deficiency as determined by extracellular proteome analysis. Proteomics *6*, 4137–4146.

Faury, D., Saidane, S., Li, H., and Morosoli, R. (2004). Secretion of active xylanase C from *Streptomyces lividans* is exclusively mediated by the Tat protein export system. Biochim. Biophys. Acta *1699*, 155–162.

Gibson, K.J., Eggeling, L., Maughan, W.N., Krumbach, K., Gurcha, S.S., Nigou, J., Puzo, G., Sahm, H., and Besra, G.S. (2003). Disruption of Cg-Ppm1, a polyprenyl monophosphomannose synthase, and the generation of lipoglycan-less mutants in *Corynebacterium glutamicum*. J. Biol. Chem. *278*, 40842–40850.

Gohlke, U., Pullan, L., McDevitt, C.A., Porcelli, I., de Leeuw, E., Palmer, T., Saibil, H.R., and Berks, B.C. (2005). The TatA component of the twin-arginine protein transport system forms channel complexes of variable diameter. Proc. Natl. Acad. Sci. U.S.A. *102*, 10482–10486.

González-Cerón, G., Miranda-Olivares, O.J., and Servín-González, L. (2009). Characterization of the methyl-specific restriction system of *Streptomyces coelicolor* A3(2) and of the role played by laterally acquired nucleases. FEMS Microbiol. Lett. *301*, 35–43.

Gurcha, S.S., Baulard, A.R., Kremer, L., Locht, C., Branch Moody, D., Muhlecker, W., Costello, C.E., Crick, D.C., Brennan, P.J., *et al.* (2002). Ppm1, a novel polyprenol monophosphomannose synthase from *Mycobacterium tuberculosis*. Biochem. J. *365*, 441–450.

Hicks, M.G., Guymer, D., Buchanan, G., Widdick, D.A., Caldelari, I., Berks, B.C., and Palmer, T. (2006). Formation of functional Tat translocases from heterologous components. BMC Microbiol. *6*, 64.

Hoffmann, C., Leis, A., Niederweis, M., Plitzko, J.M., and Engelhardt, H. (2008). Disclosure of the mycobacterial outer membrane: cryo-electron tomography and vitreous sections reveal the lipid bilayer structure. Proc. Natl. Acad. Sci. U.S.A. *105*, 3963–3967.

Hong H.-J., Hutchings, M.I., Hill, L.M., and Buttner, M.J. (2005). The role of the novel Fem protein VanK in vancomycin resistance in *Streptomyces coelicolor*. J. Biol. Chem. *280*, 13055–13061.

Hsu, T., Hingley-Wilson, S.M., Chen, B., Chen, M., Dai, A.Z., Morin, P.M., Marks, C.B., Padiyar, J., Goulding, C., Gingery, M., *et al.* (2003). The primary mechanism of attenuation of bacillus Calmette-Guerin is a loss of secreted lytic function required for invasion of lung interstitial tissue. Proc. Natl. Acad. Sci. U.S.A. *100*, 12420–12425.

Hutchings, M.I., Palmer, T., Harrington, D., and Sutcliffe, I.C. (2009). Gram-positive lipoproteins; knowing when to hold 'em, when to fold 'em. Trends Microbiol. *17*, 13–21.

Isiegas, C., Parro, V., and Mellado, R.P. (1999). *Streptomyces lividans* as a host for the production and secretion of *Escherichia coli* TEM beta-lactamase. Lett. Appl. Microbiol. *28*, 321–326.

Jack, R.L., Sargent, F., Berks, B.C., Sawers, G., and Palmer. T. (2001). Constitutive expression of *Escherichia coli tat* genes indicates an important role for the twin-arginine translocase during aerobic and anaerobic growth. J. Bacteriol. *183*, 1801–1804.

Joshi, M.V., Mann, S.G., Antelmann, H., Widdick, D. Fyans, J.K., Chandra, G., Hutchings, M.I., Toth, I., Hecker, M., Loria, R., *et al.* (2010). The Tat pathway exports multiple virulence proteins in the plant pathogen *Streptomyces scabies*. Mol. Microbiol. *77*, 252–271.

Kendall, K., and Cullum, J. (1984). Cloning and expression of an extracellular-agarase from *Streptomyces coelicolor* A3(2) in *Streptomyces lividans* 66. Gene *29*, 315–321.

Kluepfel, D., Vats-Mehta, S., Aumont, F., Shareck, F., and Morosoli, R. (1990). Purification and characterization of a new xylanase (xylanase B) produced by *Streptomyces lividans* 66. Biochem. J. *267*, 45–50.

Lara, M., Servín-González, L., Singh, M., Moreno, C., Cohen, I., Nimtz, M., and Espitia, C. (2004). Expression, secretion, and glycosylation of the 45- and 47-kDa glycoprotein of *Mycobacterium tuberculosis* in *Streptomyces lividans*. Appl. Environ. Microbiol. *70*, 679–685.

Leake, M.C., Greene, N.P., Godun, R.M., Granjon, T., Buchanan, G., Chen, S., Berry, R.M., Palmer, T., and Berks, B.C. (2008). Variable stoichiometry of the TatA component of the twin-arginine protein transport system observed by in vivo single-molecule imaging. Proc. Natl. Acad. Sci. U.S.A. *105*, 15376–15381.

Li, H., Jacques, P.E., Ghinet, M.G., Brzezinski, R., and Morosoli, R. (2005). Determining the functionality of putative Tat-dependent signal peptides in *Streptomyces coelicolor* A3(2) by using two different reporter proteins. Microbiology *151*, 2189–2198.

Lichenstein, H., Brawner, M.E., Miles, L.M., Meyers, C.A., Young, P.R., Simon, P.L., and Eckhardt, T. (1988). Secretion of interleukin-1 beta and *Escherichia coli* galactokinase by *Streptomyces lividans*. J. Bacteriol. *170*, 3924–3929.

Lommel, M., and Strahl, S. (2009). Protein O-mannosylation: conserved from bacteria to humans. Glycobiology *19*, 816–828.

Lubelski, J., Rink, R., Khusainov, R., Moll. G.N., and Kuipers, O.P. (2008). Biosynthesis, immunity, regulation, mode of action and engineering of the model lantibiotic nisin. Cell Mol. Life Sci. *65*, 455–476.

Lucana, D.O., Zou, P., Nierhaus, M., and Schrempf, H. (2005). Identification of a novel two-component system SenS/SenR modulating the production of the catalase-peroxidase CpeB and the haem-binding protein HbpS in *Streptomyces reticuli*. Microbiology *151*, 3603–3614.

Luirink, J., von Heijne, G., Houben, E., and de Gier, J.W. (2005). Biogenesis of inner membrane proteins in *Escherichia coli*. Annu. Rev. Microbiol. *59*, 329–355.

Lüke, I., Handford, J.I., Palmer, T., and Sargent, F. (2009). Proteolytic processing of *Escherichia coli* twin-arginine signal peptides by LepB. Arch. Microbiol. *191*, 919–925.

Macleod, A.M., Gilkes, N.R., Escote-Carlson, L., Warren, R.A., Kilburn, D.G., and Miller, R.C. (1992). *Streptomyces lividans* glycosylates an exoglucanase (Cex) from *Cellulomonas fimi*. Gene *121*, 143–147.

Magnet, S., Arbeloa, A., Mainardi, J.-L., Hugonnet, J.-E., Fourgeaud, M., Dubost, L., Marie, A., Delfosse, V., Mayer, C., Rice, L.B., *et al.* (2007). Specificity of L, D-transpeptidases from gram-positive bacteria producing different peptidoglycan chemotypes. J. Biol. Chem. *282*, 13151–13159.

Mainardi, J.L., Fourgeaud, M., Hugonnet, J.E., Dubost, L., Brouard, J.P., Ouazzani, J., Rice, L.B., Gutmann, L., and Arthur, M. (2005). A novel peptidoglycan cross-linking enzyme for a beta-lactam-resistant transpeptidation pathway. J. Biol. Chem. *280*, 38146–38152.

Mandlik, N., Das, A., and Ton-That-H., A. (2008). The molecular switch that activates the cell wall anchoring step of pilus assembly in gram-positive bacteria. Proc. Natl. Acad. Sci. U.S.A. *105*, 14147–14152.

Mori, H., and Cline, K. (2002). A twin arginine signal peptide and the pH gradient trigger reversible assembly of the thylakoid [Delta]pH/Tat translocase. J. Cell Biol. *157*, 205–210.

Nielsen, H., Engelbrecht, J., Brunak, S., and von Heijne, G. (1997). Identification of prokaryotic and eukaryotic signal peptides and prediction of their cleavage sites. Prot. Engineer *10*, 1–6.

Nouwen, N., and Driessen, A.J. (2005). Inactivation of protein translocation by cold-sensitive mutations in the *yajC-secDF* operon. J. Bacteriol. *187*, 6852–6855.

Onaka, H., Nakaho, M., Hayashi, K., Igarashi, Y., and Furumai, T. (2005). Cloning and characterization of the goadsporin biosynthetic gene cluster from *Streptomyces* sp. TP-A0584. Microbiology *151*, 3923-3933.

Ong, E., Kilburn, D.G., Miller, R.C., and Warren, R.A. (1994). *Streptomyces lividans* glycosylates the linker region of a beta-1, 4-glycanase from *Cellumonas fimi*. J. Bacteriol. *176*, 999–1008.

Palacin, A., de la Fuente, R., Valle, I., Rivas, L.A., and Mellado, R.P. (2003). *Streptomyces lividans* contains a minimal functional signal recognition particle that is involved in protein secretion. Microbiology *149*, 2435–2442.

Pallen, M.J., Lam, A.C., Antonio, M., and Dunbar, K. (2001). An embarrassment of sortases – a richness of substrates? Trends Microbiol. *9*, 97–102.

Pallen, M.J. (2002). The ESAT-6/WXG100 superfamily – and a new Gram-positive secretion system? Trends Microbiol. *10*, 209–212.

Palomino, C., and Mellado, R.P. (2008). Influence of a *Streptomyces lividans* SecG functional analogue on protein secretion. Int. Microbiol. *11*, 25–31.

Perlman, D., and Halvorson, H.O. (1983). A putative signal peptidase recognition site and sequence in eukaryotic and prokaryotic signal peptides. J. Mol. Biol. *167*, 391–409.

Parro, V., Schacht, S., Anné, J., and Mellado, R.P. (1999). Four genes encoding different type I signal peptidases are organized in a cluster in *Streptomyces lividans* TK21. Microbiology *145*, 2255–2263.

Pop, O.I., Westermann, M., Volkmer-Engert, R., Schulz, D., Lemke, C., Schreiber, S., Gerlach, R., Wetzker, R., and Müller, J.P. (2003). Sequence-specific binding of prePhoD to soluble TatAd indicates protein-mediated targeting of the Tat export in *Bacillus subtilis*. J. Biol. Chem. *278*, 38428–38436.

Pym, A.S., Brodin, P., Majlessi, L., Brosch, R., Demangel, C., Williams, A., Griffiths, K.E., Marchal, G., Leclerc, C., and Cole, S.T. (2003). Recombinant BCG exporting ESAT-6 confers enhanced protection against tuberculosis. Nat. Med. *9*, 533–539.

Qi, X., Jiang, R., Yao, C., Zhang, R., and Li, Y. (2008). Expression, purification, and characterization of C-terminal amidated glucagon in *Streptomyces lividans*. J. Microbiol. Biotechnol. *18*, 1076–1080.

Raghavan, S., Manzanillo, P., Chan, K., Dovey, C., and Cox, J.S. (2008). Secreted transcription factor controls *Mycobacterium tuberculosis* virulence. Nature *454*, 717–721.

Renshaw, P.S., Panagiotidou, P., Whelan, A., Gordon, S.V., Hewinson, R.G., Williamson, R.A., and Carr, M.D. (2002). Conclusive evidence that the major T-cell antigens of the *Mycobacterium tuberculosis* complex ESAT-6 and CFP-10 form a tight, 1:1 complex and characterization of the structural properties of ESAT-6, CFP-10, and the ESAT–6*CFP-10 complex. Implications for pathogenesis and virulence. J. Biol. Chem. *277*, 21598–21603.

Robichon, C., Vidal-Ingigliardi, D., and Pugsley, A.P. (2005). Depletion of apolipoprotein N-acyltransferase causes mislocalization of outer membrane lipoproteins in *Escherichia coli*. J. Biol. Chem. *280*, 974–983.

Rodrigue, A., Chanal, A., Beck, K., Müller, M., and Wu, L.F. (1999). Co-translocation of a periplasmic enzyme complex by a hitchhiker mechanism through the bacterial Tat pathway. J. Biol. Chem. *274*, 13223–13228.

Rose, R.W., Bruser, T., Kissinger, J.C., and Pohlschroder, M. (2002). Adaptation of protein secretion to extremely high-salt conditions by extensive use of the twin-arginine translocation pathway. Mol. Microbiol. *45*, 943–950.

Sääf, A., Andersson, H., Gafvelin, G., and von Heijne, G. (1995). SecA-dependence of the translocation of a large periplasmic loop in the *Escherichia coli* MalF inner membrane protein is a function of sequence context. Mol. Membr. Biol. *12*, 209–215.

Sargent, F., Bogsch, E.G., Stanley, N.R., Wexler, M., Robinson, C., Berks, B.C., and Palmer, T. (1998). Overlapping functions of components of a bacterial Sec-independent protein export pathway. EMBO J. *17*, 3640–3650.

Sargent, F., Stanley, N.R., Berks, B.C., and Palmer, T. (1999). Sec-independent protein translocation in *Escherichia coli*. A distinct and pivotal role for the TatB protein. J. Biol. Chem. *274*, 36073–36082.

Sartain, M.J., and Belisle, J.T. (2008). N-Terminal clustering of the O-glycosylation sites in the *Mycobacterium tuberculosis* lipoprotein SodC. Glycobiology *19*, 38–51.

Schaerlaekens, K., Schierova, M., Lammertyn, E., Geukens, N., Anne, J., and Van Mellaert, L. (2001). Twin-arginine translocation pathway in *Streptomyces lividans*. J. Bacteriol. *183*, 6727–6732.

Schaerlaekens, K., Van Mellaert, L., Lammertyn, E., Geukens, N., and Anne, J. (2004). The importance of the Tat-dependent protein secretion pathway in *Streptomyces* as revealed by phenotypic changes in tat deletion mutants and genome analysis. Microbiology *150*, 21–31.

Seluanov, A., and Bibi, E. (1997). FtsY, the prokaryotic signal recognition particle receptor homologue, is essential for biogenesis of membrane proteins. J. Biol. Chem. *272*, 2053–2055.

Shen, X.L., Dong, H.J., Hou, X.P., Guan, W.J., and Li, Y.Q. (2008). FtsY affects sporulation and antibiotic production by *whiH* in *Streptomyces coelicolor*. Curr. Microbiol. *56*, 61–65.

Stanley, N.R., Palmer, T., and Berks, B.C. (2000). The twin-arginine consensus motif of Tat signal peptides is involved in Sec-independent protein targeting in *Escherichia coli*. J. Biol. Chem. *275*, 11591–11596.

Stanley, S.A., Raghavan, S., Hwang, W.W., and Cox, J.S. (2003). Acute infection and macrophage subversion by *Mycobacterium tuberculosis* require a specialized secretion system. Proc. Natl. Acad. Sci. U.S.A. *100*, 13001–13006.

Strickler, J.E., Berka, T.R., Gorniak, J., Fornwald, J., Keys, R., Rowland, J.J., Rosenberg, M., and Taylor, D.P. (1992). Two novel *Streptomyces* protein protease inhibitors. Purification, activity, cloning, and expression. J. Biol. Chem. *267*, 3236–3241.

Szymanski, C.M., Yao, R., Ewing, C.P., Trust, T.J., and Guerry, P. (1999). Evidence for a system of general protein glycosylation in *Campylobacter jejuni*. Mol. Microbiol. *32*, 1022–1030.

Tarry, M.J., Schäfer, E., Chen, S., Buchanan, G., Greene, N.P., Lea, S.M., Palmer, T., Saibi, l H.R., and Berks, B.C. (2009). Structural analysis of substrate binding by the TatBC component of the twin-arginine protein transport system. Proc. Natl. Acad. Sci. U.S.A. *106*, 13284–13289.

Thomas, J.R., and Bolhuis, A. (2006). The *tatC* gene cluster is essential for viability in halophilic archaea. FEMS Microbiol. Lett. *256*, 44–49.

Thompson, S.J., Robinson, C., and Mant, A. (1999). Dual signal peptides mediate the signal recognition particle/Sec-independent insertion of a thylakoid membrane polyprotein, PsbY. J. Biol. Chem. *274*, 4059–4066.

Thompson, B.J., Widdick, D.A., Hicks, M.G., Chandra, G., Sutcliffe, I.C., Palmer, T. and Hutchings, M.I. (2010). Investigating lipoprotein biogenesis and function in the model Gram-positive bacterium *Streptomyces coelicolor*. Mol. Microbiol. *77*, 943–955.

Ton That, H., and Schneewind, O. (2004). Assembly of pili in Gram-positive bacteria. Trends Microbiol. *12*, 228–234.

Tsai, T.Y., and Lee, Y.H. (1998). Roles of copper ligands in the activation and secretion of *Streptomyces* tyrosinase. J. Biol. Chem. *273*, 19243–19250.

Tschumi A., Nai, C., Auchli, Y., Hunziker, P., Gehrig, P., Keller, P., Grau, T., and Sander, P. (2009). Identification of apolipoprotein n-acyltransferase (Lnt) in mycobacteria. J. Biol. Chem. *284*, 27146–27156.

Van den Berg, B., Clemons, Jr., W.M., Collinson, I., Modis, Y., Hartmann, E., Harrison, S.C., and Rapoport, T.A. (2004). X-ray structure of a protein-conducting channel. Nature *427*, 36–44.

von Heijne, G., and Abrahmsén, L. (1989). Species-specific variation in signal peptide design. Implications for protein secretion in foreign hosts. FEBS Lett. *244*, 439–446.

Wehmeier, S., Varghese, A.S., Gurcha, S.S., Tissot, B., Panico, M., Hitchen, P., Morris, H.R., Besra, G.S., Dell, A., and Smith, M.C.M. (2008). Glycosylation of the phosphate binding protein, PstS, in *Streptomyces coelicolor* by a pathway that resembles protein O-mannosylation in eukaryotes. Mol. Microbiol. *71*, 421–433.

Weiner, J., Bilous, P.T., Shaw, G.M., Lubitz, S.P., Frost, L., Thomas, G.H., Cole, J.A., and Turner, R.J. (1998). A novel and ubiquitous system for membrane targeting and secretion of cofactor-containing proteins. Cell *93*, 93–101.

Westermann, M., Pop, O.I., Gerlach, R., Appel, T.R., Schlörmann, W., Schreiber, S., and Müller, J.P. (2006). The TatAd component of the *Bacillus subtilis* twin-arginine protein transport system forms homo-multimeric complexes in its cytosolic and membrane embedded localisation. Biochim. Biophys. Acta *1758*, 443–451.

Westers, L., Westers, H., and Quax, W.J. (2004). *Bacillus subtilis* as cell factory for pharmaceutical proteins: a biotechnological approach to optimize the host organism. Biochim. Biophys. Acta *1694*, 299–310.

Widdick, D.A., Dilks, K., Chandra, G., Bottrill, A., Naldrett, M., Pohlschroder, M., and Palmer, T. (2006). The twin-arginine translocation pathway is a major route of protein export in *Streptomyces coelicolor*. Proc. Natl. Acad. Sci. U.S.A. *103*, 17927–17932.

Widdick, D.A., Eijlander, R.T., van Dijl, J.M., Kuipers, O.P., and Palmer, T. (2008). A facile reporter system for the experimental identification of twin-arginine translocation (Tat) signal peptides from all kingdoms of life. J. Mol. Biol. *375*, 595–603.

Willey, J.M., Willems, A., Kodani, S., and Nodwell, J.R. (2006). Morphogenetic surfactants and their role in the formation of aerial hyphae in *Streptomyces coelicolor*. Mol. Microbiol. *59*, 731–742.

Yahr, T.L., and Wickner, W.T. (2001). Functional reconstitution of bacterial Tat translocation in vitro. EMBO J. *20*, 2472–2479.

Yen, M.R., Tseng, Y.H., Nguyen, E.H., Wu, L.F., and Saier, Jr, M.H. . (2002). Sequence and phylogenetic analyses of the twin-arginine targeting (Tat) protein export system. Arch. Microbiol. *2177*, 441–450.

Yikmis, M., Arenskötter, M., Rose, K., Lange, N., Wernsmann, H., Wiefel, L., and Steinbüchel, A. (2008). Secretion and transcriptional regulation of the latex-clearing protein, Lcp, by the rubber-degrading bacterium *Streptomyces* sp. strain K30. Appl. Environ. Microbiol. *74*, 5373–5382.

Zuber, B., Chami, M., Houssin, C., Dubochet, J., Griffiths, G., and Daffé, M. (2008). Direct visualization of the outer membrane of native mycobacteria and corynebacteria. J. Bacteriol. *190*, 5672–5680.

Central Carbon Metabolic Pathways in *Streptomyces*

5

Geertje van Keulen, Jeroen Siebring and Lubbert Dijkhuizen

List of abbreviations

Glycolysis

Eno	enolase
Fba	fructose-1,6 bisphosphate aldolase
Gap	glyceraldehyde-3 phosphate dehydrogenase
Glk	glucosekinase
Pfk	6-phosphofructokinase
Pgi	phosphoglucose isomerase
Pgk	phosphoglycerate kinase
Pgm	phosphoglycerate mutase
Pyk	pyruvate kinase
Tpi	triosephosphate isomerase
1,3BGP	1,3-bisphophoglycerate
2PG	2-phosphoglycerate
3PG	3-phosphoglycerate
DHAP	dihydroxyacetone phosphate
F6P	fructose-6-phosphate
FBP	fructose-1,6-bisphosphate
G6P	glucose-6-phosphate
GAP	glyceraldehyde-3 phosphate
PEP	phosphoenolpyruvate

TCA cycle

Aco	aconitase
AKGdh	α-ketoglutarate dehydrogenase
Cs	citrate synthase
Fum	fumarase
Idh	isocitrate dehydrogenase
Mdh	malate dehydrogenase
Pdh	pyruvate dehydrogenase complex
PEPc	phosphoenolpyruvate carboxylase
Sdh	succinate dehydrogenase
Suc	succinyl CoA synthetase

Pentose phosphate pathway (PPP)

Ghd	6-phosphogluconate dehydrogenase
Pgl	6-phosphoglucono-δ-lactone
Rpe	ribose-5-phosphate epimerase
Rpi	ribose-5-phosphate isomerase
Tal	transaldolase
Tkt	transketolase
Zwf	glucose-6-phosphate dehydrogenase
6PGLU	6-phosphogluconate
E4P	erythrose-4-phosphate
Ri5P	ribose-5-phosphate
Ru5P	ribulose-5-phosphate
S7P	sedoheptulose-7 phosphate
X5P	xylulose-5-phosphate

Entner–Doudoroff pathway

KDPGal	2-keto-3-deoxy-6 phosphogluconate aldolase
Pgd	6-phosphogluconate dehydratase
KDPG	2-keto-3-deoxy-6 phosphogluconate

Gluconeogenesis

FBPase	fructose-1,6 bisphosphatase
G6Pase	glucose-6-phosphatase
PEPck	phosphoenolpyruvate carboxykinase
PEPs	phosphoenolpyruvate synthase
Pyc	pyruvate carboxylase

Abstract

Streptomyces and other actinomycetes are fascinating soil bacteria of major economic importance. They produce 70% of antibiotics known to man and numerous other pharmaceuticals for treatment of, for example, cancer, a range of infections, high cholesterol, or have immunosuppressive activity. It is not surprising that the multitude of gene clusters encoding for the biosynthesis of known and unknown secondary metabolites in genome sequences of a wide range of actinomycetes have received much attention in the last few years. In contrast, there is much less understanding of primary metabolism and its control in actinomycetes, despite its importance as supply pathways of precursors for secondary metabolite production. This review describes current information on the central carbon metabolic pathways in streptomycetes, focussing on glycolysis, pentose phosphate pathway, Entner–Doudoroff pathway, gluconeogenesis, and the source of phosphate for phosphorylation reactions. In addition, recent developments providing a greater insight into links with secondary metabolism in *Streptomyces* are reviewed.

Genomics of *Streptomyces*

S. coelicolor A3(2) is the best-studied streptomycete and a model organism for antibiotic production (Borodina *et al.*, 2008; Bentley *et al.*, 2002). In 2002 the complete sequence of the linear chromosome of *S. coelicolor* A3(2) was completed and became publicly accessible (Bentley *et al.*, 2002). In the last seven years, several other *Streptomyces* genome sequences or large parts thereof have been published [e.g. *S. avermitilis* (Ikeda *et al.*, 2003), *S. griseus* (Ohnishi *et al.*, 2008), *S. ambofaciens* (Choulet *et al.*, 2006a; Choulet *et al.*, 2006b)] or released online (e.g. *S. scabies* (Sanger Institute www.sanger.ac.uk/Projects/S_scabies), 17 further *Streptomyces* spp. including *S. albus*, *S. clavuligerus*, *S. hygroscopicus*, *S. lividans* TK24, *S. pristinaespiralis*, and *S. roseosporus* (Broad Institute, https://www.broadinstitute.org/science/data), which has enabled extensive comparative genomics studies (e.g. Ventura *et al.*, 2007; Chater and Chandra, 2006; Choulet *et al.*, 2006a; Chater and Horinouchi, 2003). From these studies it is clear that *Streptomyces* genomes in general have large numbers of genes involved in carbon uptake and utilization (see Table 5.1 for genes [putatively] involved in central carbon metabolism), and usually encode between 24 and 34 secondary metabolite gene clusters including many cryptic or orphan clusters (see also elsewhere in this book).

The yield of secondary metabolites during the production phase depends highly on the availability of biosynthetic building blocks, which are mostly drawn from primary carbon metabolic pathways (Olano *et al.*, 2008). Therefore, in depth knowledge of central carbon metabolic pathways, corresponding fluxes and key enzymes is critical for successful metabolic engineering of producer strains based on a rational approach.

Central carbohydrate metabolism in *Streptomyces*

Presence of multiple gene copies

Before discussing the present knowledge of primary metabolic pathways in streptomycetes it is important to mention a complicating factor. Multiple gene copies encoding similar or homologous gene products are a common feature of *Streptomyces* and other actinomycete genomes (Ohnishi *et al.*, 2008; Ventura *et al.*, 2007; Ikeda *et al.*, 2003; Bentley *et al.*, 2002). Table 5.1 shows an overview of central carbon metabolic enzymes and the respective gene copy number in *S. coelicolor* A3(2), *S. avermitilis*, and *S. griseus*. In bacteria, primary metabolic pathway genes are usually clustered in operons, enabling simple coordinated regulation of gene expression. Unusually, the multiple copies of central carbon metabolic genes of *Streptomyces* spp. are scattered over the core and arm regions of the chromosomes, often without displaying tight clustering in operons (Siebring *et al.*, 2010b), although some pathway genes exist that do show tight operon structure, e.g. the duplicated glycogen and trehalose operons (Schneider *et al.*, 2000) and the glycerol operon (Hindle and Smith, 1994). It is unclear how carbon metabolic gene expression at multiple loci is coordinated, adding further to the complexity of streptomycete carbon metabolism. Interestingly, extensive synteny can be found in homologous loci of central carbon metabolic genes in *S. coelicolor*, *S. avermitilis*, *S. griseus* and *S. scabies*, indicating a, yet unknown, functional relevance (Siebring *et al.*, 2010b).

The presence of multiple genes coding for a single catalytic step may become challenging

Table 5.1 Central carbon metabolic pathways and the copy numbers of respective genes annotated in the *S. coelicolor* A3(2) (SCO), *S. avermitilis* (SAV) and *S. griseus* (SGR) chromosomes, respectively

Enzyme	Full name	SCO[a]	SAV[a]	SGR[a]
Glycolysis				
Eno	Enolase	2	1	2
Fba	Fructose-1,6-bisphosphate aldolase	1	1	1
Gap	Glyceraldehyde-3-phosphate dehydrogenase	3	2	2
Glk	Glucosekinase	2	2	2
Pfk	6-Phosphofructokinase	3	3	2
Pgi	Phosphoglucose isomerase	2	2	1
Pgk	Phosphoglycerate kinase	1	1	1
Pgm	Phosphoglycerate mutase	3	2	5
Pyk	Pyruvate kinase	2	2	2
Tpi	Triosephosphate isomerase	2	1	1
TCA cycle				
Aco	Aconitase	1	1	1
Akgdh	α-Ketoglutarate dehydrogenase	2*	2*	2*
Cs	Citrate synthase	4	4	4
Fum	Fumarase	2	2	2
Idh	Isocitrate dehydrogenase	1	1	1
Mdh	Malate dehydrogenase	1	1	1
Pdh	Pyruvate dehydrogenase complex	1*	2–3*	2–3*
Pepc	Phosphoenolpyruvate carboxylase	1	1	1
Sdh	Succinate dehydrogenase	2*	2*	2*
Suc	Succinyl-CoA synthetase	2*	2*	1*
Pentose phosphate pathway				
Ghd	6-Phosphogluconate dehydrogenase	3	3	2
Pgl	6-Phosphoglucono-δ-lacton lactonase	1	1	1
Rpe	Ribose-5-phosphate epimerase	1	1	1
Rpi	Ribose-5-phosphate isomerase	3	2	1
Tal	Transaldolase	2	2	1
Tkt	Transketolase	3	4	3
Zwf	Glucose-6-phosphate dehydrogenase	2	2	1
Entner–Doudoroff pathway specific				
Kdpg (eda)	2-Keto-3-deoxy-6-phosphogluconate aldolase	4	1	1
Pgd (edd)	6-Phosphogluconate dehydratase	0	0	0
Gluconeogenesis specific				
Fbpase	Fructose-1,6-bisphosphatase	1	1	1
G6pase	Glucose-6-phosphatase	0	0	0
Pepck	Phosphoenolpyruvate carboxykinase	1	1	1
Peps	Phosphoenolpyruvate synthase	2	1	1
Pyc	Pyruvate carboxylase	1	0	0

The list was built using the KEGG database. [a]Number of (putative) gene copies annotated. *Number of annotated complexes.

for the integration of genome-scale metabolic models with metabolic flux analysis and/or gene expression data from, e.g. transcriptomics and proteomics. Similar challenges arise when a particular reaction in a pathway can be catalysed by enzymes using different cofactors. Functional redundancy of homologous genes and metabolic plasticity will need to be determined experimentally for each of these conditions.

Glucose uptake

ATP-binding cassette permeases

An *in silico* survey of ATP-binding cassette (ABC) permeases in *S. coelicolor* revealed a total of 45 solute-binding proteins, which are essential components for carbohydrate uptake via permeases (Bertram *et al.*, 2004). The putative ABC systems included proteins for transport of cellobiose and cellotriose, α-glucosides, lactose, maltose, maltodextrins, ribose, sugar alcohols, xylose, and β-xylosides (Bertram *et al.*, 2004). A novel putative bifunctional protein composed of a substrate-binding and a membrane-spanning moiety is likely to account for ribose or ribonucleoside uptake. The genome of *S. coelicolor* encodes at least 53 potential carbohydrate uptake systems when the permeases of the phosphotransferase systems are included in the analysis (Bertram *et al.*, 2004).

Phosphotransferase systems

In many bacteria the mechanism of transporting glucose into a cell is by use of PEP-dependent phosphotransferase systems (PTS), which exclusively uses PEP as the phosphoryl donor. The phosphoryl transfer chain involves the two general energy-coupling proteins, Enzyme I and HPr, as well as the sugar-specific, membrane-bound Enzyme II complexes. The Enzyme IIC permeases translocate the sugar substrates into the cytoplasmic compartment of the bacterial cell concomitant with phosphorylation (Saier Jr. and Reizer, 1994). *S. coelicolor* encodes a complete PTS, which include enzyme I, HPr, and enzyme IIA^{Crr}, encoded respectively by *ptsI*, *ptsH*, and *crr*, two genes encoding for fructose/mannitol-specific EII permeases, and four genes encoding EII permeases belonging to the glucose/sucrose family (Parche *et al.*, 2000). Unusually, *S. coelicolor* and *S. olivaceoviridis* have PTS systems that favour utilization of *N*-acetylglucosamine (Nothaft *et al.*, 2003a; Wang *et al.*, 2002). In fact, growth studies show that *S. coelicolor* prefers the C/N sources glutamate and *N*-acetylglucosamine over glucose (van Wezel *et al.*, 2009; Nothaft *et al.*, 2003a).

Glucose permease

Glucose-negative *S. coelicolor* strains that were resistant to the toxic analogue 2-deoxyglucose proved to have mutations in the gene encoding glucose kinase (Kwakman and Postma, 1994). Since no mutations were found in transporter-encoding genes in glucose-negative mutant strains, it became obvious that more than one glucose permease must exist (van Wezel *et al.*, 2007).

GlcP, coding for a major facilitator superfamily (MFS) type glucose permease, was found to be the major glucose transport system in *S. coelicolor* A3(2) (van Wezel *et al.*, 2005). Two copies of this gene were identified in the genome in two distinct loci. The two genes *glcP1* (*sco5578*) and *glcP2* (*sco7153*) remarkably differ in one base pair only, resulting in a silent mutation (van Wezel *et al.*, 2005). In contrast, *S. clavuligerus* cannot utilize glucose as sole carbon source (Garcia-Dominguez *et al.*, 1989), despite its genome coding for a functional *glcP* gene. *glcP* of *S. clavuligerus* could be functionally expressed in *S. coelicolor* from the $glcP1^{Sco}$ promoter, suggesting a lack of significant *glcP* expression in *S. clavuligerus* (Perez-Redondo *et al.*, 2010). Indeed, transcription was weak due to transcriptional regulation by formation of antisense RNA (Perez-Redondo *et al.*, 2010).

Glucose kinase and carbon catabolite repression

In unicellular bacteria such as *E. coli* and *B. subtilis,* PTS plays a dominant role in carbon catabolite repression (CCR) (Bruckner and Titgemeyer, 2002; Saier Jr and Reizer, 1994). CCR enables a cell to utilize different carbon sources in a preferential manner (van Wezel *et al.*, 2007; Titgemeyer and Hillen, 2002; Angell *et al.*, 1994). In many streptomycetes the production of extracellular hydrolytic enzymes, particularly those involved in carbohydrate degradation, is repressed by glucose (Butler *et al.*, 1999). The glucose permeases found in *S. coelicolor* are not PTS and are not involved in CCR. Instead, in *Streptomyces* glucose kinase plays a vital role in CCR by glucose (van Wezel *et al.*, 2007; Guzman *et al.*, 2005a,b; Bertram *et al.*, 2004; Nothaft *et al.*, 2003a,b; Titgemeyer and Hillen, 2002; Kwakman and Postma, 1994). Mutants lacking a functional *glkA* gene are unable to grow on glucose as the sole source of carbon and are deregulated in glucose repression of several CCR-controlled genes. Deregulated operons include those for the utilization of agar, glycerol, maltose, and galactose (Kwakman and Postma, 1994). Additionally, restoration of glucose kinase activity by introduction of a functional glucose kinase gene from *Zymomonas mobilis* failed to restore catabolite control (van Wezel *et al.*, 2007; Angell *et al.*, 1994).

The *S. coelicolor* genome encodes for two ATP-dependent glucose kinases (van Wezel *et al.*, 2007; Mahr *et al.*, 2000; Angell *et al.*, 1994). The major glucose kinase is encoded by *glcK*, while the normally inactive *glkII* can only be induced after a prolonged period of growth in the presence of glucose (van Wezel *et al.*, 2007; Angell *et al.*, 1994). Glucose kinase converts glucose into glucose-6-phosphate (G6P), which can then enter into different pathways. G6P is a substrate for glucose phosphate isomerase (glycolysis, Fig. 5.1), glucose-6-phosphate dehydrogenase [ED pathway (Fig. 5.2), pentose phosphate pathway, (Fig. 5.3)], and glucose-6-phosphatase. It can also be converted by phosphoglucomutase into G1P for subsequent usage in biosynthesis of galactose and glycogen (KEGG database, http://www.genome.jp/kegg). In conclusion, the G6P node is a key distribution point for carbon flux.

How GlcK functions as a catalytic and regulatory protein is still largely enigmatic. Western blot experiments showed that GlcK protein is produced constitutively in glucose- and mannitol-grown liquid cultures of *S. coelicolor*, whereas glucose kinase activity varied enormously dependent on carbon source and growth phases (van Wezel *et al.*, 2007). Cultures growing

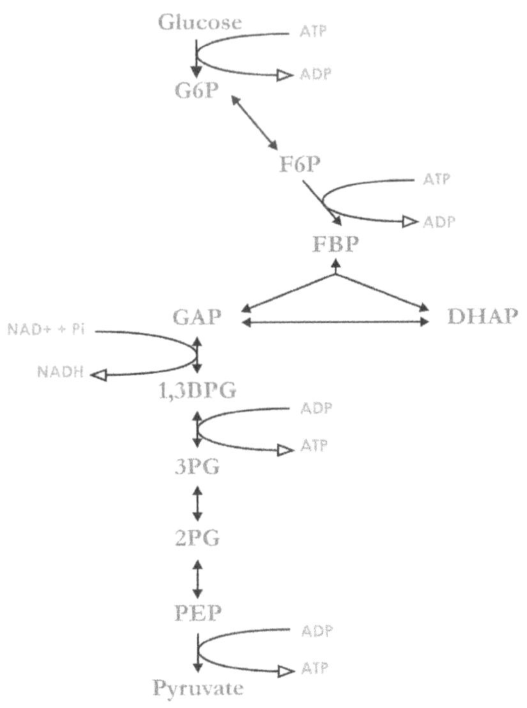

Figure 5.1 Schematic representation of glycolysis. The conversion of glucose into pyruvate is usually a sequence of 10 reactions.

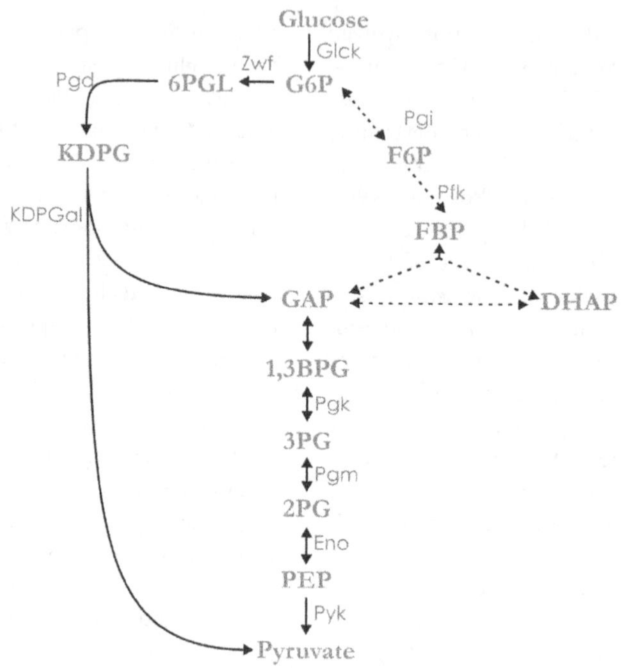

Figure 5.2 Schematic representation of the Entner–Doudoroff pathway (solid arrows) and glycolysis (non-shared reactions; dotted arrows).

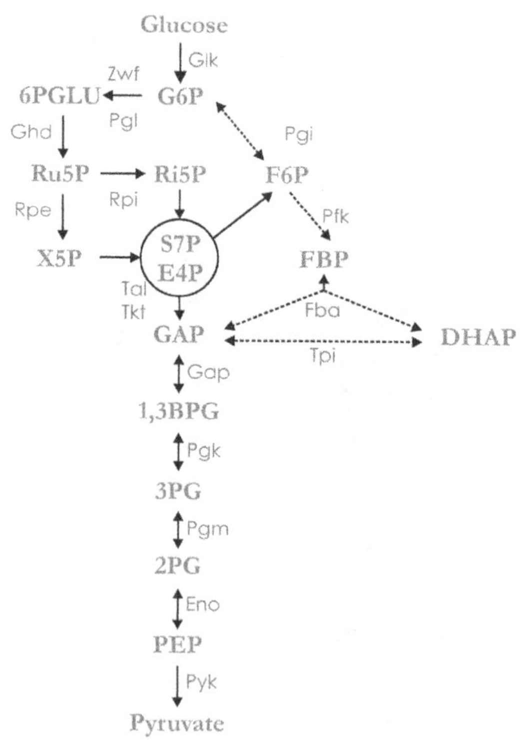

Figure 5.3 Schematic representation of the pentose phosphate pathway (solid arrows) and glycolysis (non-shared reactions; dotted arrows).

exponentially on glucose exhibited high GlcK activity, whereas mannitol-grown cultures show negligible GlcK activity. Glucose kinase activity correlated with the abundance of a faster migrating minor Glk isoform, suggesting either proteolytic processing or posttranslational modification of GlcK protein (van Wezel *et al.*, 2007). Cross-talk between GlcK and the general components of the PTS complement, with HPr as the most likely candidate, could, however, not be demonstrated (Mahr *et al.*, 2000). Interestingly, GlcK interacts with a high molecular weight-binding activity from cell extracts of glucose- and glycerol-grown cultures (Mahr *et al.*, 2000). Furthermore, GlcK must form a complex with GlcP for efficient uptake of glucose, while the free cytoplasmic GlcK responds to cellular levels of glucose, ATP, and other yet unknown metabolites (van Wezel *et al.*, 2007). In addition to GlcK–GlcP complex formation, the gene immediately upstream of *glcK*, *sco2127*, may also influence glucose uptake and CCR (Chavez *et al.*, 2009; van Wezel *et al.*, 2007; Guzman *et al.*, 2005a,b).

PTS, CCR and control of other cellular processes

Nutrient limitation normally triggers metabolic and morphological differentiation in *S. coelicolor*. A direct link between carbon utilisation, secondary metabolite production and development could be shown for null mutants of the global *pts* genes (*ptsH*, *ptsI*, and *crr*), which are developmentally arrested (van Wezel *et al.*, 2007; Rigali *et al.*, 2006; Nothaft *et al.*, 2003b). The PTS permease gene (*nagE2*) for *N*-acetylglucosamine, the monomer of chitin which is the second most abundant polymer in nature, cross-communicates with DasR, a pleiotropic regulator in *S. coelicolor* (Nothaft *et al.*, 2010; Rigali *et al.*, 2008; Rigali *et al.*, 2006). High extracellular concentrations of *N*-acetylglucosamine prevent metabolic and morphological differentiation of *S. coelicolor* under rich nutrient growth conditions, but activate these processes under poor growth conditions (van Wezel *et al.*, 2009; Rigali *et al.*, 2008; Rigali *et al.*, 2006). The glucosamine-6-phosphate-responsive regulator DasR controls these complex processes through an extensive regulon that includes *N*-acetylglucosamine transport and metabolism as well as antibiotic production and development (Nothaft *et al.*, 2010; Rigali *et al.*, 2008, reviewed in Chater *et al.*, 2010; van Wezel *et al.*, 2009). Interestingly, uptake and phosphorylation of *N*-acetylglucosamine by the PTS results in intracellular *N*-acetylglucosamine-6-phosphate, which can be quickly converted via GlcNAc-6 deacetylase and glucosamine-6-phosphate deaminase into fructose-6-phosphate (KEGG database), which links it in two steps to central carbon metabolic pathways.

Glycolysis in *Streptomyces*: phosphofructokinase

Presence and (putative) redundancy of multiple gene copies of an enzymatic activity in metabolism is a poorly understood phenomenon (Bentley *et al.*, 2002). It has been proposed that isoenzymes channel fluxes through different pathways that share enzyme activities (Moore *et al.*, 2002; Alves *et al.*, 2001; Ureta, 1991). For Pfk a few examples exist in support of this hypothesis. In the actinomycete *Amycolatopsis methanolica* two Pfk enzymes are present; a PPi-Pfk that is highly homologous to related actinomycete Pfks (Alves *et al.*, 1996) and an ATP-Pfk that is highly similar to spirochete Pfks and the result of horizontal gene transfer (Alves *et al.*, 2001). The ATP-Pfk was shown to be active during growth on C1 substrates such as methanol, functioning as part of the ribulose monophosphate (RuMP) cycle, whereas the PPi-Pfk fulfils a glycolytic role during growth on glucose (Alves *et al.*, 2001; Alves *et al.*, 1996). The idea of isoenzymes used in different pathways also appears to be upheld for the two *E. coli* Pfks Pfk-1 and Pfk-2, where Pfk-2 apparently prefers to channel F6P produced by the pentose phosphate pathway (PPP) (Fig. 5.3) (Cabrera *et al.*, 2008).

Knowledge of *pfk* expression in *S. coelicolor* and other streptomycetes is critical for understanding the specific functions of the multiple PfkA isoenzymes (see Table 5.1) in primary and secondary metabolism. The protein products of the three *S. coelicolor pfk* genes (*sco2119*, *sco5426*, and *sco1214*) are highly homologous with 56–72% identity and 69–82% similarity at the amino acid level (Siebring, 2010; Borodina *et al.*, 2008; Bapteste *et al.*, 2003; Bentley *et al.*, 2002; Siebers *et al.*, 1998). The high level of sequence similarity between *pfkA* genes has made gene expression studies extremely difficult due to extreme cross-hybridization. It proved impossible to distinguish the three *pfkA* mRNAs of *S. coelicolor* using Northern hybridization, microarray analysis or normal reverse transcription-PCR (RT-PCR) (Siebring, 2010; Borodina *et al.*, 2008). Adding to the confusion, the previously characterized ATP-dependent PfkA2 (Sco5426) enzyme activity in YEME-grown cells of *S. coelicolor* was misidentified as PfkA1 (Sco2119) due to high similarities of the PfkA N-termini (Borodina *et al.*, 2008; Alves *et al.*, 1997). *pfk* expression studies could only be performed successfully by applying a modified RNA isolation method (van Keulen *et al.*, 2004) coupled to RT-PCR followed by gene-specific restriction digestion (Siebring *et al.*, 2010b; Borodina *et al.*, 2008). The genome sequences of *S. avermitilis* and *S. scabies* also contain three highly homologous *pfk* genes, whereas *S. griseus* encodes only two *pfk* genes (Table 5.1). While *S. griseus* does not encode a *pfkA1* homologue, its corresponding gene region displays high levels of synteny with the *pfkA1* loci in the other three streptomycetes, suggesting the region has a conserved function in streptomycetes (Siebring *et al.*, 2010b). Interestingly, the *pfkA1* locus is better conserved then the *pfkA2* locus, even though PfkA2 is the major Pfk activity in *S. coelicolor* (Siebring *et al.*, 2010b; Borodina *et al.*, 2008).

By combining *pfkA* gene expression analyses with *pfkA* mutant analyses, it was shown that PfkA2 is the major Pfk activity in *S. coelicolor* with PfkA1 and PfkA3 functioning as minor, but non-redundant, PfkA activities (Siebring *et al.*, 2010b; Borodina *et al.*, 2008; Alves *et al.*, 1997). Though the three *pfkA* genes are differentially expressed in cells grown in liquid glucose medium in a growth phase-dependent manner, *pfkA2* is expressed under most of the conditions tested (Siebring *et al.*, 2010b; Hesketh *et al.*, 2002). Surprisingly, deletion of the *pfkA2* gene also resulted in loss of all measurable Pfk activity in cells grown in glucose-containing minimal media, despite the expression of *pfkA3* and a functional, though less active, glycolytic pathway as shown by metabolic flux analysis (Borodina *et al.*, 2008). As *pfkA1* and *pfkA3* expression is not induced when Pfk activity becomes limiting, it appears that PfkA1 and PfkA3 serve specific functions in *S. coelicolor*.

PfkA biochemistry and bioinformatics: ATP vs. PPi dependency

The PfkAs of *S. coelicolor* A3(2) belong to the Family A Group III Pfks (Siebring *et al.*, 2010a; Bapteste *et al.*, 2003). This group of Pfks has not been as extensively studied as the other Family A groups I and II (Bapteste *et al.*, 2003). Interestingly, some Group III Pfk enzymes display no or hardly any allosteric regulation and this group has therefore been suggested to represent the most primitive lineage of Pfk enzymes (Siebers *et al.*, 1998). Phylogenetic analysis of several PfkA sequences showed that most actinomycete PfkAs belong to the Group III PfkAs (Siebring *et al.*, 2010a). A few sequences were found outside of Group III and are considered to be the result of horizontal gene transfer events. Actinomycete species encoding these Group II PfkAs are multi-*pfk* bearing species (e.g. *A. methanolica, Propionibacterium, Frankia*) that also encode a Group III Pfk.

Measuring Pfk activity in crude extracts of several actinomycetes encoding a single Group III *pfkA* gene, resulted in detection of PPi-dependent Pfks in several species (*Salinospora tropica, Saccharopolyspora erythraea*, and *Thermobifida fusca*), while others produced an ATP-dependent Pfk (*Rhodococcus jostii* RHA1, and *Corynebacterium glutamicum*) (Siebring *et al.*, 2010a). This data was combined with Group III enzyme activities that had already been identified (*S. coelicolor* ATP-PfkA2 (Alves *et al.*, 1997), *A. methanolica* PPi-Pfk (Alves *et al.*, 1996), *Thermoproteus tenax* PPi-Pfk (Siebers *et al.*, 1998). The resulting PfkA profile of PPi- and ATP-dependency showed a pattern of scattered phosphoryl donor preference in the Group III PfkA branch, indicating that the phosphoryl donor dependency of Group III PfkAs has changed on a number of occasions during evolution (Siebring *et al.*, 2010a).

Comparison to biochemically studied PfkA enzymes of other phylogenetic Pfk groups (Meurice *et al.*, 2004; Bapteste *et al.*, 2003; Moore *et al.*, 2002; Siebers *et al.*, 1998), of which crystal structures are available (e.g. Shirakihara and

Evans, 1988), learned that most of the amino acids previously assigned as being important to phosphoryl binding are not conserved in Group III Pfks (Siebring *et al.*, 2010a). However, one particular amino acid, at position 104, is conserved in either ATP- or PPi-Pfks. In ATP-Pfks this position is occupied by glycine (G), whereas this position is occupied by an aspartic acid (D) in PPi-Pfks (Siebring *et al.*, 2010a). The sterical hindrance of the large aspartic acid residue would prevent the adenosine from ATP from entering the phosphoryl-binding pocket (Moore *et al.*, 2002; Chi and Kemp, 2000). This concept has been suggested and proven before by mutation of this amino acid in the *Entamoeba histolytica* PPi-Pfk (Chi and Kemp, 2000): the preference of phosphoryl donor changed from PPi to ATP upon mutating 104D into 104G. The preference for ATP was, however, over 10^4 times lower than for PPi. The authors suggested that this was because of specialization for PPi of the phosphoryl-binding pocket in this Group II enzyme. This suggestion was based on an observed increase in ATP preference by additional amino acid changes in this pocket. The fact that, apart from amino acid 104, no other previously assigned important amino acids appear to be highly conserved between the Group III ATP- and PPi-Pfks, suggests Group III Pfks to be less specialized. The lower degree of specialization allows higher catalytic flexibility that resulted in multiple phosphoryl donor dependency changes. Therefore, it is suggested that amino acid position 104 is the sole discriminator in these enzymes and by changing that amino acid, phosphoryl donor dependency changes. The change from glycine to aspartic acid or back can be facilitated by a single point mutation within the *pfkA* gene. This particular codon would allow actinomycetes to switch from ATP- to PPi-dependent enzymes or back during evolution (Siebring *et al.*, 2010a). Based on sequence data, we predict that *S. coelicolor* encodes one PPi-Pfk (PfkA1, 104D) in addition to two ATP-Pfks (PfkA2 and PfkA3, both 104G) enzymes (Siebring *et al.*, 2010a). *S. avermitilis* and *S. scabies* are predicted to have a similar complement of ATP- and PPi-dependent PfkAs as *S. coelicolor*, whereas *S. griseus* is predicted to have only two ATP-PfkAs and no PPi-PfkA (Siebring *et al.*, 2010a).

ATP- and PPi-Pfks are suggested to play different physiological roles (Siebring *et al.*, 2010b; Alves *et al.*, 2001). Due to the lower free energy of the phosphate bond in PPi, reverse reactions are allowed (Rao *et al.*, 2009; Costa *et al.*, 2003; Kornberg, 1995). This is in contrast to the virtually irreversible reaction catalysed by ATP-Pfks. Therefore, PPi-Pfks can also play a role in gluconeogenesis as demonstrated in *E. coli* (Kemp and Tripathi, 1993). Also, PPi-Pfks would lower the demand for ATP in a cell, which is advantageous in conditions of low ATP/ADP ratios. Low ATP/ADP ratios are more likely to occur during anaerobiosis, or in case of phosphorus deprivation (Mertens, 1993; Mertens, 1991). Anaerobiosis is frequently encountered during growth of *S. coelicolor* in static liquid cultures or inside a colony (van Keulen *et al.*, 2007; van Keulen *et al.*, 2003), where the older cells in the core of the colony are being shielded from oxygen by younger cells. Being able to switch between ATP- and PPi-dependent Pfks during anaerobic, microaerophilic and aerobic growth would provide enhanced metabolic flexibility.

Gluconeogenesis in *Streptomyces*

The Pfk-catalysed step can be bypassed in several ways, one of which is via the reverse, gluconeogenic reaction catalysed by FBPase (Berg *et al.*, 2007). In the absence of glycolytic carbon sources, *S. coelicolor* is able to grow on gluconeogenic carbon sources, e.g. asparagine and glutamine, as the sole source of carbon (Borodina *et al.*, 2005a). Indeed, the complete gluconeogenic pathway has been found encoded on the genomes of *S. coelicolor*, *S. avermitilis*, and *S. griseus*, which includes one copy each of *fbpase* (Ohnishi *et al.*, 2008; Ikeda *et al.*, 2003; Bentley *et al.*, 2002; Table 5.1).

In *S. coelicolor* anaplerotic enzymes are not only active in gluconeogenesis. When *S. coelicolor* is grown on glucose as sole carbon source, activity of phosphoenolpyruvate carboxylase (PEPc; Fig. 5.2) can be detected (Bramwell *et al.*, 1993). Under these conditions, this is the only anaplerotic activity that can be detected in *S. coelicolor*. Anaplerotic enzymes provide branch points between energy-producing and biosynthetic pathways and are therefore likely to be regulated at the switch from growth to production of secondary metabolites. Indeed, an increase in PEPc activity was found during actinorhodin production in *S. coelicolor* (Bramwell *et al.*, 1993). In addition, clavulanic acid production in *S. clavuligerus* appears dependent of anaplerotic reactions. It was proposed that carbon limitation restricted the capacity for anaplerotic metabolism, thereby minimising the potential for extensive TCA-cycle derived biosynthesis, which ultimately resulted in loss of clavulanic acid production (Kirk *et al.*, 2000).

Entner–Doudoroff pathway in actinomycetes

Several actinomycetes have active ED pathways, for example *Nonomuraea* and the streptomycete *S. tenebrarius* (Borodina *et al.*, 2005b; Gunnarsson *et al.*, 2004c). In fact, *Nonomuraea* sp. ATCC 39272 and *Mycobacterium smegmatis* predominantly metabolize glucose via the ED pathway (Gunnarsson *et al.*, 2004a,c).

S. tenebrarius is an industrially important microorganism, producing an antibiotics complex that mainly consists of the aminoglycosides apramycin, tobramycin carbamate, and kanamycin B carbamate. The fluxes through glycolysis and ED pathways in *S. tenebrarius* were almost evenly distributed during the exponential growth on glucose (Borodina *et al.*, 2005b). An active ED pathway was not observed in *S. coelicolor*, which was confirmed by the absence of the unique ED pathway gene, *pgd* (also known as *edd*, coding for 6-phosphogluconate dehydratase; Fig. 5.3) in the genome (Borodina *et al.*, 2005b; Gunnarsson *et al.*, 2004c). Further BLAST searches (Altschul *et al.*, 1997) show that the genomes of *S. avermitilis*, *S. scabies* and *S. griseus* also do not encode a *pgd* homologue. Interestingly, the genome sequences of *Sp. erythraea*, *R. jostii* RHA1, *R. opacus* B4, various *Frankia* spp., *Saccharomonospora viridis*, *Streptomyces sp.* AA4, *Tsukamurella paurometabola*, *Actinosynnema mirum* and *Nocardiopsis dassonvillei* contain high identity homologues (73–84%) of *pdg* of *S. tenebrarius*, suggesting that the use of ED pathway for metabolism of glucose may be more widespread in actinomycetes then originally considered.

Pentose phosphate pathway in *Streptomyces*

Actinomycetes generally metabolize glucose via glycolysis and PPP (Fig. 5.3) (Gunnarsson *et al.*, 2004c). In *S. lividans* the distribution of fluxes over glycolysis and PPP was affected by both the growth rate and the nature of the carbon- and energy source. Metabolic flux analysis showed that in both glucose- and gluconate-grown cells an increase in growth rate is accompanied by increased flux through PPP. Under conditions in which the growth rate was approximated zero, PPP flux was close to zero (Avignone-Rossa *et al.*, 2002). This trend was also observed in *S. coelicolor* (Obanye *et al.*, 1996). In batch fermenters with mineral phosphate-limited glucose medium, 7% of the imported glucose went into the oxidative branch of the PPP in *S. coelicolor* M145. No back-flux from the PPP into glycolysis was detected, from which it was concluded that the main role of the PPP was generation of biomass precursors (Borodina *et al.*, 2008; Obanye *et al.*, 1996). In PPP, Zwf catalyses the first step in the PPP, converting G6P to D-glucono-1,5-lactone 6-phosphate, producing one molecule of NADPH. Interestingly, in *S. coelicolor* the activity of G6PDh measured *in vitro* increased after deletion of either of the *zwf* genes, coding for Zwf (Bushell *et al.*, 2006; Fig. 5.4 and Table 5.1). When an extra copy of *actII-ORF4*, the pathway-specific activator of actinorhodin production, was introduced in either of the *zwf* mutant strains, Zwf2 activity was significantly higher than Zwf1 activity, suggesting that Zwf2 is the major activity in *S. coelicolor* (Ryu *et al.*, 2006). These findings suggest that the activity of Zwf can be modulated, which may be of use for metabolic engineering. In *S. antibioticus*, however, PPP is the dominant pathway during exponential growth (Butler *et al.*, 2002). This result was not unexpected at the time as Pfk activity could not be detected in this streptomycete. However, further analysis may be required because of the recent finding of an active glycolytic pathway without detection of a Pfk activity in the *S. coelicolor pfkA2* deletion strain (Borodina *et al.*, 2008).

Genome-scale metabolic modelling and metabolic flux analysis

Metabolic flux analysis using labelled precursor metabolites has been used extensively to characterize fluxes through primary and secondary metabolic pathways in several actinomycetes as part of strategies for improvement of producer strains (Borodina *et al.*, 2005b, 2008; Bushell *et al.*, 2006; Kim *et al.*, 2000, 2004; Gunnarsson *et al.*, 2004b,c; Bruheim *et al.*, 2002; Avignone-Rossa *et al.*, 2002; Naeimpoor and Mavituna, 2000). A limited number of metabolic reactions were incorporated into flux analyses before the availability of genomic sequence information. Genome sequence information is being used for genome-scale reconstruction of metabolic networks (GSMN), enabling prediction of a phenotype for a modified genotype and prediction of cellular behaviour under different physiological conditions, e.g. nutrient limitation (Durot *et al.*, 2009; Zamboni and Sauer, 2009; Borodina *et al.*, 2005a). Currently, three actinomycete genome sequences have been used for GSMN reconstruction: *S. coelicolor* (Borodina *et al.*, 2005a), *M. tuberculosis* H37Rv (Beste *et al.*, 2007; Jamshidi and Palsson, 2007) and *C. glutamicum* (Kjeldsen and Nielsen, 2009).

The GSMN of *S. coelicolor* consists of 500 metabolites based on 971 reactions, of which 700 are unique, with the remainder representing reactions catalysed by isoenzymes (Borodina *et al.*, 2005a). As the 711 genes that were included in the model are highly conserved between actinomycetes, the GSMN could also be used for modelling other streptomycetes, and was used for modelling of *M. tuberculosis* (Jamshidi and Palsson, 2007). Experimental data of growth of wild type and mutant strains of *S. coelicolor* on different carbon sources agreed in most cases with the GSMN predictions. In addition, annotation gaps in critical metabolic pathways, e.g. phospholipid and polyprenoid biosynthesis, were filled by adding missing reactions and re-analysing the genome sequence (Borodina *et al.*, 2005a).

Using wild type flux distribution as input, the GSMN was used to predict how fluxes would change with a decrease in flux through glycolysis based on limiting Pfk (Borodina *et al.*, 2008). Assuming that cells will undergo as few flux changes as possible to adjust to the new conditions and constraining secretion of compounds such as acetate, acetaldehyde, ethanol, formate, and proline, the model predicted pyruvate, actinorhodin, and calcium-dependent antibiotic as fermentation products (Borodina *et al.*, 2008). The decrease of flux through phosphofructokinase was predicted to correlate with decreased specific growth rate and an increased flux through the PPP. Indeed, the latter was observed experimentally in a *pfkA2* deletion strain of *S. coelicolor*, which also displayed increased production of pigmented antibiotics (Borodina *et al.*, 2008). An integrated 'Systems Biology' analysis of genome-wide transcriptomic data, metabolic flux analysis and the GSMN further revealed transcriptional changes in genes encoding redox co-factor-dependent enzymes as well as those encoding pentose phosphate pathway enzymes and enzymes involved in storage carbohydrate biosynthesis (Borodina *et al.*, 2008). The integrated approach thus revealed extensive remodelling of metabolism at the G6P node, and to a lesser extent, at the acetyl-CoA node (Borodina *et al.*, 2008). These findings show that the GSMN of *S. coelicolor* can be successfully used as a rational approach for metabolic engineering of central carbon metabolic pathways to improve secondary metabolite production.

Secondary metabolism

In primary metabolism, energy and reducing power is harnessed for direct support of synthesis of biomass. In contrast, secondary metabolism produces compounds that have no direct apparent function in cell growth or maintenance, except perhaps siderophores. Secondary metabolites may, however, display useful 'bio-activity' for the producing organism such as anti-microbial activity, or play a role in symbiosis or development (Chater *et al.*, 2010; Berdy, 2005; Challis and Hopwood, 2003; Hodgson, 2000). Secondary metabolites are typically produced after the phase of most active vegetative growth, before cells enter a dormant or reproductive stage. In *Streptomyces* the production of secondary metabolites coincides with, or slightly precedes, the development of aerial mycelium when surface-grown (Bibb, 2005; Challis and Hopwood, 2003). This point in both liquid and surface-grown cultures is marked by a metabolic reorganization, or metabolic shift, setting secondary metabolite production in motion (Nieselt *et al.*, 2010; Hesketh *et al.*, 2007, 2009; Ryu *et al.*, 2007).

The link between primary and secondary metabolism in *Streptomyces*

Precursor molecules for production of secondary metabolites are usually obtained from primary metabolism (Olano *et al.*, 2008; Gunnarsson *et al.*, 2004b; Fig. 5.5). It is therefore not surprising that changes in primary metabolism subsequently affect secondary metabolism (Borodina *et al.*, 2008; Olano *et al.*, 2008; Bibb, 2005; Gunnarsson *et al.*, 2004b; Butler *et al.*, 2002; Obanye *et al.*, 1996). E.g. clavulanic acid production by *S. clavuligerus* could be improved by deletion of *gap*1 (Fig. 5.4). This deletion resulted in an increased supply of GAP, one of the precursors required for clavulanic acid production (Li and Townsend, 2006). In *S. coelicolor*, malonyl-CoA is used as a precursor for actinorhodin biosynthesis. Overexpression

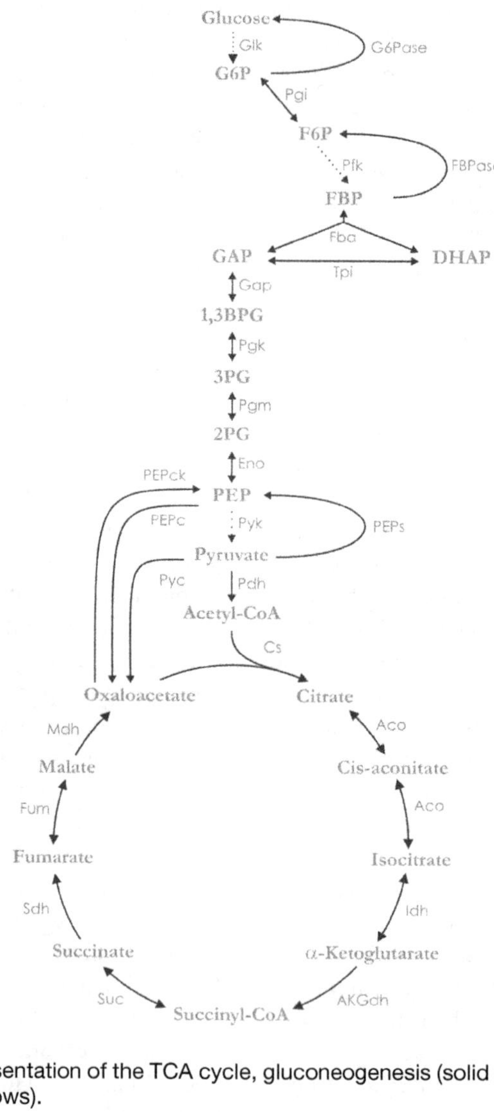

Figure 5.4 Schematic representation of the TCA cycle, gluconeogenesis (solid arrows), and glycolysis (non-shared reactions; dotted arrows).

, I., Scholler, C., Eliasson, A., and Nielsen, J. b). Metabolic network analysis of *Streptomyces arius*, a *Streptomyces* species with an active r–Doudoroff pathway. Appl. Environ. Microbiol. 94–2302.

, I., Siebring, J., Zhang, J., Smith, C.P., van n, G., Dijkhuizen, L., and Nielsen, J. (2008). iotic Overproduction in *Streptomyces coelicolor*) Mediated by Phosphofructokinase Deletion. J. Chem. *283*, 25186–25199.

, H., Nimmo, H.G., Hunter, I.S., and Coggins, 1993). Phosphoenolpyruvate carboxylase from *omyces coelicolor* A3(2): purification of the ne, cloning of the *ppc* gene and over-expression e protein in a streptomycete. J. Biochem. *293*, 136.

, R., and Titgemeyer, F. (2002). Carbon catabo- pression in bacteria: choice of the carbon source autoregulatory limitation of sugar utilization. S Microbiol. Lett. *209*, 141–148.

, P., Sletta, H., Bibb, M.J., White, J., and Levine, (2002). High-yield actinorhodin production l-batch culture by a *Streptomyces lividans* strain xpressing the pathway-specific activator gene *ORF4*. J. Ind. Microbiol. Biotechnol. *28*, 103–111.

M.E., Sequeira, S.I.P., Khannapho, C., Zhao, Chater, K.F., Butler, M.J., Kierzek, A.M., and one-Rossa, C.A. (2006). The use of genome metabolic flux variability analysis for process formulation based on an investigation of the s of the *zwf* mutation on antibiotic production *eptomyces coelicolor*. Enzyme Microb. Techn. *39*, –1353.

I.J., Bruheim, P., Jovetic, S., Marinelli, F., Postma, and Bibb, M.J. (2002). Engineering of primary n metabolism for improved antibiotic produc- n *Streptomyces lividans*. Appl. Environ. Microbiol. 731–4739.

M.J., Deutscher, J., Postma, P.W., Wilson, T.J., ier, A., and Bibb, M.J. (1999). Analysis of a ptsH logue from *Streptomyces coelicolor* A3(2). FEMS biol. Lett. *177*, 279–288.

R., Ambrosio, A.L., Garratt, R.C., Guixe, V., and l, J. (2008). Crystallographic structure of phos- uctokinase-2 from *Escherichia coli* in complex two ATP molecules. Implications for substrate ition. J. Mol. Biol. *383*, 588–602.

G.L., and Hopwood, D.A. (2003). Synergy contingency as driving forces for the evolution ultiple secondary metabolite production by tomyces species 4. Proc. Natl. Acad. Sci. U.S.A. 14555–14561.

K.F., Biro, S., Lee, K.J., Palmer, T., and Schrempf, 2010). The complex extracellular biology of *tomyces*. FEMS Microbiol. Rev. *34*, 171–198.

K.F., and Chandra, G. (2006). The evolution of opment in *Streptomyces* analysed by genome arisons. FEMS Microbiol. Rev. *30*, 651–672.

K.F., and Horinouchi, S. (2003). Signalling developmental events in two highly diverged *tomyces* species. Mol. Microbiol. *48*, 9–15.

Chavez, A., Garcia-Huante, Y., Ruiz, B., Langley, E., Rodriguez-Sanoja, R., and Sanchez, S. (2009). Cloning and expression of the *sco2127* gene from *Streptomyces coelicolor* M145. J. Ind. Microbiol. Biotechnol. *36*, 649–654.

Chi, A., and Kemp, R.G. (2000). The primordial high energy compound: ATP or inorganic pyrophosphate? J. Biol. Chem. *275*, 35677–35679.

Choulet, F., Aigle, B., Gallois, A., Mangenot, S., Gerbaud, C., Truong, C., Francou, F.X., Fourrier, C., Guerineau, M., Decaris, B., *et al.* (2006a). Evolution of the terminal regions of the *Streptomyces* linear chromosome. Mol. Biol. Evol. *23*, 2361–2369.

Choulet, F., Gallois, A., Aigle, B., Mangenot, S., Gerbaud, C., Truong, C., Francou, F.X., Borges, F., Fourrier, C., Guerineau, M., Decaris, B., *et al.* (2006b). Intraspecific variability of the terminal inverted repeats of the linear chromosome of *Streptomyces ambofaciens*. J. Bacteriol. *188*, 6599–6610.

Costa dos, S.A., Seixas da-Silva, W., de Meis, L., and Galina, A. (2003). Proton transport in maize tonoplasts supported by fructose-1, 6-bisphosphate cleavage. Pyrophosphate-dependent phosphofructokinase as a pyrophosphate-regenerating system. Plant Physiol. *133*, 885–892.

Cropp, T.A., Smogowicz, A.A., Hafner, E.W., Denoya, C.D., McArthur, H.A.I., and Reynolds, K.A. (2000). Fatty-acid biosynthesis in a branched-chain alpha-keto acid dehydrogenase mutant of *Streptomyces avermitilis*. Can. J. Microbiol. *46*, 506–514.

Durot, M., Bourguignon, P.Y., and Schachter, V. (2009). Genome-scale models of bacterial metabolism: reconstruction and applications. FEMS Microbiol. Rev. *33*, 164–190.

Firn, R.D., and Jones, C.G. (2003). Natural products – a simple model to explain chemical diversity. Nat. Prod. Rep. *20*, 382–391.

Garcia-Dominguez, M., Martin, J.F., and Liras, P. (1989). Characterization of sugar uptake in wild-type *Streptomyces clavuligerus*, which is impaired in glucose uptake, and in a glucose-utilizing mutant. J. Bacteriol. *171*, 6808–6814.

Ghorbel, S., Kormanec, J., Artus, A., and Virolle, M.J. (2006a). Transcriptional studies and regulatory interactions between the *phoR-phoP* operon and the *phoU*, *mtpA*, and *ppk* genes of *Streptomyces lividans* TK24. J. Bacteriol. *188*, 677–686.

Ghorbel, S., Smirnov, A., Chouayekh, H., Sperandio, B., Esnault, C., Kormanec, J., and Virolle, M.J. (2006b). Regulation of *ppk* expression and in vivo function of Ppk in *Streptomyces lividans* TK24. J. Bacteriol. *188*, 6269–6276.

Gunnarsson, N., Bruheim, P., and Nielsen, J. (2004a). Glucose metabolism in the antibiotic producing actinomycete *Nonomuraea* sp. ATCC 39727. Biotechnol. Bioeng. *88*, 652–663.

Gunnarsson, N., Eliasson, A., and Nielsen, J. (2004b). Control of fluxes towards antibiotics and the role of primary metabolism in production of antibiotics. Adv. Biochem. Eng. Biotechnol. *88*, 137–178.

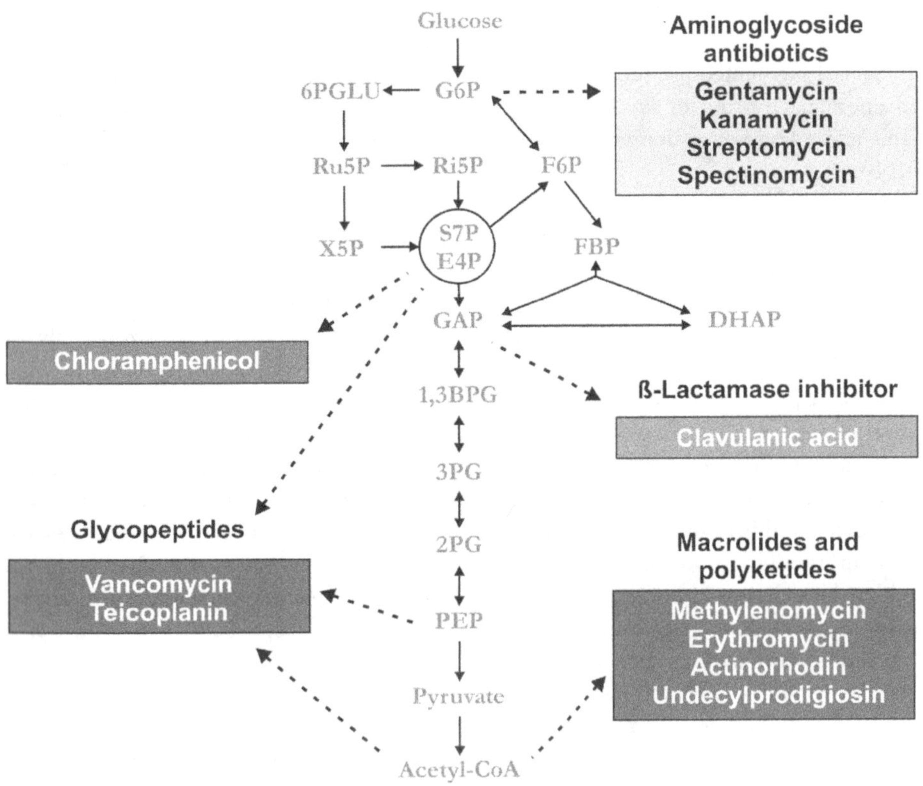

Figure 5.5 Relationship between primary metabolism and some precursors of actinomycete derived antibiotics. Modified after Alves (1997). Black arrows represent reactions of primary carbon metabolism. Dotted arrows represent pathways for several groups of secondary metabolites.

of *acc*, the malonyl-CoA-producing acetyl-CoA carboxylase encoding gene, improved the yield of actinorhodin produced by *S. coelicolor* (Ryu *et al.*, 2006). Deletion of important TCA cycle genes (citrate synthase, *citA*; aconitase, *acoA*; Table 5.1) also affected secondary metabolite production by *S. coelicolor* (Viollier *et al.*, 2001a,b). Furthermore, inactivation of *ccr*, the gene encoding crotonyl-CoA reductase, which provides the methylmalonyl-CoA precursor, in *S. cinnamonensis* resulted in 85% reduction of monensin production in oil-based fermentations (Li *et al.*, 2004).

Branched-chain amino acid metabolism provided at least 50% of the acetate incorporated into actinorhodin in a nutrient-dependent manner (Stirrett *et al.*, 2009). A *S. avermitilis* strain lacking branched-chain alpha-keto acid dehydrogenase (*bkd*) produced straight-chain fatty acids only via a fatty acid synthase with a very broad substrate specificity (Cropp *et al.*, 2000). In *S. lividans*, polyphosphate kinase (*ppk*) gene inactivation resulted in accumulation of polyphosphates and activation of actinorhodin production, which is normally silenced in this species (strain TK24) (Ghorbel *et al.*, 2006a; Ghorbel *et al.*, 2006b).

It has also been observed that the shift from growth phase to production phase in *S. coelicolor* is accompanied by a shift in flux ratio between PPP and glycolysis (Borodina *et al.*, 2008; Obanye *et al.*, 1996). Upon production of methylenomycin the relative flux through PPP increases. As described earlier, an increased flux through PPP was also observed in the *pfkA2* deletion strain of *S. coelicolor*, resulting in increased actinorhodin and undecylprodigiosin production. It was furthermore shown that NADPH levels were increased, which could further explain the increase in NADPH-dependent biosynthesis of pigmented antibiotics (Borodina *et al.*, 2008). Increased pigmented production was also observed upon inactivation of *pfkA1* or *pfkA3*, though this effect was not as dramatic (Siebring *et al.*, 2010b).

In *S. coelicolor* two copies of *zwf* genes (*zwf1* and *zwf2*) are found (Ryu *et al.*, 2006; Bentley *et al.*, 2002; Obanye *et al.*, 1996). Zwf catalyses the first step in the PPP, converting G6P to D-glucono-1,5-lactone 6-phosphate, producing one molecule of NADPH. Overall mycelial growth and actinorhodin production were enhanced by the deletion of *zwf2*, while glucose consumption was slightly delayed. The increase in actinorhodin production was, however, attributed to the increased growth of this mutant strain, while the specific actinorhodin production rate was decreased by the deletion of *zwf2*. In batch culture, glucose consumption, mycelium formation, and actinorhodin production were not greatly affected by the deletion of *zwf1* (Ryu *et al.*, 2006).

The results described for *S. coelicolor* differ from the results obtained from *S. lividans*, a species closely related to *S. coelicolor*. *S. lividans* strains used for these experiments were transformed with multicopy plasmids carrying the pathway-specific transcriptional activator genes for either the actinorhodin or undecylprodigiosin biosynthetic pathway [respectively pIJ86 (*actIII*-ORF4) and pIJ6014 (*redD*); Butler *et al.*, 2002; Avignone-Rossa *et al.*, 2002]. In *S. lividans* both *zwf* gene products contribute equally to the carbon flux through the initial PPP step. Blocking PPP activity by deletion of both *zwf* genes resulted in a marked reduction in antibiotic production but normal growth characteristics when compared with the parental producer strain. Single deletion of either *zwf* gene had no effect on mycelium growth, but resulted in an increased antibiotic production (2 to 4 times). The results suggested that, under the conditions used with *S. lividans* strains containing additional pathway-specific transcriptional activator genes, the production of actinorhodin and undecylprodigiosin is not limited by the supply of NADPH. The authors suggested that the increase in antibiotic production is due to increased carbon flow into glycolysis, resulting in more antibiotic precursors, e.g. malonyl-CoA (Butler *et al.*, 2002; Obanye *et al.*, 1996). So, in contrast, deletion of *zwf2* resulted in increased growth but a decreased specific actinorhodin production rate and increased methylenomycin production is accompanied by increased PPP flux in *S. coelicolor* (Ryu *et al.*, 2006; Obanye *et al.*, 1996).

Origins of secondary metabolite production

It has been proposed that actinorhodin or undecylprodigiosin are synthesized only when a surplus of precursor metabolites is present and that precursor availability may be a limiting factor in the synthesis of polyketides (Olano *et al.*, 2008; Avignone-Rossa *et al.*, 2002). This surplus of precursors might also imply a metabolic imbalance. Accumulation of primary metabolites, e.g. phosphorylated sugars, or a disturbed redox balance, is potentially harmful to an organism. Excess metabolites can be disposed of through the production of 'waste' products that have no immediate function (Vining, 1990). Early speculations were that this is possibly the original function of the relatively ancient secondary metabolic pathways, but this explanation failed to explain the bioactivity – a rare molecular property – of such products (Firn and Jones, 2003; Vining, 1992). Also, it fails to accommodate the employment of energetically costly pathways and the existence of relatively large gene clusters dedicated to producing such compounds (Firn and Jones, 2003; Vining, 1992; Vining, 1990). Finally, the phylogenetic clustering (as observed in the case of actinomycetes) of bioactive compound producers supports the suggestion of natural selection (Challis and Hopwood, 2003). For these reasons, secondary metabolites may have physiological functions to increase the fitness of the producing organism.

In streptomycetes, secondary metabolites are produced mostly after the transition phase, the result of nutrient exhaustion. This is accompanied by aerial mycelium formation, the breakdown of the vegetative hyphae, and the subsequent release of nutrients. Streptomycete antibiotics are proposed to defend the food source when other soil microorganisms threaten it (Challis and Hopwood, 2003). However, a single rationale is not adequate for explaining the diversity of chemical structures and biological activities

of secondary metabolites (Chater *et al.*, 2010; Challis and Hopwood, 2003; Vining, 1992).

Concluding remarks

With the onset of functional genomics, major advances have been made towards understanding cellular metabolism of *Streptomyces* spp. It has become clear through genome-scale metabolic modelling and flux analysis coupled to high-resolution 'omics' experiments on wild type and mutant genotype cultures, that a major metabolic shift occurs in central carbon metabolic pathways to accommodate secondary metabolism. Through past and ongoing experiments based on random mutagenesis, personal interest, model prediction and serendipity, it only gradually becomes apparent what the role is of each of the multiple copies of primary metabolic genes that are generally found in *Streptomyces*. Further experiments are required to distinguish the roles of multiple-copy genes, in particular under unusual laboratory growth conditions or, for example, in relation to interactions of *Streptomyces* with other organisms, e.g. fungi, plants or animals. To increase the success rate in future metabolic engineering approaches for improved production of bioactive compounds based on rational design better insights are required into how *Streptomyces* controls expression of the multiple-copy central carbon metabolic genes. This knowledge would further improve the predictive power of genome-scale metabolic models, which, in turn, would strengthen current modelling approaches of whole organisms, including modelling of intricate developmental and behaviour patterns.

Acknowledgements

The authors wish to thank Paul Dyson for giving us the opportunity to contribute to this book and for his patience. This work was supported by the EU FP6 ActinoGEN project (Contract LSHM-CT-2004–005224) to JS/LD, the HEFCW Reconfiguration and Collaboration Fund, and by a Marie Curie IEF Postdoctoral fellowship to GVK (MEIF-CT-2004–506056).

References

Altschul, S.F., Madden, T.L., Schaffer, A.A., Zhang, J., Zhang, Z., Miller, W., and Lipman, D.J. (1997). Gapped BLAST and P[…] protein database searc[…] 25, 3389–3402.

Alves, A.M. (1997). Reg[…] in the actinomycetes […] *Streptomyces coelicolor* […] of Groningen, the Net[…]

Alves, A.M., Euverink, G[…] L. (1997). Identifica[…] phofructokinase as a […] pathway of the actin[…] A3(2). Appl. Environ. […]

Alves, A.M., Euverink, G[…] L. (2001). Different […] PP(i)-dependent phos[…] the methylotrophic ac[…] *anolica*. J. Bacteriol. *18*[…]

Alves, A.M., Meijer, W.G., […] L. (1996). Character[…] *pfp* gene of *Amycolato*[…] dependent phosphofr[…] 149–155.

Angell, S., Lewis, C.G., […] (1994). Glucose repre[…] A3(2): a likely regulat[…] Gen. Genet. *244*, 135–[…]

Avignone-Rossa, C.A., W[…] P.W., Bibb, M., and Te[…] Carbon flux distributi[…] mostat cultures of *Strep*[…] 138–150.

Bapteste, E., Moreira, D[…] Rampant horizontal ge[…] change in the evolutio[…] Gene *318*, 185–191.

Bentley, S.D., Chater, K[…] Challis, G.L., Thomso[…] D.E., Quail, M.A., Kies[…] Complete genome se[…] mycete *Streptomyces c*[…] 141–147.

Berdy, J. (2005). Bioact[…] Antibiot. (Tokyo) *58*, 1[…]

Berg, J.M., Tymoczko, J[…] Biochemistry (New Y[…] Company).

Bertram, R., Schlicht, M., […] Jr., M.H., and Titgeme[…] transcriptional analysis […] tems of *Streptomyces coe*[…] 1362–1373.

Beste, D.J.V., Hooper, T., Ste[…] Rossa, C., Bushell, M., W[…] A.M., and McFadden, J[…] based genome scale net[…] *tuberculosis* metabolism. […]

Bibb, M.J. (2005). Regulat[…] in streptomycetes. Curr. […]

Borodina, I., Krabben, P[…] Genome-scale analysis o[…] metabolism. Genome R[…]

Gunnarsson, N., Mortensen, U.H., Sosio, M., and Nielsen, J. (2004c). Identification of the Entner–Doudoroff pathway in an antibiotic-producing actinomycete species. Mol. Microbiol. *52*, 895–902.

Guzman, S., Carmona, A., Escalante, L., Imriskova, I., Lopez, R., Rodriguez-Sanoja, R., Ruiz, B., Servin-Gonzalez, L., Sanchez, S., and Langley, E. (2005a). Pleiotropic effect of the *SCO$_2$127* gene on the glucose uptake, glucose kinase activity and carbon catabolite repression in *Streptomyces peucetius* var. *caesius*. Microbiology *151*, 1717–1723.

Guzman, S., Ramos, I., Moreno, E., Ruiz, B., Rodriguez-Sanoja, R., Escalante, L., Langley, E., and Sanchez, S. (2005b). Sugar uptake and sensitivity to carbon catabolite regulation in *Streptomyces peucetius* var. *caesius*. Appl. Microbiol. Biotechnol. *69*, 200–206.

Hesketh, A., Chen, W.J., Ryding, J., Chang, S., and Bibb, M. (2007). The global role of ppGpp synthesis in morphological differentiation and antibiotic production in *Streptomyces coelicolor* A3(2). Genome Biol. *8*, R161.

Hesketh, A., Kock, H., Mootien, S., and Bibb, M. (2009). The role of *absC*, a novel regulatory gene for secondary metabolism, in zinc-dependent antibiotic production in *Streptomyces coelicolor* A3(2). Mol. Microbiol. *74*, 1427–1444.

Hesketh, A.R., Chandra, G., Shaw, A.D., Rowland, J.J., Kell, D.B., Bibb, M.J., and Chater, K.F. (2002). Primary and secondary metabolism, and post-translational protein modifications, as portrayed by proteomic analysis of *Streptomyces coelicolor*. Mol. Microbiol. *46*, 917–932.

Hindle, Z., and Smith, C.P. (1994). Substrate induction and catabolite repression of the *Streptomyces coelicolor* glycerol operon are mediated through the GylR protein. Mol. Microbiol. *12*, 737–745.

Hodgson, D.A. (2000). Primary metabolism and its control in streptomycetes: a most unusual group of bacteria. Adv. Microb. Physiol. *42*, 47–238.

Ikeda, H., Ishikawa, J., Hanamoto, A., Shinose, M., Kikuchi, H., Shiba, T., Sakaki, Y., Hattori, M., and Omura, S. (2003). Complete genome sequence and comparative analysis of the industrial microorganism *Streptomyces avermitilis*. Nat. Biotechnol. *21*, 526–531.

Jamshidi, N., and Palsson, B.O. (2007). Investigating the metabolic capabilities of *Mycobacterium tuberculosis* H37Rv using the in silico strain iNJ661 and proposing alternative drug targets. BMC Syst. Biol. *1*, 26.

Kemp, R.G., and Tripathi, R.L. (1993). Pyrophosphate-dependent phosphofructo-1-kinase complements fructose 1, 6-bisphosphatase but not phosphofructokinase deficiency in *Escherichia coli*. J. Bacteriol. *175*, 5723–5724.

Kim, H.B., Smith, C.P., Micklefield, J., and Mavituna, F. (2004). Metabolic flux analysis for calcium dependent antibiotic (CDA) production in *Streptomyces coelicolor*. Metab. Eng. *6*, 313–325.

Kirk, S., Avignone-Rossa, C.A., and Bushell, M.E. (2000). Growth limiting substrate affects antibiotic production and associated metabolic fluxes in *Streptomyces clavuligerus*. Biotechnol. Letters 22, 1803–1809.

Kjeldsen, K.R., and Nielsen, J. (2009). In silico genome-scale reconstruction and validation of the *Corynebacterium glutamicum* metabolic network. Biotechnol. Bioeng. *102*, 583–597.

Kornberg, A. (1995). Inorganic polyphosphate: toward making a forgotten polymer unforgettable. J. Bacteriol. *177*, 491–496.

Kwakman, J.H., and Postma, P.W. (1994). Glucose kinase has a regulatory role in carbon catabolite repression in *Streptomyces coelicolor*. J. Bacteriol. *176*, 2694–2698.

Li, C.X., Florova, G., Akopiants, K., and Reynolds, K.A. (2004). Crotonyl-coenzyme A reductase provides methylmalonyl-CoA precursors for monensin biosynthesis by *Streptomyces cinnamonensis* in an oil-based extended fermentation. Microbiology *150*, 3463–3472.

Li, R., and Townsend, C.A. (2006). Rational strain improvement for enhanced clavulanic acid production by genetic engineering of the glycolytic pathway in *Streptomyces clavuligerus*. Metab. Eng. *8*, 240–252.

Mahr, K., van Wezel, G.P., Svensson, C., Krengel, U., Bibb, M.J., and Titgemeyer, F. (2000). Glucose kinase of *Streptomyces coelicolor* A3(2): large-scale purification and biochemical analysis. Antonie Van Leeuwenhoek *78*, 253–261.

Mertens, E. (1991). Pyrophosphate-dependent phosphofructokinase, an anaerobic glycolytic enzyme? FEBS Lett. *285*, 1–5.

Mertens, E. (1993). ATP versus pyrophosphate: glycolysis revisited in parasitic protists. Parasitol. Today *9*, 122–126.

Meurice, G., Deborde, C., Jacob, D., Falentin, H., Boyaval, P., and Dimova, D. (2004). In silico exploration of the fructose-6-phosphate phosphorylation step in glycolysis: genomic evidence of the coexistence of an atypical ATP-dependent along with a PPi-dependent phosphofructokinase in *Propionibacterium freudenreichii subsp. shermanii*. In Silico Biol. *4*, 517–528.

Moore, S.A., Ronimus, R.S., Roberson, R.S., and Morgan, H.W. (2002). The structure of a pyrophosphate-dependent phosphofructokinase from the Lyme disease spirochete *Borrelia burgdorferi*. Structure (Camb.) *10*, 659–671.

Naeimpoor, F., and Mavituna, F. (2000). Metabolic flux analysis in *Streptomyces coelicolor* under various nutrient limitations. Metab. Eng. *2*, 140–148.

Nieselt, K., Battke, F., Herbig, A., Bruheim, P., Wentzel, A., Jakobsen, O., Sletta, H., Alam, M., Merlo, M., Moore, *et al.* (2010). The dynamic architecture of the metabolic switch in *Streptomyces coelicolor*. BMC Genomics *11*, 10.

Nothaft, H., Dresel, D., Willimek, A., Mahr, K., Niederweis, M., and Titgemeyer, F. (2003a). The phosphotransferase system of *Streptomyces coelicolor* is biased for N-acetylglucosamine metabolism. J. Bacteriol. *185*, 7019–7023.

Nothaft, H., Parche, S., Kamionka, A., and Titgemeyer, F. (2003b). In vivo analysis of HPr reveals a fructose-specific phosphotransferase system that confers high-affinity uptake in *Streptomyces coelicolor*. J. Bacteriol. *185*, 929–937.

Nothaft, H., Rigali, S., Boomsma, B., Swiatek, M., McDowall, K.J., van Wezel, G.P., and Titgemeyer, F. (2010). The permease gene *nagE2* is the key to N-acetylglucosamine sensing and utilization in *Streptomyces coelicolor* and is subject to multi-level control. Mol. Microbiol. *75*, 1133–1144.

Obanye, A.I.C., Hobbs, G., Gardner, D.C.J., and Oliver, S.G. (1996). Correlation between carbon flux through the pentose phosphate pathway and production of the antibiotic methylenomycin in *Streptomyces coelicolor* A3(2). Microbiology *142*, 133–137.

Ohnishi, Y., Ishikawa, J., Hara, H., Suzuki, H., Ikenoya, M., Ikeda, H., Yamashita, A., Hattori, M., and Horinouchi, S. (2008). Genome sequence of the streptomycin-producing microorganism *Streptomyces griseus* IFO 13350. J. Bacteriol. *190*, 4050–4060.

Olano, C., Lombo, F., Mendez, C., and Salas, J.A. (2008). Improving production of bioactive secondary metabolites in actinomycetes by metabolic engineering. Metab. Eng. *10*, 281–292.

Parche, S., Nothaft, H., Kamionka, A., and Titgemeyer, F. (2000). Sugar uptake and utilisation in *Streptomyces coelicolor*: a PTS view to the genome. Antonie Van Leeuwenhoek *78*, 243–251.

Perez-Redondo, R., Santamarta, I., Bovenberg, R., Martin, J.F., and Liras, P. (2010). The enigmatic lack of glucose utilization in *Streptomyces clavuligerus* is due to inefficient expression of the glucose permease gene. Microbiology *156*, 1527–1537, DOI 10.1099/mic.0.035840-0.

Rao, N.N., Gomez-Garcia, M.R., and Kornberg, A. (2009). Inorganic Polyphosphate: Essential for Growth and Survival. Annu. Rev. Biochem. *78*, 605–647.

Rigali, S., Nothaft, H., Noens, E.E., Schlicht, M., Colson, S., Muller, M., Joris, B., Koerten, H.K., Hopwood, D.A., Titgemeyer, F., *et al.* (2006). The sugar phosphotransferase system of *Streptomyces coelicolor* is regulated by the GntR-family regulator DasR and links *N*-acetylglucosamine metabolism to the control of development. Mol. Microbiol. *61*, 1237–1251.

Rigali, S., Titgemeyer, F., Barends, S., Mulder, S., Thomae, A.W., Hopwood, D.A., and van Wezel, G.P. (2008). Feast or famine: the global regulator DasR links nutrient stress to antibiotic production by *Streptomyces*. EMBO Reports *9*, 670–675.

Ryu, Y.G., Butler, M.J., Chater, K.F., and Lee, K.J. (2006). Engineering of primary carbohydrate metabolism for increased production of actinorhodin in *Streptomyces coelicolor*. Appl. Environ. Microbiol. *72*, 7132–7139.

Ryu, Y.G., Kim, E.S., Kim, D.W., Kim, S.K., and Lee, K.J. (2007). Differential stringent responses of *Streptomyces coelicolor* M600 to starvation of specific nutrients. J. Microbiol. Biotechnol. *17*, 305–312.

Saier Jr., M.H., and Reizer, J. (1994). The bacterial phosphotransferase system: new frontiers 30 years later. Mol. Microbiol. *13*, 755–764.

Schneider, D., Bruton, C.J., and Chater, K.F. (2000). Duplicated gene clusters suggest an interplay of glycogen and trehalose metabolism during sequential stages of aerial mycelium development in *Streptomyces coelicolor* A3(2). Mol. Gen. Genet. *263*, 543–553.

Shirakihara, Y., and Evans, P.R. (1988). Crystal structure of the complex of phosphofructokinase from *Escherichia coli* with its reaction products. J. Mol. Biol. *204*, 973–994.

Siebers, B., Klenk, H.P., and Hensel, R. (1998). PPi-dependent phosphofructokinase from *Thermoproteus tenax*, an archaeal descendant of an ancient line in phosphofructokinase evolution. J. Bacteriol. *180*, 2137–2143.

Siebring, J. (2010). The phosphofructokinases of *Streptomyces coelicolor*. PhD thesis (University of Groningen, the Netherlands).

Siebring, J., van Keulen, G., Karpinski, P., and Dijkhuizen, L. (2010a). Evolution of phosphofructokinase enzymes in actinomycetes: Occurrence of multiple phosphoryl donor switches. In The phosphofructokinases of *Streptomyces coelicolor*, Siebring, J., PhD thesis (University of Groningen, the Netherlands), pp. 71–84.

Siebring, J., Weber, S., Karpinski, P., Wallin, S., Dijkhuizen, L., and van Keulen, G. (2010b). Differential expression of the highly identical *pfkA* genes in *Streptomyces coelicolor* A3(2). In The phosphofructokinases of *Streptomyces coelicolor*, Siebring, J., PhD thesis (University of Groningen, the Netherlands), pp. 87–106.

Stirrett, K., Denoya, C., and Westpheling, J. (2009). Branched-chain amino acid catabolism provides precursors for the Type II polyketide antibiotic, actinorhodin, via pathways that are nutrient dependent. J. Indust. Microbiol. Biotechnol. *36*, 129–137.

Titgemeyer, F., and Hillen, W. (2002). Global control of sugar metabolism: a gram-positive solution. Antonie Van Leeuwenhoek *82*, 59–71.

Ureta, T. (1991). The role of isozymes in metabolite channelling. J. Theor. Biol. *152*, 81–84.

van Keulen, G., Alderson, J., White, J., and Sawers, R.G. (2007). The obligate aerobic actinomycete *Streptomyces coelicolor* A3(2) survives extended periods of anaerobic stress. Environ. Microbiol. *9*, 3143–3149.

van Keulen, G., Jonkers, H.M., Claessen, D., Dijkhuizen, L., and Wosten, H.A. (2003). Differentiation and anaerobiosis in standing liquid cultures of *Streptomyces coelicolor*. J. Bacteriol. *185*, 1455–1458.

van Keulen, G., Siebring, J., Rembacz, K.P., Hoogeveen, M., Tomczynska, M., and Dijkhuizen, L. (2004). Improved method for the isolation of RNA from (standing liquid cultures of) Streptomycetes. J. Microbiol. Methods *58*, 139–142.

van Wezel, G.P., Konig, M., Mahr, K., Nothaft, H., Thomae, A.W., Bibb, M., and Titgemeyer, F. (2007). A new piece of an old jigsaw: glucose kinase is activated posttranslationally in a glucose transport-dependent manner in *Streptomyces coelicolor* A3(2). J. Mol. Microbiol. Biotechnol. *12*, 67–74.

van Wezel, G.P., Mahr, K., Konig, M., Traag, B.A., Pimentel-Schmitt, E.F., Willimek, A., and Titgemeyer, F. (2005). GlcP constitutes the major glucose uptake system of *Streptomyces coelicolor* A3(2). Mol. Microbiol. *55*, 624–636.

van Wezel, G.P., McKenzie, N.L., and Nodwell, J.R. (2009). Applying the genetics of secondary metabolism in model actinomycetes to the discovery of new antibiotics. Methods Enzymol. *458*, 117–141.

Ventura, M., Canchaya, C., Tauch, A., Chandra, G., Fitzgerald, G.F., Chater, K.F., and Van Sinderen, D. (2007). Genomics of Actinobacteria: tracing the evolutionary history of an ancient phylum. Microbiol. Mol. Biol. Rev. *71*, 495–548.

Vining, L.C. (1990). Functions of secondary metabolites. Annu. Rev. Microbiol. *44*, 395–427.

Vining, L.C. (1992). Secondary metabolism, inventive evolution and biochemical diversity – a review. Gene *115*, 135–140.

Viollier, P.H., Minas, W., Dale, G.E., Folcher, M., and Thompson, C.J. (2001a). Role of acid metabolism in *Streptomyces coelicolor* morphological differentiation and antibiotic biosynthesis. J. Bacteriol. *183*, 3184–3192.

Viollier, P.H., Nguyen, K.T., Minas, W., Folcher, M., Dale, G.E., and Thompson, C.J. (2001b). Roles of aconitase in growth, metabolism, and morphological differentiation of *Streptomyces coelicolor*. J. Bacteriol. *183*, 3193–3203.

Waksman, S. (1940). On the classification of actinomycetes. J. Bacteriol. *39*, 549–558.

Wang, F., Xiao, X., Saito, A., and Schrempf, H. (2002). *Streptomyces olivaceoviridis* possesses a phosphotransferase system that mediates specific, phosphoenolpyruvate-dependent uptake of *N*-acetylglucosamine. Mol. Genet. Gen. *268*, 344–351.

Zamboni, N., and Sauer, U. (2009). Novel biological insights through metabolomics and ^{13}C-flux analysis. Curr. Opin. Microbiol. *12*, 553–558.

Regulation of Nitrogen Assimilation in Streptomycetes and Other Actinobacteria

6

Wolfgang Wohlleben, Yvonne Mast and Jens Reuther

Abstract

Streptomycetes, as most bacteria, possess two ways to assimilate ammonium. Whereas the glutamate dehydrogenase is active under high nitrogen supply, the glutamine synthetase-glutamate synthase (GS-GOGAT) pathway is required under nitrogen limiting conditions. The major GS activity is mediated by the typical prokaryotic-type GSI enzyme. A second GS activity is conveyed by the GSII enzyme, belonging to the eukaryotic-type GS. This enzyme contributes to the overall GS activity mainly under N-limitations and during stationary growth. Three further genes (*glnA1, glnA2, glnA3*) encoding GSI-like enzymes to which no function could yet be assigned are located in all genomes of *Streptomyces* sequenced so far. The activity of GSI is post-translationally regulated by an adenylyltransferase (GlnE), which modifies the enzyme in response to nitrogen availability. In contrast to enteric bacteria, the P_{II} protein and its modifying enzyme, the adenylyltransferase GlnD, are not involved in the control of GlnE. The central transcriptional regulator GlnR which can act as activator and repressor, respectively, controls the expression of various genes involved in nitrogen supply, such as genes encoding a urease and an ammonium transporter, as well as genes of the ammonium assimilation pathways. GlnR itself can be multiply modified probably reflecting different nitrogen conditions.

Introduction

Streptomycetes require, as all organisms, nitrogen for their vegetative growth. About 50 mg of nitrogen are necessary to synthesize 1 g of cell mass. In addition, in the stationary growth phase nitrogen is needed for the synthesis of many secondary metabolites. Under standard growth conditions in the soil streptomycetes face severe nutrient limitations in a competitive environment (Hodgson, 2000). In particular nitrogen is one of the bioelements which is only poorly accessible in the natural environment.

On the other hand, if nitrogen-containing compounds become available for non-motile streptomycetes, they have to react promptly to rapid changes. Therefore, streptomycetes require complex machinery regulating nitrogen metabolism under different growth conditions as well as a number of versatile enzymes for the assimilation of nitrogen.

The preferred nitrogen source for most bacteria is ammonium. Ammonium is introduced into metabolism either by glutamate dehydrogenase (GDH) or glutamine synthetase/glutamate synthase (GS/GOGAT; Fig. 6.1). Whereas GDH (due to its relatively high K_m-value) is used under conditions of high nitrogen supply, GS is active under N-limiting conditions.

Nitrogen assimilation in *Streptomyces coelicolor* A3(2)

Glutamine synthetase

Glutamine synthetases are key enzymes in nitrogen metabolism (Paolo and Cohen, 1973). They have two essential functions: ammonium assimilation and glutamine synthesis (Tyler, 1978). A long time it was thought that

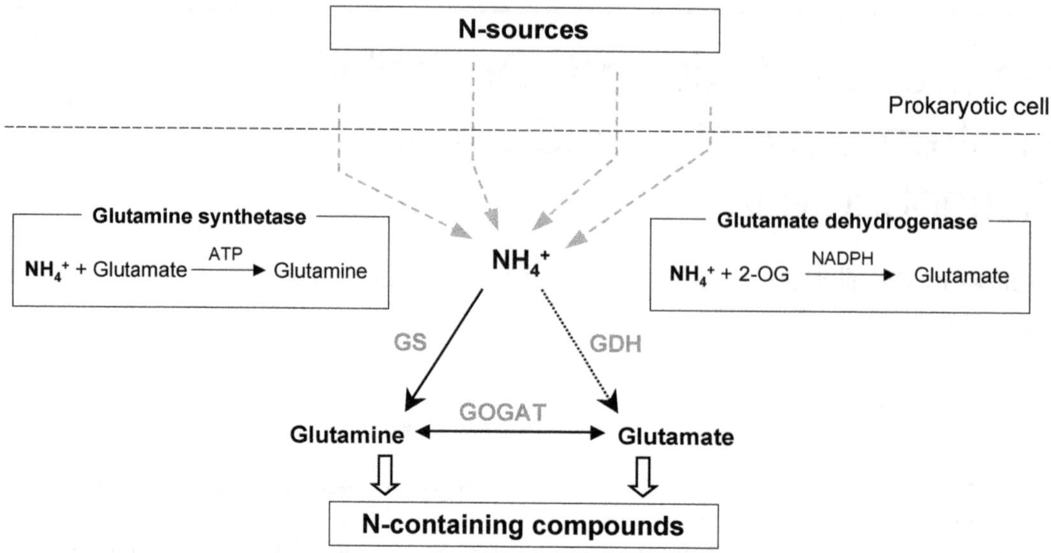

Figure 6.1 Nitrogen assimilation in prokaryotes. Different N-containing compounds can either diffuse or actively be transported into bacterial cells and are subsequently converted to ammonium (dashed arrows). Ammonium assimilation is mediated either by the glutamine synthetase (at low ammonium concentrations) or by the glutamate dehydrogenase (at high ammonium concentrations). Glutamate and glutamine then can serve as major amino donors for all cellular components. Abbreviations: GS: Glutamine synthetase; GDH: Glutamate dehydrogenase; GOGAT, glutamate synthase; OG, oxoglutarate.

prokaryotes and eukaryotes synthesize different GSs (Kumada *et al.*, 1993): GSI by bacteria and GSII by eukaryotes. However, in the course of the analysis of the biosynthesis of the GS-inhibitor phosphinothricyl-alanyl-alanine (= phosphinothricin tripeptide = PTT = bialaphos) in the producing *Streptomyces viridochromogenes* Tü494 and *Streptomyces hygroscopicus*, respectively, a GS gene (*glnII*) was isolated the gene product of which resembled significantly plant GSII enzymes (Behrmann *et al.*, 1990; Kumada *et al.*, 1990; Hillemann *et al.*, 1993). This gene was originally identified as PTT resistance gene (Behrmann *et al.*, 1990; Wohlleben *et al.*, 1991). Subsequent analysis revealed that the GSII enzyme itself is not phosphinothricin resistant but over-expression of the gene resulting in an increased amount of GS enzyme (the target of PTT) conferred PTT-resistance (Behrmann *et al.*, 1990). The occurrence of *glnII* genes is not limited to PTT producers, but it is a common feature for all streptomycetes (Behrmann *et al.*, 1990; Weiβschuh *et al.*, 2000). In the related actinomycetes, *Amycolatopsis*, *Mycobacterium* and *Corynebacterium*, however, *glnII* genes have not been detected (Reuther and Wohlleben, 2007).

In addition to the eukaryotic-like GSII all streptomycetes have a further, prokaryotic-like GSI enzyme encoded by a *glnA* gene (Wray and Fisher, 1988; Hillemann *et al.*, 1993) which turned out to be the major GS in streptomycetes as in almost all bacteria (see below). The two enzymes can be distinguished by their heat stability and the number of subunits composing the catalytically active holoenzyme: Whereas GSI is still active after heat treatment (60°C), GSII is heat labile. GSI consists of 12 identical subunits, in contrast to *S. coelicolor* GSII which is a decamer as shown by crystallographic studies (B. Schreier, unpublished).

Bacterial *glnII* genes were first identified in nitrogen-fixing bacteria like *Rhizobium*, *Agrobacterium* and *Frankia* which live in symbiotic interaction with plants. This was the reason why Carlson and Chelm (1986) speculated that *glnII* may have been acquired by these bacteria by horizontal gene transfer. However, the occurrence of *glnII* in streptomycetes as well as detailed

Table 6.1 Occurrence of glutamine synthetase genes in actinobacteria

	Prokaryotic-type GS				Eukaryotic-type GS
	glnA	*glnA2*	*glnA3*	*glnA4*	*glnII*
Amycolatopsis balhimycina	•	•	•	•	
Arthrobacter aurescens	•	•		•	
Bifidobacterium longum	•	•			
Brevibacterium linens	•	•		•	
Corynebacterium glutamicum	•	•			
Frankia alni	•	•	•	•	•
Mycobacterium tuberculosis	•	•	•	•	
Nocardia farcinica	•	•			
Propionibacterium acnes	•	•			
Rhodococcus jostii	•	•	•	•	
Streptomyces coelicolor	•	•	•	•	•
Occurrence (%) in 23 actinobacterial genera	100	96	26	52	9

phylogenetic analysis suggested that an ancient gene duplication event which may have occurred about 3500 million years ago is the reason for the presence of two evolutionary distinct GS genes in some bacteria (Kumada *et al.*, 1993).

Surprisingly, genome sequencing projects revealed (e.g. Bentley *et al.*, 1992; Ikeda *et al.*, 2003) that streptomycetes harbour three further *glnA*-type genes (*glnA2, glnA3, glnA4*), counterparts of which are also found in mycobacteria (Harth *et al.*, 2005) and in some other actinobacteria (Table 6.1). Using sequence analysis the gene products of *glnA2, glnA3 and glnA4*, respectively, can be categorized into the GSIα-group to which also the GS of *Bacillus subtilis* belongs. In contrast, the *glnA* gene product is a member of the GSIβ-group, like most of the GS enzymes of Gram-negative bacteria.

Functional analysis of all five *gln*-like genes in expression experiments using a glutamine auxotrophic *Escherichia coli* strain revealed that only *glnA* and *glnII* (after fusion to the *lac* promoter) can complement an *E. coli glnA* mutant (Rexer *et al.*, 2006) demonstrating that *glnA* and *glnII*, but not *glnA2, glnA3* and *glnA4* encode functional GS enzymes in streptomycetes. In mycobacteria, GlnA1 (which corresponds to the *Streptomyces* GlnA) is essential for growth, while the other copies (GlnA2, GlnA3, and GlnA4) are not (Harth *et al.*, 2005) indicating that none of the other copies can synthesize glutamine. However, after recombinant over-expression, GlnA3 and GlnA4 were found to catalyse the synthesis of L-glutamine, whereas GlnA2 synthesizes mainly D-glutamine and D-isoglutamine (Harth *et al.*, 2005). From sequence alignments Hayward *et al.* (2009) speculated that *glnA1* and *glnA2* were inherited from an *Actinobacteria* ancestor, whereas the *glnA3* and *glnA4* genes were sequentially acquired during *Actinobacteria* speciation.

In *S. coelicolor* the two GS-activities can replace each other since both a *glnA* as well as a *glnII* mutant can grow on minimal medium with an indistinguishable phenotype compared with the parental M145 strain. Only inactivation of both genes resulted in a glutamine auxotrophic phenotype (Grimpo, unpublished). Enzyme assays after growth of *S. coelicolor* in liquid medium under different media conditions revealed that GSI is the major GS activity: In *S. viridochromogenes* GSII contributes to about 25% (Hillemann *et al.*, 1993), in *S. coelicolor* about 5% (Fink *et al.*, 2002) of the total GS activity during vegetative growth. Relevant GSII activity can be observed at late growth phases and under nitrogen limiting conditions. This is in agreement with experiments demonstrating the presence of the GSII protein only after about 96h of growth. This observation was confirmed by reporter gene analysis in which fusions of the *glnII* promoter with the *egfp* gene resulted in detectable fluorescence only after N-limitation (Fink *et al.*, 2002).

It remains still unclear why streptomycetes contain – in addition to the prokaryotic-type GlnA – the second, eukaryotic-like GlnII. It was first speculated that GlnII may have a special function in secondary metabolite production since it was only found in those members of the *Actinobacteria* which are known to produce such compounds. However, investigations on the ability of *Streptomyces glnII* mutants to produce antibiotics led to conflicting results. But the speculations were confirmed by the observation that the genomes of corynebacteria (Burkovski, 2007) and related bacteria, which are not known to synthesize secondary metabolites, do not harbour *glnII* genes (Table 6.1). But thorough analysis of additional antibiotic producers like *Amycolatopsis mediterranei* (Peng *et al.*, 2006) and *Amycolatopsis balhimycina* (Wohlleben *et al.*, unpublished) revealed that these bacteria also are lacking a *glnII* gene. So the specific function of GlnII in streptomycetes has still to be elucidated.

Post-translational regulation

The nitrogen regulation (Ntr) system in enteric bacteria is used as a paradigm for the adaptation to nutrient supply in many textbooks (Fig. 6.2A). Genetic and biochemical analyses revealed that the GS is modified by adenylylation (and thereby inactivated) in response to high nitrogen availability (Jiang and Ninfa, 2009) by the adenylyltransferase GlnE. GlnE itself is controlled by the signalling protein P_{II} (GlnK) (Atkinson and Ninfa, 1998). The adenylylating and deadenylylating activity of GlnE is depending on the state of uridylylation of P_{II} which is mediated by the uridylyltransferase GlnD. Finally, N-availability as well as the C/N-ratio in the cell determines the level of P_{II}-modification (for details see Merrick and Edwards, 1995).

In the *S. coelicolor* genome, all genes for the relevant components of the Ntr-system are present: *glnD, glnE, glnK.* Whereas *glnD* and *glnK* form an operon together with *amtB* (encoding a putative ammonium transporter), *glnE* is located somewhere else on the chromosome (Bentley *et al.*, 2002).

Modification of GSI by adenylylation in response to high N-level was first shown by

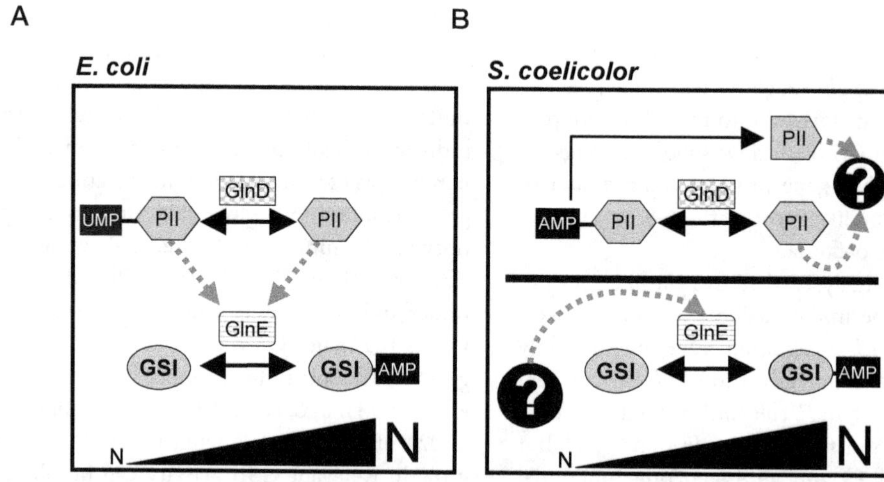

Figure 6.2 Comparison of post-translational regulation of GS in *E. coli* (A) and *S. coelicolor* (B). The GlnE activity in *E. coli* is controlled by the P_{II} protein in dependence of its uridylylation status (A). The signal which controls GlnE activity in *Streptomyces* is unknown. PII modification in *Streptomyces* includes adenylylation and N-terminal truncation; the targets of modified and unmodified P_{II} are unknown (B). GlnD: P_{II}-modifying enzyme (adenylylation in *S. coelicolor*, uridylylation in *E. coli*); GlnE, adenylyltransferase which modifies GS.

biochemical approaches for *S. cattleya* (Streicher and Tyler, 1981), *S. clavuligerus* (Bascarán *et al.*, 1989), *S. coelicolor* (Fisher and Wray, 1989), and *S. viridochromogenes* (Hillemann *et al.*, 1993). Subsequent genetic analysis could assign the adenylylating and deadenylylating activity to the gene product of *glnE* (Fink *et al.*, 1999, 2002). Investigations into possible regulation of GSII by adenylylation after ammonium shock revealed that treatment of GSII with snake venom phosphodiesterase (SVPDE) which is known to cleave the AMP moiety (Stadtman *et al.*, 1970) did not result in a change of GSII activity (Hillemann *et al.*, 1993). This is a clear indication that GSII, in contrast to GSI, is not subjected to post-translational modification by adenylylation.

All other GSI-like proteins (GlnA2, GlnA3, GlnA4) belong to the GSIα-subgroup (like *B. subtilis* GS) lacking the specific residue which is adenylylated in the GSI enzymes of Gram-negative bacteria and in GlnA of streptomycetes, and are therefore probably not prone to modification by adenylylation.

The GlnE-mediated down-regulation of GSI is probably required for short-term protection of the intracellular glutamate pool under conditions of sudden ammonium excess. This regulatory mechanism is widely distributed among bacteria. In enteric bacteria GlnE itself is controlled by GlnK (=P_{II}), GlnD and GlnB (a GlnK homologue) (Atkinson and Ninfa, 1998), although the latter is missing in streptomycetes. Surprisingly, modification of the P_{II}-protein by a GlnD-homologue does not occur by uridylylation like in *E. coli* but by adenylylation (Hesketh *et al.*, 2002). Furthermore, a specific degradation of P_{II} (removal of the three N-terminal amino acids) was observed in response to an ammonium shock (Hesketh *et al.*, 2002).

The most intriguing difference between the Ntr system in enteric bacteria and that of streptomycetes is the way these components interact: in *S. coelicolor* GlnE activity is completely uncoupled to the P_{II} signalling system: Modification of GSI by GlnE occurs in a *glnK* as well as in a *glnD* mutant in the same way as in the wild type, suggesting that GlnE requires a yet unidentified signal transmitter which responds to the N-status of the cell. On the other hand,one has to postulate alternative targets to which the PII/GlnD system transfers its signals (Fig. 6.2).

Also, *in silico* analysis of GSII did not provide any hint for the presence of possible residues for adenylylation. However, after N-upshift a rapid degradation of GSII was observed (Grimpo, unpublished). Whether this degradation is caused by an accelerated protein turnover or whether a controlled proteolytic process is responsible for the GSII breakdown has still to be investigated.

Transcriptional control

GlnR-mediated control of genes of the nitrogen metabolism

The key components of nitrogen metabolism in many bacteria are regulated at the transcriptional level: The NtrB/NtrC two component system in *E. coli* (Merrick and Edwards, 1995), NtcA in Cyanobacteria (Vega-Palas *et al.*, 1992), AmtR in Corynebacteria (Jakoby *et al.*, 2000; Burkowski, 2007), and TnrA/GlnR in *B. subtilis* (Schreier *et al.*, 1989; Wray *et al.*, 1996) are directly or indirectly involved in transcriptional regulation.

In *S. coelicolor* GlnR was first identified by mutational analysis as a transcriptional activator of the *glnA* gene (Wray *et al.*, 1991). GlnR belongs to the OmpR-family of response regulators; the corresponding sensor kinase could not yet be identified. Based on primary evidence that GlnR controls transcription not only of *glnA* but also of *glnII* and the *amtB-glnK-glnD* operon (Fink *et al.*, 2002) it was possible to identify a putative GlnR-binding box by sequence alignments. This box was more precisely defined by DNAse-footprint experiments (Tiffert *et al.*, 2008) to comprise the following sequence: gTnAc('a-site')-n_6-GaAAc('b-site')-n_6-GtnAC('a-site')-n_6-GAAAc('b-site')-n_6. This sequence was then used to screen the *S. coelicolor* genome for the presence of further GlnR-binding boxes. In total, 13 genes were identified containing such a GlnR-binding box and their functionality was proven by band shift assays (Tiffert *et al.*, 2008). RT-PCR experiments with the wild type and a *glnR* mutant confirmed that GlnR indeed controls the transcription of these 13 genes. Interestingly, GlnR does not only act as an activator of genes, the gene products of which

Gene	Protein
glnA	Glutamine synthetase I
glnII	Glutamine synthetase II
amtB	Ammonium transporter
glnK	PII protein
glnD	Adenylyl transferase
nirB	Nitrite reductase
ureA	Urease gamma subunit
gdhA	NADP-specific glutamate dehydrogenase

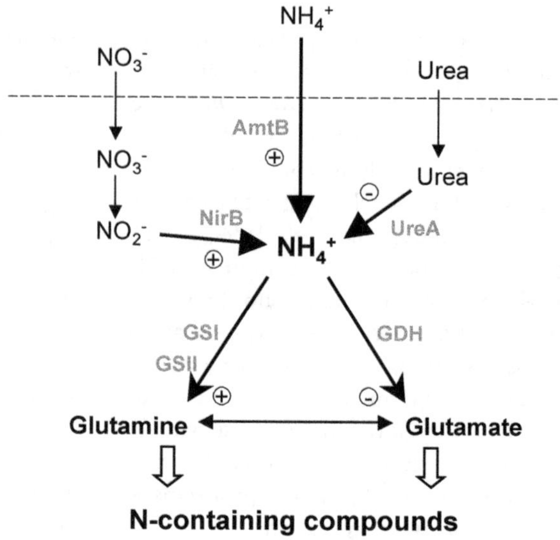

Figure 6.3 GlnR-regulated components of N-metabolism in *S. coelicolor.* At least transcription of eight genes involved in nitrogen metabolism is regulated by binding to their GlnR-box (left). Transcription is either activated (+) or repressed (–) in response to low nitrogen supply.

are required for growth under N-limiting conditions, but it also represses the transcription of *gdhA* which encodes an enzyme only active under N surplus (Fig. 6.3).

Recent analysis revealed that GlnR can also bind to a *cis*-element, similar but different to the previously described binding box (Wang and Zhao, 2009). This *cis*-element only contained the two 'a-sites' (see above) with a slightly different spacing. It was identified in front of the newly reported *nasA* (SCO2473) gene encoding an assimilatory nitrate reductase gene (Wang and Zhao, 2009).

Originally the *glnR* mutant was characterized as glutamine auxotrophic (Wray *et al.*, 1991) since it could not grow on minimal medium with nitrate as sole nitrogen source. This led to the conclusion that GlnR controls the expression of the GS gene(s) (Fisher, 1992). This is in contradiction to RT-PCR experiments in which a basal transcription of *glnA* and *glnII* in a *glnR* mutant was demonstrated (Tiffert *et al.*, 2008). Comparison of growth on different nitrogen sources finally revealed that the growth deficiency of a *glnR* mutant is due to its inability to reduce nitrite to ammonium (Tiffert *et al.*, 2008). Using the Griess-Ilosvay assay accumulation of nitrite was observed for a *glnR* mutant suggesting that nitrate can be taken up and reduced to nitrite by a *glnR* mutant. Further reduction is then blocked due to the lack of NirB (nitrite reductase large subunit) whose expression strongly depends on GlnR.

The control of *nirB* as well as of *amtB* transcription by GlnR suggested that not only functions of ammonium assimilation but also of ammonium supply are governed by this central regulator (Fig. 6.4).

Regulation of GlnR

Although GlnR is an orphan response regulator, phosphorylation of GlnR could be demonstrated. Overexpression of *glnR* in *S. coelicolor* using the expression vector pGM190 (Wohlleben *et al.*, 2009) and subsequent 2D-page analysis led to the identification of seven GlnR spots, probably representing differently phosphorylated GlnR variants (Tiffert, unpublished). Repeating these experiments with *S. coelicolor* wild type by including either enrichment steps for phosphorylated proteins or phosphatase treatment revealed the occurrence of GlnR as well as mono-, di-, and triphosphorylated GlnR.

The expression of *glnR* itself depends on the nitrogen status of the cell as shown by using a reporter system: *glnR* transcription is strongly repressed on plates containing ammonium chloride or glutamine as nitrogen source at concentrations of 1 mM or higher (van Wezel *et al.*, 2000). This observation was confirmed by the detection of different *glnR* mRNA levels when cultivated on nitrogen-rich and nitrogen-limited media, respectively (Tiffert *et al.*, 2008).

Three putative promoters were identified in front of *glnR* by primer extension experiments (Wray and Fisher, 1993). S1 nuclease protection analysis (Fink *et al.*, 2002) revealed one constitutive and two possible regulated promoters. No binding site of GlnR in its own promoter region could be detected.

It is unclear how the N-status of the cell is sensed and how the signal is transferred to end up with different *glnR* expression levels. In the same way it remains to be elucidated how different phosphorylation levels are attained and how they mediate variation in the expression of N-metabolism genes.

Recent analysis from Yang *et al.* (2009) suggested that a regulator, NdgR, might be involved in this signalling and ScbR might be one of the signal transmitters. ScbR was originally characterized to influence predominantly genes involved in secondary metabolism (Takano *et al.*, 2001). Since microarray analysis of an *scbR* mutant did not give any indication that genes involved in nitrogen metabolism are controlled by ScbR (Takano *et al.*, 2005), further studies are required to elucidate a putative role of ScbR in nitrogen regulation.

Regulation of nitrogen metabolism has not only to take into account the N-status of the cell, but it also has to reflect the C- and P- status of the bacteria. The cross-talk of C- and N- metabolism in many bacteria probably occurs via the PII:GlnD system which finally controls GS activity in response to glutamine and oxoglutarate availability in many bacteria (e.g. Jiang *et al.*, 1998; Jonsson and Nordlund, 2007; Kamberov *et al.*, 1994). Since this system does not regulate GS activity in streptomycetes (see above) one has to postulate an alternative control mechanism which couples C- and N-regulation in streptomycetes. Clear evidence that phosphate control interacts with nitrogen control was obtained by analysing binding of the phosphate regulator PhoP to upstream regions of genes involved in nitrogen metabolism (Rodriguez-Garcia *et al.*, 2009). PhoP binds to the promoters of the two functional GS genes (*glnA* and *glnII*), the *amtB-glnK-glnD* operon and – in contrast to GlnR – also to the *glnR* promoter. Footprint analysis indicated that the PhoP binding sequence overlaps the GlnR-box (Rodriguez-Garcia *et al.*, 2009) suggesting a tight interconnection between nitrogen and phosphorus regulation.

Transcriptional regulation of nitrogen metabolism genes in *S. coelicolor* is further complicated by the presence of a second transcriptional regulator: GlnRII, which has 31% identity to GlnR (Fink *et al.*, 2002). The *glnRII* gene is the third gene downstream of *glnII* suggesting a putative role in *glnII* regulation. Whereas both regulators seem to have very similar binding sites they cannot functionally replace each other. In contrast to the *glnR* mutant which cannot grow on nitrate-containing minimal medium, the *glnRII* mutant can (Fink *et al.*, 2002). Further functional differences between these two regulators are reflected by the lack of a PhoP-binding box in front of *glnRII* (Rodriguez-Garcia *et al.*, 2009).

Nitrogen regulation in S. coelicolor *grown in batch fermentation*

The regulation of nitrogen metabolism of streptomycetes and its interplay with secondary metabolite production was previously studied mainly in shaking flasks and on solid medium. A detailed analysis how nitrogen availability influences antibiotic production was carried out with the pristinamycin producer *Streptomyces pristinaespiralis* cultivated with different nitrogen sources (Voelker and Altaba, 2001). Production of pristinamycin was delayed until after exhaustion of amino acids and depletion of ammonium. However, different kinetics of utilization of carbon and nitrogen sources were observed on the different media, in particular when the amino acids valine, glutamate, or alanine, respectively, were used.

Only recently data on *S. coelicolor* cultivated in batch fermentation became available. Using protocols for highly reproducible submerged batch fermentation and custom designed oligonucleotide microarrays highly resolved transcriptomic data were generated. In bacteria undergoing the switch from exponential growth to stationary phase, the expression of a large number of genes is changed (Nieselt *et al.*, 2009). The key genes in nitrogen metabolism (such as *glnA*, *glnII*, *glnK*) show the same expression profile: High expression at early time points (20 h after inoculation), rapid decline after 24 h and a second, smaller peak at 29 h. Thereafter expression reduces to low expression levels (Battke *et al.*, 2009; Nieselt *et al.*, 2009). This profile does not correspond to the *glnR* expression course which shows a continuous slight decrease (Nieselt *et al.*, 2009).

The differences of these expression data to those obtained in shake flasks may be explained by the differences in media used. Whereas in shake flasks media were applied (such as N-Evans or other minimal media) in which a defined status of nitrogen limitation could be generated (Fink *et al.*, 2002; Tiffert *et al.*, 2009), the batch fermentations were carried out in a medium containing glutamate as sole nitrogen source. This was necessary to generate sufficient amounts of biomass required for the highly resolved time courses and for relating transcript data to antibiotic biosynthesis. Apparently, under these conditions nitrogen limitations occurs much later and *glnR* expression is not induced. The situation may be further complicated by the fact that glutamate is first used as a carbon source (Voelker and Altaba; 2001) which may lead to a cross-talk between carbon and nitrogen regulation.

Nitrogen assimilation in other actinobacteria

GlnR-like proteins in actinobacteria

In the course of investigations on the rifamycin producer *A. mediterranei* a GlnR orthologue was identified (Yu *et al.*, 2006). GlnR was shown to be essential for *glnA* transcription in *A. mediterranei*. GlnR could complement a *S. coelicolor glnR* mutant indicating functional equivalence of these two GlnR proteins. However, transformation of a *S. coelicolor* wild type with the *A. mediterranei glnR* gene resulted in severe changes to the pattern of antibiotic production and in morphological defects (poor sporulation) (Yu *et al.*, 2007), similar to the phenotype of the previously described *S. coelicolor glnR* mutant FS2 (Wray *et al.*, 1991).

In order to investigate whether regulation of nitrogen metabolism by GlnR-like proteins may be a general feature for all actinobacteria, the genome sequences of various members of the order actinomycetales were screened for the presence of GlnR orthologues. Indeed, all sequences,

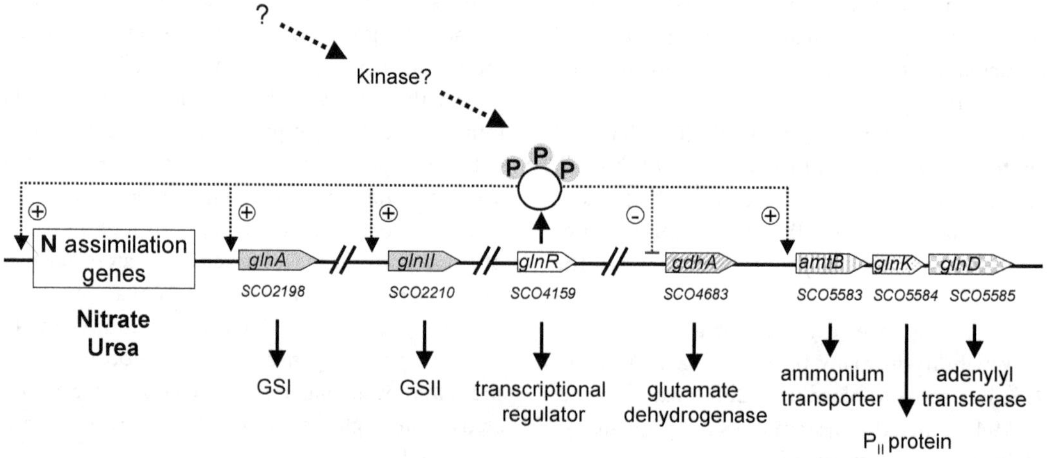

Figure 6.4 Scheme of the components involved in GlnR-mediated regulation. Chromosomal localization of genes whose transcription is controlled by GlnR.

except those of corynebacteria, contained GlnR counterparts with 60–80% identity to *S. coelicolor* GlnR (Tiffert *et al.*, 2008). In particular the C-terminal region characterized as a recognition helix (Martinez-Hackert and Stock, 1997) is conserved to 100%. Furthermore, GlnR binding sites in front of genes involved in nitrogen metabolism were identified by *in silico* analysis.

In corynebacteria central genes of nitrogen metabolism are controlled by the repressor AmtR (Jakoby *et al.*, 2000; Burkovski, 2007) which has no similarity to GlnR. Whether this special mode of nitrogen regulation reflects a yet unknown, more general difference in the life style of corynebacteria compared with that of other actinobacteria has still to be investigated. An AmtR-homologue was also found by genome comparisons in *Mycobacterium smegmatis* (Amon *et al.*, 2009) which is the only known bacterium containing both GlnR and AmtR.

Functional characterization of GlnR of Mycobacterium tuberculosis

Whether the agreement at the sequence level between streptomycetes and mycobacteria corresponds to functional equivalence was tested for *Mycobacterium tuberculosis*. A *M. tuberculosis rglnR* mutant (like a *S. coelicolor glnR* mutant) cannot grow on minimal medium with nitrate as sole nitrogen source. Complementation of this mutant by either *M. tuberculosis glnR* or *S. coelicolor glnR* restored its ability to grow. Band shift assays with promoter regions of *M. tuberculosis* genes corresponding to those *S. coelicolor* genes which are regulated by GlnR revealed that *S. coelicolor* and *M. tuberculosis* GlnR can bind to most of these DNA fragments (Malm *et al.*, 2009). This makes *S. coelicolor* a model system for studying nitrogen metabolism of the human pathogen *M. tuberculosis*. This is of special interest since GlnA1 of *M. tuberculosis* was shown to be essential for virulence (Tullius *et al.*, 2003).

Conclusions

Streptomycetes have acquired a huge armamentarium of catabolic and anabolic pathways to utilize different nitrogen sources. Due to their sessile life style they are equipped with enzymes which enable them to react to various environmental situations. Therefore, a great portion of its large genome is devoted to hydrolytic enzymes such as proteases, amylases, lipases, nucleases etc (Bentley *et al.*, 2002) to break down macromolecules which may be available occasionally in the soil. The enzymes which convert these components finally into glutamine and glutamate are basically the same as in many other bacteria with one major difference: streptomycetes have two GSs which both can use ammonium and glutamate to synthesize glutamine, but which are differently regulated. The function of three further GS-like enzymes is still unknown. What makes nitrogen metabolism in streptomycetes special is their sophisticated regulation which allows growth under extremely different conditions. In addition to transcriptional regulation a number of post-translational regulatory steps are involved which are specific for streptomycetes. There is growing evidence that the regulation is tightly connected with other cellular control processes and that in addition to its well understood elements also various two-component regulators such as AfsQ1,2 (Shu *et al.*, 2009) and even small RNAs (D'Alia *et al.*, 2010) are involved. So the complexity of the morphological differentiation program and the physiological versatility of streptomycetes is reflected by the complex nitrogen metabolism and in particular its regulation.

In future experiments a Systems Biology approach combining proteomics, transcriptomics, metabolomics as well as bioinformatics will be a powerful tool to analyse the nitrogen regulon and its integration into the overall metabolic network to get a broader view on *Streptomyces* physiology.

Acknowledgements

We thank all members of the 'nitrogen group' in the Mikrobiologie/Biotechnologie Department who contributed over the years to the results summarized in this chapter. The work was supported by the EU (FPVI ActinoGEN) and the BMBF [SysMO (GEN2006-27745-E/SYS)].

References

Amon, J., Titgemeyer, F., and Burkovski, A. (2009). A genomic view on nitrogen metabolism and nitrogen control in mycobacteria. J. Mol. Microbiol. Biotechnol. *17*, 20–29.

Atkinson, M.R., and Ninfa, A.J. (1998). Role of the GlnK signal transduction protein in the regulation of nitrogen assimilation in *Escherichia coli*. Mol. Microbiol. *29*, 431–447.

Bascarán, V., Hardisson, C., and Braña, A.F. (1989). Regulation of nitrogen catabolic enzymes in *Streptomyces clavuligerus*. J. Gen. Microbiol. *135*, 2465–2474.

Battke, F., Herbig, A., Wentzel, A., Jakobsen, Ø.M., Bonin, M., Hodgson, D.A., Wohlleben, W., Ellingsen, T.E., the STREAM Consortium and Nieselt, K. (2010). A technical platform for generating reproducible expression data from *Streptomyces coelicolor* batch cultivations. In Software Tools and Algorithms for Biological Systems, Arabina, H.R., and Tran, Q.–N., eds (Berlin, Springer Verlag), in press.

Behrmann, I., Hillemann, D., Pühler, A., Strauch, E., and Wohlleben, W. (1990). Overexpression of a *Streptomyces viridochromogenes* gene (*glnII*) encoding a glutamine synthetase similar to those of eucaryotes confers resistance against the antibiotic phosphinothricyl-alanyl-alanine. J. Bacteriol. *172*, 5326–5334.

Bentley, S.D., Chater, K.F., Cerdeño-Tárraga, A.M., Challis, G.L., Thomson, N.R., James, K.D., Harris, D.E., Quail, M.A., Kieser, H., Harper, D., *et al.* (2002). Complete genome sequence of the model actinomycete *Streptomyces coelicolor* A3(2). Nature *417*, 141–147.

Burkovski, A. (2007). Nitrogen control in *Corynebacterium glutamicum*: proteins, mechanisms, signals. J. Microbiol. Biotechnol. *17*, 187–194.

Carlson, T., and Chelm, B.K. (1986). Apparent eukaryotic origin of glutamine synthetase II from the bacterium *Bradirhizobium japonicum*. Nature *322*, 568–570.

D´Alia, D., Nieselt, K., Steigele, S., Müller, J., Verburg, I., and Takano, E. (2010). Non-coding RNA of glutamine synthetase I modulates antibiotic production in *Streptomyces coelicolor* A3(2). J. Bacteriol., 192, 1160–1164.

Fink, D., Falke, D., Wohlleben, W., and Engels, A. (1999). Nitrogen metabolism in *Streptomyces coelicolor* A3(2): modification of glutamine synthetase I by an adenylyltransferase. Microbiology *145*, 2313–2322.

Fink, D., Weiβschuh, N., Reuther, J., Wohlleben, W., and Engels, A. (2002). Two transcriptional regulators GlnR and GlnRII are involved in regulation of nitrogen metabolism in *Streptomyces coelicolor* A3(2). Mol. Microbiol. *46*, 331–347.

Fisher, S.H. (1992). Glutamine synthesis in *Streptomyces* – a review. Gene *115*, 13–17.

Fisher, S.H., and Wray, L.V. (1989). Regulation of glutamine synthetase in *Streptomyces coelicolor*. J. Bacteriol. *171*, 2378–2383.

Harth, G., Masleša-Galić, S., Tullius, M.V., and Horwitz, M.A. (2005). All four *Mycobacterium tuberculosis glnA* genes encode glutamine synthetase activities but only GlnA1 is abundantly expressed and essential for bacterial homeostasis. Mol. Microbiol. *58*, 1157–1172.

Harth, G., Clemens, D.L., and Horwitz, M.A. (1994). Glutamine synthetase of *Mycobacterium tuberculosis*: extracellular release and characterization of its enzymatic activity. Proc. Natl. Acad. Sci. U.S.A. *27*, 9342–9346.

Hayward, D., van Helden, P.D., and Wiid, I.J. (2009). Glutamine synthetase sequence evolution in the mycobacteria and their use as molecular markers for *Actinobacteria* speciation. BMC Evol. Biol. *9*, 48.

Hesketh, A., Fink, D., Gust, B., Rexer, H.U., Scheel, B., Chater, K., Wohlleben, W., and Engels, A. (2002). The GlnD and GlnK homologues of *Streptomyces coelicolor* A3(2) are functionally dissimilar to their nitrogen regulatory system counterparts from enteric bacteria. Mol. Microbiol. *46*, 319–330.

Hillemann, D., Dammann, T., Hillemann, A., and Wohlleben, W. (1993). Genetic and biochemical characterization of the two glutamine synthetases GSI and GSII of the phosphinothricyl-alanyl-alanine producer, *Streptomyces viridochromogenes* Tü494. J. Gen. Microbiol. *139*, 1773–1783.

Hodgson, D.A. (2000). Primary metabolism and its control in streptomycetes: a most unusual group of bacteria. Adv. Microb. Physiol. *42*, 47–238.

Ikeda, H., Ishikawa, J., Hanamoto, A., Shinose, M., Kikuchi, H., Shiba, T., Sakaki, Y., Hattori, M., and Omura, S. (2003). Complete genome sequence and comparative analysis of the industrial microorganism *Streptomyces avermitilis*. Nat. Biotechnol. *21*, 526–531.

Jakoby, M., Nolden, L., Meier-Wagner, J., Krämer, R., and Burkovski, A. (2000). AmtR, a global repressor in the nitrogen regulation system of *Corynebacterium glutamicum*. Mol. Microbiol. 37, 964–977.

Jiang, P., and Ninfa, A.J. (2009). Reconstitution of *Escherichia coli* glutamine synthetase adenylyltransferase from N-terminal and C-terminal fragments of the enzyme. Biochemistry *48*, 415–423.

Jiang, P., Peliska, A., and Ninfa, A.J. (1998). Enzymological characterization of the signal-transducing uridydyltransferase/uridylyl-removing enzyme (EC 2.7.7.59) of *Escherichia coli* and its interaction with the PII protein. Biochemistry *37*, 12782–12794.

Jonsson, A., and Nordlund, S. (2007). In vitro studies of the uridylylation of the three PII protein paralogs from *Rhodospirillum rubrum*: the transferase activity of *R. rubrum* GlnD is regulated by alpha-ketoglutarate and divalent cations but not by glutamine. J. Bacteriol. *189*, 3471–3478.

Kamberov, E.S., Atkinson, M.R., Feng, J., Chandran, P., and Ninfa, A.J. (1994). Sensory components controlling bacterial nitrogen assimilation. Cell Mol. Biol. Res. *40*, 175–191.

Kumada, Y., Benson, D.R., Hillemann, D., Hosted, T.J., Rochefort, D.A., Thompson, C.J., Wohlleben, W., and Tateno, Y. (1993). Evolution of the glutamine synthetase gene, one of the oldest existing and functioning genes. Proc. Natl. Acad. Sci. U.S.A. *90*, 3009–3013.

Kumada, Y., Takano, E., Nagaoka, K., and Thompson, C.J. (1990). *Streptomyces hygroscopicus* has two glutamine synthetase genes. J. Bacteriol. *172*, 5343–5351.

Malm, S., Tiffert, Y., Micklinghoff, J., Schultze, S., Joost, I., Weber, I., Horst, S., Ackermann, B., Schmidt, M., Wohlleben, W., *et al.* (2009). The roles of the nitrate reductase NarGHJI, the nitrite reductase NirBD and the response regulator GlnR in nitrate assimilation of *Mycobacterium tuberculosis*. Microbiology *155*, 1332–1339.

Martinez-Hackert, E., and Stock, A.M. (1997). The DNA-binding domain of OmpR: crystal structures of a winged helix transcription factor. Structure *5*, 109–124.

Merrick, M.J., and Edwards, R.A. (1995). Nitrogen control in bacteria. Microbiol. Rev. *59*, 604–622.

Nieselt, K., Battke, F., Herbig, A., Bruheim, P., Wentzel, A., Jakobsen, Ø.M., Sletta, H., Alam, M.T., Merlo, M.E., Moore, J., *et al.* (2010). The dynamic architecture of the metabolic switch in *Streptomyces coelicolor.* B.M.C. Genomics *11*, 10.

Peng, W.-T., Wang, J., Wu, T., Huang, J.-Q., Chiao, J.-S., and Zhao, G.-P. (2006). Bacterial type I glutamine synthetase of the rifamycin SV producing actinomycete, *Amycolatopsis mediterranei* U32, is the only enzyme responsible for glutamine synthesis under physiological conditions. Acta Biochim. Biophys. Sin. *38*, 821–830.

Reuther, J., and Wohlleben, W. (2007). Nitrogen Metabolism in *Streptomyces coelicolor*: Transcriptional and Post-Translational Regulation. J. Mol. Microbiol. Biotechnol. *12*, 139–146.

Rexer, H.U., Schäberle, T., Wohlleben, W., and Engels, A. (2006). Investigation of the functional properties and regulation of three glutamine synthetase-like genes in *Streptomyces coelicolor* A3(2). Arch. Microbiol. *186*, 447–458.

Rodríguez-García, A., Sola-Landa, A., Apel, K., Santos-Beneit, F., and Martín, J.F. (2009). Phosphate control over nitrogen metabolism in *Streptomyces coelicolor*: direct and indirect negative control of *glnR, glnA, glnII* and *amtB* expression by the response regulator PhoP. Nucleic Acids Res. *37*, 3230–3242.

Schreier, H.J., Brown, S.W., Hirschi, K.D., Nomellini, J.F., and Sonenshein, A.L. (1989). Regulation of *Bacillus subtilis* glutamine synthetase gene expression by the product of the *glnR* gene. J. Mol. Biol. *210*, 51–63.

Shu, D., Chen, L., Wang, W., Yu, Z., Ren, C., Zhang, W., Yang, S., Lu, Y., and Jiang, W. (2009). *afsQ1-Q2-sigQ* is a pleiotropic but conditionally required signal transduction system for both secondary metabolism and morphological development in *Streptomyces coelicolor*. Appl. Microbiol. Biotechnol. *81*, 149–160.

Stadtman, E.R., Ginsburg, A., Ciardi, J.E., Yeh, J., Hennig, S.B., and Shapiro, B.M. (1970). Multiple molecular forms of glutamine synthetase produced by enzyme catalysed adenylylation and deadenylylation reactions. Adv. Enzyme Reg. *8*, 99–118.

Streicher, S.L., and Tyler, B. (1981). Regulation of glutamine synthetase activity by adenylylation in the Gram-positive bacterium *Streptomyces cattleya*. Proc. Natl. Acad. Sci. U.S.A. *78*, 229–233.

Takano, E., Chakaraburtty, R., G., Nihira, T., Yamada, Y., and Bibb, M.J. (2001). A complex role for the role of the γ-butyrolactone SCB1 in regulating antibiotic production in *Streptomyces coelicolor* A3(2). Mol. Microbiol. *41*, 1015–1028.

Takano, E., Kinoshita, H., Mersinias, V., Bucca, G., Hotchkiss, G., Nihira, T., Smith, C.P., Bibb, M., Wohlleben, W., and Chater, K. (2005). A bacterial hormone (the SCB1) directly controls the expression of a pathway-specific regulatory gene in the cryptic type I polyketide biosynthetic gene cluster of *Streptomyces coelicolor*. Mol. Microbiol. *56*, 465–379.

Tiffert, Y., Supra, P., Wurm, R., Wohlleben, W., Wagner, R., and Reuther, J. (2008). The *Streptomyces coelicolor* GlnR regulon: identification of new GlnR targets and evidence for a central role of GlnR in nitrogen metabolism in actinomycetes. Mol. Microbiol. *67*, 861–880.

Truffa-Bachi, P., and Cohen, G.N. (1973). Amino acid metabolism. Annu. Rev. Biochem. *42*, 113–134.

Tullius, M.V., Harth, G., and Horwitz, M.A. (2003). Glutamine synthetase GlnA1 is essential for growth of *Mycobacterium tuberculosis* in human THP-1 macrophages and guinea pigs. Infect. Immun. *71*, 3927–3936.

Tyler, B. (1978). Regulation of the assimilation of nitrogen compounds. Annu. Rev. Biochem. *47*, 1127–1162.

van Wezel, G.P., White, J., Hoogvliet, G., and Bibb, M.J. (2000). Application of *redD*, the transcriptional activator gene of the undecylprodigiosin biosynthetic pathway, as a reporter for transcriptional activity in *Streptomyces coelicolor* A3(2) and *Streptomyces lividans.* J. Mol. Microbiol. Biotechnol. *2*, 551–556.

Vega-Palas, M.A., Flores, E., and Herrero, A. (1992). NtcA, a global nitrogen regulator from the cyanobacterium *Synechococcus* that belongs to the Crp family of bacterial regulators. Mol. Microbiol. *6*, 1853–1859.

Voelker, F., and Altaba, S. (2001). Nitrogen source governs the patterns of growth and pristinamycin production in *'Streptomyces pristinaespiralis'*. Microbiology *147*, 2447–2459.

Wang, J., and Zhao, G.-P. (2009). GlnR positively regulates *nasA* transcription in *Streptomyces coelicolor*. Biochem. Biophys. Res. Comm. *386*, 77–81.

Weiβschuh, N., Fink, D., Vierling, S., Bibb, M.J., Wohlleben, W., and Engels, A. (2000). Transcriptional analysis of the gene for glutamine synthetase II and two upstream genes in *Streptomyces coelicolor* A3(2). Mol. Gen. Genet. *264*, 461–469.

Wohlleben, W., Arnold, W., Behrmann, I., Broer, I., Hillemann, D., Pühler, A., and Strauch, E. (1991). Genetic analysis of different resistance mechanisms against the herbicidal antibiotic phosphinothricyl-alanyl-alanine. In Genetics and Product Formation in *Streptomyces*, Baumberg, S., Krügel, H., Noack, D., eds (Plenum Press, New York), pp. 171–183.

Wohlleben, W., Stegmann, E., and Süssmuth, R.D. (2009). Molecular genetic approaches to analyze glycopeptide biosynthesis. Methods Enzymol. *458*, 459–486.

Wray, Jr., L.V., Atkinson, M.R., and Fisher, S.H. (1991). Identification and cloning of the *glnR* locus, which is required for transcription of the *glnA* gene in *Streptomyces coelicolor* A3(2). J. Bacteriol. *173*, 7351–7360.

Wray, Jr., L.V., Ferson, A.E., Rohrer, K., and Fisher, S.H. (1996). TnrA, a transcription factor required for global nitrogen regulation in *Bacillus subtilis*. Proc. Natl. Acad. Sci. U.S.A. *93*, 8841–8845.

Wray, L.V., and Fisher, S.H. (1988). Cloning and nucleotide sequence of the *Streptomyces coelicolor* gene encoding glutamine synthetase. Gene *71*, 247–256.

Wray, L.V., and Fisher, S.H. (1993). The *Streptomyces coelicolor glnR* gene encodes a protein similar to other bacterial response regulators. Gene *130*, 145–150.

Yang, Y.H., Song, E., Kim, E.J., Lee, K., Kim, W.S., Park, S.S., Hahn, J.S., and Kim, B.G. (2009). NdgR, an IclR-like regulator involved in amino-acid-dependent growth, quorum sensing, and antibiotic production in *Streptomyces coelicolor*. Appl. Microbiol. Biotechnol. *82*, 501–511.

Yu, H., Peng, W.-T., Liu, Y., Wu, T., Yao, Y.-F., Cui, M.-X., Jiang, W.-H., and Zhao, G.-P. (2006). Identification and characterization of *glnA* promoter and its corresponding *trans*-regulatory protein GlnR in the rifamycin SV producing actinomycete, *Amycolatopsis mediterranei* U32. Acta Biochim. Biophys. Sin. *38*, 831–843.

Yu, H., Yao, Y., Liu, Y., Jiao, R., Jiang, W., and Zhao, G.-P. (2007). A complex role of *Amycolatopsis mediterranei* GlnR in nitrogen metabolism and related antibiotics production. Arch. Microbiol. *188*, 89–96.

Network Mechanisms of Phosphate Control of Primary and Secondary Metabolism

7

Juan F. Martín, Alberto Sola-Landa, Fernando Santos-Beneit and Antonio Rodríguez-García

Abstract

The biosynthesis of hundreds of antibiotics and other secondary metabolites is repressed by inorganic phosphate, in spite of the fact that those metabolites are synthesized by quite different pathways. Derepression of many antibiotic biosynthesis genes occurs when the concentration of inorganic phosphate becomes limiting for growth. Phosphate control of primary and secondary metabolism in *Streptomyces* species is mediated by the two component system PhoR-PhoP. The genes under direct control of PhoP (the *pho* regulon) have been identified in *S. coelicolor*. They are involved in a variety of cellular functions, most notably in phosphate scavenging and phosphate transport. Phosphate control of secondary metabolism genes is not exerted by direct binding of PhoP to the promoters of those genes. Rather, PhoP controls directly other intermediate regulators in a regulatory cascade that in turn modulate expression of pathway-specific regulators. Indeed PhoP regulates expression of *afsS* in *S. coelicolor* that in turn triggers expression of *act*II-ORF4 and *redD* controlling actinorhodin and undecylprodigiosin biosynthesis. PhoP binds to a PHO box in the *afsS* promoter that overlaps with the AfsR binding sequence. Phosphate control also interacts with nitrogen regulation. PhoP binds to the promoters of *glnR*, *glnA* and other key genes of the nitrogen metabolism. This article provides new insights into the mechanisms underlying systems biology in *Streptomyces*.

Introduction

The biosynthesis of different classes of secondary metabolites is under phosphate control. Bacteria have evolved regulatory systems to respond to nutrient scarcity. The concentration of soluble inorganic orthophosphate (Pi), the preferred source of phosphorous, is usually low in the soil environment where most *Streptomyces* species live.

Over the last four decades an impressive number of antibiotics and other secondary metabolites have been reported to be produced by different bacteria (particularly actinobacteria) and filamentous fungi (Bérdy, 2005; Clardy *et al.*, 2006). The biosynthesis of many of these compounds is regulated negatively by phosphate (Masuma *et al.*, 1986; Martín, 1989; Liras *et al.*, 1990) but, surprisingly, the molecular mechanism of phosphate control has remained obscure for years (Martín and Demain, 1980; Martín *et al.*, 1994) in spite of its basic and industrial relevance. It is interesting to note that inorganic phosphate in the culture media controls the synthesis of a large number of secondary metabolites belonging to different chemical classes, e.g. macrolides, tetracyclines, anthracyclines, polyether compounds, aminoglycosides, amino acid-derived metabolites such as clavulanic acid, among others (Romero *et al.*, 1984; Dekeva *et al.*, 1985; Doull and Vining, 1990; Hobbs *et al.*, 1990; Lounes *et al.*, 1996). From a biosynthetic point of view, these groups of metabolites have very little in common, except that they all are dispensable 'secondary' metabolites (Vining, 1992; Martín *et al.*, 2000). Organic phosphates are also important nutrients for the biosynthesis of secondary metabolites and the release of phosphate from these compounds is under strict control by the inorganic phosphate level.

Martín and Demain (1980) proposed that phosphate control is used as a mechanism that triggers secondary metabolite biosynthesis when phosphate in the environment is depleted and, therefore, growth of the microorganisms cannot proceed at a normal rate (Demain and Vaishnav, 2004). This hypothesis has been supported by recent transcriptomic studies (Rodríguez-García *et al.*, 2007). When phosphate concentration in the culture media decreases below a threshold level, bacteria increase their production of a variety of metabolites that might serve as direct antagonists to other microorganisms (Vögtli *et al.*, 1994) or as biochemical cross-talk signals (Horinouchi and Beppu, 1992; Vining, 1992; Kaiser and Losick, 1993) to enhance survival under harsh nutritional conditions (Mapplestone *et al.*, 1992).

In 2003, we reported that phosphate control of antibiotic biosynthesis in *Streptomyces lividans* is mediated by the two-component PhoR-PhoP system (Sola-Landa *et al.*, 2003) that also controls the alkaline phosphatase gene (*phoA*) and other *phoA*-related genes (Apel *et al.*, 2007). The *S. coelicolor* and *S. lividans* PhoRP systems belong to class IIIA of two component systems (Fabret *et al.*, 1999; Hutchings *et al.*, 2004). The PhoR protein is a standard membrane sensor protein kinase, and PhoP is a member of the OmpR family of DNA-binding response regulators. In *Escherichia coli* and *Bacillus subtilis*, the phosphorylated PhoP protein (PhoP~P) or its homologue in *E. coli* (PhoB~P), in response to phosphate-starvation, activates expression of the *pho* regulon genes by binding to PHO boxes in their promoter regions (Torriani-Gorini, 1994; Hulett, 1996). In *S. lividans* expression of *phoA* is induced by phosphorylated PhoP (PhoP~P) and mutants lacking *phoP* (or the *phoR-phoP* cluster) do not form PhoA. These mutants overproduce large amounts of actinorhodin and undecylprodigiosin (Sola-Landa *et al.*, 2003). Similar mutants of *S. coelicolor* behave in a slightly different manner. Although actinorhodin and undecylprodigiosin are under phosphate control, the *S. coelicolor* Δ*phoP* mutants do not overproduce these pigmented metabolites (Santos-Beneit *et al.*, 2009a).

PhoP, as other members of the response regulator OmpR family, contains the DNA binding domain (DBD) in the C-terminal region (amino acids 187 to 201) (Sola-Landa *et al.*, 2005). In *E. coli* it is known that when the N-terminal region of PhoB (homologous to the *Streptomyces* PhoP) is deleted, including part or all of the α5 helix, the truncated protein binds constitutively to the PHO boxes with higher affinity than the unmodified protein (Ellison and McCleary, 2000; Allen *et al.*, 2001). The C-terminal *S. coelicolor* PhoP domain (hereafter named $PhoP^{DBD}$) was able to recognize PHO boxes with similar affinity than intact PhoP.

Genes controlled by PhoP in *S. coelicolor*: a global view

Genes under direct control of PhoP constitute, by definition, the *pho* regulon. As presented in this review, several PhoP operators have been characterized by protein–DNA binding assays (EMSA and DNase I footprinting). The total number of genes involved in the response to Pi-limitation is expected to be considerably larger than those regulated by PhoP, as indicated in the transcriptome and proteome analysis of *S. coelicolor* wild-type and Δ*phoP* mutant strains after Pi depletion (Rodríguez-García *et al.*, 2007). This is not a surprise since phosphate has essential metabolic, structural and regulatory roles. Phosphate-containing metabolites take part in central and energy metabolism, in the synthesis of membrane lipids and nucleic acids, and in the posttranslational regulation of many proteins *via* phosphorylation (Wanner, 1996). The PhoP regulated genes are involved in different cellular metabolic processes as detailed in the following paragraphs.

Phosphate scavenging and transport

The primary response to Pi starvation involves the induction of extracellular enzymes to obtain Pi from organic phosphates, and the activation of genes encoding Pi transporters. Thus, the *phoA* and *phoD* genes encoding the secreted alkaline phosphatase and phospholipase D proteins, respectively, are activated by PhoP (Apel *et al.*, 2007). The alkaline phosphatase is a non-specific phosphomonoesterase which is commonly induced in response to Pi starvation in bacteria (Wanner, 1996). The genome of *S. coelicolor* contains another alkaline phosphatase gene, *phoC*. This gene, despite its poor expression, is induced by low Pi concentration in a PhoP-independent fashion (Apel *et al.*, 2007). Several *Streptomyces* phospholipase D genes have been sequenced because of their biotechnological applications (Uesugi and Hatanaka, 2009). These enzymes release phosphatidic acids from the hydrolysis of phospholipids. Phospholipases are well-known virulence factors; they participate in Pi scavenging in both pathogenic and non-pathogenic bacteria (Titball, 1993; Oh *et al.*, 2007).

At least other three secreted enzymes are encoded by *pho* regulon genes. These are the glycerophosphodiester phosphodiesterases (GDPD) genes *glpQ1* and *glpQ2* (Santos-Beneit *et al.*, 2009b) and the putative phytase gene SCO7697 (Rodríguez-García *et al.*, 2007; Sola-Landa *et al.*, 2008). Glycerophosphodiesters are released from phospholipids by deacylating phospholipases, and then are hydrolysed by GDPD to yield sn-glycerol-3-phosphate and an alcohol. Specific transporters for the uptake of glycerol-3-phosphate have been characterized in Gram-positive and Gram-negative bacteria. It is expected that homologue transporters exist in the *S. coelicolor* genome (Bentley *et al.*, 2002).

The main identified organic phosphorus compounds in soil are phospholipids, inositol phosphates and nucleic acids. Inositol phosphates (phytates), which serve as phosphorous storage compounds in plants, frequently represent up to 80% of the total soil organic phosphorous (Quiquampoix and Mousain, 2005). Thus, it makes sense that PhoP activates phytase genes in response to phosphate starvation in *Streptomyces* (Rodríguez-García *et al.*, 2007; Sola-Landa *et al.*, 2008) as occurs also in *Bacillus* (Hoi *et al.*, 2006; Makarewicz *et al.*, 2006). Finally, other genes encoding putative phosphate scavengers are SCO1906, SCO3790 and SCO4152. Both SCO1906 and SCO3790 encode putative secreted phosphatases (other than PhoA and PhoC) and their promoters are bound by PhoP (Rodríguez-García *et al.*, 2007; Sola-Landa *et al.*, 2008). SCO4152 encodes a putative secreted 5′-nucleotidase that is up-regulated in the wild type transcriptome after phosphate shiftdown, but not in the Δ*phoP* mutant. Therefore, this gene is a candidate member of the *pho* regulon (Rodríguez-García *et al.*, 2007).

The released Pi from organic sources can enter the cell by means of (at least) three Pi transporters. These are the high-affinity Pst system, and the low-affinity high-velocity PitH1 and PitH2 systems (Díaz *et al.*, 2005; Santos-Beneit *et al.*, 2008). The *pitH1* promoter is active in the Δ*phoP* mutant, whereas the promoters of the *pstS* gene (the first component of the *pstSCAB* cluster) and of the *pitH2* gene are totally dependent on PhoP binding to become active. After Pi-depletion the highest activation of the *pitH2* promoter precedes that of the *pstS* promoter and then drops. Since the low affinity Pit systems are driven by the proton motive force, the high-affinity transporter (Pst) is energetically more expensive to the cell because it consumes ATP. Nevertheless, Pst is more effective in conditions of very low Pi concentrations because of its high affinity for Pi. These factors explain the sequential pattern of expression of these genes observed in flask cultures of *S. coelicolor* (Santos-Beneit *et al.*, 2008).

Phosphate storage and saving

Microorganisms accumulate polyphosphate as energy and Pi reserves. In *S. lividans* polyphosphates are synthesized from ATP by the kinase encoded by the *ppk* gene. The polyphosphate kinase can also catalyse the reverse reaction to regenerate the ATP pool in excess of ADP (Chouayekh and Virolle, 2002). When exogenous Pi is limiting for growth, the Ppk protein is synthesized in the wild type strain but not in a mutant *phoR-phoP* (Ghorbel *et al.*, 2006). Interestingly, the *S. coelicolor* Δ*phoP* mutant is able to consume Pi from culture medium to the same extent as the wild type, but its growth is impaired (Santos-Beneit *et al.*, 2008). This mutant strain seems to be unable to use the endogenous Pi source (e.g. polyphosphate). The lack of *ppk* activation in the Δ*phoP* mutant might explain, at least in part, the growth phenotype of this strain.

Teichoic acids are phosphate-rich polymers found in the cell wall of Gram-positive bacteria. It is known since a long time that in phosphate-limited cultures of *B. subtilis* teichoic acids are partially replaced by teichuronic acids. Their biosynthesis genes are repressed and activated, respectively, by PhoP (reviewed in Bhavsar *et al.*, 2004). It is very likely that in *Streptomyces* cell-wall teichoic acids also function as a Pi storage material. The replacement with Pi-free teichuronic acids would allow the scarce Pi to be directed to basic metabolic processes. This is indicated by the PhoP-dependent transcriptional activation of some members of a gene cluster that is probably involved in the biosynthesis of teichuronic acids (Rodríguez-García *et al.*, 2007). According to Charaniya *et al.* (2007), this cluster is organized in the transcriptional units SCO4872–3 (chromosomal strand −1), SCO4874–6 (+1), SCO4877–8 (−1), and SCO4879–82 (+1). PhoP has a predicted binding site in the bidirectional promoter region SCO4873-SCO4874 (Rodríguez-García *et al.*, 2007), and has two experimentally validated operators in the bidirectional region SCO4878-SCO4879 (Sola-Landa *et al.*, 2008).

PhoP as a global regulator of intermediary metabolism and macromolecule biosynthesis

In shiftdown cultures of *S. coelicolor*, Pi depletion up-regulates or down-regulates the expression of genes belonging to several metabolic processes (Rodríguez-García *et al.*, 2007). These include core metabolic pathways such as central and energy metabolism, transcription and translation. Other differentially expressed genes participate in stress responses, nitrogen assimilation, nucleotide biosynthesis, differentiation and secondary metabolism. Biologically, most of these changes make sense, such as the lower transcription of ribosomal genes when Pi is limiting the growth.

Some genes were differentially expressed in the Δ*phoP* mutant with respect to the wild type strain, indicating a direct or indirect regulation by PhoP. To date, the only PhoP role investigated in detail is the negative regulation of nitrogen assimilation (see below). It seems that there is a 'general response mechanism' to nutrient scarcity that activates assimilatory pathways and causes the transcriptional increase of the nitrogen genes in the Δ*phoP* mutant.

The short time at which samples were taken from shiftdown cultures (Rodríguez-García *et al.*, 2007) reflected the first step of the expression response to nutrient scarcity. According to Matin *et al.* (1989), the first steps are the cell efforts to escape Pi starvation and to maintain growth. In differentiating bacteria the second step (after the inability to continue growing) is sporulation. In *Streptomyces*, both morphological differentiation and secondary metabolism are often co-regulated. We have observed a PhoP-dependent down-regulation of both differentiation and prodiginine biosynthesis genes. PhoP might be signalling that 'it is not yet the moment to sporulate'. In *S. coelicolor* the production of prodiginine, however, is lower in the Δ*phoP* mutant than in the wild type at difference of what occurs in *S. lividans*. These different *Streptomyces* species respond in a somewhat different manner to PhoP.

Binding of PhoP to PHO boxes: different types of operator sequences

The control of *pho* regulon genes in *Streptomyces* is mediated by the two-component system PhoR-PhoP (Sola-Landa *et al.*, 2003; 2005; Mendes *et al.*, 2007). The transcriptional regulation of these genes is controlled by binding of the response regulator to the PHO boxes. The *Streptomyces* PHO box, as in other bacteria, is formed by direct repeat units (DRus) of eleven nucleotides composed of seven well conserved and four less conserved nucleotides (Blanco *et al.*, 2002; Suzuki *et al.*, 2004; Sola-Landa *et al.*, 2005; Wösten *et al.*, 2006; Yuan *et al.*, 2006; Schaaf and Bott, 2007). In *S. coelicolor*, the consensus sequence of the most conserved region (GTTCACC) was initially deduced from the alignment of the PhoP binding sites of *pstS*, *phoRP-phoU* (Sola-Landa *et al.*, 2005), *phoA*, *phoD* (Apel *et al.*, 2007) and *pitH2* (Santos-Beneit *et al.*, 2008).

With the aim of finding new PhoP binding sites, the *S. coelicolor* genome was searched using an informatics matrix made with these sequences. Fourteen new sequences were determined. With the information of these accumulated nineteen PhoP binding sequences a new model of the PhoP binding site, called model I, was constructed (Sola-Landa *et al.*, 2008). This model takes into account the number of retarded complexes formed in EMSA experiments, the protected sequences obtained in DNase I footprinting assays and the conservation value of each DRu calculated using an informatics program based on the 'information theory' described by Schneider (1997). According to this theory, the conservation of each base and its contribution to the information content of the DRu sequences (*Ri* value) is measured in bytes. So the maximum information content for a base position in DNA sequence is 2 (see Fig. 7.1).

The premises of model I are (i) a PhoP monomer interacts with a sequence of 11 nucleotides through specific contacts between the amino acids and the nitrogenous bases of a DRu (Blanco *et al.*, 2002; Yamane *et al.*, 2008), (ii) a minimum of two 11 nucleotides DRus are needed for PhoP binding (Sola-Landa *et al.*, 2008), (iii) the binding of monomers to consecutive sequences is favoured by protein–protein interactions (Birck *et al.*, 2003; Chen *et al.*, 2003; Perron-Savard *et al.*, 2005; Sinha *et al.*, 2008), and (iv) the highest conserved sequences bind monomers with higher affinity.

According to the previous premises, model I enables the classification of the DRus that form the PhoP operators into two different types: the ones constituting the core of the operator (called C) and the others involved in the extension (called E) of the core region. The core is the simplest structure required for proper binding of PhoP to DNA and is formed by two or three DRus, depending on the higher or lower conservation of each. In the EMSA analyses, binding of PhoP to the core results in a single retarded band (all the DRus are bound by PhoP at the same time). On the other hand, the extension DRus in a PHO box are occupied only when the protein concentration increases, leading to extra retarded bands in the EMSA experiments. Three mechanisms explain this sequential binding of PhoP to the DNA: (1) the extension DRu (E) localized adjacent to the core has a low degree of conservation; (2) there is a non-conserved DRu, called unstable DRu (E_U) between the core and a conserved DRu (which receives the name of extension support DRu [E_S] because it makes possible the binding of two monomers to both, E_U and E_S, at the same time); and (3) there is a separation of one or two nucleotides between two DRu that makes protein–protein interactions difficult. As they also have a supporting function, they are called E_S. An example of each of these

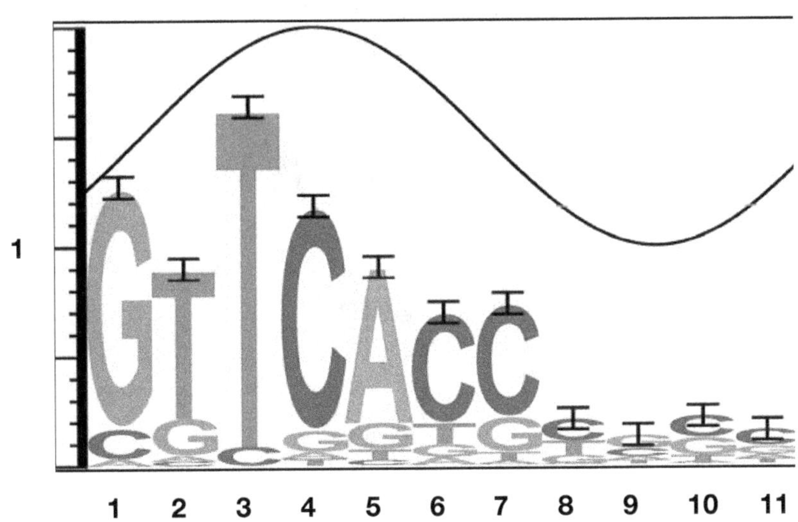

Figure 7.1 Sequence logo of the 11-nucleotide direct repeats units that form the PhoP binding site in *S. coelicolor*, according to model I (modified from Sola-Landa *et al.*, 2008).

mechanisms is shown in Fig. 7.2. The cooperative effect of several PhoP molecules when binding to the DRu is supported by evidence of PhoP–PhoP monomers interaction in *B. subtilis* (Eder *et al.*, 1999).

The sequence conservation, number and organization of the DRus that form the different PhoP operators show high variability. Using model I, these PhoP operators have been classified into three different classes. The first and second classes include operators with two or three DRus of type C, respectively. Class III operators are composed of a core region associated with extension DRus. Different structures are possible in this class and they can be formulated using the letters that symbolized each type of DRu. For example, the PhoP operator with higher number of DRus, *pitH2,* which forms four retarded complexes in the EMSA experiments, is formulated in the following way: E_S[1] E_SE_UCC[2]E_S (Santos-Beneit *et al.*, 2008). The numbers in square brackets represent the separation in base pairs between contiguous DRus. See Table 7.1 for all the operator structures characterized to date.

Recently, seven new PhoP binding sites have been fully characterized (Santos-Beneit *et al.*, 2009a,b; Rodríguez-García *et al.*, 2009). Six of them have been classified according to model I as class III.

The fact that most of the PhoP binding sites (70%) belong to class III (complex operators) may indicate that PhoP does not operate like a simple on-off switch. On the contrary, PhoP seems to exert several levels of induction or repression, or even both. Thus, PhoP can activate the expression of one promoter when it binds to a DRu localized in the –35 region and

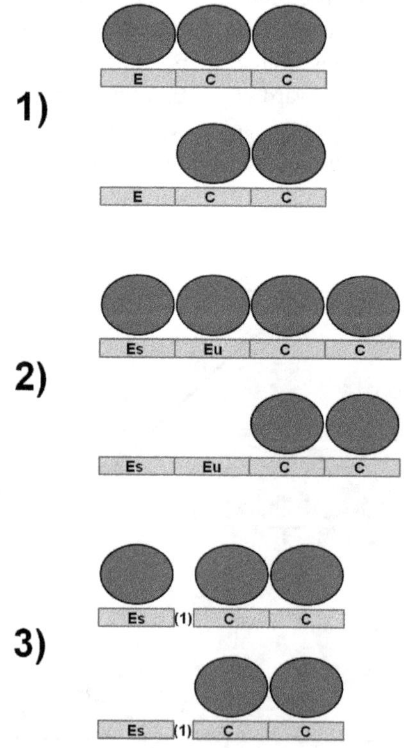

Figure 7.2 Scheme summarizing the three mechanisms that account for the sequential binding of PhoP to the DNA according to model I.

Table 7.1 PhoP binding sites characterized in *S. coelicolor* to date

	Gene	Operator structure	Class	Distance to ATG (nt)	Regulation	Reference
1	*phoA*	CC	I	128	Activation	Apel *et al.* (2007)
2	*afsS*	CC	I	162	Repression	Santos-Beneit *et al.* (2009a)
3	SCO3790	CC	I	79	Activation	Sola-Landa *et al.* (2008)
4	SCO6169	CC	I	195	ND	Sola-Landa *et al.* (2008)
5	SCO0034	CCC	II	127	Activation	Sola-Landa *et al.* (2008)
6	SCO1906	CCC	II	92	ND	Sola-Landa *et al.* (2008)
7	SCO4261	CCC	II	43	Repression	Sola-Landa *et al.* (2008)
8	*pstS*	EECC	III	114	Activation	Sola-Landa *et al.* (2005)
9	*phoU*	CCEuEs	III	109	Activation	Sola-Landa *et al.* (2005)
10	*pitH2*	Es[1]EsEuCC[2]Es	III	123	Activation	Santos-Beneit *et al.* (2008)
11	*phoD*	CCC/Es[1]EE	III	52/10	Activation	Apel *et al.* (2007)
12	*hrdA*	ECC	III	80	Repression	Sola-Landa *et al.* (2008)
13	*glpQ1*	Es[1]CCE	III	49	Activation	Santos-Beneit *et al.* (2009b)
14	*glpQ2*	Es[2]CCEuEs	III	48	Activation	Santos-Beneit *et al.* (2009b)
15	*amtB*	ECC	III	149	Repression	Rodríguez-García *et al.* (2009)
16	*glnR*	CCEE	III	108	Repression	Rodríguez-García *et al.* (2009)
17	*glnII*	CCEEuEs	III	87	Repression	Rodríguez-García *et al.* (2009)
18	*glnA*	Es[1]CC	III	103	Repression	Rodríguez-García *et al.* (2009)
19	SCO1196	CCC[2]Es	III	83	Activation	Sola-Landa *et al.* (2008)
20	SCO1394	ECC	III	30	Activation	Sola-Landa *et al.* (2008)
21	SCO2262	ECC	III	70	Repression	Sola-Landa *et al.* (2008)
22	SCO2878	CCCEuEs	III	42	Activation	Sola-Landa *et al.* (2008)
23	SCO5447	ECCE	III	214	Activation	Sola-Landa *et al.* (2008)
24	SCO7697	CCEE	III	122	Activation	Sola-Landa *et al.* (2008)
25	SCO4878	4 DRu (ND)	ND	40	ND	Sola-Landa *et al.* (2008)
26	SCO4879	5 DRu (ND)	ND	30	Activation	Sola-Landa *et al.* (2008)

EMSA and DNase I footprinting assays have been performed in all genes. The operator structure of *phoD* is localized both in the coding and non-coding strands. ND; not determined.

repress the expression when it binds to a DRu in the −10 region. This dual character of PhoP has been also reported for the control of the *phyC* gene of *Bacillus amyloliquefaciens* (Makarewicz *et al.*, 2006), and could also be the case with the *Streptomyces phoD* and *pitH2* genes (Apel *et al.*, 2007; Santos-Beneit *et al.*, 2008).

Phosphate control of secondary metabolites: cascades of regulation

The biosynthesis of secondary metabolites in *Streptomyces* and other bacteria is under the control of both global and pathway specific regulators. The global regulators respond to a variety of nutritional or environmental stress signals, i.e. phosphate or nitrogen starvation, heat shock or pH stress.

The lack of PHO boxes and direct binding of PhoP to genes encoding the structural biosynthetic enzymes for secondary metabolites suggest that its regulatory effect may be exerted through interaction of PhoP with other regulators.

The biosynthesis of actinorhodin and undecylprodigiosin is controlled by the global regulator AfsR, a member of the large-size SARP family (Tanaka *et al.*, 2007). Floriano and Bibb (1996) found that the stimulatory effect of AfsR on antibiotic production is exerted through the activation of the pathway-specific regulatory genes *actII*-ORF4 and *redD*. The only known target of the AfsR regulator is a small gene located downstream of *afsR*, named *afsS* in *S. coelicolor* and *afsR2* in *S. lividans*. Amplification of *afsR2/afsS* on a high-copy-number plasmid conferred overproduction of actinorhodin and undecylprodigiosin in both *S. coelicolor* and *S. lividans* (Vögtli *et al.*, 1994; Matsumoto *et al.*, 1995; Floriano and Bibb, 1996). The binding sequence of AfsR to the *afsS* promoter region was identified by Lee *et al.* (2002); our analysis of this sequence identified a putative PHO box overlapping with the AfsR binding sequence suggesting that both PhoP and AfsR might interact in the regulation of *afsS*.

Electrophoretic mobility shift assays showed that $PhoP^{DBD}$ does not bind directly to the *actII-ORF4, redD* and *atrA* promoters, but it binds to the *afsS* promoter, in a region overlapping with the *afsR* operator. DNase I footprinting studies revealed a PhoP protected region of 26 nucleotides (two direct repeats of 11 nucleotides) that overlaps with the AfsR recognition sequence. Binding experiments indicated a competition between AfsR and PhoP; increasing concentrations of $PhoP^{DBD}$ resulted in the disappearance of the AfsR–DNA complex (Santos-Beneit *et al.*, 2009a).

Expression studies using the reporter *luxAB* gene coupled to the *afsS* promoter showed that PhoP downregulates *afsS* expression probably by a competition with the AfsR activator. Interestingly, AfsR binds to other PhoP-regulated promoters including those of *pstS* (a member of the phosphate transport system) and *phoRP* (encoding the two component system itself). Analysis of the AfsR-protected sequences in each of these promoters allowed us to distinguish the AfsR binding sequence from the overlapping PHO box (Santos-Beneit *et al.*, 2009a).

Interaction of phosphate, carbon and nitrogen regulation

Carbon and nitrogen sources in addition to phosphate play an important role in the regulation of primary and secondary metabolism in almost all bacteria including *Streptomyces* (Martín and Demain, 1980; Hobbs *et al.*, 1990; Merrick and Edwards, 1995; Martín, 2004; Rigali *et al.*, 2008). The different mechanisms of control have been widely studied in an individual form, but the connections existing between them have not been investigated. Only a few examples of genes regulated simultaneously by different nutritional sources in *Streptomyces* species have been described up until now. The *glpQ1* and *glpQ2* genes of *S. coelicolor* code for two extracellular glycerophosphodiester phosphodiesterases required for the utilization of glycerophosphodiesters as phosphate and carbon sources. Both genes are directly regulated by PhoP (Rodríguez-García *et al.*, 2007; Santos-Beneit *et al.*, 2009b) and their expression is also affected, in a different manner, by the carbon source (Santos-Beneit *et al.*, 2009b). A similar interaction of phosphate and carbon sources has been described for the *pstS* gene in *S. lividans* and *S. coelicolor*. PstS is the phosphate-binding protein belonging to the high-affinity phosphate-specific transport system (*pst*), a member of the *pho* regulon (Sola-Landa *et al.*, 2005; Rodríguez-García *et al.*, 2007). PstS is highly accumulated in the supernatant of liquid cultures when the bacteria are grown in the presence of different sugars under phosphate limitation conditions (Díaz *et al.*, 2005; Esteban *et al.*, 2008). An interaction between phosphate and carbon sources has been shown in other microorganisms, e.g. in *Vibrio vulnificus*, the patterns of alkaline phosphatase and phospholipase expression are affected by both phosphate concentration and carbon source (Oh *et al.*, 2007). In *B. subtilis* it is known that the carbon catabolite protein-A (CcpA) exerts a control over the expression of *phoRP* genes, although there is a discrepancy about whether such regulation is exerted through the *cre* sequence (Puri-Taneja *et al.*, 2006) or indirectly (Choi and Saier, 2005).

The relationship between the phosphate and nitrogen source regulatory pathways in *Streptomyces* is more complex because it affects not only the expression of structural genes but also the direct regulators. GlnR is the main nitrogen regulator and controls the expression of several genes involved in nitrogen metabolism at the transcriptional level (Wray and Fisher, 1993; Reuther and Wohlleben, 2007). In response to nitrogen limitation GlnR activates the transcription of numerous genes (Tiffert *et al.*, 2008), including that of *glnA* and *glnII* (encoding two different types of glutamine synthetases) and the operon *amtB-glnK-glnD*, involved in ammonium transport and metabolism (Reuther and Wohlleben, 2007). GlnRII is a second regulator similar to GlnR that also binds to the promoters of *glnA*, *glnII* and *amtB* genes (Fink *et al.*, 2002). However, GlnRII is not totally homologous to GlnR and its role in nitrogen regulation is not yet clear (Reuther and Wohlleben, 2007). Under phosphate limitation the expression of some nitrogen metabolism genes like *glnA*, *glnII*, *amtB* and *glnK* is increased in a *S. coelicolor* Δ*phoP* mutant (Rodríguez-García *et al.*, 2007). The control of PhoP over these genes is exerted indirectly through the binding to the promoter of the *glnR* gene, responsible for their activation. In addition, there is also a direct control exerted by the binding of PhoP to the promoters of *glnA*, *glnII*, and the *amtB-glnK-glnD* operon (Rodríguez-García *et al.*, 2009). In the regulation of *glnA* and probably also in the control of *glnII* there is a direct competition between PhoP and GlnR proteins (Sola-Landa *et al.*, unpublished), which recognize overlapping sequences in its promoter (Rodríguez-García *et al.*, 2009). In the *amtB* promoter the PhoP and GlnR binding sequences are not overlapping, but PhoP appears to act as a 'road-block', preventing the correct formation of the RNA polymerase complex needed for the transcription. This double mechanism ensures a fine control of *glnA*, *glnII* and *amtB-glnK-glnD* by PhoP. A possible explanation for this control is that under phosphate limitation, the cells need to channel their energy resources to obtain phosphate and decrease the expression of the genes for nitrogen assimilation in order to equilibrate the P/N ratio in the cells.

Interestingly, PhoP binds to the promoter of *glnR* gene but not to the promoter of *glnRII*, suggesting that there is a 'second' nitrogen regulatory system independent of PhoP (Rodríguez-García *et al.*, 2009). The cross-regulation between phosphate and nitrogen metabolism is not symmetric since GlnR does not bind to promoters of the *pho* regulon (Sola-Landa *et al.*, unpublished results).

A negative effect of phosphate starvation over the expression of nitrogen metabolism genes has been described in *E. coli* and *Sinorhizobium meliloti* (VanBogelen *et al.*, 1996; Krol and Becker, 2004).

Future outlook

Taking into account all available information, it is clear that the two-component PhoR-PhoP system plays an important role on the control of primary and secondary metabolism in *Streptomyces*. The structure of the PhoP binding sites is now clearly established. One important lesson that we have learnt is that binding of PhoP monomers to the 11-nucleotide DRus occurs in a cooperative manner, and this allows considerable variability in the consensus sequence of the 11-nucleotide DRus. Further characterization of the location of the binding sites with regards to the transcription start point is required to explain the dual regulatory effect exerted by PhoP. In most cases PhoP acts as an activator of transcription of the PhoP-regulated genes. This is exerted by helping to recruit the RNA polymerase complex. However, in a few cases, PhoP exerts a negative regulation (i.e. it acts as transcriptional repressor) and this correlates with binding to the −10 region or even inside the ORF of the regulated gene.

Another important aspect of phosphate control is the interaction with other wide-domain regulators such as GlnR or AfsR. Multiple interaction or competition mechanisms are likely to occur in the cell, as a way of maintaining the equilibrium between different nutrients for optimal growth. The overlapping binding of the PhoP and GlnR regulators needs to be investigated in more detail.

Finally, the well-known phosphate control of the biosynthesis of secondary metabolites is exerted through PhoR-PhoP, but not exclusively. PhoP binds to the promoter of *afsS*; and this explains its effect on the expression of *actII*-ORF4 and *redD* that in turn control actinorhodin and undecylprodigiosin biosynthesis. Other pleiotropic regulators, yet unknown, are likely to be regulated by PhoP.

However, a significant part of phosphate-regulated genes are PhoP-independent (Rodríguez-García *et al.*, 2007). In these cases, alternative mechanisms of control are likely to occur. *Streptomyces* species contain many two-component systems and orphan response regulators. The possible involvement of some two-component systems other than PhoR-PhoP, or orphan regulators, in phosphate control of secondary metabolism has not been explored yet. The PhoR-PhoP system is conserved in many bacteria, and its role in phosphate control is also conserved, but this fact does not exclude that some other two component systems may also be involved in phosphate signal transduction, e.g. SenX3-RegX3, have been described in mycobacteria to play related roles (Glover *et al.*, 2007).

References

Allen, M.P., Zumbrennen, K.B., and McCleary, W.R. (2001). Genetic evidence that the alpha5 helix of the receiver domain of PhoB is involved in interdomain interactions. J. Bacteriol. *183*, 2204–2211.

Apel, A.K., Sola-Landa, A., Rodríguez-García, A., and Martín, J.F. (2007). Phosphate control of *phoA*, *phoC* and *phoD* gene expression in *Streptomyces coelicolor* reveals significant differences in binding of PhoP to their promoter regions. Microbiology *153*, 3527–3537.

Bentley, S.D., Chater, K.F., Cerdeno-Tarraga, A.M., Challis, G.L., Thomson, N.R., James, K.D., Harris, D.E., Quail, M.A., Kieser, H., Harper, D., *et al.* (2002). Complete genome sequence of the model actinomycete *Streptomyces coelicolor* A3(2). Nature *417*, 141–147.

Bérdy, J. (2005). Bioactive microbial metabolites, a personal view. J. Antibiot. *58*, 1–26.

Bhavsar, A.P., Erdman, L.K.; Schertzer, J.W., and Brown, E.D. (2004). Teichoic acid is an essential polymer in *Bacillus subtilis* that is functionally distinct from teichuronic acid. J. Bacteriol. *186*, 7865–7873.

Birck, C., Chen, Y., Hulett, F.M., and Samama, J.P. (2003). The crystal structure of the phosphorylation domain in PhoP reveals a functional tandem association mediated by an asymmetric interface. J. Bacteriol. *185*, 254–261.

Blanco, A.G., Solà, M., Gomis-Rüth, F.X., and Coll, M. (2002). Tandem DNA recognition by PhoB, a two-component signal transduction transcriptional activator. Structure *10*, 701–713.

Charaniya, S., Mehra, S., Lian, W., Jayapal, K.P., Karypis, G., and Hu, W.-S. (2007). Transcriptome dynamics-based operon prediction and verification in *Streptomyces coelicolor*. Nucleic Acids Res. *35*, 7222–7236.

Chen, Y., Birck, C., Samama, J.P., and Hulett, F.M. (2003). Residue R113 is essential for PhoP dimerization and function: a residue buried in the asymmetric PhoP dimer interface determined in the PhoPN three-dimensional crystal structure. J. Bacteriol. *185*, 262–273.

Choi, S.K., and Saier, Jr, M.H. (2005). Regulation of *pho* regulon gene expression by the carbon control protein A, CcpA, in *Bacillus subtilis*. J. Mol. Microbiol. Biotechnol. *10*, 40–50.

Chouayekh, H., and Virolle, M.-J. (2002). The polyphosphate kinase plays a negative role in the control of antibiotic production in *Streptomyces lividans*. Mol. Microbiol. *43*, 919–930.

Clardy, J., Fischbach, M.A., and Walsh, C.T. (2006). New antibiotics from bacterial natural products. Nat. Biotechnol. *24*, 1541–1550.

Dekeva, M.L., Titus, J.A., and Strohl, W.R. (1985). Nutrient effects on anthracycline production by *Streptomyces peucetius* in a defined medium. Can. J. Microbiol. *31*, 287–294.

Demain, A.L., and Vaishnav, P. (2004). Secondary metabolism in microbes and its control by phosphate and metals. S. I. M. News *54*, 104–113.

Díaz, M., Esteban, A., Fernández-Abalos, J.M., and Santamaría, R.I. (2005). The high-affinity phosphate-binding protein PstS is accumulated under high fructose concentrations and mutation of the corresponding gene affects differentiation in *Streptomyces lividans*. Microbiology *151*, 2583–2592.

Doull, J.L., and Vining, L.C. (1990). Nutritional control of actinorhodin production by *Streptomyces coelicolor* A3(2): suppressive effects of nitrogen and phosphate. Appl. Microbiol. Biotechnol. *32*, 449–454.

Eder, S., Liu, W., and Hulett, F.M. (1999). Mutational analysis of the *phoD* promoter in *Bacillus subtilis*: implications for PhoP binding and promoter activation of Pho regulon promoters. J. Bacteriol. *181*, 2017–2025.

Ellison., D.W., and McCleary, W.R. (2000). The unphosphorylated receiver domain of PhoB silences the activity of its output domain. J. Bacteriol. *182*, 6592–6597.

Esteban, A., Díaz, M., Yepes, A., and Santamaría, R.I. (2008). Expression of the *pstS* gene of *Streptomyces lividans* is regulated by the carbon source and is partially independent of the PhoP regulator. BMC Microbiol. *8*, 201.

Fabret, C., Feher, V.A., and Hoch, J.A. (1999). Two-component signal transduction in *Bacillus subtilis*: how one organism sees its world. J. Bacteriol. *181*, 1975–1983.

Fink, D., Weiβschuh, N., Reuther, J., Wohlleben, W., and Engels, A. (2002). Two transcriptional regulators GlnR and GlnRII are involved in regulation of nitrogen metabolism in *Streptomyces coelicolor* A3(2). Mol. Microbiol. *46*, 331–347.

Floriano, B., and Bibb, M. (1996). *afsR* is a pleiotropic but conditionally required regulatory gene for antibiotic production in A3(2). Mol. Microbiol., 385–396.

Ghorbel, S., Smirnov, A., Chouayekh, H., Sperandio, B., Esnault, C., Kormanec, J., and Virolle, M.-J. (2006). Regulation of *ppk* expression and in vivo function of Ppk in *Streptomyces lividans* TK24. J. Bacteriol. *188*, 6269–6276.

Glover, R.T., Kriakov, J., Garforth, S.J., Baughn, A.D., and Jacobs, Jr, W.R. (2007). The two-component regulatory system *senX3-regX3* regulates phosphate-dependent gene expression in *Mycobacterium smegmatis*. J. Bacteriol. *189*, 5495–5503.

Hobbs, G., Frazer, C.M., Gardner, D.C.J., Flett, F., and Oliver, S.G. (1990). Pigmented antibiotic production by *Streptomyces coelicolor* A3(2): kinetics and the influence of nutrients. J. Gen. Microbiol. *136*, 2291–2296.

Hoi, L.T., Voigt, B., Jürgen, B., Ehrenreich, A., Gottschalk, G., Evers, S., Feesche, J., Maurer, K.-H., Hecker, M., and Schweder, T. (2006). The phosphate-starvation response of *Bacillus licheniformis*. Proteomics *6*, 3582–3601.

Horinouchi, S., and Beppu, T. (1992). Autoregulatory factors and communication in actinomycetes. Annu. Rev. Microbiol. *46*, 377–398.

Hulett, F.M. (1996). The signal-transduction network for Pho regulation in *Bacillus subtilis*. Mol. Microbiol. *19*, 933–939.

Hutchings, M.I., Hoskisson, P.A., Chandra, G., and Buttner, M.J. (2004). Sensing and responding to diverse extracellular signals? Analysis of the sensor kinases and response regulators of *Streptomyces coelicolor* A3(2). Microbiology *150*, 2795–2806.

Kaiser, D., and Losick, D. (1993). How and why bacteria talk to each other. Cell 73, 873–885.

Krol, E., and Becker, A. (2004). Global transcriptional analysis of the phosphate starvation response in *Sinorhizobium meliloti* strains 1021 and 2011. Mol. Genet. Genomics 272, 1–17.

Lee, P.C., Umeyama, T., and Horinouchi, S. (2002). *afsS* is a target of AfsR, a transcriptional factor with ATPase activity that globally controls secondary metabolism in *Streptomyces coelicolor* A3(2). Mol. Microbiol. *43*, 1413–1430.

Liras, P., Asturias, J.A., and Martín, J.F. (1990). Phosphate control sequences involved in transcriptional regulation of antibiotic biosynthesis. Trends Biotechnol. *8*, 184–189.

Lounes, A., Lebrihi, A., Benslimane, C., Lefebvre, G., and Germain, P. (1996). Regulation of spiramycin synthesis in *Streptomyces ambofaciens*: Effects of glucose and inorganic phosphate. Appl. Microbiol. Biotechnol. *45*, 204–211.

Makarewicz, O., Dubrac, S., Msadek, T., and Borriss, R. (2006). Dual role of the PhoP~P response regulator: *Bacillus amyloliquefaciens* FZB45 phytase gene transcription is directed by positive and negative interactions with the *phyC* promoter. J. Bacteriol. *188*, 6953–6965.

Mapplestone, R.A., Stone, M.J., and Williams, D.H. (1992). The evolutionary role of secondary metabolites – a review. Gene *115*, 151–157.

Martín, J.F. (1989). Molecular mechanism for the control by phosphate of the biosynthesis of antibiotic and secondary metabolites. In Regulation of Secondary Metabolism in Actinomycetes, Shapiro, S., ed. (Boca Raton, CRC Press), pp. 213–236.

Martín, J.F. (2004). Phosphate control of the biosynthesis of antibiotics and other secondary metabolites is mediated by the PhoR-PhoP system: an unfinished story. J. Bacteriol. *186*, 5197–5201.

Martín, J.F., and Demain, A. (1980). Control of antibiotic biosynthesis. Microbiol. Rev. *44*, 230–251.

Martín, J.F., Marcos, A.T., Martín, A., Asturias, J.A., and Liras, P. (1994). Phosphate control of antibiotic biosynthesis at the transcriptional level. In Phosphate in Microorganisms, Torriani-Gorini, A., Yagil, E., and Silver, S., eds (ASM Press, Washington), pp. 140–147.

Martín, J.F., Gutiérrez, S., and Aparicio, J.F. (2000). Secondary metabolites. In Encyclopedia of Microbiology, vol. 4, 2nd edn., Lederberg, J., ed. (Academic Press, San Diego), pp. 213–237.

Masuma, R., Tanaka, Y., Tanaka, H., and Omura, S. (1986). Production of nanaomycin and other antibiotics by phosphate-depressed fermentation using phosphate-trapping agents. J. Antibiotics *39*, 1557–1564.

Matin, A., Auger, E.A., Blum, P.H., and Schultz, J.E. (1989). Genetic basis of starvation survival in non-differentiating bacteria. Annu. Rev. Microbiol. *43*, 293–316.

Matsumoto, A., Ishizuka, H., Beppu, T., and Horinouchi, S. (1995). Involvement of a small ORF downstream of the *afsR* gene in the regulation of secondary metabolism in *Streptomyces coelicolor* A3(2). Actinomycetologica 9, 37–43.

Mendes, M.V., Tunca, S., Antón, N., Recio, E., Sola-Landa, A., Aparicio, J.F., and Martín, J.F. (2007). The two-component *phoR-phoP* system of *Streptomyces natalensis*: Inactivation or deletion of *phoP* reduces the negative phosphate regulation of pimaricin biosynthesis. Metabol. Engineer. 9, 217–227.

Merrick, M.J., and Edwards, R.A. (1995). Nitrogen control in bacteria. Microbiol. Rev. *59*, 604–622.

Oh, W.S., Im, Y.S., Yeon, K.Y., Yoon, Y.J., and Kim, J.W. (2007). Phosphate and carbon source regulation of alkaline phosphatase and phospholipase in *Vibrio vulnificus*. J. Microbiol. *45*, 311–317.

Perron-Savard, P., De Crescenzo, G., and Le Moual, H. (2005). Dimerization and DNA binding of the *Salmonella enterica* PhoP response regulator are phosphorylation independent. Microbiology *151*, 3979–3987.

Puri-Taneja, A., Paul, S., Chen, Y., and Hulett, F.M. (2006). CcpA causes repression of the *phoPR* promoter through a novel transcription start site, P_{A6}. J. Bacteriol. *188*, 1266–1278.

Quiquampoix, H., and Mousain, D. (2005). Enzymatic hydrolysis of organic phosphorus. In Organic Phosphorus in the Environment, Turner, B.L., Frossard, E., and D.S Baldwin, eds. (CABI Publishing, Cambridge, MA, USA,), pp. 89–112.

Reuther, J., and Wohlleben, W. (2007). Nitrogen metabolism in *Streptomyces coelicolor*: transcriptional and post-translational regulation. J. Mol. Microbiol. Biotechnol. *12*, 139–146.

Rigali, S., Titgemeyer, F., Barends, S., Mulder, S., Thomae, A.W., Hopwood, D.A., and van Wezel, G.P. (2008). Feast or famine: the global regulator DasR links nutrient stress to antibiotic production by *Streptomyces*. EMBO Rep. *9*, 670–675.

Rodríguez-García, A., Barreiro, C., Santos-Beneit, F., Sola-Landa, A., and Martín, J.F. (2007). Genome-wide transcriptomic and proteomic analysis of the primary response to phosphate limitation in *Streptomyces coelicolor* M145 and in a Δ*phoP* mutant. Proteomics *7*, 2410–2429.

Rodríguez-García, A., Sola-Landa, A., Apel, K., Santos-Beneit, F., and Martín, J.F. (2009). Phosphate control over nitrogen metabolism in *Streptomyces coelicolor*: direct and indirect negative control of *glnR*, *glnA*, *glnII* and *amtB* expression by the response regulator PhoP. Nucleic Acids Res. *37*, 3230–3242.

Romero, J., Liras, P., and Martín, J.F. (1984). Dissociation of cephamycin and clavulanic acid biosynthesis in *Streptomyces clavuligerus*. Appl. Microbiol. Biotechnol. *20*, 318-325.

Santos-Beneit, F., Rodríguez-García, A., Franco-Domínguez, E., and Martín, J.F. (2008). Phosphate-dependent regulation of the low- and high-affinity transport systems in the model actinomycete *Streptomyces coelicolor*. Microbiology *154*, 2356–2370.

Santos-Beneit, F., Rodríguez-García, A., Sola-Landa, A., and Martín, J.F. (2009a). Cross-talk between two global regulators in *Streptomyces*: PhoP and AfsR interact in the control of *afsS*, *pstS* and *phoRP* transcription. Mol. Microbiol. 72, 53–68.

Santos-Beneit, F., Rodríguez-García, A., Apel, A.K., and Martín, J.F. (2009b). Phosphate and carbon source regulation of two PhoP-dependent glycerophosphodiester phosphodiesterase genes of *Streptomyces coelicolor*. Microbiology *155*, 1800–1811.

Schaaf, S., and Bott, M. (2007). Target genes and DNA-binding sites of the response regulator PhoR from *Corynebacterium glutamicum*. J. Bacteriol. *189*, 5002–5011.

Schneider, T.D. (1997). Information content of individual genetic sequences. J. Theor. Biol. *189*, 427–441.

Sinha, A., Gupta, S., Bhutani, S., Pathak, A., and Sarkar, D. (2008). PhoP–PhoP interaction at adjacent PhoP binding sites is influenced by protein phosphorylation. J. Bacteriol. *190*, 1317–1328.

Sola-Landa, A., Moura, R.S., and Martín, J.F. (2003). The two-component PhoR-PhoP system controls both primary metabolism and secondary metabolite biosynthesis in *Streptomyces lividans*. Proc. Natl. Acad. Sci. U.S.A. *100*, 6133–6138.

Sola-Landa, A., Rodríguez-García, A., Franco-Domínguez, E., and Martín, J.F. (2005). Binding of PhoP to promoters of phosphate-regulated genes in *Streptomyces coelicolor*: identification of PHO boxes. Mol. Microbiol. *56*, 1373–1385.

Sola-Landa, A., Rodríguez-García, A., Apel, A.K., and Martín, J.F. (2008). Target genes and structure of the direct repeats in the DNA binding sequences of the response regulator PhoP in *Streptomyces coelicolor*. Nucleic Acids Res. *36*, 1358–1368.

Suzuki, S., Ferjani, A., Suzuki, I., and Murata, N. (2004). The SphS-SphR two component system is the exclusive sensor for the induction of gene expression in response to phosphate limitation in *Synechocystis*. J. Biol. Chem. *279*, 13234–13240.

Tanaka, A., Takano, Y., Ohnishi, Y., and Horinouchi, S. (2007). AfsR recruits RNA polymerase to the *afsS* promoter: a model for transcriptional activation by SARPs. J. Mol. Biol. *369*, 322–333.

Tiffert, Y., Supra, P., Wurm, R., Wohlleben, W., Wagner, R., and Reuther, J. (2008). The *Streptomyces coelicolor* GlnR regulon: identification of new GlnR targets and evidence for a central role of GlnR in nitrogen metabolism in actinomycetes. Mol. Microbiol. *67*, 861–880.

Titball, R.W. (1993). Bacterial phospholipases C. Microbiol. Rev. *57*, 347–366.

Torriani-Gorini, A. (1994). Introduction: the Pho regulon of *Escherichia coli*. In Phosphate in Microorganisms. Torriani-Gorini, A., Yagil, E., and Silver, S., eds (ASM Press, Washington), pp.1–4.

Uesugi, Y., and Hatanaka, T. (2009). Phospholipase D mechanism using *Streptomyces* PLD. Biochim. Biophys. Acta. *1791*, 962–969.

VanBogelen, R.A., Olson, E.R., Wanner, B.L., and Neidhardt, F.C. (1996). Global analysis of proteins synthesized during phosphorus restriction in *Escherichia coli*. J. Bacteriol. *178*, 4344–4366.

Vining, L.C. (1992). Secondary metabolism, inventive evolution and biochemical diversity – a review. Gene *115*, 135–140.

Vögtli, M., Chang, P.C., and Cohen, S.N. (1994). afsR2: a previously undetected gene encoding a 63-amino acid protein that stimulates antibiotic production in *Streptomyces lividans*. Mol. Microbiol. *14*, 643–653.

Wanner, B.L. (1996). Phosphorus assimilation and control of the phosphate regulon. In *Escherichia coli* and *Salmonella*, 2nd edn., Neidhardt, F.C., Ingraham, J.L., Lin, E.C.C., Low, K.B., Magasanik, B., Reznikoff, W.S., Riley, M., Schaechter, M., and Umbarger, H.E., eds (American Society for Microbiology, Washington D.C.), pp. 1357–1381.

Wösten, M.M., Parker, C.T., van Mourik. A., Guilhabert, M.R., van Dijk. L., and van Putten, J.P. (2006). The *Campylobacter jejuni* PhosS/PhosR operon represents a non-classical phosphate-sensitive two-component system. Mol. Microbiol. *62*, 278–291.

Wray, L.V., and Fisher, S.H. (1993). The *Streptomyces coelicolor glnR* gene encodes a protein similar to other bacterial response regulators. Gene *130*, 145–150.

Yamane, T., Okamura, H., Ikeguchi, M., Nishimura, Y., and Kidera, A. (2008). Water-mediated interactions between DNA and PhoB DNA-binding/transactivation domain: NMR-restrained molecular dynamics in explicit water environment. Proteins *71*, 1970–1983.

Yuan, Z.C., Zaheer, R., Morton, R., and Finan, T.M. (2006). Genome prediction of PhoB regulated promoters in *Sinorhizobium meliloti* and twelve proteobacteria. Nucleic Acids Res. *34*, 2686–2697.

γ-Butyrolactones and their Role in Antibiotic Regulation

8

Marco Gottelt, Stefan Kol and Eriko Takano

Abstract

γ-Butyrolactones are small signalling molecules regulating antibiotic production and sometimes also morphological development in streptomycetes. The related regulatory systems are complex, and some are species specific. Focusing on well-studied examples in *S. griseus*, *S. virginiae* and *S. coelicolor*, we depict common basic features and differences between the γ-butyrolactone systems. Biosynthesis and structural diversity of the γ-butyrolactones are described, as well as the mode of action and structure–function relationship of the bacterial 'hormone receptors'. We also report the recent discovery of a previously undescribed antibiotic compound in *S. coelicolor* by deletion of a γ-butyrolactone homologue, and the identification and characterization of a second form of the major γ-butyrolactone receptor in the model streptomycete *S. coelicolor*.

Bacterial communication and 'microbial hormones'

Collective behaviour in bacteria is often governed by small diffusible signalling molecules. The information relayed by these molecules synchronizes the activities of large groups of cells, and thus allows bacteria to modulate gene expression on a population-wide scale. This synchronization is known as quorum sensing, while the signalling molecules are often referred to colloquially as 'bacterial hormones' or more formally as autoinducers. The population density dependent concentration of such autoinducers affects, for example, bioluminescence in *Vibrio fischeri* (Fuqua *et al.*, 2001), and virulence and biofilm formation in *Pseudomonas aeruginosa* (Williams and Camara, 2009). Two examples of small-molecule autoinducers are *N*-acyl-homoserine lactones (AHLs) in Gram-negative and modified oligopeptides in Gram-positive bacteria (Camilli and Bassler, 2006) (Fig. 8.1).

γ-Butyrolactone regulatory systems in *Streptomyces*

In actinomycetes, the most commonly used signalling molecules are γ-butyrolactones (or butanolides). Diffusible and effective in nanomolar concentrations, butanolides induce and synchronize antibiotic production and, in some cases, also morphological differentiation in the mycelial colonies of streptomycetes (Bibb, 2005; Horinouchi, 2002).

In this chapter, we depict the general characteristics of the γ-butyrolactone-dependent regulatory systems in *Streptomyces*, focusing on the three best-studied species: the A-factor system in *S. griseus*, the virginiae butanolide system in *S. virginiae* and, in particular, the *S. coelicolor* butanolide system. Differences and similarities among these and further *Streptomyces* species will be discussed to demonstrate the diversity and complexity of this system.

The A-factor system in *Streptomyces griseus*

The best studied *Streptomyces* γ-butyrolactone system is the A-factor system (Fig. 8.2) regulating streptomycin and grixazone production, as well as

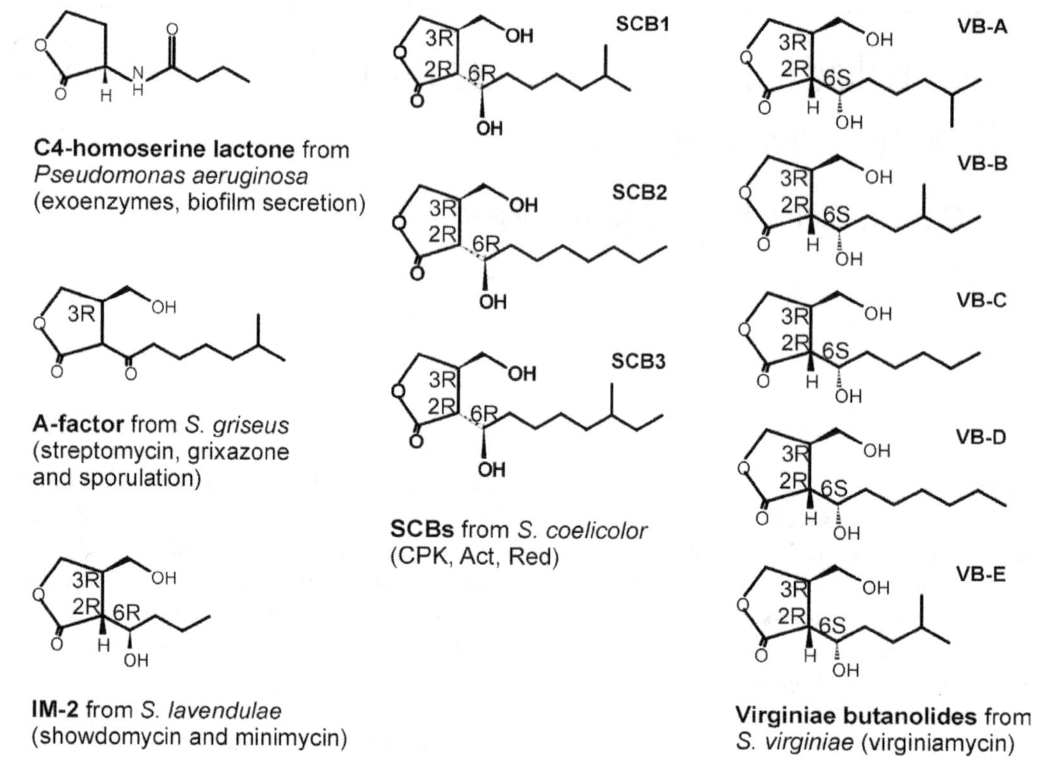

Figure 8.1 Chemical structures of γ-butyrolactones from *Streptomyces* and C4-homoserine lactone from *Pseudomonas aeruginosa* (adapted from Takano, 2006). The name of the signalling molecule appears in bold type, and the antibiotic that it affects or its other functions are listed in brackets.

sporulation in *S. griseus*; several comprehensive reviews on this system have been published (e.g. in Bibb, 2005; Horinouchi, 2007).

Pioneering work by Khokhlov and co-workers (Khokhlov *et al.*, 1967) originally identified a diffusible factor that was able to induce sporulation and streptomycin production. Subsequent work by Hara and Beppu using mutants deficient in streptomycin production and sporulation (Hara and Beppu, 1982), led to a re-evaluation of Khokhlov's work and finally to a detailed description of the A-factor regulatory cascade and its biosynthesis (Horinouchi, 2007).

The A-factor butanolide is synthesized by AfsA from a glycol derivative and a β-keto acid (derived from fatty acid biosynthesis) (Ando *et al.*, 1997; Kato *et al.*, 2007). A-factor binds to the cytoplasmic transcriptional repressor and A-factor receptor protein ArpA. ArpA consequently dissociates from the promoter region of *adpA* encoding a pleiotropic transcriptional activator. The A-factor dependent protein AdpA binds to the promoter regions of many genes required for morphological development and secondary metabolism. For streptomycin biosynthesis, AdpA activates *strR*, which encodes the pathway-specific transcriptional activator for the streptomycin biosynthesis gene cluster (Horinouchi, 2007). AdpA was also reported to auto-repress its own expression (Kato *et al.*, 2005) and seems to impair A-factor synthesis (Kato *et al.*, 2004). Consistent with the regulatory cascade described, an *afsA* mutant lost A-factor productivity (Hara and Beppu, 1982), and an *arpA* mutant showed precocious antibiotic production (Miyake *et al.*, 1990).

The virginiae butanolide system in *Streptomyces virginiae*

Streptomyces virginiae produces the antibiotics virginiamycin M (VM) and virginiamycin S (VS), both of which belong to the family of streptogramin antibiotics. The biosynthesis of VM and VS in *S. virginiae* is initiated by the production of virginiae butanolide (VB) (Fig. 8.1). The γ-butyrolactone binds to the VB-specific receptor protein BarA, a DNA-binding transcriptional repressor (Kinoshita *et al.*, 1997). The flanking regions of *barA* contain other important genes that are involved in VB biosynthesis (*barX*, *barS1* and *barS2*) and mechanisms of resistance to VM and VS (*varM*, *varR* and *varS*) (Kawachi *et al.*, 2000a; Namwat *et al.*, 2001).

The *Streptomyces coelicolor* signalling system

The butanolide system

In the *S. coelicolor* γ-butyrolactone regulatory system (Figs. 8.2 and 8.3) (reviewed in Takano, 2006), *Streptomyces coelicolor* butanolides (SCBs) are produced at late transition phase dependent on the AfsA homologue ScbA (Takano *et al.*, 2005). The structures of three distinct SCB compounds, SCB1, 2 and 3 (Fig. 8.1), have been determined (Takano *et al.*, 2000) and neither was produced in a *scbA* mutant (Takano *et al.*, 2001). From bioassays, at least six γ-butyrolactones were predicted in *S. coelicolor* (Anisova *et al.*, 1984; Kawabuchi *et al.*, 1997). The SCBs bind to the ArpA receptor homologue ScbR and prevent DNA binding of the transcriptional regulator. *ScbR*, *scbA* and *cpkO* have been experimentally identified as target genes of ScbR (Takano *et al.*, 2001; Takano *et al.*, 2005). CpkO is the pathway specific regulator of the *cpk* gene cluster (formerly called *kas* gene cluster (Pawlik *et al.*, 2007). Thus, in addition to its own expression, ScbR also regulates the production of the signalling molecules of the *S. coelicolor* butanolide system, as well as an secondary metabolite biosynthesis gene cluster. ScbR was also shown to affect two pigmented *S. coelicolor* antibiotics, the blue-coloured actinorhodin (Act) (Bystrykh *et al.*, 1996; Rudd and Hopwood, 1979), and the red-pigmented prodigiosins (Red) (Feitelson *et al.*, 1985) (Takano *et al.*, 2001).

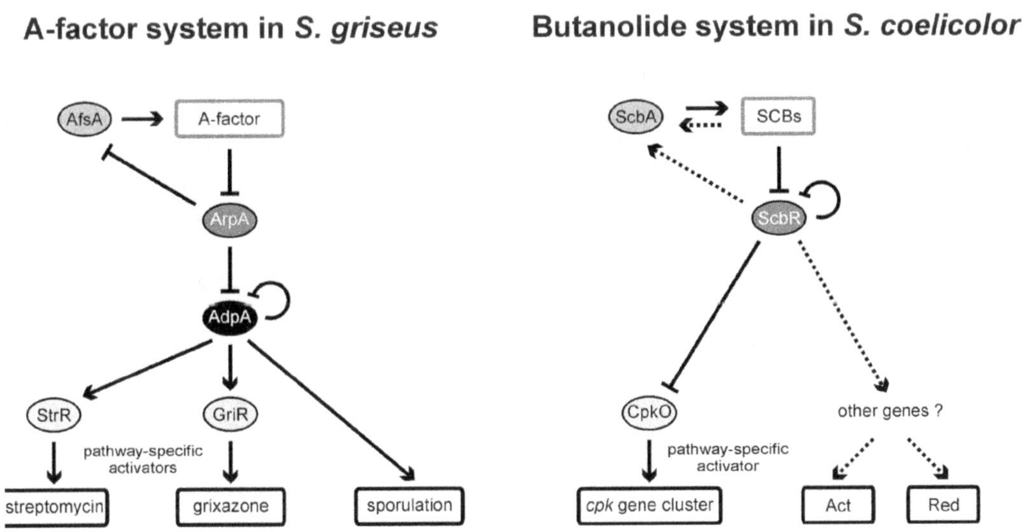

Figure 8.2 Schematic representations of the A-factor regulatory cascade of *Streptomyces griseus* and the butanolide system of *Streptomyces coelicolor*. Arrows denote activation and lines with a bar denote repression. Dotted lines represent effects currently under investigation. Homologues or homologous functions are indicated. Detailed descriptions are given in the text. For the *S. coelicolor* system, also see Fig. 8.3.

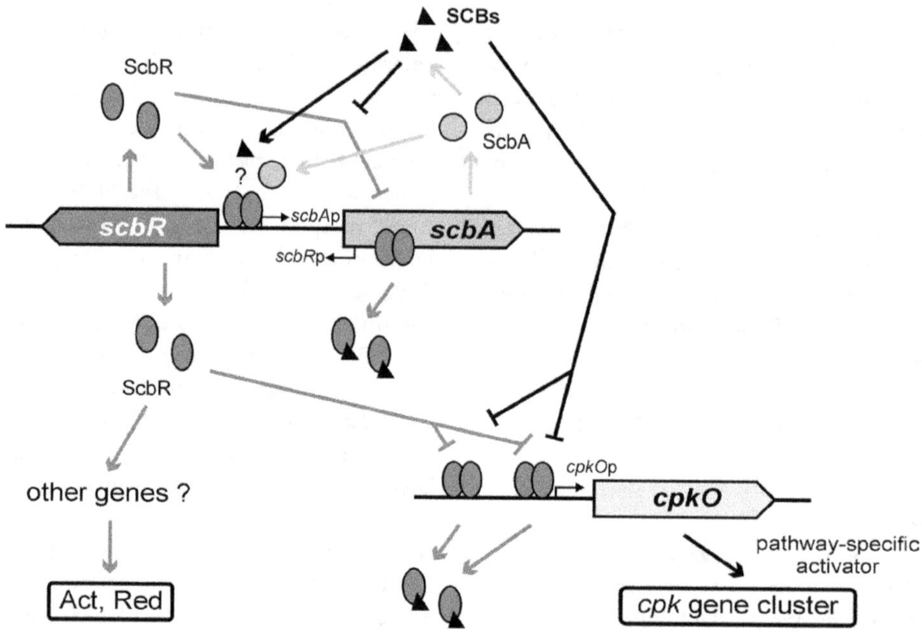

Figure 8.3 Detailed schematic model of the butanolide system in *Streptomyces coelicolor* (adapted from Takano, 2006). ScbR (ovals) binds as a dimer to 4 targets sites. One repressor site is in front of its own promoter (*scbR*p) and two other repressor sites are in front of the *cpkO* (pathway-specific activator) promoter (*cpkO*p). In all cases, the γ-butyrolactones (SCBs; triangles) bind to ScbR (ovals with triangle) and relieves the repression. ScbA (circles) is involved in production of SCBs and also is required along with SCBs and ScbR for its own expression (*scbA*p, the *scbA* promoter). The precise mechanism is under investigation (shown as '?'). ScbR might regulate other genes which could in turn regulate the production of Act and Red. CpkO actively regulates the transcription of the *cpk* gene cluster. Arrows denote activation and lines with a bar denote repression.

In total, four binding sites were identified in front of the three ScbR target genes: Site R in front of *scbR*, site A in front of *scbA* and site O_A and site O_B in front of *cpkO* (Fig. 8.3) (Takano, 2006). Site R and site O_A show an identical inverted repeat sequence, whereas site O_B and site A are less congruent with the deduced ScbR binding consensus sequence. *In silico* analysis predicted a single additional fully conserved ScbR binding site in the *S. coelicolor* chromosome in front of *orfB* (Takano *et al.*, 2005). *OrfB* is located only three genes downstream of *scbR*, separated by *scbA* and *scbB*. Binding of ScbR in front of the additional target gene was also shown experimentally (Takano, unpublished). The function of the predicted histidine kinase OrfB, however, is not known. Interestingly, only for the conserved binding sites, site R and site O_A, ScbR has been confirmed as transcriptional repressor. ScbR binding at site A was suggested to possibly promote activation (Takano *et al.*, 2001), and preliminary *in vitro* data indicate a corresponding function of ScbR also at site O_B (Takano, unpublished). However, the putative dualistic function of ScbR at its multiple binding sites (repressor at site R and site O_A, activator at site A and site O_B) is not consistent with the position of the sites. Site A and site O_A overlap with the –10 and -35 region of the transcriptional start site of *scbA* and *cpkO*, respectively, whereas site R and site O_B are located –42 nt to –68 nt and -222 nt to –244 nt upstream of *scbR* and *cpkO*, respectively. Given corresponding functions also *in vitro*, this indicates that the regulatory processes taking place at the ScbR target sites involve additional regulatory proteins and consequently are even more complex.

Other signals in S. coelicolor

Recently, in addition to the butanolide system in *S. coelicolor*, a new class of signalling molecules, the methylenomycin furans (MMFs; 2-alkyl-4-hydroxymethylfuran-3-carboxylic acids) regulating production of the plasmid-encoded antibiotic methylenomycin, was reported (Corre *et al.*, 2008). The MMF biosynthesis genes are also located on the autonomous *Streptomyces coelicolor* plasmid SCP1, which is missing in the sequenced strain M145, and MMF production was shown to be independent from the chromosomally encoded γ-butyrolactone biosynthetic pathway (O'Rourke *et al.*, 2009).

S. coelicolor butanolide system target: the CPK antibiotic biosynthesis gene cluster

There is evidence that the butanolide system in *S. coelicolor* primarily provides regulation for the *cpk* cluster. The previously observed effects on the production of Act and Red seem to be downstream effects only, e.g. by competition for common precursors (Takano, 2006), since direct regulation, as shown for the *cpk* gene cluster via CpkO, could not be shown for the *act* and *red* clusters (Takano *et al.*, 2001). Two novel *S. coelicolor* secondary metabolites, the yellow pigment yCPK (yellow *coelicolor* polyketide) and the antibacterial compound abCPK (antibiotic *coelicolor* polyketide), related to the hitherto orphan *cpk* biosynthesis gene cluster were described recently (Gottelt *et al.*, 2010).

Fig. 8.4 shows a comprehensive hypothetical model of the *cpk* biosynthetic pathway and its regulation. The model summarizes the results presented in Gottelt *et al.* (submitted) and in other relevant publications, as well as amino acid sequence (homology) analysis of the *cpk* gene products using BLASTP (http://blast.ncbi.nlm.nih.gov/Blast.cgi, http://www.expasy.org/tools/blast/; Gasteiger *et al.*, 2003), Tatfind (http://signalfind.org/tatfind.html; Rose *et al.*, 2002), SignalP (http://www.cbs.dtu.dk/services/SignalP/; Bendtsen *et al.*, 2004; Pawlik *et al.*, 2007) and computational prediction of the functions of Cpk proteins in cellular processes of *S. coelicolor* [using KEGG (Kanehisa *et al.*, 2009) and BRENDA (Chang *et al.*, 2009)].

CpkPα and CpkPβ are predicted to form together the branched-chain α-keto acid dehydrogenase (BCDH) complex. The CpkPαβ holoenzyme is supposed to catalyse the conversion of 2-oxoglutarate into succinyl-CoA in the TCA cycle of *S. coelicolor* (KEGG). Glutamate was shown to increase the production of yCPK and abCPK (Gottelt *et al.*, 2010). Interestingly, homologues of the putative class-III aminotransferase CpkG were predicted to form 2-oxoglutarate from glutamate (BLASTP, ExPASy, BRENDA). Thus, CpkG may catalyse a direct anaplerotic reaction of the TCA cycle. However, both reactions are reversible and, thus, CpkPαβ and CpkG could also provide glutamate from TCA cycle intermediates. Glutamate and other ω-amino acids could also be used by CpkG as donor substrate in a direct transamination (Pawlik *et al.*, 2007; Yonaha *et al.*, 1992) of a metabolite of the *cpk* biosynthesis pathway. A corresponding hypothetical role for CpkG is discussed in detail later and is denoted in Fig. 8.4. Any of these proposed enzymatic functions may be involved in the stimulating effect of glutamate on the production of the *cpk* products reported in Gottelt *et al.* (submitted). Clearly, more work is needed to elucidate the dependency of yCPK and abCPK production on primary metabolism and on specific nitrogen sources in particular.

During antibiotic production the acetyl-CoA flux may be redirected from entering the TCA cycle into the biosynthesis of the polyketide precursor malonyl-CoA (de Carvalho Lima Lobato *et al.*, 2007). The putative AccA1/CpkK Acyl-CoA carboxylase complex (Pawlik *et al.*, 2007) is predicted to form malonyl-CoA from acetyl-CoA and other acyl-CoA species (Rodriguez and Gramajo, 1999). In this case, homeostasis of the TCA cycle may be impaired and its metabolic balance then depends on 3-phosphoglycerate and anaplerotic reactions (de Carvalho Lima Lobato *et al.*, 2007), such as the one suggested for CpkG.

In total, six malonyl-CoA units are used as starter and extender units in the formation of a first aliphatic, unsaturated C_{12} CPK polyketide intermediate by the Type I polyketide synthase (PKS) subunits CpkA, CpkB and CpkC (Pawlik *et al.*, 2007). The thioesterase ScoT most probably removes acyl residues different from malonyl-CoA that block the PKS extension modules in CpkABC (Kotowska *et al.*, 2002; Kotowska *et al.*, 2009). The primary polyketide may subsequently be further modified by CpkE,

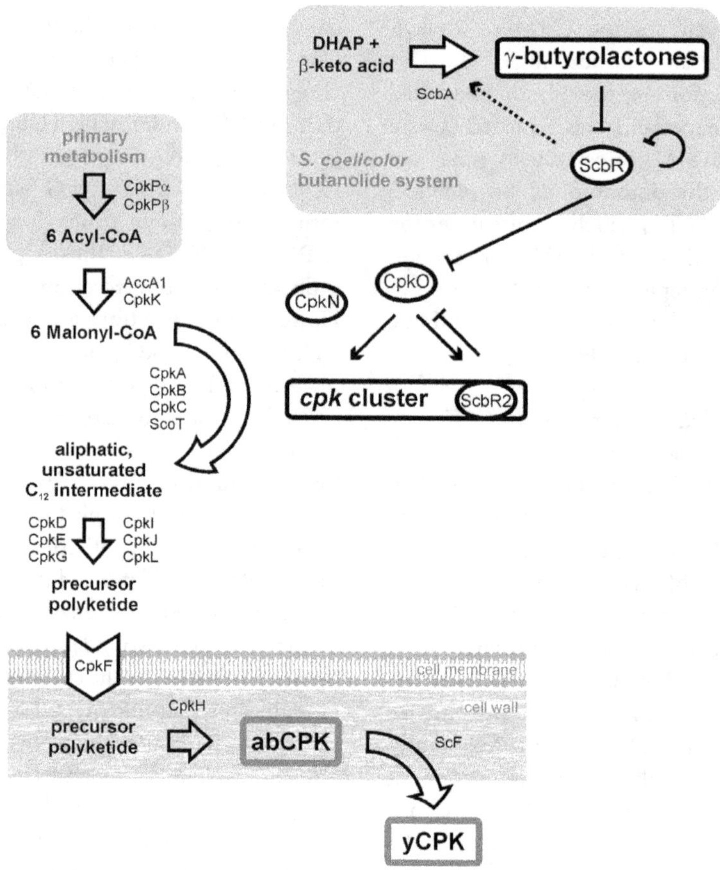

Figure 8.4 Hypothetical model of the CPK biosynthetic pathway and its regulation (explanation in the text).

a predicted epoxide hydrolase, CpkI, a putative 3-oxoacyl-ACP reductase, CpkJ, a nucleoside-diphosphate-sugar epimerase homologue, the hypothetical protein CpkL, and by CpkD (Pawlik *et al.*, 2007). Although CpkD is predicted to be a secreted FAD-binding protein in view of a Twin-Arginine motif in its putative signal peptide (Signal P; Pawlik *et al.*, 2007), experimentally, it could not be verified as substrate for the Tat secretion system (Widdick *et al.*, 2006).

The tailored *cpk* precursor polyketide could then be transported out of the cell via the putative transmembrane efflux protein CpkF (Pawlik *et al.*, 2007). CpkH and ScF are Tat-dependent, cell wall-associated (Widdick *et al.*, 2006) FAD-binding oxidoreductase homologues (Pawlik *et al.*, 2007). Based on the phenotype of a *scF* mutant that produces abCPK, but not yCPK, we propose that the precursor is first converted into the antibiotic abCPK by CpkH, and then subsequently modified to the yellow pigment yCPK by ScF. In *S. coelicolor* liquid cultures, abCPK could be extracted best from the cells, in contrast to the secreted yCPK. This is in agreement with the localization of CpkH and ScF in the cell wall: although also found diffusing into solid media (Gottelt *et al.*, 2010), the intermediate abCPK may mainly stay attached to the cell before being further converted to yCPK.

ScF and CpkH show high sequence homology (BLASTP) and protein structure similarity (M.W. Fraaije, personal communication) to AknOx, a secreted oxidoreductase in *S. galilaeus* involved in aclacinomycin biosynthesis. AknOx catalyses the modification of the sugar residue of this antibiotic (Alexeev *et al.*, 2007). Also, pyridoxal phosphate (PLP)-dependent transaminases, such as CpkG, are involved in the synthesis of amino sugars that subsequently serve as glycone moiety in *Streptomyces* secondary metabolites (Liu and Thorson, 1994), e.g. DnmJ in daunosamine biosynthesis in the doxorubicin producer *S. peuceticus* (Hutchinson and Colombo, 1999). Interestingly, with FdtB from *Aneurinibacillus thermoaerophilus*, among several amino group donors tested, including glutamine, alanine, and aspartate, only glutamate resulted in a turnover (Pfoestl *et al.*, 2003). This suggests a putative role of glutamate and CpkG in amino sugar synthesis in the *cpk* biosynthetic pathway. This would be consistent with a function of CpkG in the direct transamination of a *cpk*-related metabolite proposed above. Furthermore, also the epimerase CpkJ is predicted to act on sugars (Pawlik *et al.*, 2007). A likely function of CpkJ is the conversion of a C-1 activated nucleoside diphospho-sugar, typically the TDP-D-glucose derivative TDP-4-keto-6-deoxy-D-glucose. Usually, the resulting activated deoxy-L-hexose isomer is then subsequently incorporated in glycosylated antibiotics (Walsh *et al.*, 2003). All these findings suggest the glycosylation of an intermediate of the *cpk* biosynthetic pathway and the presence of a sugar moiety attached to the polyketide aglycone of yCPK and abCPK.

On the other hand, the *cpk* gene cluster does not provide various additional gene functions needed for, e.g. glycon synthesis, and also no glycosyltransferase is predicted in the *cpk* gene cluster (Pawlik *et al.*, 2007). Generally, the genes encoding the dedicated glycosyltransferases are found in the same gene cluster that contains the genes to make the particular activated deoxysugars and the aglycon (Walsh, 2002).

However, we cannot exclude that the missing enzymatic functions for the putative glycosylation of the *cpk* product are encoded elsewhere on the *S. coelicolor* chromosome, outside of the *cpk* cluster. In the elloramycin producer *Streptomyces olivaceus* it has been demonstrated that a small glycon biosynthesis gene cluster and the polyketide aglycon biosynthesis gene cluster are present on separated chromosomal loci (Ramos *et al.*, 2008). In *Saccharopolyspora spinosa*, only one set of biosynthetic genes is present, providing rhamnose for both primary structural components (cell walls) and a secondary metabolite (spinosyn). Also, in *S. spinosa* the sugar biosynthesis genes are found far away from the spinosyn aglycon gene cluster (Madduri *et al.*, 2001). Biosynthesis of the heterocyst envelope polysaccharide (HEP) in the filamentous cyanobacterium *Anabaena sp.* is mainly encoded in the HEP genetic island. Glycosyltransferases essential for HEP production, however, are located outside the HEP island (Wang *et al.*, 2007). The given examples suggest that also in yCPK and abCPK production yet unidentified gene functions may be encoded outside of the *cpk* gene cluster, involved in synthesis and attachment of a putative glycon moiety. CpkG, CpkJ, and possibly also glutamate, may be involved in deoxy-sugar biosynthesis; CpkH and ScF may successively modify the glycon resulting in abCPK and yCPK, respectively. Structure elucidation of the *cpk* metabolites may eventually reveal whether they have one or more sugars attached, or whether a sugar-like structure is present in an unglycosylated *cpk* polyketide.

CpkN is a predicted *Streptomyces* antibiotic regulatory protein (SARP) and overexpression of CpkN leads to enhanced production of yCPK (Pawlik *et al.*, poster 47, Biology of *Streptomyces*, Münster, Germany, 2009). Its function may therefore resemble that of CpkO, the other SARP regulator and pathway-specific activator of the gene cluster. A third, presumably negative, regulatory gene, ScbR2, together with CpkO might contribute to negative feedback regulation of the *cpk* genes (Gottelt *et al.*, 2010). As yet, DNA binding sites and direct target genes have been characterized for none of these transcriptional regulators.

Via CpkO that is under the direct control of ScbR (Takano *et al.*, 2005), regulation of the *cpk* gene cluster is linked to the *S. coelicolor* butanolide system (Fig. 8.4). The results presented in Gottelt *et al.* (submitted), thus provide for the first time a continuous regulatory cascade from the bacterial hormones (the γ-butyrolactones) to secondary metabolites (abCPK and yCPK) in *S. coelicolor*. Furthermore, the identification of abCPK is the first successful example of the induced production of a hitherto undescribed antibiotic in a streptomycete by the manipulation of a regulatory gene (*scbR2*). This concept could be applied not only to the *cpk* gene cluster, but to virtually all orphan gene clusters; not only in *S. coelicolor* but essentially in all streptomycetes and even beyond the genus. Hence, awakening these sleeping genes affords a promising chance for the discovery of countless novel secondary metabolites and potentially new antibiotics.

γ-Butyrolactone regulatory system diversity in *Streptomyces*

Although both the composition and the molecular mode of action of the key players is studied best with the A-factor system in *S. griseus*, it may not serve as a general model for *Streptomyces* γ-butyrolactones regulatory systems.

Genes involved in the synthesis and binding of the γ-butyrolactones, respectively, are often located in close vicinity of each other and, furthermore, in or close to the target antibiotic biosynthesis gene cluster. ScbA and ScbR in *S. coelicolor* are neighbours and are situated only a few genes upstream from the CPK gene cluster (Fig. 8.2). In contrast, AfsA and ArpA in *S. griseus* are ~100 kb away from each other and ArpA, involved in the regulation of streptomycin production, is not co-located with the cognate gene cluster (Bibb, 2005; Takano, 2006).

AdpA is the central transcriptional activator in *S. griseus*. The *S. coelicolor* butanolide system lacks a counterpart with an analogous function. Nevertheless, we find *adpA* homologues in *S. coelicolor*. With 84% amino acid identity, the one that most closely resembles *adpA* from *S. griseus* is part of the *Bld* cascade related to development and its deletion results in the *bldH* mutant phenotype. Development is suggested not to be influenced by γ-butyrolactones in *S. coelicolor* and, even more important, in contrast to A-factor regulation of *adpA* in *S. griseus*, transcription of *S. coelicolor adpA* is not affected by the SCBs (Fig. 8.4) (Chater and Horinouchi, 2003).

Differing from A-factor, and similar to the SCBs, most other *Streptomyces* γ-butyrolactones appear to be devoted only to the regulation of secondary metabolism (Bibb, 2005). For example, virginiae butanolides (VBs; Fig. 8.3) control virginiamycin production in *S. virginiae* (Okamoto *et al.*, 1995), and IM-2 (Fig. 8.3) controls showdomycin and minimycin production in *S. lavendulae* (Waki *et al.*, 1997). Furthermore, in contrast to ArpA from *S. griseus*, the γ-butyrolactone receptor proteins of the other three species (ScbR from *S. coelicolor*, BarA from *S. virginiae*, FarA from *S. lavendulae*) all seem to be required for normal production levels of their respective ligands (Nakano *et al.*, 1998 and Takano *et al.*, 2001, respectively).

All these atypical features of the A-factor system raise doubts about its representativeness for *Streptomyces* γ-butyrolactone regulatory systems. However, despite partial similarities between other γ-butyrolactone regulatory systems, e.g. in *S. coelicolor*, *S. virginiae* and *S. lavendulae*, due to differences concerning other major aspects of the system, there may not be a feasible 'general' model for these complex systems above the level of the molecular mode of action of the homologous proteins involved. The more examples are studied, the clearer the diversity and the complexity of *Streptomyces* γ-butyrolactone systems becomes (Chater and Horinouchi, 2003; Takano, 2006).

A yet unidentified γ-butyrolactone is presumed to be involved in the regulation of tylosin production in *S. fradiae*. Besides three other regulatory genes, the unique regulatory cascade contains two γ-butyrolactone receptor homologues, *tylP* and *tylQ*, with TylP regulating *tylQ* transcription (reviewed in Cundliffe, 2008). *S. ambofaciens* (Bunet *et al.*, 2008; Pang *et al.*, 2004) and *S. virginiae* (Matsuno *et al.*, 2004) are other examples of *Streptomycetes* with multiple γ-butyrolactone receptors.

S. coelicolor possesses, in addition to ScbR, at least four other chromosomal ArpA-like receptor proteins: CprA, CprB, SCO6323 and ScbR2 (Takano *et al.*, 2005). Interestingly, ScbR2 is located in the ScbR-controlled *cpk* gene cluster (Pawlik *et al.*, 2007). The detailed functions of CprA, CprB, SCO6323 and ScbR2 are unknown. In particular, binding to a cognate ligand could not be shown for any of them, and no DNA target sequences were identified yet (Takano, 2006). Based on detailed amino acid sequence analysis, CprA, CprB and ScbR2, together with e.g. TylQ from *S. fradiae* and BarB from *S. virginiae*, make up a subfamily within the γ-butyrolactone receptors, which is suggested to comprise 'pseudo'-receptors with a function as transcriptional regulators, but without ligand binding ability (Takano *et al.*, 2001). More work will be needed to resolve the role of this group of regulators in *Streptomyces*.

γ-Butyrolactone biosynthesis

The 2,3-disubstituted lactone ring structure of the γ-butyrolactones resembles that of acyl homoserine lactones (AHLs) from Gram-negative bacteria, except the carbon side chains (Fig. 8.3). Length, branching and stereochemistry of the fatty acid side chain differ between particular γ-butyrolactones (Fig. 8.3) (Horinouchi, 2002; Takano, 2006) and seem to determine the highly specific binding of the different γ-butyrolactones to their cognate receptor molecules (Hsiao *et al.*, 2009). Currently, biosynthesis of the small signalling molecules is of strong interest. AfsA was shown to be the key enzyme in A-factor biosynthesis, and the putative biosynthesis pathway in *S. griseus* was recently reported (Ando *et al.*, 1997; Kato *et al.*, 2007). Corresponding functions were shown for BarX, an AfsA homologue in *S. virginiae*, and ScbA from *S. coelicolor* (Hsiao *et al.*, 2007; Kawachi *et al.*, 2000; Lee *et al.*, 2009). In addition to this enzymatic role, ScbA and perhaps BarX may also possess a regulatory function (Hsiao *et al.*, 2007; Kawachi *et al.*, 2000; Lee *et al.*, 2009).

The origin of the β-keto acid derivate, whose structure determines the variation in the C-2 side chain of the γ-butyrolactones, has been determined for the novel class of autoregulators, the methylenomycin furans. Corre and colleagues reported that MmfL, the AfsA homologue encoded on the *S. coelicolor* plasmid SCP1, plays an important role in MMF biosynthesis (Corre *et al.*, 2008). When MmfL was solely expressed in *E. coli*, only MMFs with linear C-2 side chains were produced whereas in *S. coelicolor*, with wild-type *scbA*, five MMFs with both linear and branched C-2 side chains were produced. This observation is consistent with the fact that *E. coli* only produces straight-chain fatty acids, while *Streptomyces* produces both straight- and branched-chain fatty acids. This indicates that the precursors of MMFs are derived from fatty acid biosynthesis.

Structure–function relationship in γ-butyrolactone receptors

Including ArpA, Horinouchi and co-workers reported 37 ArpA-like γ-butyrolactone receptor homologues in actinomycetes (Nishida *et al.*, 2007). As a general mode of action, the ligand-free γ-butyrolactone receptors bind to specific DNA sequences upstream of their target genes, typically repressing transcription in the absence of the signalling molecules. The γ-butyrolactones inhibit DNA binding of the receptor protein, and the target genes are expressed (reviewed in Takano, 2006).

The cytoplasmic γ-butyrolactone receptors belong to the TetR family of transcriptional repressors (Ramos *et al.*, 2005). Representative for the γ-butyrolactone receptors, the crystal structure of CprB, one of the ArpA homologues in *S. coelicolor* with a yet unknown function, was determined to be a dimer with an Ω shape (Natsume *et al.*, 2004) (Fig. 8.5). A monomer is composed of two domains, the DNA binding domain, containing a helix–turn–helix (HTH) motif, and a regulatory domain. The latter comprises a dimerization domain and a hydrophobic cavity probably serving as ligand-binding pocket (Natsume *et al.*, 2004; Fig. 8.5). This domain structure was confirmed by site-directed mutagenesis yielding a mutant ArpA protein that lacks DNA binding ability but retains γ-butyrolactone binding ability (Val41Ala in the HTH motif) (Sugiyama *et al.*, 1998) or that binds to DNA but does not bind A-factor (Pro115Ser in the regulatory domain) (Onaka *et al.*, 1997). Also, tryptophan residue 119 was shown to be essential for ligand binding (Trp119Ala in the ligand-binding pocket) (Sugiyama *et al.*, 1998). It is suggested that the binding of γ-butyrolactones induces the relocation of helix α4, which links the binding pocket with the DNA binding domain. As a result of the relocation of the DNA binding domain, the γ-butyrolactone receptor dissociates from its DNA target sequences (Horinouchi, 2007; Natsume *et al.*, 2004). The computer-modelled protein structure shown in Fig. 8.5 uses crystallographic data of CprB (Horinouchi, 2007; Natsume *et al.*, 2004) to illustrate the general structure and functional domains of ArpA-like

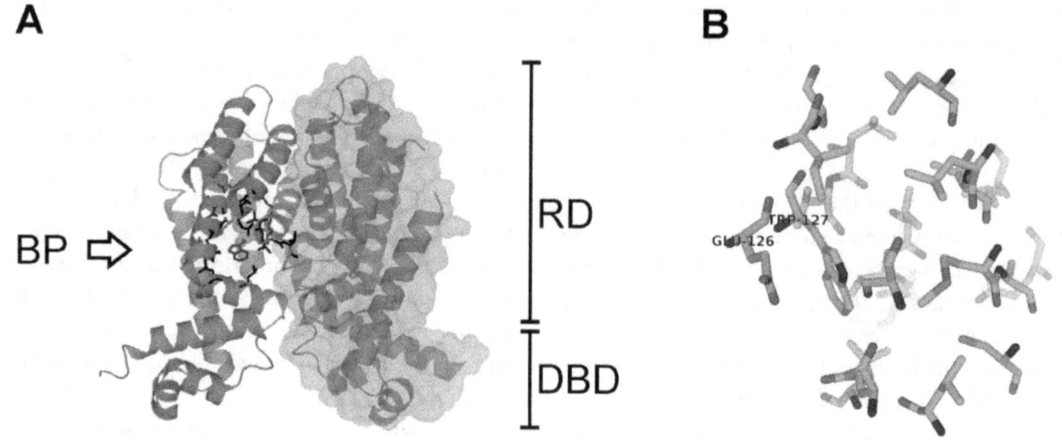

Figure 8.5 Protein structure of CprB. (A) Overall structure of CprB. The γ-butyrolactone receptor homologue constitutes a dimer, each subunit of which contains an N-terminal DNA binding domain (DBD) with a helix-turn-helix motif and a C-terminal regulatory domain (RD) including the dimerization domain and the ligand-binding pocket (BP). The latter is located in a cavity of each regulatory domain and indicated by an arrow. For one monomer, amino acid residues involved in forming the BP (Natsume *et al.*, 2004) are shown. (B) Close-up of the BP from another angle. Trp127 (corresponding to Trp119 in ArpA and Trp121 in ScbR) and Glu126 (corresponding to Arg120 in ScbR), that is not part of the BP, are labelled. The structure model was newly created using structure information from the RCSB Protein Data Bank (CprB (form Ia), accession code 1UI5 (Natsume *et al.*, 2004) and PyMOL (DeLano Scientific) software.

γ-butyrolactone receptors (Horinouchi, 2007; Natsume *et al.*, 2004).

ScbR$_{M600}$ – discovery and investigation of a second form of the *S. coelicolor* γ-butyrolactone receptor

After decades of *S. coelicolor* research, a family tree of strains is available, deriving from the original isolate (Waksman strain W3443) (Weaver *et al.*, 2004). *S. coelicolor* M145 is the typical laboratory strain that was fully sequenced (Bentley *et al.*, 2002), but also strain M600 is often used in many research groups.

When analysing the proteome of these two *S. coelicolor* strains by 2D gel electrophoresis, we identified in M600 a variant of ScbR (ScbR$_{M600}$) that is mutated in the vicinity of the γ-butyrolactone-binding pocket compared with ScbR from M145 (ScbR$_{M145}$) (Gottelt *et al.*, submitted). A single amino acid change, Arg120Ser, in ScbR$_{M600}$ was revealed in MALDI-TOF analysis. A corresponding point mutation was confirmed by DNA sequence analysis in *scbR*$_{M600}$ of independent isolates of strain M600 provided by several laboratories. The same mutation has also been found in the *scbR* gene of strain A3(2) N2.

Effects of the ScbR point mutation

Possible effects of the mutation in ScbR$_{M600}$ were examined *in vitro* and *in vivo*. Contrary to our expectations, *in vitro* characterization using *E. coli* crude cell extract showed that the mutation resulted in an instable DNA-binding activity of the transcriptional regulator. This occurs despite the mutation's location near the ligand-binding pocket and not in the vicinity of the DNA-binding domain. Surprisingly, binding of the γ-butyrolactones, on the other hand, was not affected in ScbR$_{M600}$.

From the protein structure of the ScbR homologue CprB (Horinouchi, 2007; Natsume *et al.*, 2004), amino acid 120 in ScbR (Glu126 in CprB; Fig. 8.5) was deduced to be part of α-helix 7 in the regulatory region of the protein (Natsume *et al.*, 2004). The residue faces the protein surface and is distant from the DNA binding and the dimerization domain; its only functional feature seems to be that it is the neighbouring residue to Trp121 in ScbR (Trp127 in CprB; Fig. 8.5), a residue crucial for ligand binding (Sugiyama *et al.*, 1998; D. Linke, personal communication, based on Natsume *et al.*, 2004). Therefore, the nature of the structural effect of the mutation in ScbR$_{M600}$ leading to the quick loss of DNA binding *in vitro*, remains unclear. Protein crystallization of the γ-butyrolactone receptor homologues is difficult, and consequently the structure of CprB remains the only one available. Co-crystallization of both variants of ScbR with its cognate ligand and also with its DNA target region would lead to a better understanding of the structure–function relationship of these proteins and is therefore of great interest in future work.

For *in vivo* characterization, and to avoid the influence of a different chromosomal background and the effects of putative further mutations in the genome of *S. coelicolor* M600, the mutated *scbR* locus in a M145Δ*scbR*$_{M145}$ in-frame deletion mutant (M752) (Takano *et al.*, 2001) was replaced by *scbR*$_{M600}$ and *scbR*$_{M145}$ yielding strains LW33 and LW34, respectively. LW33 therefore encodes ScbR$_{M600}$ in the M145 genetic background; LW34 is the control strain expressing the native ScbR$_{M145}$. *In vivo*, in agreement with our *in vitro* results, the DNA-binding activity of ScbR$_{M600}$ was impaired. Quantitative RT-PCR analysis showed delayed transcription of the pathway-specific regulatory genes of the *act*, *red* and the *cpk* biosynthesis gene clusters in strain LW33. However, production levels of the related secondary metabolites did not differ significantly between strains LW33 and LW34. Also, expression of *scbA* involved in the synthesis of the γ-butyrolactones was delayed in LW33, and consistently a γ-butyrolactone bioassay indicated a minor decrease in the production of the signalling molecules.

Regulation of the *act*, *red* and *cpk* gene clusters by the butanolide system is intertwined with the complex regulatory network of secondary metabolism in *S. coelicolor*. The *cpk* gene cluster, for example, was shown to be also regulated by the RapA1/A2 two-component system (Lu *et al.*, 2007) and the global regulator DasR (Rigali *et al.*, 2008). Production of other, known and unknown, secondary metabolites in *S. coelicolor* may impair the synthesis of Act, Red and the *cpk* end product, e.g. by competition for common precursors from primary metabolism. Precursor competition would affect production of Act, Red and the *cpk* product even when the related biosynthesis genes are highly expressed, thereby overriding the transcriptional regulation of the *act*, *red* and *cpk* gene clusters. All this may explain why the comparably small changes in gene expression of their pathway-specific regulatory genes caused by the mutation in ScbR$_{M600}$ did not result in an apparent change in the antibiotic phenotype.

Presence of ScbR in different stages of development

ScbR$_{M600}$ was initially identified in 2D gel analysis in the proteome of spores of *S. coelicolor* M600. Surprisingly, ScbR$_{M600}$ was never detected in mycelial samples of M600, whereas in M145 ScbR$_{M145}$ is present in the mycelium, but was never found in spores (A. Hesketh, personal communication). The absence of ScbR$_{M600}$ in mycelial cells was confirmed by Western analysis. This striking difference is most probably not caused by the mutation in ScbR$_{M600}$, but is due to the M600 genetic background, since ScbR$_{M600}$ was present in the mycelium when expressed in a M145 genetic background in LW33. Intracellular conditions in the dormant spores may differ from those in growing mycelial cells. Although studied in vegetative mycelium, the ScbR$_{M600}$ specific protein characteristics may therefore represent an adaptation of the protein to the spores where expression of ScbR$_{M600}$ was detected and thus to its natural cellular environment. *In vivo* activity of ScbR$_{M600}$ in spores has not been shown yet, and one could further speculate that the role of ScbR$_{M600}$ in the spores, where no antibiotics are produced, differs from that described for ScbR$_{M145}$ in the mycelium. A M600Δ*scbR*$_{M600}$ mutant, as well as the expression of *scbR*$_{M145}$ in a M600Δ*scbR*$_{M600}$ genetic background would be useful to address these hypotheses in future work. Putative effects on development, especially on sporulation and germination, but also changed properties of the spores (e.g. heat resistance) would be of special interest.

Evolutionary relevance of the ScbR point mutation

In the lineage of the *S. coelicolor* laboratory strains, two major changes in the genome are most apparent: the loss of the two naturally occurring *S. coelicolor* plasmids, the linear SCP1 and the circular SCP2, and the duplication of a 1.06 Mb region from the left chromosome end. The duplication contains 1005 genes, among them four gene clusters known or predicted to be related to secondary metabolites (eicosapentaenoic acid, isorenieratene, deoxysugar synthases/ glycosyl transferases, and coelichelin) (Bentley *et al.*, 2002). Transcript levels of the duplicated genes, however, are similar in strains with or without the duplication, and few obvious phenotypic differences exist (Weaver *et al.*, 2004). On the *S. coelicolor* plasmids, we find secondary metabolite genes, i.e. the methylenomycin (Mm) biosynthesis gene cluster on SCP1 (Bentley *et al.*, 2004). Mm production is controlled by the above mentioned MMFs, with analogous functions to the γ-butyrolactone regulatory molecules. Also, ScbR-like receptor and transcriptional regulator proteins (MmyR and MmfR) were identified (Corre *et al.*, 2008). Given the encoded functions, the chromosomal duplication and the presence or absence of SCP1 plasmid may influence antibiotic production and regulation. Therefore, one could speculate that the mutation in $ScbR_{M600}$ is related to these genomic alterations. *S. coelicolor* M145 and M600 lost the plasmids; strain A3(2) N2 possesses SCP1 and SCP2. Interestingly, strains M600 and A3(2) N2 ($scbR_{M600}$) possess the duplication, whereas M145 ($scbR_{M145}$) is the only one without the prolonged chromosome. A putative correlation between the duplication and the mutant $scbR_{M600}$, however, could not be confirmed, since in 7 additional *S. coelicolor* strains with the duplication (Weaver *et al.*, 2004), $scbR_{M145}$ was identified (Gottelt *et al.*, submitted).

Although unknown, a natural relevance for the amino acid change Arg120Ser in ScbR of strain M600 and A3(2) N2 seems likely, since this was the only variant form of the protein found and based on the lineage of *S. coelicolor* (Weaver *et al.*, 2004) the mutation occurred independently in the two strains. Also, strains M600 and A3(2) did not undergo any mutagenic treatment (Weaver *et al.*, 2004).

References

Alexeev, I., Sultana, A., Mantsala, P., Niemi, J., and Schneider, G. (2007). Aclacinomycin oxidoreductase (AknOx) from the biosynthetic pathway of the antibiotic aclacinomycin is an unusual flavoenzyme with a dual active site. Proc. Natl. Acad. Sci. U.S.A. *104*, 6170–6175.

Ando, N., Matsumori, N., Sakuda, S., Beppu, T., and Horinouchi, S. (1997). Involvement of *afsA* in A-factor biosynthesis as a key enzyme. Antibiot. J. (Tokyo) *50*, 847–852.

Anisova, L.N., Blinova, I.N., Efremenkova, O.V., Koz'min, I., and Onoprienko, V.V. (1984). [Regulators of development in *Streptomyces coelicolor* A3(2)]. Izv Akad Nauk SSSR Biol. 98–108.

Bendtsen, J.D., Nielsen, H., von, H.G., and Brunak, S. (2004). Improved prediction of signal peptides: SignalP 3.0. J. Mol. Biol. *340*, 783–795.

Bentley, S.D., Brown, S., Murphy, L.D., Harris, D.E., Quail, M.A., Parkhill, J., Barrell, B.G., McCormick, J.R., Santamaria, R.I., Losick, R., *et al.* (2004). SCP1, a 356, 023 bp linear plasmid adapted to the ecology and developmental biology of its host, *Streptomyces coelicolor* A3(2). Mol. Microbiol. *51*, 1615–1628.

Bentley, S.D., Chater, K.F., Cerdeno-Tarraga, A.M., Challis, G.L., Thomson, N.R., James, K.D., Harris, D.E., Quail, M.A., Kieser, H., Harper, D., *et al.* (2002). Complete genome sequence of the model actinomycete *Streptomyces coelicolor* A3(2). Nature *417*, 141–147.

Bibb, M.J. (2005). Regulation of secondary metabolism in streptomycetes. Curr. Opin. Microbiol. *8*, 208–215.

Bunet, R., Mendes, M.V., Rouhier, N., Pang, X., Hotel, L., Leblond, P., and Aigle, B. (2008). Regulation of the synthesis of the angucyclinone antibiotic alpomycin in *Streptomyces ambofaciens* by the autoregulator receptor AlpZ and its specific ligand. J. Bacteriol. *190*, 3293–3305.

Bystrykh, L.V., Fernandez-Moreno, M.A., Herrema, J.K., Malpartida, F., Hopwood, D.A., and Dijkhuizen, L. (1996). Production of actinorhodin-related 'blue pigments' by *Streptomyces coelicolor* A3(2). J. Bacteriol. *178*, 2238–2244.

Camilli, A., and Bassler, B.L. (2006). Bacterial small-molecule signaling pathways. Science *311*, 1113–1116.

Chang, A., Scheer, M., Grote, A., Schomburg, I., and Schomburg, D. (2009). BRENDA, AMENDA and FRENDA the enzyme information system: new content and tools in 2009. Nucleic Acids Res. *37*, D588-D592.

Chater, K.F., and Horinouchi, S. (2003). Signalling early developmental events in two highly diverged *Streptomyces* species. Mol. Microbiol. *48*, 9–15.

Corre, C., Song, L., O'Rourke, S., Chater, K.F., and Challis, G.L. (2008). 2-Alkyl-4-hydroxymethylfuran-3-carboxylic acids, antibiotic production inducers discovered by *Streptomyces coelicolor* genome mining. Proc. Natl. Acad. Sci. U.S.A. *105*, 17510–17515.

Cundliffe, E. (2008). Control of tylosin biosynthesis in *Streptomyces fradiae*. J. Microbiol. Biotechnol. *18*, 1485–1491.

de Carvalho Lima Lobato, A.K., Munoz-Hernandez, R., Araujo Furlan, R.L., Padilla, G., Garrido, L.M., Ribeiro de Macedo, G., and Mavituna, F. (2007). BioSysBio 2007, Systems Biology, Bioinformatics and Synthetic Biology Manchester, UK. 11–13 January 2007. Abstracts. I(Suppl I):P43. BMC Syst. Biol. *1*, S1–S15.

Feitelson, J.S., Malpartida, F., and Hopwood, D.A. (1985). Genetic and biochemical characterization of the red gene cluster of *Streptomyces coelicolor* A3(2). J. Gen. Microbiol. *131*, 2431–2441.

Fuqua, C., Parsek, M.R., and Greenberg, E.P. (2001). Regulation of gene expression by cell-to-cell communication: acyl-homoserine lactone quorum sensing. Annu. Rev. Genet. *35*, 439–468.

Gasteiger, E., Gattiker, A., Hoogland, C., Ivanyi, I., Appel, R.D., and Bairoch, A. (2003). ExPASy: The proteomics server for in-depth protein knowledge and analysis. Nucleic Acids Res. *31*, 3784–3788.

Gottelt, M., Kol, S., Gomez-Escribano, J.P., Bibb, M., and Takano, E. (2010) Deletion of a regulatory gene within the *cpk* gene cluster reveals novel antibacterial activity in *Streptomyces coelicolor* A3(2). Microbiology *156*, 2343–2353.

Hara, O., and Beppu, T. (1982). Mutants blocked in streptomycin production in *Streptomyces griseus* – the role of A-factor. Antibiot, J. (Tokyo) *35*, 349–358.

Horinouchi, S. (2002). A microbial hormone, A-factor, as a master switch for morphological differentiation and secondary metabolism in *Streptomyces griseus*. Front Biosci. 7, d2045-d2057.

Horinouchi, S. (2007). Mining and polishing of the treasure trove in the bacterial genus *Streptomyces*. Biosci. Biotechnol. Biochem. *71*, 283–299.

Hsiao, N.H., Nakayama, S., Merlo, M.E., de, V.M., Bunet, R., Kitani, S. *et al.* (2009). Analysis of two additional signaling molecules in *Streptomyces coelicolor* and the development of a butyrolactone-specific reporter system. Chem. Biol. *16*, 951–960.

Hsiao, N.H., Soding, J., Linke, D., Lange, C., Hertweck, C., Wohlleben, W., and Takano, E. (2007). ScbA from *Streptomyces coelicolor* A3(2) has homology to fatty acid synthases and is able to synthesize gamma-butyrolactones. Microbiology *153*, 1394–1404.

Hutchinson, C.R., and Colombo, A.L. (1999). Genetic engineering of doxorubicin production in *Streptomyces peucetius*: a review. J. Ind. Microbiol. Biotechnol. *23*, 647–652.

Kanehisa, M., Goto, S., Furumichi, M., Tanabe, M., and Hirakawa, M. (2009). KEGG for representation and analysis of molecular networks involving diseases and drugs. Nucleic Acids Res. *38*, D355-D360.

Kato, J.Y., Funa, N., Watanabe, H., Ohnishi, Y., and Horinouchi, S. (2007). Biosynthesis of gamma-butyrolactone autoregulators that switch on secondary metabolism and morphological development in *Streptomyces*. Proc. Natl. Acad. Sci. U.S.A. *104*, 2378–2383.

Kato, J.Y., Miyahisa, I., Mashiko, M., Ohnishi, Y., and Horinouchi, S. (2004). A single target is sufficient to account for the biological effects of the A-factor receptor protein of *Streptomyces griseus*. J. Bacteriol. *186*, 2206–2211.

Kato, J.Y., Ohnishi, Y., and Horinouchi, S. (2005). Autorepression of AdpA of the AraC/XylS family, a key transcriptional activator in the A-factor regulatory cascade in *Streptomyces griseus*. J. Mol. Biol. *350*, 12–26.

Kawabuchi, M., Hara, Y., Nihira, T., and Yamada, Y. (1997). Production of butyrolactone autoregulators by *Streptomyces coelicolor* A3(2). FEMS Microbiol. Lett. *157*, 81–85.

Kawachi, R., Akashi, T., Kamitani, Y., Sy, A., Wangchaisoonthorn, U., Nihira, T., and Yamada, Y. (2000). Identification of an AfsA homologue (BarX) from *Streptomyces virginiae* as a pleiotropic regulator controlling autoregulator biosynthesis, virginiamycin biosynthesis and virginiamycin M1 resistance. Mol. Microbiol. *36*, 302–313.

Khokhlov, A.S., Tovarova, I.I., Borisova, L.N., Pliner, S.A., Shevchenko, L.N., Kornitskaia, E.I., Ivkina, N.S., and Rapoport, I.A., *et al.* (1967). [The A-factor, responsible for streptomycin biosynthesis by mutant strains of Actinomyces streptomycini]. Dokl. Akad. Nauk. SSSR *177*, 232–235.

Kotowska, M., Pawlik, K., Butler, A.R., Cundliffe, E., Takano, E., and Kuczek, K. (2002). Type II thioesterase from *Streptomyces coelicolor* A3(2). Microbiology *148*, 1777–1783.

Kotowska, M., Pawlik, K., Smulczyk-Krawczyszyn, A., Bartosz-Bechowski, H., and Kuczek, K. (2009). Type II thioesterase ScoT, associated with *Streptomyces coelicolor* A3(2) modular polyketide synthase Cpk, hydrolyzes acyl residues and has a preference for propionate. Appl. Environ. Microbiol. *75*, 887–896.

Lee, Y.J., Kitani, S., and Nihira, T. (2009). Null mutation analysis of an *afsA*-family gene, *barX*, that is involved in biosynthesis of the gamma-butyrolactone autoregulator in *Streptomyces virginiae*. Microbiology *156*, 206–210.

Liu, H.W., and Thorson, J.S. (1994). Pathways and mechanisms in the biogenesis of novel deoxysugars by bacteria. Annu. Rev. Microbiol. *48*, 223–256.

Lu, Y., Wang, W., Shu, D., Zhang, W., Chen, L., Qin, Z., Yang, S., and Jiang, W. (2007). Characterization of a novel two-component regulatory system involved in the regulation of both actinorhodin and a type I polyketide in *Streptomyces coelicolor*. Appl. Microbiol. Biotechnol. *77*, 625–635.

Madduri, K., Waldron, C., and Merlo, D.J. (2001). Rhamnose biosynthesis pathway supplies precursors for primary and secondary metabolism in *Saccharopolyspora spinosa*. J. Bacteriol. *183*, 5632–5638.

Matsuno, K., Yamada, Y., Lee, C.K., and Nihira, T. (2004). Identification by gene deletion analysis of *barB* as a negative regulator controlling an early process of virginiamycin biosynthesis in *Streptomyces virginiae.* Arch. Microbiol. *181,* 52–59.

Miyake, K., Kuzuyama, T., Horinouchi, S., and Beppu, T. (1990). The A-factor-binding protein of *Streptomyces griseus* negatively controls streptomycin production and sporulation. J. Bacteriol. *172,* 3003–3008.

Nakano, H., Takehara, E., Nihira, T., and Yamada, Y. (1998). Gene replacement analysis of the *Streptomyces virginiae barA* gene encoding the butyrolactone autoregulator receptor reveals that BarA acts as a repressor in virginiamycin biosynthesis. J. Bacteriol. *180,* 3317–3322.

Natsume, R., Ohnishi, Y., Senda, T., and Horinouchi, S. (2004). Crystal structure of a gamma-butyrolactone autoregulator receptor protein in *Streptomyces coelicolor* A3(2). J. Mol. Biol. *336,* 409–419.

Nishida, H., Ohnishi, Y., Beppu, T., and Horinouchi, S. (2007). Evolution of gamma-butyrolactone synthases and receptors in *Streptomyces.* Environ. Microbiol. *9,* 1986–1994.

O'Rourke, S., Wietzorrek, A., Fowler, K., Corre, C., Challis, G.L., and Chater, K.F. (2009). Extracellular signalling, translational control, two repressors and an activator all contribute to the regulation of methylenomycin production in *Streptomyces coelicolor.* Mol. Microbiol. *71,* 763–778.

Okamoto, S., Nakamura, K., Nihira, T., and Yamada, Y. (1995). Virginiae butanolide binding protein from *Streptomyces virginiae.* Evidence that VbrA is not the virginiae butanolide binding protein and reidentification of the true binding protein. J. Biol. Chem. *270,* 12319–12326.

Onaka, H., Sugiyama, M., and Horinouchi, S. (1997). A mutation at proline-115 in the A-factor receptor protein of *Streptomyces griseus* abolishes DNA-binding ability but not ligand-binding ability. J. Bacteriol. *179,* 2748–2752.

Pang, X., Aigle, B., Girardet, J.M., Mangenot, S., Pernodet, J.L., Decaris, B., and Leblond, P. (2004). Functional angucycline-like antibiotic gene cluster in the terminal inverted repeats of the *Streptomyces ambofaciens* linear chromosome. Antimicrob. Agents Chemother. *48,* 575–588.

Pawlik, K., Kotowska, M., Chater, K.F., Kuczek, K., and Takano, E. (2007). A cryptic Type I polyketide synthase (*cpk*) gene cluster in *Streptomyces coelicolor* A3(2). Arch. Microbiol. *187,* 87–99.

Pfoestl, A., Hofinger, A., Kosma, P., and Messner, P. (2003). Biosynthesis of dTDP-3-acetamido-3, 6-dideoxy-alpha-D-galactose in *Aneurinibacillus thermoaerophilus* L420-91T. J. Biol. Chem. *278,* 26410–26417.

Ramos, A., Lombo, F., Brana, A.F., Rohr, J., Mendez, C., and Salas, J.A. (2008). Biosynthesis of elloramycin in *Streptomyces olivaceus* requires glycosylation by enzymes encoded outside the aglycon cluster. Microbiology *154,* 781–788.

Ramos, J.L., Martinez-Bueno, M., Molina-Henares, A.J., Teran, W., Watanabe, K., Zhang, X., Gallegos, M.T., Brennan, R., Tobes, R. (2005). The TetR family of transcriptional repressors. Microbiol. Mol. Biol. Rev. *69,* 326–356.

Rigali, S., Titgemeyer, F., Barends, S., Mulder, S., Thomae, A.W., Hopwood, D.A., and van Wezel, G.P. (2008). Feast or famine: the global regulator DasR links nutrient stress to antibiotic production by *Streptomyces.* EMBO Rep. *9,* 670–675.

Rodriguez, E., and Gramajo, H. (1999). Genetic and biochemical characterization of the alpha and beta components of a propionyl-CoA carboxylase complex of *Streptomyces coelicolor* A3(2). Microbiology *145,* 3109–3119.

Rose, R.W., Bruser, T., Kissinger, J.C., and Pohlschroder, M. (2002). Adaptation of protein secretion to extremely high-salt conditions by extensive use of the twin-arginine translocation pathway. Mol. Microbiol. *45,* 943–950.

Rudd, B.A., and Hopwood, D.A. (1979). Genetics of actinorhodin biosynthesis by *Streptomyces coelicolor* A3(2). J. Gen. Microbiol. *114,* 35–43.

Sugiyama, M., Onaka, H., Nakagawa, T., and Horinouchi, S. (1998). Site-directed mutagenesis of the A-factor receptor protein: Val-41 important for DNA-binding and Trp-119 important for ligand-binding. Gene *222,* 133–144.

Takano, E. (2006). Gamma-butyrolactones: *Streptomyces* signalling molecules regulating antibiotic production and differentiation. Curr. Opin. Microbiol. *9,* 287–294.

Takano, E., Chakraburtty, R., Nihira, T., Yamada, Y., and Bibb, M.J. (2001). A complex role for the gamma-butyrolactone SCB1 in regulating antibiotic production in *Streptomyces coelicolor* A3(2). Mol. Microbiol. *41,* 1015–1028.

Takano, E., Kinoshita, H., Mersinias, V., Bucca, G., Hotchkiss, G., Nihira, T., Smith, C.P., Bibb, M., Wohlleben, W., and Chater, K. (2005). A bacterial hormone (the SCB1) directly controls the expression of a pathway-specific regulatory gene in the cryptic Type I polyketide biosynthetic gene cluster of *Streptomyces coelicolor.* Mol. Microbiol. *56,* 465–479.

Takano, E., Nihira, T., Hara, Y., Jones, J.J., Gershater, C.J., Yamada, Y., and Bibb, M. (2000). Purification and structural determination of SCB1, a gamma-butyrolactone that elicits antibiotic production in *Streptomyces coelicolor* A3(2). J. Biol. Chem. *275,* 11010–11016.

Waki, M., Nihira, T., and Yamada, Y. (1997). Cloning and characterization of the gene (*farA*) encoding the receptor for an extracellular regulatory factor (IM-2) from *Streptomyces sp.* strain FRI-5. J. Bacteriol. *179,* 5131–5137.

Walsh, C.T. (2002). Combinatorial biosynthesis of antibiotics: challenges and opportunities. Chembiochem *3,* 125–134.

Walsh, C.T., Losey, H.C., and Freel Meyers, C.L. (2003). Antibiotic glycosyltransferases. Biochem. Soc. Trans. *31,* 487–492.

Wang, Y., Lechno-Yossef, S., Gong, Y., Fan, Q., Wolk, C.P., and Xu, X. (2007). Predicted glycosyl transferase genes located outside the HEP island are required for formation of heterocyst envelope polysaccharide in *Anabaena sp.* strain PCC 7120. J. Bacteriol. *189*, 5372–5378.

Weaver, D., Karoonuthaisiri, N., Tsai, H.H., Huang, C.H., Ho, M.L., Gai, S., Patel, K.G., Huang, J., Cohen, S.N., Hopwood, D.A., *et al.* (2004). Genome plasticity in *Streptomyces*: identification of 1 Mb TIRs in the *S. coelicolor* A3(2) chromosome. Mol. Microbiol. *51*, 1535–1550.

Widdick, D.A., Dilks, K., Chandra, G., Bottrill, A., Naldrett, M., Pohlschroder, M., and Palmer, T. (2006). The twin-arginine translocation pathway is a major route of protein export in *Streptomyces coelicolor*. Proc. Natl. Acad. Sci. U.S.A. *103*, 17927–17932.

Williams, P., and Cámara, M. (2009). Quorum sensing and environmental adaptation in Pseudomonas aeruginosa: a tale of regulatory networks and multifunctional signal molecules. Curr. Opin. Microbiol. *12*, 182–191.

Yonaha, K., Nishie, M., and Aibara, S. (1992). The primary structure of omega-amino acid:pyruvate aminotransferase. J. Biol. Chem. *267*, 12506–12510.

Clavulanic Acid and Clavams Biosynthesis and Regulation

9

Paloma Liras, Irene Santamarta and Rosario Pérez-Redondo

Abstract

The (3*R* 5*R*) clavulanic acid and (5*S*) clavam molecules share a structure formed by a four member β-lactam and a five member oxazolidine ring and have several initial common steps in their biosynthesis pathways. The precursors of the molecules are glyceraldehyde-3-phosphate and arginine, condensed by the carboxyethyl-arginine synthase (CeaS). The next steps in the pathway occur by the subsequent action of the β-lactam synthetase (β-LS) forming the β-lactam ring, a proclavaminic acid guanidine hydrolase (Pah) and the clavaminate synthase (CAS), which forms the two rings clavam structure of clavaminic acid. Modifications of this compound result in late step intermediates for clavulanic acid biosynthesis, N-glycylclavaminic acid or clavaldehyde, and in the clavams structures. In addition to the clavulanic acid gene cluster, two additional clusters containing paralogous genes for clavulanic acid biosynthesis and clavam biosynthesis have been located in *S. clavuligerus*. Biochemical characterization of the clavam non producer mutants will clarify the biosynthetic pathway of these compounds.

Introduction

The basic chemical clavam structure of clavulanic acid (CA) and the clavams is a bicyclic molecule. It is formed by a four-member β-lactam ring and a five-member, oxygen-containing, oxazolidine ring (Fig. 9.1). Clavulanic acid is a β-lactamase inhibitor of industrial relevance and possesses a (3*R*, 5*R*) stereochemistry. The clavams are a group of different compounds with (5*S*) stereochemistry, which do not possess β-lactamase inhibitory properties, but have antibacterial or antifungal activity.

The industrial clavulanic acid producer is *Streptomyces clavuligerus* but this compound has been reported to be produced also by *Streptomyces jumonjinensis* and *Streptomyces katsurahamanus*. Clavams, but not CA, are produced by *Streptomyces antibioticus* and *Streptomyces hygroscopicus* (for reviews see Baggaley *et al.*, 1997; Liras and Rodríguez-García, 2000). The current abundance of genome sequences in the databases permits the annotation of multiple genes for clavulanic acid in the thermophile *Anoxybacillus flavithermus* and in *Saccharomonospora viridis,* although no production studies confirm this observation.

Clavulanic acid is measured by the growth inhibition, due to penicillin G, on the β-lactamase producing *Klebsiella pneumoniae*. In addition, imidazole derivatized CA can be quantified by HPLC (Liras and Martín, 2005). However, HPLC quantification and studies on clavams production are hampered by the lack of available clavam standards.

Most of the research in the literature deals with CA biochemistry and genetics in *Streptomyces clavuligerus.* Knowledge of clavam biosynthesis derives mainly on the amino acid-deduced sequences of genes involved in clavams formation. The relation between CA and clavam biosynthesis is not completely elucidated, although the initial steps of their pathways appear to be common.

^{13}C-NMR spectroscopy studies of the labelling pattern of different ^{13}C-labelled precursors and

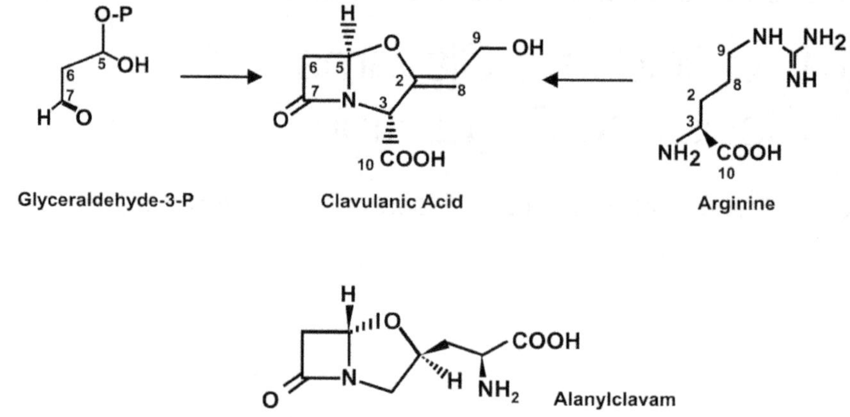

Figure 9.1 Structure of clavulanic acid (above) and alanylclavam (below). The contribution of the precursors arginine and glyceraldehyde-3-phosphate to the clavulanic acid structure is shown.

incorporation of ^{14}C-labelled precursors suggested that CA was formed from a C3 molecule, probably derived from glycerol or pyruvate and either ornithine or arginine (Townsend and Ho, 1985a, Townsend and Ho, 1985b). Later studies revealing the best substrate for the first enzyme of the pathway (CeaS2), and the use of arginine auxotroph mutants, demonstrated that glycerol-3-phosphate and arginine are the direct precursors of CA (Khaleeli *et al.*, 1999; Valentine *et al.*, 1993)

In fact, an increase of the glyceraldehyde-3-P intracellular pool due to disruption of *gap1*, encoding the glyceraldehyde-3-phosphate dehydrogenase (Li and Townsend, 2006) or to amplification of the copy number of *gylD1* and *gylK1*, for glycerol utilization (Baños *et al.*, 2009), results in higher CA formation. The same occurs in *S. clavuligerus* transformants carrying additional copies of arginine biosynthesis genes (Rodríguez-García *et al.*, 1995).

Biosynthesis enzymes and genes

The clavulanic acid biosynthesis pathway (Fig. 9.2B) has been partially elucidated due to the isolation of several intermediates (carboxyethylarginine, proclavaminic acid, deoxyguanidine proclavaminic acid, clavaldehyde and N-glycylclavaminic acid) from the broth of high producing strain cultures, and to the purification of biosynthetic enzymes, most of them as recombinant proteins in *E. coli*. However, some steps of the pathway are not completely understood, and the role of several proteins encoded by essential genes for CA production (*oppA2, oppA1, cyp-fd, orf12, orf14*), is unclear.

In this article we will comment the steps of the pathway and will relate them to the corresponding biosynthetic enzymes and their encoding genes. These genes form a cluster of about 30 kb, named the CA gene cluster (Fig. 9.2A), located side by side to the cephamycin biosynthesis gene clusters in *S. clavuligerus, S. jumonjinensis* and *S. katsurahamanus* (Ward and Hodgson, 1993).

In addition, in *S. clavuligerus* there is a separated paralogous clavulanic acid gene cluster formed by the duplicated genes *ceaS1-bls1-pah1-oat1* (Jensen *et al.*, 2000), while the duplicated *cas1* gene, encoding the clavaminate synthase 1 isoenzyme, is located in the clavams gene cluster (Mosher *et al.*, 1999) at a different location in the genome. The role of these additional gene clusters in CA biosynthesis remains unclear.

The condensation of glyceraldehyde-3-phosphate and arginine is carried out by the N2-(2-carboxyethyl)arginine synthase (CeaS2), the first enzyme of the CA pathway, encoded by *ceaS2*. Using cell-free extracts of an *E. coli* transformant carrying the *ceaS2* gene, Khaleeli *et al.* (1999) followed the condensation of U-^{14}C-arginine and different glycolytic intermediates into carboxyethylarginine (CEA), a compound previously isolated from the fermentation broth of *S. clavuligerus* and *bls*-disrupted mutants

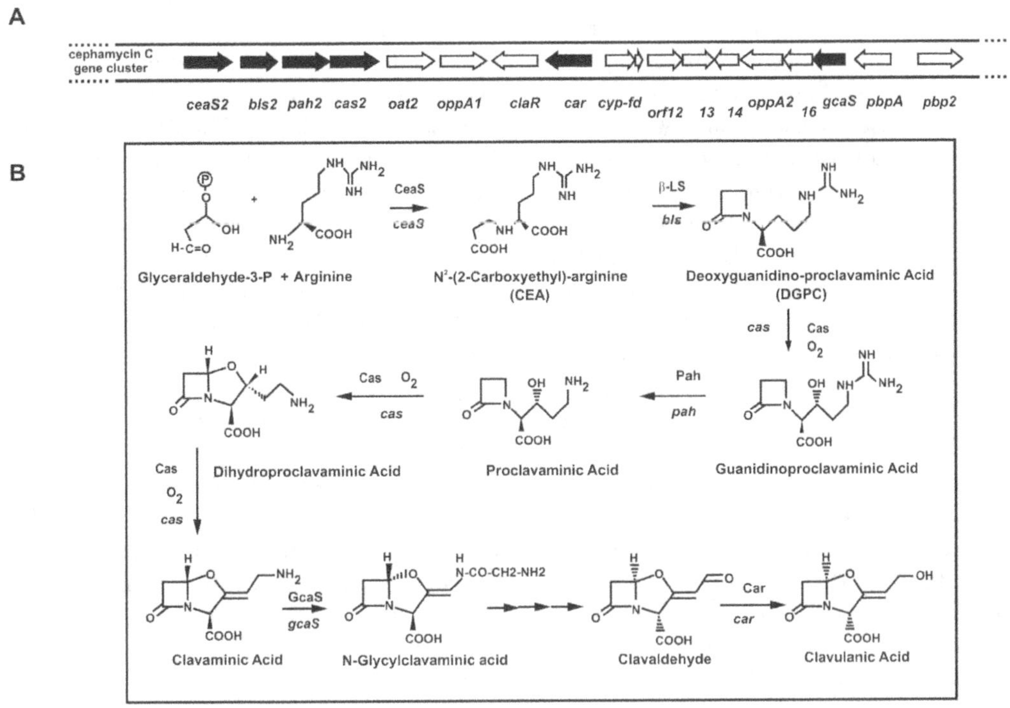

Figure 9.2 Clavulanic acid gene cluster and biosynthetic pathway. (A) Clavulanic acid biosynthesis gene cluster. Genes encoding enzymes of the pathway are indicated in black. (B) Clavulanic acid biosynthesis pathway. Note that the pathway between clavaminic acid and clavaldehyde is not completely understood.

cultures (see below). Highly purified CeaS2 used glyceraldehyde-3-phosphate as the most efficient substrate for the reaction. The enzyme contains a thiamine recognizing domain and requires thiamine diphosphate and Mg^{2+} ions for activity. The deduced amino acid sequence for the 60.9 kDa monomer possesses 30% identity to C–C bond forming, thiamine-dependent, acetolactate synthetases of different origins. CeaS2 replaces a C–O bond in glyceraldehyde-3-phosphate by a rare C–N bond in CEA and forms an acryloyl-thiazolium intermediate with thiamine (Merski and Townsend, 2007). The crystal structure of the enzyme complexed with thiamine diphosphate and Mg^{2+} (Caines *et al.*, 2004), shows that CeaS2 is formed by a tetramer of two associated dimers, with three domains α, β and γ in each subunit, a general folding similar to that of pyruvate decarboxylases, which provides relative flexibility in the amino acid binding site (Caines *et al.*, 2009).

The *ceaS2* gene is essential for CA formation, although some residual CA is formed by *S. clavuligerus ceaS2*-disrupted mutants in glycerol supplemented media (Pérez-Redondo *et al.*, 1999) probably due to the activity of the paralogous *ceaS1* gene. Since *ceaS2* encodes the first enzyme of the pathway, expression of the *ceaS2-bls2-pah2-cas2* polycistronic transcript is an essential site for regulation (see below).

CEA is the substrate for the next enzyme in the pathway encoded by *bls2*, in the CA gene cluster. Disruption of this gene results in complete loss of CA production. Cultures of the *bls2*-disrupted mutant accumulate large amounts of a guanidine compound that was identified by NMR spectroscopy as CEA. The *bls2* mutant can be complemented by feeding the cultures with chemically synthesized deoxyguanidine proclavaminic acid (DGPC), a compound with a four member β-lactam ring. The *bls2* encoded enzyme, purified as recombinant protein in *E. coli,* forms DGPC from CEA in presence of ATP and Mg^{2+} ions (Bachmann *et al.,* 1998). The amino acid sequence of the 54 kDa protein has 33% identity to the highly conserved class B amidotransferases (AS-Bs), which are asparagine synthetases. This was the first reported enzyme forming a four membered β-lactam ring without dioxygenase activity, and was named β-lactam synthetase (β-LS). Crystallization to 1.95Å resolution and comparison with asparagine synthetases show that the β-LS shares many secondary and tertiary structural features and has the overall fold structure of class B amidotransferases. These enzymes possess an N-terminal domain able to hydrolyse glutamine to glutamic acid and ammonia, and a C-terminal domain for the adenylation of aspartic acid. Both domains (amino acids 1 to 210 and 210 to 513) are present in the β-LS, although the N-terminal domain is less conserved and lacks a cysteine residue present in all AS-Bs (Miller *et al.,* 2001, 2002). The kinetic and thermodynamic features of substrate capture by the β-LS have been followed by modifying the pH, temperature and viscosity of the reaction for the natural and site-directed mutagenesis K443R variant of the enzyme. The β-LS catalyses the formation of DGPC in two steps: it first carries out the acyl-adenylation of CEA and then makes the intramolecular nucleophilic addition of the α-amine of CEA to form the β-lactam ring (Raber *et al.,* 2009).

The β-lactam synthetase is 26% identical to CarA, the enzyme that forms (3*S* 5*S*) carbapenem-3-carboxylic acid, for carbapenem biosynthesis in *Erwinia carotovora* (Miller *et al.,* 2003) and 31% identical to ThnM, a putative, non purified, β-lactam synthetase for thienamycin biosynthesis in *S. cattleya* (Núñez *et al.,* 2003).

A duplicate gene, *bls1,* is present in the paralogous gene cluster in *S. clavuligerus.* It encodes a protein with 49% amino acids identity to β-LS2. Disruption of *bls1* results in mutants able to produce different levels of clavulanic acid and clavams depending on medium and culture time (Tahlan *et al.,* 2004a,b). Only disruption of both *bls1* and *bls2* results in lack of clavulanic acid or 5*S*-clavams formation.

The next step in the pathway is carried out by the clavaminate synthase (CAS). This was the first enzyme activity measured in *S. clavuligerus* cell-extracts, and converts proclavaminic acid into clavaminic acid, the first bicyclic compound in the pathway. The clavaminate synthase was purified from *S. clavuligerus* and the enzyme activity was associated to two polypeptides (CAS1 and CAS2) with similar molecular weight and kinetic properties (Salowe *et al.,* 1990; Marsh *et al.,* 1992). A third enzyme, CAS3, performing the same oxidative reactions and with similar kinetic properties to CAS1 and CAS2 was purified from *S. antibioticus,* a clavam producer (Janc *et al.,* 1995).

CAS1 and CAS2 are non-haem iron requiring dioxygenases that convert α-ketoglutarate and oxygen into succinate, CO_2 and H_2O. The reaction is similar to the oxidative ring expansion of penicillin to cephalosporins encoded by the deacetoxycephalosporin C synthase (Liras and Demain, 2009).

Later it was discovered that the first CAS-catalysed reaction is the hydroxylation of deoxyguanidinoproclavaminic acid (DGPC) to guanidinoproclavaminic acid. The last compound is not a substrate for CAS activity and requires the hydrolysis of the guanidine group by the proclavaminic acid hydrolase (PAH, see below) to form proclavaminic acid. This compound is cyclized to form a stable CAS-bound intermediate, dihydroclavaminic acid, which is desaturated to form clavaminic acid (Salowe *et al.,* 1991). In the absence of proclavaminic acid, CAS catalyses an uncoupled decarboxylation reaction in which α-ketoglutarate forms succinate and CO_2 with formation of oxygen peroxide and partial enzyme inactivation (Zhou *et al.,* 2001).

The crystal structure of recombinant CAS1 and CAS2 complexed to Fe^{2+}, α-ketoglutarate and either proclavaminic acid or the substrate analogue N-α-acetyl-L-arginine has been analysed. The enzyme possesses a jelly roll β-barrel core similar to those of the isopenicillin N synthase (IPNS) and deacetoxycephalosporin C synthase (DAOCS), two similar dioxygenases involved in penicillin and cephalosporin biosynthesis. This jelly roll motif, not present in other members of the α-ketoglutarate oxygenases family, suggests that CAS, IPNS and DAOCS are evolutively divergent enzymes and constitute a separate subfamily of oxygenases (Zhang *et al.*, 2000). The His-145 and His-280 residues of the protein are the iron ligands for CAS as shown by directed mutagenesis (Doan *et al.*, 2000; Khaleeli *et al.*, 2000).

Two genes with 87% identity, *cas2* and *cas1*, are located in the clavulanic acid and the clavam gene cluster, respectively, and encode the clavaminate synthetase isozymes of *S. clavuligerus*. Disruption of *cas2* results in a strain unable to produce CA in defined SA medium but still producing the compound in complex soy medium (Paradkar and Jensen, 1995).

As indicated above, to complete the oxidative steps leading to clavaminic acid the guanidine group of DGPC has to be removed. This hydrolysis is carried out by proclavaminic acid amidino hydrolase (PAH) to give proclavaminic acid and urea. The PAH activity was purified from *S. clavuligerus* and N-terminal sequencing permitted the location of *pah2*, upstream of *cas2* (Aidoo *et al.*, 1994). PAH is a member of the arginase family of proteins and the 33 kDa protein has a 63% amino acid sequence homology to agmatine ureidohydrolases (forming putrescine and urea from agmatine), and in the order of 55% with arginases of different origins forming ornithine and urea from arginine (Wu *et al.*, 1995).

PAH was purified as N-terminal modified recombinant protein in *E. coli*. The enzyme activity, measured through urea formation, is activated by Mn^{2+}, Co^{2+} and Ni^{2+} ions. It has an optimal pH of 9.0 and a K_m of 10^{-15} mM for DGPC. The PAH is not able to release urea from D- or L-arginine or *N*-acetyl-L-arginine (a substrate for the clavaminate synthase). Recombinant PAH from *S. clavuligerus* is hexameric, and the crystal structure has been characterized at 1.75 Å and 2.45 Å resolution and found to be formed by two trimers (Elkins *et al.*, 2002). The proclavaminic acid formed by PAH is again a substrate for CAS2 which forms the two ring clavaminic acid molecule (Fig. 9.2).

The last part of the pathway is unclear. Disruption of *oppA2* (*orf15*) or *orf16* in the CA gene cluster blocks clavulanic acid production and results in the accumulation of *N*-acetyl-glycyl-clavaminic acid; this compound and *N*-glycyl-clavaminic acid were previously isolated from fermentation broths (Jensen *et al.*, 2004; Elson *et al.*, 1988). The protein encoded by *orf17*, an essential gene for CA production, was found to be involved in the formation of these pathway intermediates. The amino acid sequence of ORF17 shows that it corresponds to a member of the ATP-grasp fold superfamily. The protein possesses an ATP-binding site but otherwise shows only limited conservation with the members of the ATP-grasp superfamily and the lack of a glycine-rich phosphate binding motif suggests that it belongs to a different subfamily in the group. The shorter sequence for *orf17*, proposed by Jensen *et al.* (2004), was expressed in *E. coli*. An assay using the purified recombinant protein, with ATP, clavaminic acid and several acyl-donors reveals that, only in the presence of glycine, the compound identified as *N*-glycyl-clavaminic acid was formed. *N*-acetyl-glycine is not accepted as substrate by the enzyme that appears to be a rather specific *N*-glycyl-clavaminic acid synthetase (GCAS) using glycine as the carboxylate acceptor and clavaminic acid as amino donor (Arulanantham *et al.*, 2006).

Several steps of the CA pathway are still unknown. An epimerization step must occur, since CA has a (*3R, 5R*) stereochemistry. The 9-amino group present in N-glycyl-clavaminic acid has to be modified to a hydroxyl group in which the oxygen derives from molecular oxygen (Townsend and Krol, 1988). This type of reaction requires an aldehyde or an ester as preferred substrates. A C9-aldehyde, clavaldehyde, was chemically synthesized and later extracted from acid culture broths. The (*3R, 5R*) clavulanate-9-aldehyde is relatively unstable and is slowly decarboxylated giving a *5S* clavam structure. *In vitro* transformation of clavaldehyde into CA by *S. clavuligerus* cell-extracts occurs in the presence of NADPH (but no NADH) as hydrogen donor. The protein carrying the reaction has an Mr of 28 kDa, shows homology to NADPH-dependent dehydrogenases and was named clavaldehyde dehydrogenase (Nicholson *et al.*, 1994). As deduced from the N-terminal amino acid sequence of this enzyme, is encoded by a gene, *car*, located upstream of *claR* in the CA cluster (Pérez-Redondo *et al.*, 1998).

Enzymes and genes of unknown function in clavulanic acid formation

Several genes located in the CA gene cluster are essential or affect clavulanic acid production but their function in biosynthesis is unclear.

Downstream of *cas2*, the *oat2* (*orf6*) gene encodes a protein of 41.6 kDa with high similarity to ornithine acetyltransferases. There is a parallel gene in the genome, *argJ*, that encodes the ornithine N-acetyltransferase for arginine biosynthesis. ArgJ shares 31% amino acids identity with Oat2. Both enzymes catalyse the transfer of an acetyl group from L-N-acetylornithine to L-glutamate to form ornithine and N-acetylglutamate. When L-arginine is the substrate, these enzymes produce N-acetylarginine, a substrate for CAS; however, the efficiency of this reaction is in the order of 13–24% in relation to that using L-glutamate. Both enzymes have similar kinetics and Km values. Disruption of *oat2* decreases clavulanic acid production about 40% but, surprisingly, the total cellular OAT activity is not affected. Oat2, as ArgJ and other OATs, is formed by a $\alpha_2\beta_2$ heterotetramer, which is synthesized as a precursor and undergoes autocatalytic cleavage between the alanine and threonine residues of the sequence KGXGMXXPX(M/L)AT(M/L)L generating the α and β subunits (Kershaw *et al.*, 2002; Fuente *et al.*, 2004). The *oat2* gene belongs to the arginine regulon. It has an ARG-box upstream of the promoter, which binds the repressor protein ArgR, and is strongly expressed in an ArgR-disrupted mutant. Although Oat2 is not essential for CA formation, the *oat2*-disrupted mutant produces more CA when the media is supplemented with 2.5 to 20 mM arginine, concentrations which inhibits up to 50% CA formation in the wild type strain, in spite of being a CA precursor. It has been postulated that Oat2 channels glutamic acid towards arginine and clavulanic acid but it might also act as a clavulanic acid biosynthesis modulator at high arginine concentrations (Fuente *et al.*, 2004). The role of this enzyme in the formation of N-acetyl metabolites derived from clavaminic acid has never been tested. A second gene, named *oat1*, is present in the CA paralogous gene cluster but the double mutant *oat1/oat2* still produces clavulanic acid and clavams.

Three contiguous *orf's* (*orf10*, *orf11* and *orf12*), are located downstream of *car*. They encode a cytochrome P450 (*cyp*), an associated [3Fe–4S] ferredoxin (*fd*) and a protein similar to the *Mycobacterium tuberculosis* LpqF lipoprotein, of unknown function (*orf12*) (Li *et al.*, 2000; Mellado *et al.*, 2002). A highly conserved SDN motif in *orf12* includes this protein in the β-lactamase superfamily but it lacks other characteristic residues of β-lactamases. Disruption of either *orf10* or *orf12*, which are separately transcribed, results in strains unable to produce clavulanic acid but producing cephamycin C at higher levels than the control strain (Li *et al.*, 2000; Lorenzana, 2003). The ability to synthesize CA, as well as normal levels of cephamycin C, is regained by complementing the mutants with either of the genes, indicating that the lack of CA production is not due to a polar effect. Pure recombinant cytochrome P450 is not active on chemically synthesized ^{14}C-N-α-acetyl-L-arginine nor increases the efficiency of recombinant clavaminate synthase to produce N-α-acetyl-L-hydroxyarginine from that substrate (Lorenzana, 2003). Therefore, it has been suggested that the P450 cytochrome has a role in the late oxidative reactions between clavaminic acid and clavaldehyde. The ferredoxin encoded by *fd* has, probably, low specificity and can be disrupted with only a 25% decrease in CA formation.

A large DNA region upstream of *car* was sequenced by Mellado *et al.* (2002). It contains five additional orf's: *orf13* to *orf18*. The *orf13* gene encodes a membrane protein similar to the O-acetyl-L-serine, N-acetyl-L-serine and cysteine *E. coli* exporter. It has been suggested that this protein might be involved in the secretion of N-acyl metabolites derived from clavaminic acid. Disruption of *orf13* results in a strong decrease (about 80%) of CA and clavams production.

The *orf14* encodes a protein of the GNAT superfamily, with a clear CoA binding domain, able to transfer acetyl-CoA to primary amines. The effect of *orf14*-disruption on CA formation, in extended multisampled fermentation in three different media, is a decrease in the order of 35%

relative to the wild type, and the same decrease is found for cephamycin C when defined media are used (Lorenzana, 2003; Mellado *et al.*, 2002), although other authors reported a more drastic CA decrease (Jensen *et al.*, 2004). The amino acid sequence similarity to acetyltransferases responsible for tabtoxin or phosphinothricin resistance suggests that Orf14 might be involved in the detoxification of some intermediate of the pathway, i.e. the formation of N-acyl-clavaminic acid derived metabolites, tentatively exported by *orf13* (or perhaps by *oat2*). No accumulation of acyl-metabolites is observed in this mutant.

Two genes, *oppA1* and *oppA2* (*orf7* and *orf15*), located 8 kb apart in the CA gene cluster encode proteins with 48% amino acid identity. Their amino acid sequences show them to be periplasmic dipeptide or oligopeptide-binding proteins. They have bacterial extracellular solute-binding motifs and are functional, as shown by the lack of growth of *oppA*-disrupted mutants on bradykinin as sole nitrogen source, and by their resistance to the toxic tripeptide bialaphos. These oligopeptide transporters are essential for CA production, and might transport peptides acting as signal for clavulanic acid triggering, or alternatively, arginine-modified intermediates of the pathway. In spite of the high amino acid similarity of *oppA1* and *oppA2*-encoded proteins, they are not able to complement each other, and probably exert different functions. The *oppA2*-disrupted mutant, forms bald colonies, produces the yellow antibiotic holomycin and accumulates N-acetylglycyl-clavaminic acid (Jensen *et al.*, 2004), phenotypes not present in the *oppA1*-disrupted mutant. The effect of *oppA2*-disruption is not due to a polar effect on *orf14*, which is independently transcribed, and transformation with *oppA2-orf14* does not improve the effect of complementing with only *oppA2* (Lorenzana, 2003). Using a peptide-binding assay recombinant glutathione transferase-fused OppA1 and OppA2 were tested against a set of arginine-containing peptides (Mackenzie *et al.*, 2010). Each protein binds the peptides with different specificity. The OppA2 protein has been crystallized as free form and bound to bradykinin (an arginine-containing peptide). The protein resembles other oligopeptide-binding proteins in having two lobes formed by three domains of mixed α/β secondary structure. Arginine (or bradykinin) binds the cleft between both lobes (Mackenzie *et al.*, 2010). These results confirm the proposal that OppA1 and OppA2 act as peptide transporters, but the lack of CA production by the *oppA* mutants suggests that nutrient provision is not the only function of these proteins.

The *orf16* gene encodes a hypothetical protein with no homologues in other microorganisms but in *Saccharomonospora viridis*. Disruption of *orf16* results in complete loss of CA formation. The mutant accumulates N-acetylglycyl-clavaminic acid, which is not due to a polar effect on *oppA2* transcription.

Two genes, *orf18* and *19*, encoding penicillin-binding proteins, are present at the right side of the CA cluster. They show similarities to type A and type 2 PBPs and therefore have been named *pbpA* and *pbp2*. *pbpA* possesses four of the five conserved motifs of β-lactamases and a ATP/GTP binding motif. Both PBPs have been expressed in *E. coli* and the MIC of the transformants in relation to penicillin G and clavulanic acid was tested. The PBPs proteins have very low affinity to clavulanic acid and *pbp2* possesses lower affinity to penicillin G than *pbpA* (Ishida *et al.*, 2006). In spite of their location, these genes might not belong to the CA cluster as shown by the lack of effect of *pbp2* disruption (Jensen *et al.*, 2004).

Regulatory genes

A regulatory gene, *ccaR*, located in the cephamycin gene cluster of *S. clavuligerus* is required for CA and cephamycin biosynthesis as shown by the non-producing phenotype of *ccaR*-disrupted mutants, which are complemented by *ccaR* (Pérez-Llarena *et al.*, 1997). CcaR is a SARP-type antibiotic biosynthesis autoregulatory activator that binds the *cefD–cmcI* bidirectional promoter region and triggers expression of cephamycin biosynthetic genes (Santamarta *et al.*, 2002). However, no binding sites for CcaR have been detected so far in the CA biosynthetic gene cluster, suggesting that the effect on CA production depends on a regulatory cascade. Transcription of *ccaR* has been followed by fusion to the green fluorescent protein. Expression of *ccaR*, in TSB cultures inoculated with spores, starts at 20–24 h, reaches the maximal intensity at 47–55 h and is uniformly distributed in the mycelium (Kyung *et al.*, 2001). This expression is dependent on a putative anti-anti-sigma factor encoded by *bldG* (Bignell *et al.*, 2005).

Northern experiments have shown that *claR* and *car* (for the last enzyme of the pathway) are not expressed in a *ccaR*-disrupted mutant (Pérez-Redondo *et al.*, 1998; Paradkar *et al.*, 1998). Quantitative RT-PCR indicates that *ccaR*-disrupted mutants transcribe the polycistronic *ceaS2-bls2-pah2-cas2* mRNA at 1000 to 6000-fold lower level that the parental strain but express the late genes of the pathway with about a 15-fold reduced efficiency (López-García, M.T., personal communication).

A second regulatory protein is encoded by *claR*. This gene is located upstream of *car*. It encodes a LysR-type transcriptional regulator of molecular weight 47 kDa. A clear HTH motif for DNA binding is present at the N-terminal end (aa 3–51), and a second HTH motif can be found at the C-terminal end (aa 338–393) (Pérez-Redondo *et al.*, 1998; Paradkar *et al.*, 1998). Similar LysR-type regulators are present in the thienamycin biosynthetic gene cluster (ThnI, 32% amino acids identity), in the clavulanic acid paralogous gene cluster (*orfI*) and in other antibiotic biosynthesis clusters.

claR-disrupted mutants do not produce clavulanic acid but produce about twice as much cephamycin C than the wild type strain, and overexpression of *claR* in the transformant *S. clavuligerus* (pClaR) results in increased clavulanic acid and alanylclavam formation but no production of cephamycin C. HPLC analysis shows that the *claR*-disrupted mutants accumulate clavaminic acid and S1 analysis shows lack of transcription of *orf7, orf9* and *orf10* in the mutant. Quantitative RT-PCR analysis of the CA biosynthesis genes in cultures grown in TSB medium confirm that the most drastically affected genes are *car, cyp, orf12–13–14* and *oppA2*. ClaR does not appear to regulate its own transcription but negatively controls the transcription of *ccaR*, as shown by Q-RT-PCR experiments (López-García, unpublished results).

Three *orfs* (*orf21 to orf23*) at the far right end of the CA gene cluster encode a SigL-type of RNA polymerase sigma factor, a putative kinase and a response regulator, respectively. Disruption of *orf21* or *orf22* has no effect on CA production, but disruption of *orf23* slightly decreases the volumetric production, affects morphological differentiation and cell growth. A very small decrease in *claR* transcription is observed in the *orf23* mutant (Song *et al.*, 2009) but quantitative RT-PCR and specific production of CA should be studied before it can be claimed that the gene belongs to the CA biosynthetic cluster.

More details on clavulanic acid and cephamycin regulation in *S. clavuligerus* can be obtained in recent revisions (Liras *et al.*, 2008).

Clavams biosynthesis and genetics

The proposal of a shared pathway for clavulanic acid and clavams is based on the incorporation of [2,3-$^{13}C_2$]-D,L,-proclavaminic acid by *S. antibioticus* into valclavam and hydroxyethylclavam. Incorporation of proclavaminic acid in clavam-2-carboxylate has been also demonstrated in *S. clavuligerus* (Janc *et al.*, 1993). Mutants blocked in *ccaR* are unable to produce CA and the clavams in SA but produce clavams in complex soy medium. However, strains blocked in the early steps of the pathway (*pah2, cas2*) still produce clavams in some culture media as shown by HPLC analysis of the culture broth (Jensen *et al.*, 2000). Small amounts of CA are also produced by *ceaS2*-disrupted strains when glycerol is present in the media (Pérez-Redondo *et al.*, 1999). This could be explained by the presence of a second set of paralogous genes encoding proteins with 50–66% amino acid identity to some of those encoded in the CA cluster; these genes have been named *ceaS1, pah1, bls1,* and *oat1* and are regulated in a different manner that the homologous genes in the CA cluster. Double disruption of *ceaS1/ceaS2, bls1/bls2* or *pah1/pah2* completely block CA and clavams (Tahlan *et al.*, 2004b).

Genes for clavam biosynthesis have been found around the paralogous *cas1* (Mosher *et al.*, 1999); this cluster is separate from the CA and from the paralogous gene clusters in *S. clavuligerus* genome and was further enlarged to a sequence encompassing 16 genes (Tahlan *et al.*, 2007). The *cvm1*- and *cvm5*-disrupted mutants are unable to produce 5S-clavams but the *cvm5*-disrupted strain accumulates a novel compound characterized as 2-carboxymethylideclavam. Disruption of *cvm2* gene results in mutants severely compromised in alanylclavam or 2-hydroxymethyl-clavam and totally blocked in clavam-2-carboxylate formation. Disruption of other genes of the cluster (each *cvm9* to *cvm13, cvmP, cvmG*) has no effect on clavulanic acid or clavams formation. The putative proteins encoded by the genes essential for clavam biosynthesis are shown in Fig. 9.3C.

Upstream of *ceaS1,* in the CA paralogous gene cluster, a large DNA fragment containing twelve genes *(orfA* to *orfL)* has been sequenced (Zelyas *et al.,* 2008). Individual disruption of these genes by transposon mutagenesis did not result in significant effects on CA or cephamycin production. However, mutants in *orfA, orfB, orfC* or *orfD* were defective in alanylclavam formation and mutants *orfC-D* accumulated a new metabolite characterized by electrospray MS as 8-hydroxy-alanylclavam.

Upstream of *oat1* (Fig. 9.3B) are located two genes, *cvm6P* and *cvm7P*. They encode proteins

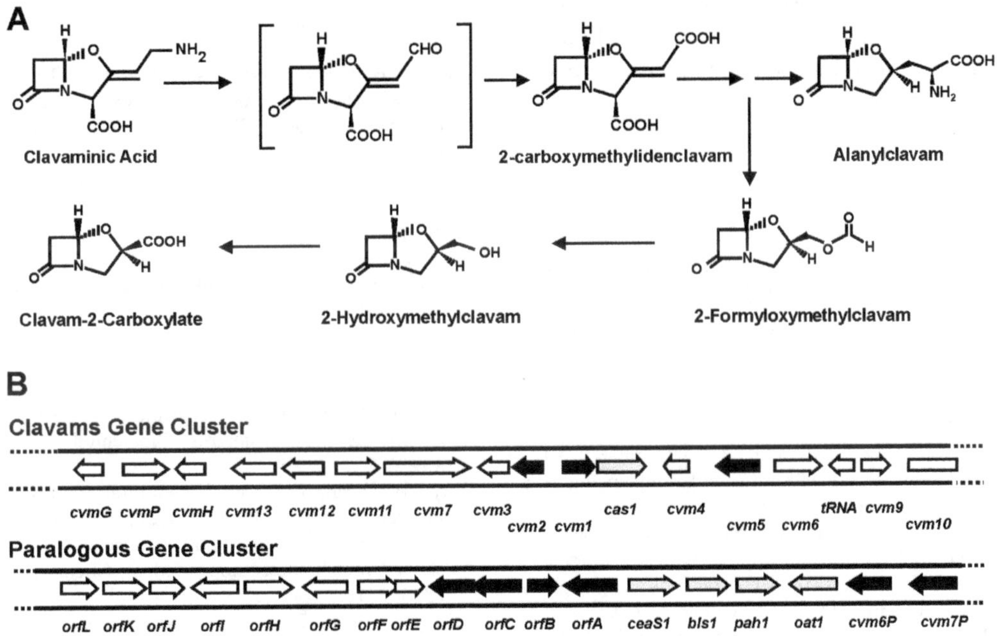

C

Clavams Gene cluster

Gene	Similarity/Identity
cvm1	Aldo/keto reductase
cvm2	Hypothetical isomerase
cvm5	Oxidoreductase

"Paralogous" gene cluster

Gene	Similarity/Identity
cvm6P	Aminotransferase
cvm7P	Transcriptional regulator Lal-type
orfA	Hidroxymethyltransferase
orfB	Aminoacid biosynthesis regulator
orfC	Aspartic/aromatic aminotransferase
orfD	Threonine dehydratase

Figure 9.3 Gene clusters involved in clavam biosynthesis. (A) Proposed pathway for clavams biosynthesis in *S. clavuligerus*. (B) Gene organization of the 'paralogous' and the clavams gene cluster. Genes involved in clavam biosynthesis are indicated in black, the paralogous genes to the clavulanic acid gene cluster are marked in grey. (C) Similarity of the deduced proteins involved in clavam biosynthesis to other proteins in the databases.

homologous to those encoded by *cvm6* and *cvm7* (located in the clavam gene cluster) which are not required for clavam biosynthesis. The Cvm6P protein resembles ArgD, encoding an aminotransferase, while Cvm7P is a transcriptional regulator of the LAL family. Disruption of either *cvm6P* or *cvm7P* results in mutants with normal CA formation but unable to produce any 5S clavam. Therefore, genes for clavam biosynthesis are located in two separate locations in the genome, what is unusual in antibiotic biosynthesis genes and thus more information for clavam biosynthesis is present in the paralogous gene cluster than in the so called clavam gene cluster.

On the basis of the knowledge obtained from mutants altered in clavam formation, a putative clavams pathway has been proposed, although biochemical analyses will be required to support the proposal.

Acknowledgements

This work was supported by grants of the European Community (Actinogen LSHMCT-2004-005224), the Spanish Ministry of Science and Technology (BIO2003-03274, BIO2006-14853) and the Regional Castilla and León Government (LE 001A07).

References

Aidoo, K.A., Wong, A., Alexander, D.C., Rittamer. R.A.R., and Jensen, S.E. (1994). Cloning, sequencing and disruption of a gene from *Streptomyces clavuligerus* involved in clavulanic acid biosynthesis. Gene *147*, 41–46.

Arulanantham, H., Kershaw, N.J., Hewitson, K.S., Hughes, C.E., Thirkettle, J.E., and Schofield, C.J. (2006). ORF17 from the clavulanic acid biosynthesis gene cluster catalyzes the ATP-dependent formation of N-glycyl-clavaminic acid. J. Biol. Chem. *281*, 279–287.

Bachmann, B.O., Li, R., and Townsend, C.A. (1998). β-Lactam synthetase: a new biosynthetic enzyme. Proc. Natl. Acad. Sci. U.S.A. *95*, 9082–9086.

Baggaley, K.H., Brown, A.G., and Schofield, C.J. (1997). Chemistry and biosynthesis of clavulanic acid and other clavams. Natl. Prod. Rep. *14*, 309–333.

Baños, S., Pérez-Redondo, R., Koekman, B., and Liras, P. (2009). Glycerol utilization gene cluster in *Streptomyces clavuligerus*. Appl. Environ. Microbiol. *75*, 2991–2995.

Bignell, D.R., Tahlan, K., Colvin, K.R., Jensen, S.E., and Leskiw, B.K. (2005). Expression of *ccaR*, encoding the positive activator of cephamycin C and clavulanic acid production in *Streptomyces clavuligerus*, is dependent on *bldG*. Antimicrob. Agents Chemother. *49*, 1529–1541.

Caines, M.E., Elkins, J.M., Hewitson, K.S., and Schofield, C.J. (2004). Crystal structure and mechanistic implications of N2-(2-carboxyethyl)arginine synthase, the first enzyme in the clavulanic acid biosynthesis pathway. J. Biol. Chem. *279*, 5685–5692.

Caines, M.E., Sorensen, J.L., and Schofield, C.J. (2009). Structural and mechanistic studies on N(2)-(2-carboxyethyl)arginine synthase. Biochem. Biophys. Res. Commun. *385*, 512–517.

Doan, L.X., Hassan, A., Lipscomb, S.J., Dhanda, A., Zhang, Z., and Schofield, C.J. (2000). Mutagenesis studies on the iron binding ligands of clavaminic acid synthase. Biochem. Biophys. Res. Commun. *279*, 240–244.

Elkins, J.M., Clifton, I.J., Hernández, H., Doan, L.X., Robinson, C.V., Schofield, C.J., and Hewitson, K., S. (2002). Oligomeric structure of proclavaminic acid amidino hydrolase: evolution of a hydrolytic enzyme in clavulanic acid biosynthesis. Biochem. J. *366*, 423–434.

Elson, W.E., Gillet, J., Nicholson, N.H., and Tyler, J.W. (1988). N-acyl derivatives of clavaminic acid produced by a mutant of *Streptomyces clavuligerus*. J. Soc. Chem. Commun. *1988*, 979–980.

Fuente, A. de la, Martín, J.F., Rodríguez-García, A., and Liras, P. (2004). Two proteins with ornithine acetyltransferase activity show different functions in *Streptomyces clavuligerus*: Oat2 modulates clavulanic acid biosynthesis in response to arginine. J. Bacteriol. *186*, 6501–6507.

Ishida, K., Hung, T.V., Liou, K., Lee, H.C., Shin, C.H., and Sohng, J.K. (2006). Characterization of *pbpA* and *pbp2* encoding penicillin-binding proteins located on the downstream of clavulanic acid gene cluster in *Streptomyces clavuligerus*. Biotechnol. Lett. *28*, 409–417.

Janc, J.W., Egan, L.A., and Townsend, C.A. (1993). Emerging evidence for a shared biosynthetic pathway among clavulanic acid and the structural diverse clavam metabolites. Bioorg. Med. Chem. Lett. *3*, 2313–2316.

Janc, J.W., Egan, L.A., and Townsend, C.A. (1995). Purification and characterization of clavaminate synthase from *Streptomyces antibioticus*. A multifunctional enzyme of clavam biosynthesis. J. Biol. Chem. *270*, 5399–5404.

Jensen, S.E., Elder, K.J., Aidoo, K.A., and Paradkar, A.S. (2000). Enzymes catalyzing the early steps of clavulanic acid biosynthesis are encoded by two sets of paralogous genes in *Streptomyces clavuligerus*. Antimicrob. Agents Chemother. *44*, 720–726.

Jensen, S.E., Paradkar, A.S., Mosher, R.H., Anders, C., Beatty, P.H., Brumlik, M.J., Griffin, A., and Barton, B. (2004). Five additional genes are involved in clavulanic acid biosynthesis in *Streptomyces clavuligerus*. Antimicrob. Agents Chemother. *48*, 192–202.

Kershaw, N.J., McNaughton, H.J., Hewitson, K.S., Hernández, H., Griffin, J., Hughes, C., Greaves, P., Barton, B., Robinson, C.V., and Schofield, C.J.

(2002). ORF6 from the clavulanic acid gene cluster of *Streptomyces clavuligerus* has ornithine acetyltransferase activity. Eur. J. Biochem. *269*, 2052–2059.

Khaleeli, N., Busby, R.W., and Townsend, C.A. (2000). Site-directed mutagenesis and biochemical analysis of the endogenous ligands in the ferrous active site of clavaminate synthase. The His-3 variant of the 2-His-1-carboxylate model. Biochemistry *39*, 8666–8673.

Khaleeli, N., Li, R., and Townsend, C.A. (1999). Origin of the β-lactam carbons in clavulanic acid from an unusual thiamine pyrophosphate-mediated reaction. J. Am. Chem. Soc. *121*, 9223–9224.

Kyung, Y.S., Hu, W.S., and Sherman, D.H. (2001). Analysis of temporal and spatial expression of the CcaR regulatory element in the cephamycin C biosynthetic pathway using green fluorescent protein. Mol. Microbiol. *40*, 530–541.

Li, R., and Townsend, C.A. (2006). Rational strain improvement for enhanced clavulanic acid production by genetic engineering of the glycolytic pathway in *Streptomyces clavuligerus*. Metab. Eng. *8*, 240–252.

Li, R., Khaleeli, N., and Townsend, C.A. (2000). Expansion of the clavulanic acid gene cluster: identification and in vivo functional analysis of three new genes required for biosynthesis of clavulanic acid by *Streptomyces clavuligerus*. J. Bacteriol. *182*, 4087–4095.

Liras P., Gómez-Escribano, J.P., and Santamarta, I. (2008). Regulatory mechanisms controlling antibiotic production in *Streptomyces clavuligerus*. J. Ind. Microbiol. Biotechnol. *35*, 667–676.

Liras, P., and Rodríguez-García, A. (2000). Clavulanic acid, a β-lactamase inhibitor: Biosynthesis and molecular genetics. Applied Microbiol. Biotechnol. *54*, 467–475.

Liras, P., and Demain, A.L. (2009). Enzymology of β-lactam compounds with cephem structure produced by Actinomycetes. In Methods in Enzymology Complex Enzymes in Microbial Natural Product Biosynthesis, Hopwood, D.A., ed. (Academic Press, New York), vol. 458, pp. 401–429.

Liras, P., and Martín, J.F. (2005). Assay methods for detection and quantification of antimicrobial metabolites produced by *Streptomyces clavuligerus*. In Methods in Biotechnology, Barredo, J.L., ed. (The Humana Press Inc., Totowa, NJ), vol. 18, pp. 149–163.

Lorenzana, L.M., Pérez-Redondo, R., Santamarta, I., Martín, J.F., and Liras, P. (2004). Two oligopeptide-permease-encoding genes in the clavulanic acid cluster of *Streptomyces clavuligerus* are essential for production of the beta-lactamase inhibitor. J. Bacteriol. *186*, 3431–3438.

Lorenzana, L.M. (2003). Agrupación de genes de biosíntesis de ácido clavulánico en *Streptomyces clavuligerus*. Análisis de la región corriente arriba del gen *car*. PhD Thesis (University of León, Spain).

Mackenzie, A.K., Valegård, K., Iqbal, A., Caines, M.E.C., Kershaw, N.J., Jensen, S.E., Schofield, C.J., and Andersson, I. (2010). Crystal structures of an oligopeptide-binding protein from the biosynthetic pathway of the β-lactamase inhibitor clavulanic acid. J. Mol. Biol. *396*, 332–344.

Marsh, N.E., Chang, M.D-T., and Townsend, C.A. (1992). Two isozymes of clavaminate synthase central to clavulanic acid formation: cloning and sequencing of both genes from *Streptomyces clavuligerus*. Biochemistry *31*, 12648–12657.

Mellado, E., Lorenzana, L.M., Rodríguez-Sáiz, M., Díez, B., Liras, P., and Barredo, J.L. (2002). The clavulanic acid biosynthetic cluster of *Streptomyces clavuligerus*: genetic organization of the region upstream of the *car* gene. Microbiology *148*, 1427–1438.

Merski, M., and Townsend, C.A. (2007). Observation of an acryloyl-thiamin diphosphate adduct in the first step of clavulanic acid biosynthesis. J. Am. Chem. Soc. *129*, 15750–15751.

Miller, M.T., Bachmann, B.O., Townsend, C.A., and Rosenzweig, A.C. (2001). Structure of beta-lactam synthetase reveals how to synthesize antibiotics instead of asparagine. Nat. Struct. Biol. *8*, 684–689.

Miller, M.T., Bachmann, B.O., Townsend, C.A., and Rosenzweig, A.C. (2002). The catalytic cycle of beta-lactam synthetase observed by x-ray crystallographic snapshots. Proc. Natl. Acad. Sci. U.S.A. *99*, 14752–14757.

Miller, M.T., Gerratana, B., Stapon, A., Townsend, C.A., and Rosenzweig, A.C. (2003). Crystal structure of carbapenem synthetase (CarA). J. Biol. Chem. *278*, 40996–41002.

Mosher, R.H., Paradkar, A.S., Anders, C., Barton, B., and Jensen, S.E. (1999). Genes specific for the biosynthesis of clavam metabolites antipodal to clavulanic acid are clustered with the gene for clavaminate synthase 1 in *Streptomyces clavuligerus*. Antimicrob. Agents Chemother. *43*, 1215–1224.

Nicholson, N.H., Baggaley, K.H., Cassels, R., Davison, M., Elson, S.W., Fulston, M., Tyler, J.W., and Woroniecki, S.R. (1994). Evidence that the immediate biosynthetic precursor of clavulanic acid is its N-aldehyde analogue. J. Chem. Soc. Chem. Commun. *1994*, 1281–1282.

Núñez, L.E., Méndez, C., Braña, A.F., Blanco, G., Salas, J.A. (2003). The biosynthetic gene cluster for the beta-lactam carbapenem thienamycin in *Streptomyces cattleya*. Chem. Biol. *10*, 301–311.

Paradkar, A.S., Aidoo, K.A., and Jensen, S.E. (1998). A pathway-specific transcriptional activator regulates late steps of clavulanic acid biosynthesis in *Streptomyces clavuligerus*. Mol. Microbiol. *27*, 831–843.

Paradkar, A.S., and Jensen, S.E. (1995). Functional analysis of the gene encoding the clavaminate synthase 2 isoenzyme involved in clavulanic acid biosynthesis in *Streptomyces clavuligerus*. J. Bacteriol. *177*, 1307–1314.

Pérez-Llarena, F., Liras, P., Rodríguez-García, A., and Martín, J.F. (1997). A regulatory gene (*ccaR*) required for cephamycin and clavulanic acid production in *Streptomyces clavuligerus*: amplification results in overproduction of both β-lactam compounds. J. Bacteriol. *179*, 2053-2059.

Pérez-Redondo, R., Rodríguez-García, A., Martín, J.F., and Liras, P. (1998). The *claR* gene of *Streptomyces clavuligerus*, encoding a LysR-type regulatory protein controlling clavulanic acid biosynthesis, is linked to the clavulanate-9-aldehyde reductase (*car*) gene. Gene *211*, 311–321.

Pérez-Redondo, R., Rodríguez-García, A., Martín, J.F., and Liras, P. (1999). Deletion of the *pyc* gene blocks clavulanic acid biosynthesis except in glycerol-containing medium: evidence for two different genes in formation of the C3 unit. J. Bacteriol. *181*, 6922–6928.

Raber, M.L., Freeman, M.F., and Townsend, C.A. (2009). Dissection of the stepwise mechanism to beta-lactam formation and elucidation of a rate-determining conformational change in beta-lactam synthetase. J. Biol. Chem. *284*, 207–217.

Rodríguez-García, A., Martín, J.F., and Liras, P. (1995). The *argG* gene of *Streptomyces clavuligerus* has low homology to unstable *argG* from other actinomycetes: effect of amplification on clavulanic acid biosynthesis. Gene *167*, 9–15.

Salowe, S.P., Krol, W.J., Iwata-Reuyl, D., and Townsend, C.A. (1991). Elucidation of the order of oxidations and identification of an intermediate in the multistep clavaminate synthase reaction. Biochemistry *30*, 2281–2292.

Salowe, S.P., Marsh, E.N., and Townsend, C.A. (1990). Purification and characterization of clavaminate synthase from *Streptomyces clavuligerus*: An unusual oxidative enzyme in natural product biosynthesis. Biochemistry *29*, 6499–6508.

Santamarta, I., Rodríguez-García, A., Pérez-Redondo, R., Martín, J.F., and Liras, P. (2002). CcaR is an autoregulatory protein that binds to the *ccaR* and the *cefD-cmcI* promoters of the cephamycin C-clavulanic acid cluster in *Streptomyces clavuligerus*. J. Bacteriol. *184*, 3106–3113.

Song, J.Y., Kim, E.S., Kim, D.W., Jensen, S.E., and Lee, K.J. (2009). A gene located downstream of the clavulanic acid gene cluster in *Streptomyces clavuligerus* ATCC 27064 encodes a putative response regulator that affects clavulanic acid production. J. Ind. Microbiol. Biotechnol. *36*, 301–311.

Tahlan, K., Anders, C., and Jensen, S.E. (2004a). The paralogous pairs of genes involved in clavulanic acid and clavam metabolite biosynthesis are differently regulated in *Streptomyces clavuligerus*. J. Bacteriol. *186*, 6286–6297.

Tahlan, K., Park, H.U., Wong, A., Beatty, P.H., and Jensen, S.E. (2004b). Two sets of paralogous genes encode the enzymes involved in the early stages of clavulanic acid and clavam metabolite biosynthesis in *Streptomyces clavuligerus*. Antimicrob. Agents Chemother. *48*, 930–939.

Tahlan, K., Anders, C., Wong, A., Mosher, R.H., Beatty, P.H., Brumlik, M.J., Griffin, A., Hughes, C., Griffin, J., Barton, B., *et al.* (2007). 5S clavam biosynthetic genes are located in both the clavam and paralog gene clusters in *Streptomyces clavuligerus*. Chem. Biol. *14*, 131–142.

Townsend, C.A., and Krol, W.J. (1988). The role of molecular oxygen in clavulanic acid biosynthesis: Evidence for a bacterial oxidative deamination. J. Chem. Soc. Chem. Commun, 1234–1236.

Townsend, C.A., and Ho, M. (1985a). Biosynthesis of clavulanic acid: Origin of the C5 unit. J. Am. Chem. Soc. *107*, 1065–1066.

Townsend, C.A., and Ho, M. (1985b). Biosynthesis of clavulanic acid: Origin of the C3 unit. J. Am. Chem. Soc. *107*, 1066–1068.

Valentine, B.P., Bailey, C.R., Doherty, A., Morris, J., Elson, S.W., Baggaley, K.H., and Nicholson, N.H. (1993). Evidence that arginine is a later metabolic intermediate than ornithine in the biosynthesis of clavulanic acid by *Streptomyces clavuligerus*. J. Chem. Soc. Chem. Commun. *1993*, 1210–1211.

Ward, J.M., and Hodgson, J.E. (1993). The biosynthetic genes for clavulanic acid and cephamycin production occur as a 'super-cluster' in three *Streptomyces*. FEMS Microbiol. Lett. *110*, 239–242.

Wu, T.-K., Busby, R.W., Houston, T.A., McIlwaine, D.B., Egan, L.A., and Townsend, C.A. (1995). Identification, cloning, sequencing and overexpression of the gene encoding proclavaminate amidino hydrolase and characterization of protein function in clavulanic acid biosynthesis. J. Bacteriol. *177*, 3714–3720.

Zelyas, N.J., Cai, H., Kwong, T., and Jensen, S.E. (2008). Alanylclavam biosynthetic genes are clustered together with one group of clavulanic acid biosynthetic genes in *Streptomyces clavuligerus*. J. Bacteriol. *190*, 7957–7965.

Zhang, Z., Ren, J., Stammers, D.K., Baldwin, J.E., Harlos, K., and Schofield, C.J. (2000). Structural origins of the selectivity of the trifunctional oxygenase clavaminic acid synthase. Nat. Struct. Biol. *7*, 127–133.

Zhou, J., Kelly, W.L., Bachmann, B.O., Gunsior, M., Townsend, C.A., and Solomon, E.I. (2001). Spectroscopic studies of substrate interactions with clavaminate synthase 2, a multifunctional alpha-KG-dependent non-heme iron enzyme: correlation with mechanisms and reactivities. J. Am. Chem. Soc. *123*, 7388–7398.

Genome-guided Exploration of *Streptomyces ambofaciens* Secondary Metabolism

10

Bértrand Aigle, Robert Bunet, Christophe Corre, Amélie Garénaux, Laurence Hotel, Sheng Huang, Luisa Laureti, Sylvie Lautru, Marta Vaz Mendes, Šárka Nezbedová, Hoang Cuong Nguyen, Lijiang Song, Jaroslav Weiser, Gregory Challis, Pierre Leblond and Jean-Luc Pernodet

Abstract

Members of the *Streptomyces* genus are among the most prolific microorganisms producing secondary metabolites with wide uses in medicine and in agriculture. Sequencing of the genome of the model *Streptomyces*, *Streptomyces coelicolor*, has highlighted an unexpected feature, i.e. that the potential of these organisms to synthesise secondary metabolites has been largely underestimated. They indeed possess many more gene clusters encoding natural product-like biosynthetic pathways than there are known natural products. Similar observations have since been made for other bacterial or fungal genomes. Thus, it has become clear that microbial secondary metabolism has been seriously underestimated and that genome-based approaches are very promising in the search for new bioactive compounds.

Here, we present an overview of the secondary metabolite biosynthetic potential of *Streptomyces ambofaciens*, a species known for decades as producer of the macrolide spiramycin and the pyrrolamide congocidine. Interestingly, genome analysis has revealed that despite of the close phylogenetic relatedness of *S. coelicolor* and *S. ambofaciens*, most of its secondary metabolite gene clusters are species-specific.

Introduction

In the race for the discovery of new antibacterial compounds, there will be definitely a pre- and post-sequencing era in the *Streptomyces* exploitation. Genome sequencing has indeed disclosed unsuspected capacities to encode various secondary metabolites and has revived the expectations for new active metabolites from streptomycetes.

Before that quantum leap, each *Streptomyces* species was known to produce only a few molecules. Based on screening for antibacterial activities, metabolites were isolated and used in the war against infectious diseases, and for other pharmacological activities: antiproliferative, immunosuppressant, antiviral. Even in the well-studied model *Streptomyces coelicolor*, only five secondary metabolites were identified (four antibiotics and a pigment, none of them being of economical importance) before the genome sequence project (Bentley *et al.*, 2002). In *S. griseus*, the famous aminoglycoside antibiotic streptomycin producer, only six biosynthetic pathways were previously identified (Ohnishi *et al.*, 2008). In general, each *Streptomyces* species was exploited for a single active compound (e.g. *S. avermitilis* for the anthelmintic avermectins). In the macrolide spiramycin producer *Streptomyces ambofaciens*, our model bacterium, a second metabolite, the pyrrolamide congocidine, was previously known with, however, no therapeutic use.

Analysis of the complete genome sequences revealed the presence of an impressive number of cryptic genes encoding yet unidentified secondary metabolites. These genes usually direct the synthesis of a single metabolite and are clustered at a specific locus together with the resistance determinants. The *S. coelicolor* genome encodes

25 such gene clusters (Bentley *et al.*, 2002), *S. avermitilis* on its side has 30 clusters (Ikeda *et al.*, 2003) while *S. griseus* possesses 34 clusters (Ohnishi *et al.*, 2008). These clusters represent 4–6% of the total genome information underlining the importance of secondary metabolism for these micro-organisms. It became clear that streptomycete secondary metabolism has been seriously underestimated, and that genomic-based approaches are promising for discovering new interesting metabolites.

An additional feature supports the development of genomic-based approaches. Thus, the majority of the clusters identified are not randomly scattered along the linear genome, but are more frequently in the non essential part of the linear chromosome, i.e. the terminal regions. About half of the clusters are indeed included in the last quarter of the chromosome with an average of two-fold enrichment. Second, most of the clusters are species specific (e.g. 14 of the 34 clusters in *S. griseus*).

This is mostly the consequence of the specific genetic organization and evolution of the terminal regions of the chromosome. This organization was first revealed in *S. ambofaciens* by the study of genome plasticity showing that the terminal quarter of the linear chromosome was dispensable in laboratory conditions, and that antibiotic biosynthesis was prone to genetic instability (Dary *et al.*, 1992; Leblond *et al.*, 1991). Further studies also showed that amplifiable regions overlap or include genes involved in secondary metabolite biosynthesis showing that these regions are tolerant to DNA rearrangements in laboratory conditions (Aigle *et al.*, 1996). Besides, it goes along with the idea that the reiterative structure of some biosynthetic pathways (mostly type I PKS) may derive and evolve from DNA amplification (Jenke-Kodama and Dittmann, 2009). Finally, the content and variability of the ends of the *S. ambofaciens* genome was studied through genome sequencing and comparison with closely and distantly related species (Choulet *et al.*, 2006a,b). While the central part of the chromosome is well conserved, the terminal regions are highly variable and specific. In addition, some large genomic islands can interrupt the central conserved regions (12 in *S. coelicolor* and 4 in *S. ambofaciens*). The less the species are related, the more the terminal specific region is large (more than 650 kb on each arm between *S. coelicolor* and *S. ambofaciens*, 1 Mb between *S. avermitilis* and *S. ambofaciens*, (Choulet *et al.*, 2006a). Even between strains belonging to the same species, specific regions were identified in the distal part of the terminal inverted repeats (TIRs) characteristic of the invertronic structure of the *Streptomyces* genome (Choulet *et al.*, 2006b).

Genome sequencing of *S. ambofaciens* has revealed the presence of numerous cryptic secondary metabolite gene clusters located in or corresponding to specific regions. Here, we present an overview of these different clusters and of the spiramycin cluster. Among the new pathways, we investigated new type I and II PKS clusters to identify original secondary metabolites. In addition, we searched for the genes responsible for biosynthesis of the known antibiotic congocidine whose biosynthesis mechanism might reveal original enzymatic mechanisms.

Gene clusters for secondary metabolites

Initially, our sequencing project of the *S. ambofaciens* genome focused on the terminal regions of the linear chromosome i.e. on about 3 Mb for a chromosome whose size has been estimated at about 8 Mb. These regions have been fully sequenced with 1544 kb for the left chromosomal arm (accession number AM238663) and 1367 kb for the right chromosomal arm (accession number AM238664; Choulet *et al.*, 2006a,b) (http://www.weblgm.scbiol.uhp-nancy.fr/ambofaciens/). The sequencing of the central region, also designated as core region, is going to be completed soon but the partial sequence has already been analysed for identification of secondary metabolite gene clusters.

The analysis of the terminal regions has unveiled the presence of 14 secondary metabolite gene clusters (Fig. 10.1 and Table 10.1). Five of them are located on the right chromosomal arm and nine on the left one. Among these 14 clusters, two are identical and correspond to the cluster located in the TIRs of the linear chromosome (see below, the cluster *alp*). The partially sequenced central region contains at least seven

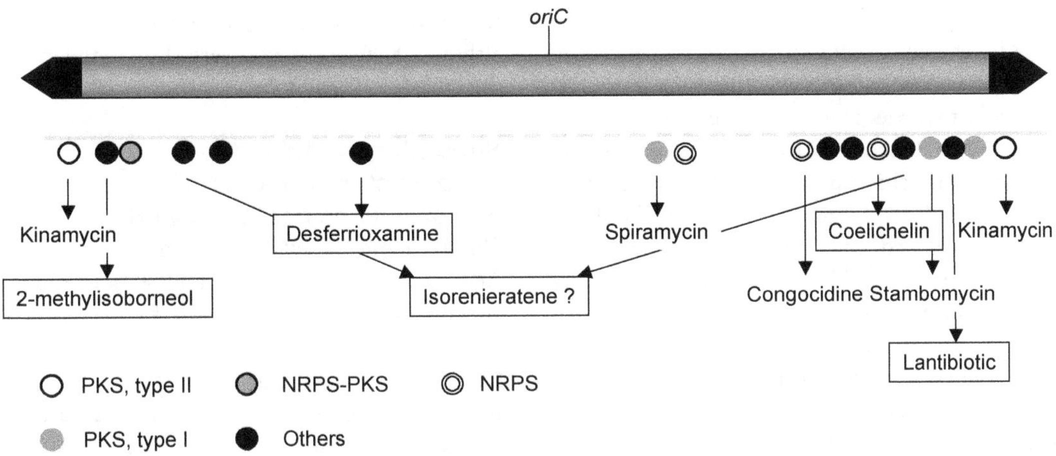

Figure 10.1 Chromosomal map of *S. ambofaciens* ATCC23877 and location of the secondary metabolite gene clusters. The known or predicted metabolites produced by some of the clusters are indicated. The metabolites framed are common to *S. ambofaciens* and *S. coelicolor*. The black arrows on the chromosomal map symbolize the TIRs. The bars represent the terminal regions completely sequenced while the hashed bar symbolizes the central region partially sequenced.

Table 10.1 Secondary metabolite gene clusters identified in the terminal regions of the *S. ambofaciens* chromosome

	Type of cluster (product*)	Limits of the cluster	Size (kb)
TIR	Type II PKS (kinamycin)	*SAMT0158-SAMT0185*	34
Left arm	Terpene synthase (2-methylisoborneol)	*SAML0357-SAML0358*	2.2
	NRPS/PKS	*SAML0370-SAML0377*	19
	Carotenoid synthase (isorenieratene?)	*SAML0730-SAML0732*	3.5
	Terpene synthase	*SAML0739-SAML0744*	5
Right arm	Type I PKS	*SAMR0267-SAMR0278*	22
	Lantibiotic	*SAMR0342-SAMR0344*	5
	Type I PKS (stambomycin)	*SAMR0465-SAMR0487*	146
	Carotenoid synthase (isorenieratene?)	*SAMR0510-SAMR0517*	9
	NRPS (coelichelin)	*SAMR0548-SAMR0559*	24
	3-Oxoacyl-(ACP) synthase	*SAMR0594-SAMR0609*	18
	Terpene synthase	*SAMR0831-SAMR0836*	6
	NRPS (congocidine)	*SAMR0900-SAMR0921*	29

*Characterized or predicted metabolites. *SAMLXXXX*: genes located in the left chromosomal arm; *SAMRXXXX*: genes located in the right chromosomal arm; *SAMTXXXX*: genes located in the terminal inverted repeats (TIRs).

other clusters in addition to the *des* cluster responsible for the production of desferrioxamine, the spiramycin cluster and an adjacent non-ribosomal peptide synthetase (NPRS) cluster (unpublished data).

Although completion of the genome sequence could yet reveal additional genes involved in secondary metabolite biosynthesis, it is already apparent that the majority of these biosynthetic gene clusters occurs outside the conserved core region of the chromosome, reinforcing the idea that the variable subtelomeric regions of linear chromosomes contribute to the metabolic diversity of *Streptomyces.* This is also reinforced by the observation that six of the 14 clusters located in the chromosomal arms are in the last 619 kb (left arm) or 660 kb (right arm) of the chromosome that have been shown to be species-specific when compared with the phylogenetically close relative *S. coelicolor* (not more than 1% of 16S ribosomal sequence divergence; Choulet *et al.*, 2006a; Table 10.1). It should nevertheless be noted that two of these clusters, the lantibiotic and the 2-methylisoborneol clusters (see below), are conserved in *S. coelicolor*, in which they are also located in the species-specific regions. However, we have identified so far only five clusters conserved between these two closely related species.

Mining the *S. ambofaciens* genome has thus highlighted that this species represents a source for discovering new secondary metabolites although *S. ambofaciens* is very close to the model *Streptomyces, S. coelicolor.*

Secondary metabolite gene clusters common to *Streptomyces coelicolor* and *Streptomyces ambofaciens*

At least five secondary metabolite gene clusters are common to *S. ambofaciens* and *S. coelicolor.*

Desferrioxamines and coelichelin

In soil, iron is mainly present in the form of insoluble hydroxide complexes and is therefore not readily bioavailable to micro-organisms. To overcome iron limitation, micro-organisms synthesize small metabolites termed siderophores that bind tightly Fe^{3+} and capture iron from the environment. *S. ambofaciens* produces the same three siderophores, desferrioxamines B and E and coelichelin as *S. coelicolor* (Barona-Gomez *et al.*, 2006). The *des* gene cluster, directing the biosynthesis of desferrioxamines B and E and extremely well conserved among *Streptomyces* species, is located in a region of the chromosome

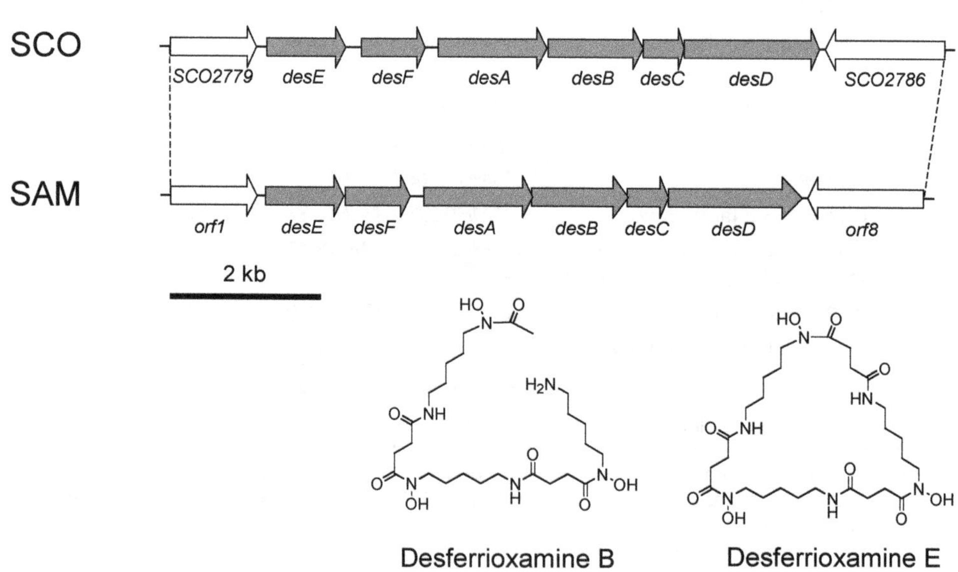

Figure 10.2 Desferrioxamine biosynthetic gene clusters in *S. coelicolor* (SCO) and *S. ambofaciens* (SAM). The product of *orf1* is homologous to the product of *SCO2779* while the product of *orf8* is homologous to the product of *SCO2786*.

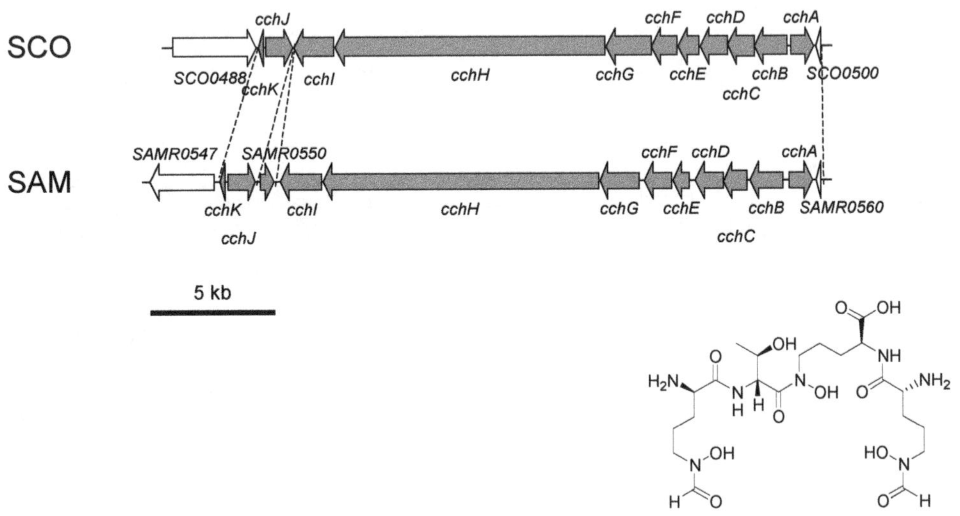

Coelichelin

Figure 10.3 Coelichelin biosynthetic gene clusters in *S. coelicolor* (SCO) and *S. ambofaciens* (SAM). The product of *SAM*R0560 is homologous to the product of *SCO0500*. The product of *SAMR0547* is homologous to the product of *SCO0854*.

of *S. ambofaciens* highly syntenic with the chromosome of *S. coelicolor* (Fig. 10.2) (unpublished data). All of the Cch proteins encoded in the *S. coelicolor* coelichelin biosynthetic gene cluster have homologues in *S. ambofaciens*. However, the *S. ambofaciens cch* gene cluster contains an extra gene, *SAMR0550*, coding for a putative integral membrane protein. The function of this protein remains to be determined (Fig. 10.3). Contrary to the *des* gene cluster, the *cch* gene cluster in *S. ambofaciens* is located in one of the variable extremities of the chromosome and therefore not in a region syntenic with the chromosome of *S. coelicolor*.

2-Methylisoborneol

2-Methylisoborneol (2-MIB) and geosmin are volatile odorous terpenoid metabolites produced by many actinomycetes. Recently, the genes directing the biosynthesis of 2-MIB have been identified in several *Streptomyces* species, including *S. ambofaciens* and *S. coelicolor* (Komatsu *et al.*, 2008; Wang and Cane, 2008). They consist of a monoterpene cyclase (SAML0357/SCO7700) and of a C-methyltransferase (SAML0358/SCO7701). The organization of the 2-MIB gene cluster is identical in *S. ambofaciens* and *S. coelicolor* (Fig. 10.4). However, as with coelichelin, the *S. ambofaciens* 2-MIB gene cluster is not located in a region of the chromosome syntenic with a region of the *S. coelicolor* chromosome. Geosmin is also known to be produced by both *Streptomyces* species, but the gene cluster directing the biosynthesis of this metabolite in *S. ambofaciens* has not been isolated yet.

Carotenoid

Carotenogenesis is widespread among both eukaryotes and prokaryotes. In non-photoautotrophic organisms, carotenoids protect the cells from photooxidative stress caused by illumination (Krugel *et al.*, 1999). A carotenoid biosynthesis gene cluster, responsible for the biosynthesis of isorenieratene, has been identified in the complete genome sequence of *S. coelicolor* A3(2) (Bentley *et al.*, 2002). It is composed of two convergent biosynthesis operons, *crtEIBV* and *crtYTU*. These operons are flanked by a regulatory region of two divergent operons, *litRQ* and *litSAB* (*lit* = light induced) (Takano *et al.*, 2005). In *S. ambofaciens*, a carotenoid-like biosynthetic region was found on the right arm (*SAMR0510–0517*).

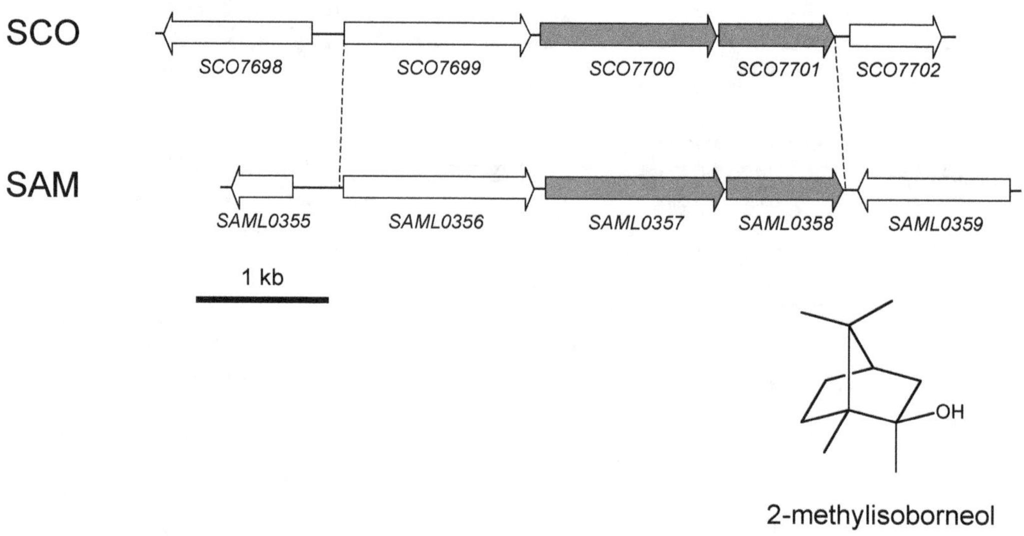

Figure 10.4 2-Methylisoborneol biosynthetic gene clusters in *S. coelicolor* (SCO) and *S. ambofaciens* (SAM). The product of *SAML0356* is homologous to the product of *SCO7699*.

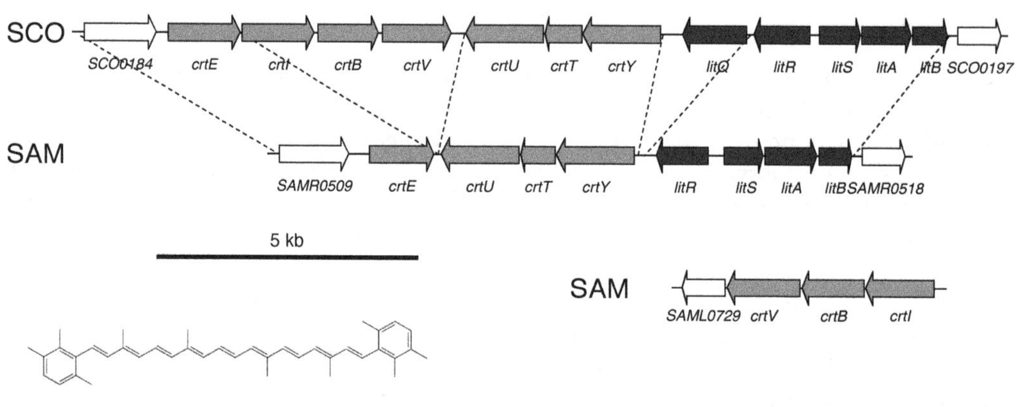

Figure 10.5 Chemical structure of isorenieratene and organization of the genes constituting the isorenieratene biosynthetic region in *S. coelicolor* (SCO) and of their homologues in *S. ambofaciens* (SAM). In grey are genes belonging to biosynthesis operons and in black genes belonging to regulatory operons.

It is not strictly identical to that of *S. coelicolor*, as the *crtEIBV* operon is split into two parts. Only the *crtE* gene remains close to the convergent *crtYTU* operon. The rest of the operon (*crtIBV*) is located on the left arm (*SAML0730–0732*). Concerning the *S. ambofaciens* regulatory region, the *litQ* oxidoreductase is absent (Fig. 10.5). In *S. coelicolor*, the role of *litQ* has not been assigned (Takano *et al.*, 2005). It may be dispensable for carotenoid biosynthesis.

Lantibiotic

Lantibiotics (=lanthionine-containing antibiotics) are peptides produced by many Gram-positive bacteria and contain unusual amino acids like lanthionine or dehydrated

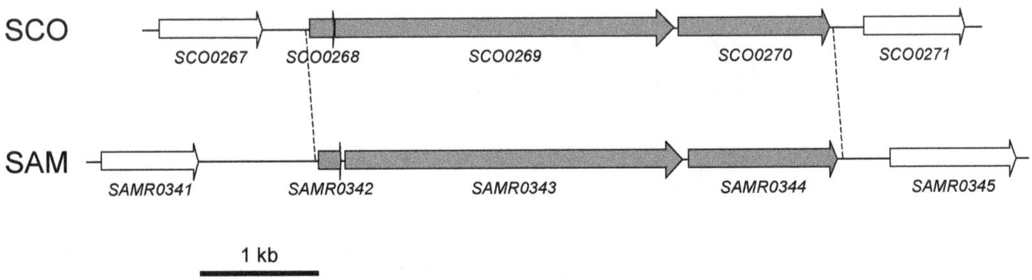

Figure 10.6 Organization of the genes constituting the putative lantibiotic-like biosynthetic region in *S. coelicolor* (SCO) and of their homologues in *S. ambofaciens* (SAM).

amino acids. Their biosynthesis begins with the synthesis of a prepeptide by ribosomes. It is composed of an N-terminal leader peptide and of a C-terminal propeptide. The latter undergoes post-translational modifications to form the bioactive lantibiotic (Nagao *et al.*, 2006). A cluster resembling a lantibiotic biosynthetic cluster is present in *S. ambofaciens* and *S. coelicolor*. It consists of a gene coding for the prepeptide (*SCO0268* and *SAMR0342*), a lantibiotic dehydratase (*SCO0269* and *SAMR0343*) and a putative lanthionine synthase (*SCO0270* and *SAMR0344*) (Fig. 10.6). This cluster seems to be absent from all the other *Streptomyces* genomes available to date. This is consistent with the fact that *S. ambofaciens* and *S. coelicolor* are phylogenetically close (Choulet *et al.*, 2006a).

Secondary metabolite gene clusters unique to *Streptomyces ambofaciens*

Spiramycin

S. ambofaciens is cultivated by the pharmaceutical industry for the production of spiramycin, a macrolide antibiotic used in human medicine as an antibacterial agent and for the treatment of *Toxoplasma* infections. The spiramycin molecule is composed of a 16-membered polyketide lactone ring (platenolide) on which two amino sugars (mycaminose and forosamine) and one neutral sugar (mycarose) are attached (Fig. 10.7). Spiramycin is produced as a mixture of three major compounds differing by acyl substitutions at carbon 3 (Fig. 10.7). Spiramycin biosynthetic genes (*srm* genes) have been isolated by several groups (Geistlich *et al.*, 1992; Karray *et al.*, 2007; Kuhstoss *et al.*, 1996; Richardson *et al.*, 1990) and the sequence of the complete *srm* cluster is now available. The involvement of the *srm* genes in spiramycin biosynthesis has been established by inactivating them (Richardson *et al.*, 1990; Geistlich *et al.*, 1992; Karray *et al.*, 2007; Nguyen, 2009).

The *srm* gene cluster is not present in the terminal parts of the chromosome for which the sequence is available, but it was mapped on the physical map of the *S. ambofaciens* chromosome previously established (Leblond *et al.*, 1996; Fig. 10.1). The *srm* cluster is not present in *S. coelicolor*; moreover it is flanked, downstream of *srm1*, by a large cryptic NRPS cluster which is also not conserved in *S. coelicolor* (unpublished data). Therefore, the region containing these two clusters should correspond to a large break of synteny between the two strains, in a part of the chromosome that is otherwise syntenic.

Comparison of the products of the *srm* genes with protein databases allowed assignment of putative functions in spiramycin biosynthesis to most of the Srm proteins. Several *srm* genes were inactivated by gene replacement; the effects of these inactivations on spiramycin biosynthesis were studied and, when relevant, the biosynthetic intermediates accumulated by the mutant strains were identified. Platenolide is synthesized by the type I PKS encoded by the five *srmG* genes (Kuhstoss *et al.*, 1996). The PKS incorporates malonyl, methylmalonyl, ethylmalonyl and methoxymalonyl into the 16-membered lactone

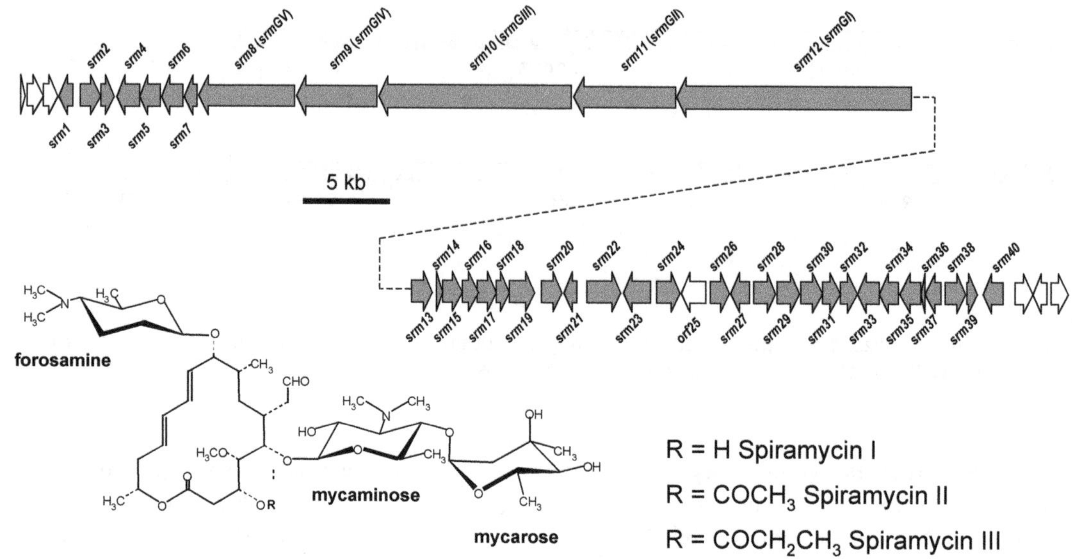

Figure 10.7 Spiramycin biosynthetic gene cluster and chemical structure of the spiramycins. Genes involved in spiramycin biosynthesis, resistance or regulation thereof are represented by grey arrows; genes not involved are represented by white arrows.

ring. Several *srm* genes are involved in the provision of extender units for the PKS: *srm4* encodes a putative crotonyl-CoA reductase which might contribute to the flux of ethylmalonyl-CoA; five genes (*srm3, srm34, srm35, srm36,* and *srm37*) are believed to be involved in the biosynthesis of methoxymalonyl-acyl carrier protein. All the genes required for the synthesis of the three deoxysugars from glucose-1-phosphate are present in the cluster. Concerning sugar attachment, four *srm* genes are predicted to encode glycosyltransferases, but only three of them (*srm5, srm29* and *srm38*) are involved in the biosynthesis (Nguyen *et al.*, 2010). The glycosyltransferase responsible for the attachment of each of the three sugars has been identified. Moreover, two *srm* genes (*srm6* and *srm28*) encode putative glycosyltransferase auxiliary proteins, which are required for the full activity of some glycosyltransferases. One of the auxiliary proteins could interact efficiently with the two glycosyltransferases transferring mycaminose and forosamine, while the other auxiliary protein interacts only with the mycaminosyltransferase. Two genes of the cluster are involved in spiramycin resistance: *srm1* (= *srmD*) encodes a 23S rRNA methyltransferase (our unpublished results) and *srm23* (= *srmB*) encode the cytoplasmic ATP-binding component of an ABC transporter (Richardson *et al.*, 1987; Schoner *et al.*, 1992). The regulation of spiramycin biosynthesis involves the two genes *srm22* (= *srmR*) and *srm40*. SrmR is an activator required for the transcription of *srm40*. Srm40 is also a transcriptional activator acting on most if not all the *srm* biosynthetic genes (Geistlich *et al.*, 1992; Karray, 2005; Karray *et al.*, 2010).

Congocidine

Congocidine belongs to the family of pyrrole-amide antibiotics (Fig. 10.8). It has been known to be produced by *S. ambofaciens* since 1952, but has also been isolated from *S. netropsis* (netropsin). Congocidine binds the DNA minor groove with some sequence specificity (a succession of four A/T bases). This confers on the molecule antimicrobial activities but also renders it too cytotoxic for clinical use. Nonetheless, molecules from the pyrrole-amide family continue to attract considerable attention as they could potentially be used to regulate the expression of genes involved in diseases such as cancer.

enoylreductase (ER) domain (Keatinge-Clay, 2007; Kwan *et al.*, 2008). The linear polyketide backbone of the product was predicted to have an approximate mass of ~1127Da and the molecular formula $C_{61}H_{107}O_{18}$. On the basis of the thioesterase domain analysis, the linear backbone was expected to be cyclized. The presence of a glycosyltransferase gene and of a set of five genes predicted to be potentially involved in the conversion of an α-D-glucose-1-phosphate into a TDP-D-mycaminose residue strongly suggested that the final product of this type PKS gene cluster was a macrolide. The possible mass and molecular formula of the macrolide were expected to be ~1318Da and $C_{69}H_{124}NO_{22}$, respectively, and according to the chemical data base, this molecule was predicted to be novel (Laureti, 2010; Laureti *et al.*, submitted).

The strategy applied to identify this new metabolite was based on the ectopic expression of a pathway-specific LAL (Large ATP-binding LuxR family) regulator encoded by a gene located in the cluster (*SAMR0484*). Indeed, the PKS encoding genes and most of the genes in the cluster were silent under different tested growth conditions. The *SAMR0484* gene was placed under the control of a strong and constitutive promoter (*ermEp**) in the conjugative and integrative plasmid pIB139 (Wilkinson *et al.*, 2002). The overexpression of the LAL regulatory gene in *S. ambofaciens* finally triggered the expression of all the silent genes. By applying a comparative LC-MS profiling approach, two peaks were identified in the LAL regulator overexpressing strain which were absent in the control strain (wt + pIB139). These peaks were detected only from the mycelium extract and not from the medium showing that the corresponding metabolites were not secreted at least in the growth conditions used. The mass of the metabolites and their molecular formulas ($C_{72}H_{131}NO_{22}$ and $C_{73}H_{133}NO_{22}$) established by high resolution mass spectrometry were consistent with our previous predictions (Laureti, 2010; Laureti *et al.*, submitted).

The structural characterization of these compounds revealed four forms of a 51-membered macrolide we named stambomycin: the isomers A and B and the isomers C and D (Fig. 10.10). Stambomycin compounds contain the largest lactone ring ever described. This large size is due to a novel cyclization process involving a cytochrome P450 (SAMR0479; encoded by the cluster) which hydroxylates the first precursor when the polyketide chain is still attached to the last module. Another originality of the stambomycin compounds is the long alkyl chain due to the incorporation of an unusual alkylmalonyl-CoA extender unit. This extender unit is proposed to be synthesized by an acyl-CoA synthase (SAMR0482) and an acyl-CoA carboxylase (SAMR0483) encoded by the cluster.

The stambomycin A/B and C/D mixtures have antimicrobial activities against Gram-positive bacteria except *Mycobacterium smegmatis* but not against Gram-negative bacteria and fungi. More interestingly, they have a very promising antiproliferative activity. Indeed, they inhibit the proliferation of the human adenocarcinoma (HT29) cell lines with an efficiency comparable to the therapeutic drug doxorubicin.

The discovery of the novel macrolide stambomycin emphasizes that the strategy of awaking sleeping gene clusters by ectopic expression of a pathway-specific regulator is a promising approach towards the identification of new metabolites from silent secondary metabolite gene clusters identified by genome mining in *Streptomyces* and more generally in actinomycetes, the main producers of microbial natural products. A similar approach was previously only described in the fungi *Aspergillus nidulans* for a PKS–NRPS hybrid gene cluster shown to be responsible for the production of aspyridone (Bergmann *et al.*, 2007).

Perspectives

Genome sequencing has revealed that *Streptomyces* genomes contain far more secondary metabolite biosynthetic gene clusters than previously estimated based on the number of compounds that had been isolated: there exist 'cryptic' or 'orphan' clusters, for which no product has been identified yet (Challis, 2008). To date, *S. ambofaciens* is known to produce four secondary metabolites with antibacterial activity (spiramycin, congocidine, kinamycin/alpomycin and stambomycin). The capacity of *S. ambofaciens* to produce spiramycin and congocidine had been known for a long time before their biosynthetic pathways were elucidated (Karray *et al.*, 2007; Juguet *et al.*, 2009). Since then, sequencing of the chromosomal arms has allowed the alpomycin/kinamycin and stambomycin gene clusters to be discovered by genome mining (Choulet *et al.*, 2006; Pang *et al.*, 2004; Bunet *et al.*, 2008 and submitted, Laureti *et al.*, submitted). Concerning stambomycin, the analysis of the cluster organization and the identification of regulators has led to the expression of a cryptic pathway and the discovery of a new molecule with numerous bioactivities. For further exploitation of the potential for production of secondary metabolites by *S. ambofaciens*, a mutant no longer producing the four known antibiotics has been constructed. It will simplify the detection of new activities, as precursors will be more available and as HPLC and bioassay analyses will be facilitated. Activation of cryptic pathways will then be possible by playing with regulation and by changing culture conditions. With these tools and using genome mining, we can assume that the complete genome sequence of *S. ambofaciens* will lead to the discovery of new molecules with potential bioactivities or to specific features in the biosynthetic pathways that could be useful for combinatorial biochemistry.

References

Aigle, B., Schneider, D., Morilhat, C., Vandewiele, D., Dary, A., Holl, A.C., Simonet, J.M., and Decaris, B. (1996). An amplifiable and deletable locus of *Streptomyces ambofaciens* RP181110 contains a very large gene homologous to polyketide synthase genes. Microbiology *142*, 2815–2824.

Aigle, B., Pang, X., Decaris, B., and Leblond, P. (2005). Involvement of AlpV, a new member of the Streptomyces antibiotic regulatory protein family, in regulation of the duplicated type II polyketide synthase *alp* gene cluster in *Streptomyces ambofaciens*. J. Bacteriol. *187*, 2491–2500.

Barona-Gomez, F., Lautru, S., Francou, F.X., Leblond, P., Pernodet, J.L., and Challis, G.L. (2006). Multiple biosynthetic and uptake systems mediate siderophore-dependent iron acquisition in *Streptomyces coelicolor* A3(2) and *Streptomyces ambofaciens* ATCC 23877. Microbiology *152*, 3355–3366.

Bate, N., Butler, A.R., Gandecha, A.R., and Cundliffe, E. (1999). Multiple regulatory genes in the tylosin biosynthetic cluster of *Streptomyces fradiae*. Chem. Biol. *6*, 617–624.

Bentley, S.D., Chater, K.F., Cerdeno-Tarraga, A.M., Challis, G.L., Thomson, N.R., James, K.D., Harris, D.E., Quail, M.A., Kieser, H., Harper, D., *et al.* (2002). Complete genome sequence of the model actinomycete *Streptomyces coelicolor* A3(2). Nature *417*, 141–147.

Bergmann, S., Schumann, J., Scherlach, K., Lange, C., Brakhage, A.A., and Hertweck, C. (2007). Genomics-driven discovery of PKS-NRPS hybrid metabolites from *Aspergillus nidulans*. Nat. Chem. Biol. *3*, 213–217.

Bunet, R., Mendes, M.V., Rouhier, N., Pang, X., Hotel, L., Leblond, P., and Aigle, B. (2008). Regulation of the synthesis of the angucyclinone antibiotic alpomycin in *Streptomyces ambofaciens* by the autoregulator receptor AlpZ and its specific ligand. J. Bacteriol. *190*, 3293–3305.

Challis, G.L. (2008) Mining microbial genomes for new natural products and biosynthetic pathways. Microbiology *154*, 1555–1569.

Choulet, F., Aigle, B., Gallois, A., Mangenot, S., Gerbaud, C., Truong, C., Francou, F.X., Fourrier, C., Guerineau, M., Decaris, B., *et al.* (2006a) Evolution of the terminal regions of the *Streptomyces* linear chromosome. Mol. Biol. Evol. *23*, 2361–2369.

Choulet, F., Gallois, A., Aigle, B., Mangenot, S., Gerbaud, C., Truong, C., Francou, F.X., Borges, F., Fourrier, C., Guerineau, M., *et al.* (2006b) Intraspecific variability of the terminal inverted repeats of the linear chromosome of *Streptomyces ambofaciens*. J. Bacteriol. *188*, 6599–6610.

Dary, A., Bourget, N., Girard, N., Simonet, J.M., and Decaris, B. (1992). Amplification of a particular DNA sequence in *Streptomyces ambofaciens* RP181110 reversibly prevents spiramycin production. Res. Microbiol. *143*, 99–112.

Decker, H., and Haag, S. (1995). Cloning and characterization of a polyketide synthase gene from *Streptomyces fradiae* Tu2717, which carries the genes for biosynthesis of the angucycline antibiotic urdamycin A and a gene probably involved in its oxygenation. J. Bacteriol. *177*, 6126–6136.

Faust, B., Hoffmeister, D., Weitnauer, G., Westrich, L., Haag, S., Schneider, P., Decker, H., Kunzel, E., Rohr, J., and Bechthold, A. (2000). Two new tailoring enzymes, a glycosyltransferase and an oxygenase, involved in biosynthesis of the angucycline antibiotic urdamycin A in *Streptomyces fradiae* Tu2717. Microbiology *146*, 147–154.

Geistlich, M., Losick, R., Turner, J.R., and Rao, R.N. (1992). Characterization of a novel regulatory gene governing the expression of a polyketide synthase gene in *Streptomyces ambofaciens*. Mol. Microbiol. *6*, 2019–2029.

Han, L., Yang, K., Ramalingam, E., Mosher, R.H., and Vining, L.C. (1994). Cloning and characterization of polyketide synthase genes for jadomycin B biosynthesis in *Streptomyces venezuelae* ISP5230. Microbiology *140*, 3379–3389.

Hasinoff, B.B., Wu, X., Yalowich, J.C., Goodfellow, V., Laufer, R.S., Adedayo, O., and Dmitrienko, G.I. (2006). Kinamycins A and C, bacterial metabolites that contain an unusual diazo group, as potential new anticancer agents: antiproliferative and cell cycle effects. Anticancer Drugs *17*, 825–837.

Ikeda, H., Ishikawa, J., Hanamoto, A., Shinose, M., Kikuchi, H., Shiba, T., Sakaki, Y., Hattori, M., and Omura, S. (2003). Complete genome sequence and comparative analysis of the industrial microorganism *Streptomyces avermitilis*. Nat. Biotechnol. *21*, 526–531.

Jenke-Kodama, H., and Dittmann, E. (2009). Evolution of metabolic diversity: insights from microbial polyketide synthases. Phytochemistry *70*, 1858–1866.

Juguet, M., Lautru, S., Francou, F.X., Nezbedova, S., Leblond, P., Gondry, M., and Pernodet, J.L. (2009). An iterative nonribosomal peptide synthetase assembles the pyrrole-amide antibiotic congocidine in *Streptomyces ambofaciens*. Chem. Biol. *16*, 421–431.

Karray, F. (2005). Etude de la biosynthèse de l'antibiotique spiramycine par *Streptomyces ambofaciens*. PhD thesis. (Université Paris-Sud 11, France).

Karray, F., Darbon, E., Oestreicher, N., Dominguez, H., Tuphile, K., Gagnat, J., Blondelet-Rouault, M.H., Gerbaud, C., and Pernodet, J.L. (2007). Organization of the biosynthetic gene cluster for the macrolide antibiotic spiramycin in *Streptomyces ambofaciens*. Microbiology *153*, 4111–4122.

Karray, F., Darbon, E., Nguyen, H.C., Gagnat, J., and Pernodet, J.-L. (2010). Regulation of the biosynthesis of the macrolide antibiotic spiramycin in *Streptomyces ambofaciens*. J. Bacteriol. 2010 3 September [Epub ahead of print] doi:10.1128/JB.00712-10.

Keatinge-Clay, A.T. (2007) A tylosin ketoreductase reveals how chirality is determined in polyketides. Chem. Biol. *14*, 898–908.

Komatsu, M., Tsuda, M., Omura, S., Oikawa, H., and Ikeda, H. (2008). Identification and functional analysis of genes controlling biosynthesis of 2-methylisoborneol. Proc. Natl. Acad. Sci. U.S.A. *105*, 7422–7427.

Krugel, H., Krubasik, P., Weber, K., Saluz, H.P., and Sandmann, G. (1999). Functional analysis of genes from *Streptomyces griseus* involved in the synthesis of isorenieratene, a carotenoid with aromatic end groups, revealed a novel type of carotenoid desaturase. Biochim. Biophys. Acta *1439*, 57–64.

Kuhstoss, S., Huber, M., Turner, J.R., Paschal, J.W., and Rao, R.N. (1996). Production of a novel polyketide through the construction of a hybrid polyketide synthase. Gene *183*, 231–236.

Kwan, D.H., Sun, Y., Schulz, F., Hong, H., Popovic, B., Sim-Stark, J.C., Haydock, S.F., and Leadlay, P.F. (2008) Prediction and manipulation of the stereochemistry of enoylreduction in modular polyketide synthases. Chem. Biol. *15*, 1231–1240.

Laureti, L. (2010). Activation of a silent type I PKS gene cluster in *Streptomyces ambofaciens* ATCC23877: isolation and characterization of a novel giant macrolide. PhD thesis (Université Henri Poncaré, France).

Leblond, P., Demuyter, P., Simonet, J.M., and Decaris, B. (1991). Genetic instability and associated genome plasticity in *Streptomyces ambofaciens*: pulsed-field gel electrophoresis evidence for large DNA alterations in a limited genomic region. J. Bacteriol. *173*, 4229–4233.

Leblond, P., Fischer, G., Francou, F.X., Berger, F., Guerineau, M., and Decaris, B. (1996). The unstable region of *Streptomyces ambofaciens* includes 210 kb terminal inverted repeats flanking the extremities of the linear chromosomal DNA. Mol. Microbiol. *19*, 261–271.

Mochizuki, S., Hiratsu, K., Suwa, M., Ishii, T., Sugino, F., Yamada, K., and Kinashi, H. (2003). The large linear plasmid pSLA2-L of *Streptomyces rochei* has an unusually condensed gene organization for secondary metabolism. Mol. Microbiol. *48*, 1501–1510.

Nagao, J., Asaduzzaman, S.M., Aso, Y., Okuda, K., Nakayama, J., and Sonomoto, K. (2006). Lantibiotics: insight and foresight for new paradigm. J. Biosci. Bioeng. *102*, 139–149.

Nguyen, H.-C. (2009). Study of the spiramycin biosynthesis and its regulation in *Streptomyces ambofaciens*. PhD thesis (Université Paris-Sud 11, France).

Nguyen, H.-C., Karray, F., Lautru, S., Gagnat, J., Lebrihi, A., Ho Huynh, T.D., and Pernodet, J.-L. (2010). Glycosylation steps during spiramycin biosynthesis in Streptomyces ambofaciens: involvement of three glycosyltransferases and their interplay with two auxiliary proteins. Antimicrob. Agents Chemother. *54*, 2830–2839.

Ohnishi, Y., Ishikawa, J., Hara, H., Suzuki, H., Ikenoya, M., Ikeda, H., Yamashita, A., Hattori, M., and Horinouchi, S. (2008). Genome sequence of the streptomycin-producing microorganism *Streptomyces griseus* IFO 13350. J. Bacteriol. *190*, 4050–4060.

Pang, X., Aigle, B., Girardet, J.M., Mangenot, S., Pernodet, J.L., Decaris, B., and Leblond, P. (2004). Functional angucycline-like antibiotic gene cluster in the terminal inverted repeats of the *Streptomyces ambofaciens* linear chromosome. Antimicrob. Agents Chemother. *48*, 575–588.

Richardson, M.A., Kuhstoss, S., Solenberg, P., Schaus, N.A., and Rao, R.N. (1987). A new shuttle cosmid vector, pKC505, for streptomycetes: its use in the cloning of three different spiramycin-resistance genes from a *Streptomyces ambofaciens* library. Gene *61*, 231–241.

Richardson, M.A., Kuhstoss, S., Huber, M.L., Ford, L., Godfrey, O., Turner, J.R., and Rao, R.N. (1990). Cloning of spiramycin biosynthetic genes and their use in constructing *Streptomyces ambofaciens* mutants defective in spiramycin biosynthesis. J. Bacteriol. *172*, 3790–3798.

Schoner, B., Geistlich, M., Rosteck, P. Jr, Rao, R.N., Seno, E., Reynolds, P., Cox, K., Burgett, S., and Hershberger, C. (1992) Sequence similarity between macrolide-resistance determinants and ATP-binding transport proteins. Gene *115*, 93–96.

Stumpp, T., Himbert, S., and Altenbuchner, J. (2005). Cloning of the netropsin resistance genes from *Streptomyces flavopersicus* NRRL 2820. J. Basic Microbiol. *45*, 355–362.

Takano, H., Obitsu, S., Beppu, T., and Ueda, K. (2005). Light-induced carotenogenesis in *Streptomyces coelicolor* A3(2): identification of an extracytoplasmic function sigma factor that directs photodependent transcription of the carotenoid biosynthesis gene cluster. J. Bacteriol. *187*, 1825–1832.

Wang, C.M., and Cane, D.E. (2008). Biochemistry and molecular genetics of the biosynthesis of the earthy odorant methylisoborneol in *Streptomyces coelicolor*. J. Am. Chem. Soc. *130*, 8908–8909.

Westrich, L., Domann, S., Faust, B., Bedford, D., Hopwood, D.A., and Bechthold, A. (1999). Cloning and characterization of a gene cluster from *Streptomyces cyanogenus* S136 probably involved in landomycin biosynthesis. FEMS Microbiol. Lett. *170*, 381–387.

Wilkinson, C. J., Hughes-Thomas, Z. A., Martin, C. J., Bohm, I., Mironenko, T., Deacon, M., Wheatcroft, M., Wirtz, G., Staunton, J. and Leadlay, P.F. (2002). Increasing the efficiency of heterologous promoters in actinomycetes. J. Mol. Microbiol. Biotechnol. *4*, 417–426.

Gene Clusters for Bioactive Natural Products in Actinomycetes and their Use in Combinatorial Biosynthesis

11

Carlos Olano, Carmen Méndez and José A. Salas

Abstract

During the last 25 years the isolation and characterization of gene clusters involved in the biosynthesis of actinomycete secondary metabolites has permitted the elucidation of the biochemical steps involved in the production of different structural classes of bioactive compounds. The characterization of these clusters has represented a great source of genes for the generation of novel 'unnatural natural' compounds by using combinatorial biosynthesis. The development of more effective methods for DNA sequencing, the improvement of targeted inactivation and heterologous host expression systems has strengthened the effectiveness of combinatorial biosynthesis. For these reasons combinatorial DNA technology has become during the last decade one of the most important approaches for generating chemical structural diversity and for increasing the number of potential useful compounds.

Introduction

Actinomycetes are Gram-positive bacteria belonging to the phylum *Actinobacteria* that represents one of the largest taxonomic groups among the 18 major lineages currently recognized within the domain Bacteria (Ventura *et al.*, 2007). Filamentous actinomycetes are characterized by a complex life cycle of morphological differentiation which begins with a spore that germinates to give a substrate mycelium and later an aerial mycelium, followed then by a sporulation stage (Chater, 1998). Actinobacteria are widely distributed in terrestrial, especially soil, and aquatic ecosystems (McCarthy and Williams, 1992; Stach and Bull, 2005). They are important for soil biodegradation and humus formation by decomposing and recycling complex mixtures of polymers in dead plant, animal and fungal materials using extracellular enzymes (McCarthy and Williams, 1992). They also exhibit diverse physiological and metabolic properties, such as the production of volatile substances (Gust *et al.*, 2003) and a wide variety of secondary metabolites (Bérdy, 2005). Some strains have been identified as pathogenic to plants, animals and humans (Locci, 1994; McNeil and Brown, 1994).

Actinomycetes are prolific producers of bioactive compounds (Fig. 11.1), with about 45% of all bioactive microbial metabolites discovered (over 10,000) produced by these bacteria, and about 76% produced by *Streptomyces* species (Bérdy, 2005). However, the rapid development of oceanographic studies in the last decade, has led to the isolation of new prolific bioactive metabolite-producing actinomycetes from marine sources such as *Salinispora* (Jensen *et al.*, 2007). Many of these secondary metabolites are potent antibiotic, antifungal, antitumour, immunosuppressant, antiviral and antiparasitic agents and also enzyme-inactivating compounds that have practical importance in medicine, veterinary medicine and agriculture. This has turned actinomycete species into the primary bioactive metabolite-producing organisms exploited by the pharmaceutical industry and has prompted the study of these microorganisms at all levels: taxonomy, genetics and physiology (Hopwood, 1999).

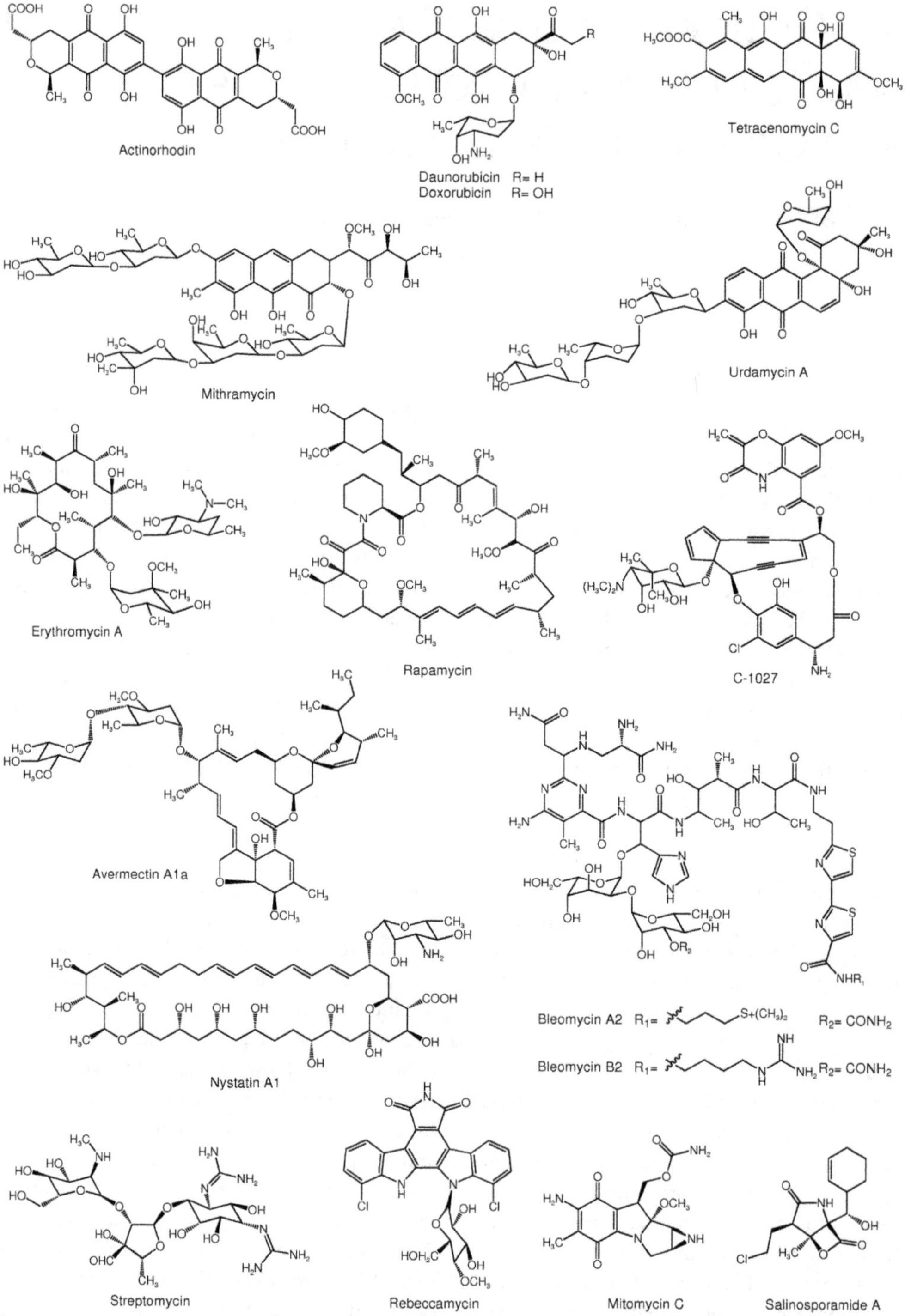

Figure 11.1 Structures of some bioactive compounds produced by actinomycetes.

The search for novel drugs has been, and still is today, a priority goal for the treatment of microbial, fungal and parasitic infections due to the rapid development of resistance to multiple chemotherapeutic drugs and to the emergence of new infectious microorganisms. In addition, cancer therapy is another field with an increased demand for novel antitumour drugs active against untreatable tumours, with fewer side effects or with greater therapeutic efficiency. Recent progress has been made on drug discovery from actinomycetes by searching uncommon actinomycetes in special niches, mining genomes for cryptic pathways, and combinatorial biosynthesis to generate novel secondary metabolites related to existing pharmacophores (Baltz, 2008a). Since the development of recombinant DNA technology an increasing number of secondary metabolite biosynthesis gene clusters have been isolated and characterized from actinomycetes, which has in turn raised the importance of combinatorial biosynthesis as an alternative approach for developing novel bioactive compounds by engineering biosynthetic pathways. In this review, we focus on the different approaches used for the isolation and characterization of secondary metabolite biosynthesis gene clusters from actinomycetes and their subsequent use for the development of novel derivatives through combinatorial biosynthesis as well as for the improvement of production titres.

Isolation of gene clusters

The isolation of gene clusters from microorganisms producing bioactive compounds is the first step for the generation of novel derivatives. Identifying the genes involved in the biosynthesis of these compounds permits the elucidation of how these metabolites are biosynthesized and lays the foundation stone for the rational design of novel derivatives. Several approaches have been reported for the isolation of gene clusters going from the early complementation of mutants to most recent whole genome sequencing (Fig. 11.2). As shown below, in some cases, several strategies have been applied for the isolation of a single biosynthesis cluster (Table 11.1). Many of these strategies are based on some initial assumptions (later on confirmed): (i) structural biosynthesis genes are generally clustered; (ii) resistance genes are normally part of the clusters;

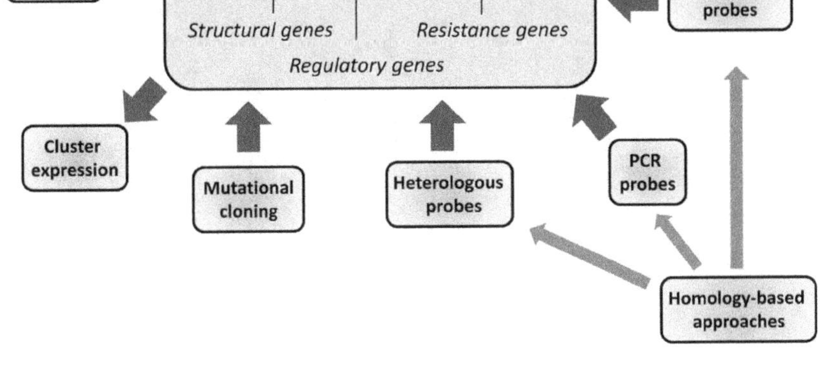

Figure 11.2 Approaches used for the identification and isolation of secondary metabolites biosynthesis gene clusters.

Table 11.1 Gene clusters isolated and characterized from actinomycetes and methods used for its identification

Compound	Structural type	Producer strain	Method for cloning	References
Aclacinomycin	Polyketide	*S. galilaeus* ATCC31615	Homology-based approaches	Räty *et al.* (2000, 2002)
			Complementation of mutants	Räty *et al.* (2000)
Actinomycin	Polypeptide	*S. chrysomallus*	Homology-based approaches	Schauwecker *et al.* (1998)
Actinorhodin	Polyketide	*S. coelicolor* A3(2)	Complementation of mutants, gene cluster expression	Malpartida and Hopwood (1984)
Amphotericin B	Polyketide	*S. nodosus* ATCC14899	Homology-based approaches	Caffrey *et al.* (2001)
Ansamitocin P-3	Polyketide	*Ac. pretiosum* subsp. *auranticum*	Homology-based approaches	Yu *et al.* (2002)
Anthramycin	Benzodiazepine	*S. refuineus* var. *thermotolerans*	Genome mining	Hu *et al.* (2007)
Aranciamycin	Polyketide	*S. echinatus* Tü303	Homology-based approaches	Luzhetskyy *et al.* (2007b)
Aureothin	Polyketide	*S. thioluteus* HKI-227	Homology-based approaches	He and Hertweck (2003)
Avermectin	Polyketide	*S. avermitilis* K139	Complementation of mutants	Streicher *et al.* (1989), Ikeda *et al.* (1998)
			Homology-based approaches	Ikeda *et al.* (1999)
Avilamycin	Oligosaccharide	*S. viridochromogenes* Tü57	Homology-based approaches	Gaisser *et al.* (1997)
Azinomycin B	Polyketide	*S. sahachiroi* NRRL2485	Homology-based approaches	Zhao *et al.* (2008)
Benastatin	Polyketide	*Streptomyces* sp. A2991200	Homology-based approaches, gene cluster expression	Xu *et al.* (2007)
Bialaphos	Polypeptide	*S. hygroscopicus* ATCC21705	Complementation of mutants, identification of resistance genes	Murakami *et al.* (1986)
Bleomycin	Glycopeptide	*S. verticillus* ATCC15003	Identification of resistance genes	Sugiyama *et al.* (1994)
Borrelidin	Polyketide	*S. parvulus* Tü4055	Homology-based approaches	Olano *et al.* (2004a)
Brasilicardin A	Isoprenoid	*N. terpenica*	Homology-based approaches	Hayashi *et al.* (2008)
Candicidin	Polyketide	*S. griseus* IMRU3570	Complementation of mutants	Gil and Hopwood (1983)
			Homology-based approaches	Campelo and Gil (2002)
Carbapenem	β-Lactam	*S. fulvoviridis*	Complementation of mutants	Nakata *et al.* (1989)
Carbomycin	Polyketide	*S. thermotolerans* ATCC11416	Identification of resistance genes	Epp *et al.* (1987)
Cephamycin C	β-Lactam	*S. cattleya* NRRL8057	Gene cluster expression	Chen *et al.* (1988)
Chalcomycin	Polyketide	*S. bikiniensis* NRRL2737	Homology-based approaches	Ward *et al.* (2004)

Chartreusin	Polyketide	*S. chartreusis* HKI-249	Homology-based approaches, gene cluster expression	Xu *et al.* (2005)
Chlorothricin	Polyketide	*S. antibioticus* DSM40725	Homology-based approaches	Shao *et al.* (2006), Jia *et al.* (2006)
Chromomycin A3	Polyketide	*S. griseus* subsp. *griseus*	Homology-based approaches	Menéndez *et al.* (2004)
Clavulanic acid	β-Lactam	*S. clavuligerus*	Complementation of mutants	Bailey *et al.* (1984)
Clorobiocin	Aminocoumarin	*S. roseochromogenes* DS12976	Homology-based approaches	Pojer *et al.* (2002)
Coelichelin	Polypeptide	*S. coelicolor* A3(2)	Genome mining	Challis and Ravel (2000)
		S. ambofaciens ATCC23877	Genome mining	Barona-Gómez *et al.* (2006)
Concanamycin A	Polyketide	*S. neyagawaensis* ATCC27449	Genome mining	Haydock *et al.* (2005)
Cosmomycin	Polyketide	*S. olindensis* DAUFPE5622	Homology-based approaches	Garrido *et al.* (2006)
Coumermycin A_1	Aminocoumarin	*S. rishiriensis* DSM40489	Homology-based approaches	Wang *et al.* (2000)
Daptomycin	Polypeptide	*S. roseosporus* NRRL11379	Gene cluster expression	Miao *et al.* (2005)
Daunorubicin	Polyketide	*S. peucetius* ATCC29050	Homology-based approaches Complementation of mutants	Stutzman-Engwall *et al.* (1989) Otten *et al.* (1990)
Desferrioxamine	Hydroxamate	*S. coelicolor* A3(2)	Genome mining	Barona-Gómez *et al.* (2004)
		S. ambofaciens ATCC23877	Genome mining	Barona-Gómez *et al.* (2006)
Doxorubicin	Polyketide	*S. peucetius* subsp. *caesius*	Homology-based approaches, complementation of mutants	Otten *et al.* (1990)
ECO-02301	Polyketide	*S. aizunensis* NRRLB-11277	Genome mining	McAlpine *et al.* (2005)
Elaiophylin	Polyketide	*Streptomyces* sp. DSM4137	Homology-based approaches	Haydock *et al.* (2004)
Elloramycin	Polyketide	*S. olivaceus* Tü2353	Homology-based approaches	Decker *et al.* (1995), Rafanan *et al.* (2001), Blanco *et al.* (2001), Ramos *et al.*(2008)
Endophenazine A	Phenazine	*S. cinnamonensis* DSM1042	Homology-based approaches	Haagen *et al.* (2006)
Enduracidin	Polypeptide	*S. fungicidicus* ATCC21013	Homology-based approaches	Yin and Zabriskie (2006)
Enterocin	Polyketide	*S. maritimus*	Homology-based approaches, gene cluster expression	Piel *et al.* (2000)
Erythromycin	Polyketide	*Sac. erythraea* NRRL2338	Complementation of mutants	Vara *et al.* (1989)
			Identification of resistance genes	Thompson *et al.* (1982)

Table 11.1 continued

Compound	Structural type	Producer strain	Method for cloning	References
FR900098	Phosphonate	*S. rubellomurinus* ATCC31215	Homology-based approaches, gene cluster expression	Eliot *et al.* (2008)
Fredericamycin	Polyketide	*S. griseus* ATCC49344	Homology-based approaches	Wendt-Pienkowski *et al.* (2005)
Fridamycin E	Polyketide	*S. rimosus* NRRL3016	Homology-based approaches, gene cluster expression	Hong *et al.* (1997)
Furanonapthoquinone I	Isoprenoid	*S. cinnamonensis* DSM1042	Homology-based approaches	Haagen *et al.* (2006)
Furaquinocin A	Isoprenoid	*Streptomyces* sp. KO-3988	Homology-based approaches, gene cluster expression	Kawasaki *et al.* (2006)
Geldanamycin	Polyketide	*S. hygroscopicus* var. *geldanus*	Homology-based approaches	Rascher *et al.* (2003)
Gentamicin	Aminoglycoside	*M. purpurea*	Identification of resistance genes	Kelemen *et al.* (1991)
		M. echinospora ATCC15835	Homology-based approaches	Unwin *et al.* (2004)
Gilvocarcin V	Polyketide	*S. griseoflavus* Gö3592	Homology-based approaches, gene cluster expression	Fischer *et al.* (2003)
Granaticin	Polyketide	*S. violaceoruber* Tü22	Homology-based approaches	Sherman *et al.* (1989)
			Gene cluster expression	Ichinose *et al.* (1998)
Griseorhodin A	Polyketide	*Streptomyces* sp. JP95	Homology-based approaches	Li and Piel (2002)
Halstoctacosanolides	Polyketide	*S. halstedii* HC34	Genome mining	Tohyama *et al.* (2004, 2006)
Hedamycin	Polyketide	*S. griseoruber* ATCC15422	Homology-based approaches	Bililign *et al.* (2004)
Hygromycin A	Glycoside	*S. hygroscopicus* NRRL2388	Homology-based approaches	Palaniappan *et al.* (2006)
Isopenicillin N	β-Lactam	*S. clavuligerus*	Homology-based approaches, complementation of mutants	Leskiw *et al.* (1988)
Kanamycin	Aminoglycoside	*S. kanamyceticus* ISP1375	Identification of resistance genes	Demydchuk *et al.* (1998)
			Homology-based approaches	Yanai and Murakami (2004)
Kijanimicin	Polyketide	*A. kijaniata* SCC1256	Homology-based approaches	Zhang *et al.* (2007)
Lactonamycin	Polyketide	*S. rishiriensis* MJ773-88K4	Homology-based approaches	Zhang *et al.* (2008a)
Lactonamycin Z	Polyketide	*S. sanglieri* AK 623	Homology-based approaches	Zhang *et al.* (2008a)
Landomycin	Polyketide	*S. cyanogenus* S136	Homology-based approaches	Westrich *et al.* (1999)

Lasalocid	Polyketide	*S. lasaliensis* ATCC35851	Homology-based approaches	Migita *et al.* (2009)
Leinamycin	Polyketide	*S. atroolivaceus* S-140	Homology-based approaches	Cheng *et al.* (2002)
Leptomycin	Polyketide	*Streptomyces* sp. ATCC39366	Genome mining	Hu *et al.* (2005)
Lincomycin	Lincosamide	*S. lincolnensis* 78-11	Identification of resistance genes	Zhang *et al.* (1992)
			Homology-based approaches	Peschke *et al.* (1995)
α-Lipomycin	Polyketide	*S. aureofaciens* Tü117	Homology-based approaches	Bihlmaier *et al.* (2006)
Medermycin	Polyketide	*Streptomyces* sp. AM-7161	Homology-based approaches	Ichinose *et al.* (2003)
Megalomicin	Polyketide	*M. megalomicea*	Homology-based approaches	Volchegursky *et al.* (2000)
Meilingmycin	Polyketide	*S. nanchangensis* NS3226	Homology-based approaches	Sun *et al.* (2003a)
Meridamycin	Polyketide	*Streptomyces* sp. NRRL30748	Homology-based approaches	He *et al.* (2006)
		Streptomyces sp. DSM4137	Homology-based approaches	Sun *et al.* (2006)
Methylenomycin A	Polyketide	*S. coelicolor* A3(2)	Mutational cloning	Chater and Bruton (1983)
Mithramycin	Polyketide	*S. argillaceus* ATCC12956	Identification of resistance genes	Fernández *et al.* (1996)
			Homology-based approaches	Lombó *et al.* (1996, 1997)
Mitomycin C	Benzoquinone	*S.lavendulae* NRRL2564	Identification of resistance genes	August *et al.* (1994)
			Homology-based approaches	Mao *et al.* (1999)
Monensin A	Polyketide	*S. cinnamonensis* ATCC15413	Homology-based approaches	Oliynyk *et al.* (2003)
Mycinamicin	Polyketide	*M. griseorubida* A11725	Identification of resistance genes	Inouye *et al.* (1994)
Nanchangmycin	Polyketide	*S. nanchangensis* NS3226	Homology-based approaches	Sun *et al.* (2002)
Napthocyclinone	Polyketide	*S. arenae* DSM40737	Homology-based approaches	Brünker *et al.* (1999)
Neocarzilin	Polyketide	*S. carzinostaticus* var. F-41	Homology-based approaches	Otsuka *et al.* (2004)
Neocarzinostatin	Polyketide	*S. carzinostaticus* ATCC15944	Homology-based approaches	Liu *et al.* (2005)
Niddamycin	Polyketide	*S. caelestis* NRRL2821	Homology-based approaches	Kakavas *et al.* (1997)
Nigericin	Polyketide	*Streptomyces* sp. DSM4137	Homology-based approaches	Harvey *et al.* (2007)
Nogalamycin	Polyketide	*S. nogalater* ATCC27451	Homology-based approaches, complementation of mutants	Ylihonko *et al.* (1996)
Nonactin	Polyketide	*S. griseus* DSM40695	Identification of resistance genes	Plater and Robinson (1992)
Novobiocin	Aminocoumarin	*S. spheroides* NCIB11891	Identification of resistance genes	Thiara and Cundliffe (1988)

Table 11.1 continued

Compound	Structural type	Producer strain	Method for cloning	References
			Homology-based approaches	Steffensky *et al.* (2000)
Nystatin	Polyketide	*S. noursei* ATCC11455	Homology-based approaches	Brautaset *et al.* (2000)
Oleandomycin	Polyketide	*S. antibioticus* ATCC11891	Identification of resistance genes	Rodríguez *et al.* (1993)
			Homology-based approaches	Swan *et al.* (1994), Aguirrezabalaga *et al.* (2000)
Oviedomycin	Polyketide	*S. antibioticus* ATCC11891	Homology-based approaches, gene cluster expression	Méndez *et al.* (2002)
Oxazolomycin	Polyketide	*S. albus* JA3453	Homology-based approaches	Zhao *et al.* (2006)
Oxytetracycline	Polyketide	*S. rimosus*	Identification of resistance genes	Binnie *et al.* (1989)
			Complementation of mutants	Butler *et al.* (1989), Binnie *et al.* (1989)
Pactamycin	Polyketide	*S. pactum* NBRC13433	Homology-based approaches	Kudo *et al.* (2007)
Pentalenolactone	Isoprenoid	*S. avermitilis* ATCC31267	Genome mining	Tetzlaff *et al.* (2006)
PD116740	Polyketide	*Streptomyces* sp. WP4669	Homology-based approaches, gene cluster expression	Hong *et al.* (1997)
Phenalinolactone	Isoprenoid	*Streptomyces* sp. Tü6071	Homology-based approaches	Dürr *et al.* (2006)
Pikromycin	Polyketide	*S. venezuelae* ATCC15439	Homology-based approaches	Xue *et al.* (1998a)
Pimaricin	Polyketide	*S. natalensis* ATCC27448	Homology-based approaches	Aparicio *et al.* (1999)
Pladienolide B	Polyketide	*S. platensis* Mer-11107	Homology-based approaches	Machida *et al.* (2008a)
Pradimicin	Polyketide	*A. hibisca* P157-2	Homology-based approaches	Kim *et al.* (2007a)
Puromycin	Aminonucleoside	*S. alboniger* ATCC12461	Identification of resistance genes	Vara *et al.* (1985)
			Homology-based approaches, gene cluster expression	Lacalle *et al.* (1992)
Rapamycin	Polyketide	*S. hygroscopicus* NRRL5491	Homology-based approaches	Schwecke *et al.* (1995)
Rebeccamycin	Indolocarbazole	*L. aerocolonigenes* ATCC39243	Homology-based approaches, gene cluster expression	Sánchez *et al.* (2002)
Rifamycin	Polyketide	*Am. mediterranei* S699	Detection of enzymatic activities	Kim *et al.* (1998)
			Homology-based approaches	August *et al.* (1998), Schupp *et al.* (1998)

Saframycin	Polypeptide	*S. lavendulae* NRRL11002	Homology-based approaches	Li *et al.* (2008a)
Salinosporamide	Polyketide	*Sa. tropica* CNB-440	Genome mining	Udwary *et al.* (2007)
Salinomycin		*S. albus* ATCC21838	Homology-based approaches	Izumikawa *et al.* (2003)
	Polyketide	*S. albus* CCM4719	Homology-based approaches	Knirschová *et al.* (2007)
Sibiromycin	Benzodiazepine	*St. sibiricum*	Homology-based approaches	Li *et al.* (2009a)
Simocyclinone D8	Polyketide	*S. antibioticus* Tü6040	Homology-based approaches	Trefzer *et al.* (2002a), Galm *et al.* (2002)
Sisomicin	Aminoglycoside	*M. inyoensis* TS388	Identification of resistance genes	Goldberg *et al.* (1990)
			Homology-based approaches	Hong *et al.* (2009)
Spinosyn	Polyketide	*Sac. spinosa*	Homology-based approaches, complementation of mutants	Waldron *et al.* (2001)
Staurosporine	Indolocarbazole	*Streptomyces* sp. TP-A0274	Homology-based approaches, gene cluster expression	Onaka *et al.* (2002)
Steffimycin	Polyketide	*S. steffisburgensis* NRRL3193	Mutational cloning	Gullón *et al.* (2006)
Streptomycin	Aminoglycoside	*S. griseus* ATCC10137	Complementation of mutants	Distler *et al.* (1985)
			Identification of resistance genes	Ohnuki *et al.* (1985)
Tautomycetin	Polyketide	*Streptomyces* sp. CK4412	Homology-based approaches	Choi *et al.* (2007)
Tautomycin	Polyketide	*S. spiroverticillatus*	Homology-based approaches	Li *et al.* (2006)
Tetracenomycin C	Polyketide	*S. glaucescens*	Complementation of mutants, identification of resistance genes, gene cluster expression	Motamedi and Hutchinson (1987)
Tetrocarcin	Polyketide	*M. chalcea* NRRL11289	Homology-based approaches	Fang *et al.* (2008)
Tetronomycin	Polyketide	*Streptomyces* sp. NRRL11266	Homology-based approaches	Demydchuk *et al.* (2008)
Thienamycin	β-Lactam	*S. cattleya* NRRL8057	Homology-based approaches	Núñez *et al.* (2003)
Thiocoraline	Polypeptide	*Micromonospora* sp. ML1	Homology-based approaches	Lombó *et al.* (2006)
Thiostrepton	Polypeptide	*S. laurentii* ATCC31255	Genome mining	Kelly *et al.* (2009)
Tomaymycin	Benzodiazepine	*S. achromogenes*	Homology-based approaches	Li *et al.* (2009b)
Tylosin	Polyketide	*S. fradiae*	Detection of enzymatic activities	Bauer *et al.* (1988)
			Homology-based approaches	Fishman *et al.* (1987), Fouces *et al.* (1999)
			Complementation of mutants	Fishman *et al.* (1987)

Table 11.1 continued

Compound	Structural type	Producer strain	Method for cloning	References
Undecylprodigiosin	Oligopyrrole	*S. coelicolor* A3(2)	Complementation of mutants	Feitelson and Hopwood (1983), Feitelson *et al.* (1985)
			Gene cluster expression	Malpartida *et al.* (1990)
Urdamycin A	Polyketide	*S. fradiae* Tü2717	Homology-based approaches	Decker and Haag (1995)
Validamycin	Aminocyclitol	*S. hygroscopicus* subsp. *jinggangensis*	Homology-based approaches. ,utational cloning	Yu *et al.* (2005)
Vicenistatin	Polyketide	*S. halstedii* HC34	Homology-based approaches	Ogasawara *et al.* (2004)
Zorbamycin	Polypeptide	*S. flavoviridis* ATCC21892	Homology-based approaches	Galm *et al.* (2009)

A., *Actinomadura*; *Ac.*, *Actinosynnema*; *Am.*, *Amycolatopsis*; *L.*, *Lechevaleria*; *M.*, *Micromonospora*; *N.*, *Nocardia*; *S.*, *Streptomyces*; *Sa.*, *Salinispora*; *Sac.*, *Saccharopolispora*; *St.*, *Streptosporangium.*

and (iii) genes from biosynthetically related pathways show cross-hybridization with each other.

Complementation of non-producing mutants

This method was one of the first used to clone biosynthesis genes and it takes advantage of the availability of non-producing mutants. The *p*-aminobenzoic acid synthetase *pabAB* involved in the biosynthesis of candicidin was cloned from a candicidin producer and expressed in *S. lividans* restoring prototrophy to a *pab* auxotroph (Gil and Hopwood, 1983). Complementation of *S. coelicolor* mutants defective in the production of undecylprodigiosin also led to the identification of the *redE* gene coding for an *O*-methyltransferase required for a late step in the biosynthesis of this red-pigmented antibiotic (Feitelson and Hopwood, 1983). Additional work led to the cloning of *redA, redB, redE,* and *redF* genes by complementing several mutant strains (Feitelson *et al.*, 1985). *S. coelicolor* produces other compounds such as actinorhodin, a blue pigmented aromatic polyketide, which gene cluster was isolated by complementing several non-producing mutants and further expression in *S. parvulus* (Malpartida and Hopwood, 1984). The streptomycin cluster was identified from *S. griseus* using a cloned DNA fragment that complemented one out of three classes of streptomycin-idiotrophic mutants which required streptidine to produce streptomycin (Distler *et al.*, 1985). Genes involved in the biosynthesis of antibiotics bialaphos, clavulanic acid, carbapenem, isopenicillin and tetracenomycin were also identified using this method (Bailey *et al.*, 1984; Leskiw *et al.*, 1988; Nakata *et al.*, 1989; Murakami *et al.*, 1986; Motamedi and Hutchinson, 1987). Genes that govern the formation of deoxysugars or their attachment during the biosynthesis of erythromycin were identified by complementing mutants accumulating intermediates erythronolide B or 3-alpha-mycarosyl erythronolide B (Vara *et al.*, 1989). The avermectin gene cluster was also identified by complementing several non-producing mutants particularly one lacking avermectin 5-O-methyltransferase activity (Streicher *et al.*, 1989; Ikeda *et al.*, 1998).

In other cases, like tylosin, daunorubicin, nogalamycin, aclacinomycin and spinosyn, the initial identification of the gene cluster was done by homology-based approaches followed by complementation of non-producing mutants (Fishman *et al.*, 1987; Stutzman-Engwall *et al.*, 1989; Otten *et al.*, 1990; Ylihonko *et al.*, 1996; Räty *et al.*, 2000; Waldron *et al.*, 2001). The oxytetracycline gene cluster was identified initially by isolating genomic DNA fragments which conferred resistance to oxytetracycline that were used for complementing *S. rimosus* non-producing mutants (Butler *et al.*, 1989).

Detection of enzymatic activities

Some gene clusters have been initially identified by detecting specific enzymatic activities. This approach was used for the identification of phenoxazinone synthase activity initially presumed to catalyse the condensation of two units of 4-methyl-3-hydroxyanthranilic acid (4-MHA) to form the chromophore of actinomycin. DNA fragments from the actinomycin producer *S. antibioticus* were cloned in *S. lividans* and the transformants were screened by enzymatic assay for their ability to produce the enzyme phenoxazinone synthase PhsA (Jones and Hopwood, 1984). However, further inactivation of the *phsA* gene led to an actinomycin-producer *S. antibioticus* mutant, which confirmed *phsA* was not involved in actinomycin biosynthesis (Jones, 2000). As illustrated below, the actinomycin gene cluster has been isolated following other methods from *S. chrysomallus* (Schauwecker *et al.*, 1998).

Biosynthesis of rifamycin involves the use of 3-amino-5-hydroxybenzoic acid (AHBA) as starter unit. The enzymatic activity responsible for the last step in the biosynthesis of this compound was purified and characterized from a rifamycin producer. A partial amino acid sequence of this enzyme was used for the construction of oligodeoxynucleotide primers to clone the AHBA synthase gene (Kim *et al.*, 1998). The same approach was used for the identification of macrocin-O-methyltransferase gene *tylF* that catalyses the final enzymatic step in the biosynthesis of tylosin (Bauer *et al.*, 1988).

Identification of resistance genes

As mentioned before, structural and resistance genes are usually clustered thus facilitating the identification of biosynthesis gene clusters by cloning resistance genes in an appropriate sensitive host. The identification of genes conferring resistance to erythromycin, *ermE* (Thompson *et al.*, 1982), oxytetracycline, *otrA* and *otrB* (Binnie *et al.*, 1989), carbomycin, *carA* and *carB* (Epp *et al.*, 1987), nonactin, *nonR* (Plater and Robinson, 1992), oleandomycin, *oleB* (Rodríguez *et al.*, 1993), mycinamicin, *myrA* and *myrB* (Inouye *et al.*, 1994), and mithramycin, *mtrA* and *mtrB* (Fernández *et al.*, 1996), led to the isolation of the clusters governing the biosynthesis of these polyketides. This approach has been used, in addition, for identifying the gene clusters involved in biosynthesis of other antibiotics such as the aminonucleoside puromycin (Vara *et al.*, 1985), aminocoumarin antibiotic novobiocin (Thiara and Cundliffe, 1988), glycopeptide bleomycin (Sugiyama *et al.*, 1994), aminobenzoquinone mitomycin C (August *et al.*, 1994), aminoglycosides kanamycin, gentamicin or sisomicin (Demydchuk *et al.*, 1998; Kelemen *et al.*, 1991; Goldberg *et al.*, 1990), and lincosamide antibiotic lincomycin (Zhang *et al.*, 1992).

In other cases, this approach has been used in combination with complementation of non-producing mutants for the isolation of several clusters. The bialaphos *bar* gene encoding an acyltransferase and tetracenomycin *tcmA* gene encoding a transport protein, both conferring resistance to the respective metabolites, were identified by their expression in *S. lividans* (Murakami *et al.*, 1986; Motamedi and Hutchinson, 1987). The streptomycin resistance determinant *strA* that encodes a 6-phosphotransferase was identified using an *S. griseus* sensitive strain (Distler *et al.*, 1985; Ohnuki *et al.*, 1985).

Homology-based approaches

This method combines agarose gel electrophoresis for size separation of DNA with methods to transfer the DNA to a filter membrane for probe hybridization. It is routinely used to check for the presence of specific sequences in a DNA sample using homologous or heterologous labelled DNA probes. This method is usually preceded by colony hybridization to identify initial positive clones.

Homologous probes

In this approach, genes previously isolated from gene clusters are used as DNA probes for the isolation of hybridizing clones. This approach was used for the isolation of avermectin, spinosyn, candicidin, rifamycin or rebeccamycin gene clusters using the homologous probes: avermectin 5-O-methyltransferase gene *aveD* (Ikeda *et al.*, 1999), a 400-kb SpeI fragment absent in some mutants unable to bioconvert the spinosyn aglycon to spinosyn and a ketosynthase gene (Waldron *et al.*, 2001), candicidin *p*-aminobenzoic acid synthetase gene *pabAB* (Criado *et al.*, 1993; Campelo and Gil, 2002), rifamycin AHBA synthase gene *rifK* (August *et al.*, 1998), and rebeccamycin *N*-glycosyltransferase gene *ngt* gene (Sánchez *et al.*, 2002), respectively. The rebeccamycin *ngt* gene was initially identified by expression in *S. lividans* and bioconversion of indolocarbazole J-104303 into its glycosylated derivative J-109384 (Ohuchi *et al.*, 2000). Other biosynthetic genes used as homologous probes were neocarzinostatin apo-protein gene *ncsA* (Liu *et al.*, 2005) or sisomicin L-glutamine 2-deoxy-scyllo-inosose aminotransferase gene *sis10* (Hong *et al.*, 2009).

In other cases, degenerate oligonucleotides were used for homology-based approaches. These probes were designed from amino acid sequences obtained from purified biosynthetic enzymes such as tylosin macrocin-O-methyltransferase TylF (Fishman *et al.*, 1987; Fouces *et al.*, 1999), actinomycin peptide synthetases ACMS I and ACMS II (Schauwecker *et al.*, 1998), or isopenicillin synthetase PcbC (Leskiw *et al.*, 1988). In addition, some resistance genes have been also used as homologous probes like puromycin *pac* (Lacalle *et al.*, 1992), gentamicin *gmrA* (Unwin *et al.*, 2004), kanamycin *kmr* (Yanai and Murakami, 2004) or lincomycin *lmrA*, *lmrB* and *lmrC* (Peschke *et al.*, 1995).

PCR-designed probes

Another way to obtain homologous probes for its use in homology-based approaches is the amplification of PCR products using degenerate oligoprimers based on the homology with previously known gene products. Several degenerate oligoprimers pairs have been designed for the isolation of genes involved in deoxyhexose biosynthesis like dNDP-D-glucose-4,6-dehydratase (NGDH; Decker *et al.*, 1996), dNDP-glucose synthases (NGS; Hyun *et al.*, 2000) and glycosyltransferases (Luzhetskyy *et al.*, 2007a); PKSs such as type I (PKS-I; Ginolhac *et al.*, 2004; Chuck *et al.*, 2006; Savic *et al.*, 2006), type II (PKS-II; Metsä-Ketelä *et al.*, 1999; Wawrik *et al.*, 2005) and enediyne PKSs (Liu *et al.*, 2003); or NRPSs (Sosio *et al.*, 2000; Ayuso-Sacido and Genilloud, 2005).

Degenerate oligonucleotides for the amplification of an internal fragment of NGDH encoding genes have been used for the isolation of gene clusters involved in the biosynthesis of deoxysugar containing compounds: oligosaccharide avilamycin (Gaisser *et al.*, 1997), aminocoumarin novobiocin (Steffensky *et al.*, 2000), aromatic polyketides landomycin, aclacinomycin, gilvocarcin, medermycin, hedamycin and elloramycin (Westrich *et al.*, 1999; Räty *et al.*, 2000; Fischer *et al.*, 2003; Ichinose *et al.*, 2003; Bililign *et al.*, 2004; Ramos *et al.*, 2008), macrocyclic lactam vicenistatin (Ogasawara *et al.*, 2004), terpene glycoside phenalinolactone (Dürr *et al.*, 2006) and pyrrolobenzodiazepine sibiromycin (Li *et al.*, 2009a). Other genes involved in deoxysugar biosynthesis and attachment have been used for designing specific degenerated oligoprimers in the search for biosynthesis clusters. Amplification of dNDP-hexose-2,3-dehydratases was used for the isolation of vicenistatin (Ogasawara *et al.*, 2004), lactonamycin (Zhang et a., 2008a) and simocyclinone (Trefzer *et al.*, 2002a) clusters. In the former case also coumarate ligase genes obtained by PCR amplification were used. Highly conserved motifs of GDP-D-mannose-4,6-dehydratases were used for designing specific primers that led to the identification of *hyg5* gene involved in hygromycin A biosynthesis (Palaniappan *et al.*, 2006). Finally, degenerate oligoprimers have been used for the identification of genes encoding glycosyltransferases from elloramycin and hedamycin gene clusters (Blanco *et al.*, 2001; Bililign *et al.*, 2004).

The isolation of gene clusters involved in the biosynthesis of polyketide compounds has taken advantage of highly conserved motifs found in PKSs. Degenerate oligonucleotides for the amplification of PKS-I, especially acyl-carrier protein (ACP), ketosynthase (KS) and acyltransferase (AT) domains, have been used for the isolation of clusters directing the biosynthesis of niddamycin (Kakavas *et al.*, 1997), nystatin (Brautaset *et al.*, 2000), aureothin (He and Hertweck, 2003), geldanamycin (Rascher *et al.*, 2003), meilingmycin (Sun *et al.*, 2003a), salinomycin (Izumikawa *et al.*, 2003; Knirschová *et al.*, 2007), chalcomycin (Ward *et al.*, 2004), neocarzilin (Otsuka *et al.*, 2004), α-lipomycin (Bihlmaier *et al.*, 2006), meridamycin (He *et al.*, 2006), tautomycetin (Choi *et al.*, 2007), or lasalocid (Migita *et al.*, 2009). For the isolation of some of those clusters, additional degenerate oligonucleotides were used for the amplification of genes involved in the biosynthesis of starter units or tailoring steps: aureothin PABA synthase (He and Hertweck, 2003), geldanamycin AHBA synthase and carbamoyltransferase (Rascher *et al.*, 2003) or meilingmycin cytochrome P450s and ketoreductases (KR) (Sun *et al.*, 2003a). The amplification of β-hydroxyacyl-CoA dehydrogenase (HADH) genes *ttmA* and *ozmG*, and acyl-CoA dehydrogenase (ADH) genes *ttmB* and *ozmD*, involved in the biosynthesis of extender unit methoxymalonyl-ACP led to the isolation of tautomycin and oxazolomycin clusters (Li *et al.*, 2006; Zhao *et al.*, 2006). The utilization of conserved motifs that are unique to iterative PKS-I for designing degenerate primers led to isolate *chlB1* gene involved in the biosynthesis of chlorothricin (Shao *et al.*, 2006; Jia *et al.*, 2006) and later to isolate mixed polyketide-peptide azinomycin cluster (Zhao *et al.*, 2008). Amplification of PKS-II ketosynthases (KSα and KSβ) has been used for the identification of clusters involved in the biosynthesis of aromatic polyketides such as griseorhodin (Li and Piel, 2002), medermycin (Ichinose *et al.*, 2003), hedamycin (Bililign *et al.*, 2004), chartreusin (Xu *et al.*, 2005), fredericamycin (Wendt-Pienkowski *et al.*, 2005), cosmomycin (Garrido *et al.*, 2006), or pradimicin (Kim *et al.*, 2007a).

Degenerate oligonucleotides based on conserved regions from several domains of modular NRPSs have been also used for identifying clusters involved in the biosynthesis of peptide antibiotics. The isolation of peptide-polyketide leinamycin cluster was achieved by using degenerated primers to amplify cyclization and oxidization domains from a NRPS coding gene involved in the formation of thiazole moieties that led to the identification of *lmnI* gene encoding PKS and NRPS modules (Cheng *et al.*, 2002). The same oligoprimers were later used for isolating the zorbamycin cluster (Galm *et al.*, 2009). Amplification of adenylation domains using some of their highly conserved motifs was used for the identification of meridamycin (He *et al.*, 2006), thiocoraline (Lombó *et al.*, 2006) and tomaymycin (Li *et al.*, 2009b) clusters, while oligoprimers for the amplification of adenylation and thiolation domains were used for the identification of enduracidin (Yin and Zabriskie, 2006) and saframycin (Li *et al.*, 2008a) clusters.

Other specific degenerate oligoprimers have been designed for the amplification of different genes like rebeccamycin $FADH_2$-dependent halogenase *rebH* (Sánchez *et al.*, 2002), staurosporine chromopyrrolic acid synthase *staD* (Onaka *et al.*, 2002), thienamycin β-lactam synthetase *thnM* (Núñez *et al.*, 2003), chromomycin ABC transporter ATP-binding protein domain *cmrA* (Menéndez *et al.*, 2004), enduracidin prephenate dehydrogenase *orf23* (Yin and Zabriskie, 2006), pactamycin Fe-S containing radical SAM methyltransferases (Kudo *et al.*, 2007), FR900098 phosphoenolpyruvate mutase *frbD* (Eliot *et al.*, 2008) or brasilicardin polyprenyl diphosphate synthase *bra2* (Hayashi *et al.*, 2008).

Heterologous probes

Once the first biosynthesis clusters were isolated, characterized genes were used as heterologous probes for identifying new clusters involving equivalent biosynthetic pathways. The identification of the actinorhodin cluster and the demonstration that it can be used as a hybridization probe for the identification of other biosynthesis clusters (Malpartida *et al.*, 1987) boosted the utilization of *actI* (KSα) and *actIII* (KR) genes for the isolation of pathways involved in the biosynthesis of several aromatic compounds such as granaticin (Sherman *et al.*, 1989; Ichinose *et al.*, 1998), mithramycin (Lombó *et al.*, 1996), PD116740 and fridamycin E (Hong *et al.*, 1997), naphthocyclinone (Brünker *et al.*, 1999), enterocin (Piel *et al.*, 2000), aclacinomycin (Räty *et al.*, 2002), oviedomycin (Méndez *et al.*, 2002; Lombó *et al.*, 2004), gilvocarcin V (Fischer *et al.*, 2003), or aranciamycin (Luzhetskyy *et al.*, 2007b). Once new PKS-II encoding genes were available, they were also used for the identification of other clusters. Thus, the daunorubicin-doxorubicin cluster was isolated using actinorhodin and tetracenomycin genes *actI*/*actIII* and *tcmIa* (Stutzman-Engwall *et al.*, 1989; Otten *et al.*, 1990) and the nogalamycin cluster was identified using *actI* and aclacinomycin PKS *acm* genes as probes (Ylihonko *et al.*, 1996). Tetracenomycin PKS genes *tcmK* and *tcmL* were used as probes for the isolation of urdamycin and elloramycin pathways (Decker and Haag, 1995; Decker *et al.*, 1995). In addition, *tcmG* gene coding for a NADPH-dependent oxygenase in the biosynthesis of tetracenomycin was used for the identification of *elmGHIJ* genes involved in cyclization and oxygenation steps during elloramycin biosynthesis (Rafanan *et al.*, 2001). The chromomycin cluster was isolated also taking advantage of the mithramycin PKS genes *mtmP* and *mtmK* (Menéndez *et al.*, 2004). A gene library from the benastatin producer was screened with griseorhodin KSα gene *grhA* as probe (Li and Piel, 2002) to isolate the pentangular polyphenol pathway (Xu *et al.*, 2007). The simocyclinone D8 cluster was identified using as probes landomycin PKS *lanB, lanC, lanD,* and *lanP,* urdamycin glycosyltransferase *urdGT2* and novobiocin cytochrome P450 *novI* genes (Galm *et al.*, 2002; Trefzer *et al.*, 2002a).

Also, PKS-I genes have been frequently used as heterologous probes for the isolation of a large number of gene clusters. In particular, erythromycin PKS genes were the first used as probes for the isolation of PKS-I containing pathways such as those involved in the biosynthesis of oleandomycin (Swan *et al.*, 1994), rapamycin (Schwecke *et al.*, 1995) megalomicin (Volchegursky *et al.*, 2000), amphotericin B (Caffrey *et al.*, 2001), candicidin (Campelo and Gil, 2002), nanchangmycin (Sun *et al.*, 2002; Sun *et al.*, 2003b), monensin (Oliynyk *et al.*, 2003), or tetronomycin (Demydchuk *et al.*, 2008). The borrelidin gene cluster was isolated using as probe a portion of *orfB* gene encoding the third subunit of the oleandomycin PKS (Olano *et al.*, 2004a). Rapamycin PKS genes were used for the isolation of pimaricin, together with candicidin PKS probes (Aparicio *et al.*, 1999; Aparicio *et al.*, 2000), elaiophylin (Haydock *et al.*, 2004), meridamycin (Sun *et al.*, 2006), or nigericin (Harvey *et al.*, 2007) clusters. The cluster directing the biosynthesis of macrolide antibiotics pikromycin and methymycin was identified using a PKS-I probe in addition to the *tylAI* gene encoding a NGS from the tylosin pathway (Xue *et al.*, 1998a). Pikromycin-methymycin PKS gene *pikAI* and cytochrome P450 gene *pikC* were then used for the isolation of pladienolide cluster (Machida *et al.*, 2008a). Other gene clusters like those involved in the biosynthesis of rifamycin (Schupp *et al.*, 1998), α-lipomycin (Bihlmaier *et al.*, 2006) and tetrocarcin (Fang *et al.*, 2008) were isolated using as probes genes encoding for the *Sorangium cellulosum* soraphen PKS, simocyclinone PKS and thioesterase, and chlorotricin PKS, respectively.

Several genes directing early steps during the biosynthesis of the deoxysugar moieties have been used as probes for the isolation of clusters involved in the biosynthesis of glycosylated compounds: streptomycin *strMDE* genes for oleandomycin (Aguirrezabalaga *et al.*, 2000) and mithramycin (Lombó *et al.*, 1997); mithramycin *mtmDE* genes for cosmomycin (Garrido *et al.*, 2006); and tylosin *tylA2* and novobiocin *novT* genes for coumermycin (Wang *et al.*, 2000), clorobiocin (Pojer *et al.*, 2002) or kijanimicin (Zhang *et al.*, 2007) clusters. Additional probes for the isolation of coumermycin (Wang *et al.*, 2000) and clorobiocin (Pojer *et al.*, 2002) were novobiocin resistance gene *gyrB* and novobiocic acid synthetase gene *novL*, respectively. Finally, *lct36* gene encoding a lactonamycin glycosyltransferase was the probe used for the identification of the lactonamycin Z biosynthesis cluster (Zhang *et al.*, 2008a).

Other genes used as heterologous probes have been rifamycin AHBA synthase *rifK*, phenazine biosynthesis genes *phzA* and *phzB* from *Pseudomonas fluorescens*, terpentecin HMG-CoA reductase *hmgg* from *S. griseolosporeus* MF730-N6, or acarbose C7-cyclitol synthase *acbC* from *Actinoplanes* sp. SE50/110 for isolating mitomycin C and ansamitocin (Yu *et al.*, 2002; Mao *et al.*, 1999), furanonapthoquinone and endophenazine (Haagen *et al.*, 2006), furaquinocin (Kawasaki *et al.*, 2006), and aminocyclitol validamycin (Yu *et al.*, 2005) biosynthesis gene clusters, respectively.

Mutational cloning

Another approach for the identification of gene clusters is the direct inactivation of genes involved in the biosynthesis of a particular compound. This method was successfully used for the identification of the methylenomycin cluster. Chromosomal DNA fragments were cloned in a phage derived vector carrying a viomycin resistance gene. These constructs were use for integration in *S. coelicolor* DNA by homologous recombination thus disrupting methylenomycin biosynthesis. The search for viomycin-resistant transductants unable to produce methylenomycin allowed the identification of vectors containing genes involved in the pathway that finally led to the isolation of the cluster (Chater and Bruton, 1983). Subsequent studies revealed that genes involved in methylenomycin production, regulation and resistance are clustered together (Chater and Bruton, 1985) and located in the 356-kb linear plasmid SCP1 (Kinashi *et al.*, 1987).

The gene cluster responsible for the biosynthesis of validamycin was identified using a similar approach. In this case, a cosmid clone showing a weak hybridization signal with the *acbC* gene from *Actinoplanes* sp. SE50/110 was used to replace 30 kb of the insert containing the *acbC* homologue with an apramycin resistance gene. The apramycin-resistant thiostrepton-sensitive exconjugants obtained were unable to produce validamycin leading to the identification of the cluster (Yu *et al.*, 2005). The isolation of the steffimycin cluster was achieved by gene disruption of a gene encoding a KSα available from the databases. Using this information a PCR fragment was amplified from steffimycin producer chromosomal DNA and used to generate a non-producing mutant that allowed the identification of the cluster (Gullón *et al.*, 2006).

Gene cluster expression

This method for isolating gene clusters involves the introduction of chromosomal DNA fragments into a heterologous host, expression of the cluster and detection of the desired compound. This method takes advantage of the availability of a number of *S. coelicolor* and *S. lividans* strains that can be used as heterologous hosts, including some engineered strains lacking the actinorhodin cluster for the expression of PKS-containing pathways (Kieser *et al.*, 2000; Fujii, 2009). Another frequently used heterologous host is *S. albus* J1074, a good recipient for transformation with *E. coli* derived DNA since it is defective in the SalGI restriction-modification system (Kieser *et al.*, 2000). Following this approach, the cephamycin cluster was identified in a 29-kb DNA fragment able to direct the biosynthesis of the antibiotic in *S. lividans* and to increase its production in other cephamycin producing strains (Chen *et al.*, 1988). The same approach was used for the identification of the cyclic lipopeptide daptomycin biosynthetic cluster located in a 128-kb DNA fragment that was cloned into a bacterial artificial chromosomal shuttle vector and expressed in *S. lividans* (Miao *et al.*, 2005).

However, in most of the cases, the expression of chromosomal DNA fragments in heterologous hosts has been used as a complementary approach, previous to the sequencing of large DNA regions, to confirm the presence of biosynthesis genes in regions identified by other methods such as homology-based approaches or isolation of resistance genes. This was used with actinorhodin (Malpartida and Hopwood, 1984) or undecylprodigiosin (Malpartida *et al.*, 1990) clusters expressed in *S. parvulus*, the tetracenomycin cluster expressed in *S. ambofaciens*, *S. lasaliensis* and *S. lividans* (Motamedi and Hutchinson, 1987), and the puromycin cluster expressed in *S. lividans* and *S. griseofuscus* (Lacalle *et al.*, 1992). The entire granaticin cluster was cloned and upon expression in *S. coelicolor* led to the accumulation of granaticin, granaticin B, dihydrogranaticin and dihydrogranaticin B (Ichinose *et al.*, 1998). The expression of the enterocin cluster in *S. lividans* allowed the production of enterocin and wailupemcyin D (Piel *et al.*, 2000). Genes required for the biosynthesis of FR900098 were expressed in *S. lividans* leading to the production of this antimalarial compound (Eliot *et al.*, 2008). Expression of cosmid clones in *S. lividans* allowed the identification of staurosporine and benastatin clusters (Onaka *et al.*, 2002; Xu *et al.*, 2007). *S. lividans* was also used for the expression of a 25-kb DNA region that harboured a mevalonate pathway involved in the biosynthesis of furaquinocin A. The incubation of *S. lividans* cell extracts, carrying the mevalonate pathway, with mevalonolactone led to identification of the compounds mevalonate-5-phosphate, mevalonate diphosphate, and isopentenyl diphosphate showing that at least the genes that coded for mevalonate kinase, phosphomevalonate kinase, and mevalonate diphosphate decarboxylase were included in the DNA fragment employed (Kawasaki *et al.*, 2006). Fridamycin, PD116740 and gilvocarcin gene clusters were expressed in *S. lividans* leading to the accumulation of tetrangulol, tetrangomycin, and fridamycin (Hong *et al.*, 1997), tetrangulol, 19-hydroxytetrangulol, 8-O-methyltetrangulol, 19-hydroxy-8-O-methyltetrangulol and PD116740 (Hong *et al.*, 1997) and gilvocarcin V and M (Fischer *et al.*, 2003), respectively. *S. albus* was used as host for the identification of cosmid clones containing rebeccamycin (Sánchez *et al.*, 2002) and oviedomycin (Méndez *et al.*, 2002) clusters, leading to the production of their respective antibiotics. Cosmid clones harbouring oviedomycin gene cluster were also successfully expressed in *S. lividans* (Méndez *et al.*, 2002). In addition, *S. albus* and *S. lividans* were used as hosts for expressing cosmids containing PKS-II loci during the search for the chartreusin cluster. The expression of cosmids containing the *pks3* cluster in *S. albus* produced chartreusin, thus confirming the nature of the encoded cluster (Xu *et al.*, 2005).

Genome mining

Whole genome sequencing of avermectin, actinorhodin, erythromycin, salinosporamide and streptomycin producers, have revealed the presence in their chromosomes of 20 to 34 clusters involved in the biosynthesis of known or predicted secondary metabolites belonging to different structural classes (Omura *et al.*, 2001; Bentley *et al.*, 2002; Oliynyk *et al.*, 2007; Udwary *et al.*,

2007; Ohnishi *et al.*, 2008). The fact that several clusters involved in the biosynthesis of unknown metabolites (orphan pathways) are present in microbial genomes has prompted the search for strategies to confirm the function of clusters putatively involved in the biosynthesis of predicted metabolites (Zazopoulos *et al.*, 2003; Gross, 2007; Challis, 2008). These approaches have led to the identification of new compounds such as the siderophores coelichelin and desferrioxamine (Challis and Ravel, 2000; Barona-Gómez *et al.*, 2004 and 2006), and sesquiterpenoid pentalenolactone (Tetzlaff *et al.*, 2006). Genome mining has also led to predict the structure and isolate previously unknown compounds such as the linear polyene ECO-02301 (McAlpine *et al.*, 2005) and macrolactones halstoctacosanolides (Tohyama *et al.*, 2004, 2006). The gene clusters involved in the biosynthesis of the benzodiazepine alkaloid anthramycin (Hu *et al.*, 2007) and polypeptide thiostrepton (Kelly *et al.*, 2009) were identified also by genome scanning. In other cases the identification of gene clusters was achieved by analogous methods such as sequencing cosmids ends or random sequencing of a cosmid library that led to the isolation of macrolide concanamycin (Haydock *et al.*, 2005) and leptomycin (Hu *et al.*, 2005) clusters, respectively.

Validation of gene clusters

Isolation of gene clusters usually needs their validation to confirm the involvement of a particular cluster in the biosynthesis of the compound under study. The validation process is especially important for clusters identified by homology-based approaches using both homologous or heterologous probes and genome mining. The isolation of clusters through complementation of non-producing mutants, gene disruption or cluster expression implies its self-validation during the identification process and normally there is no need for further confirmation. The validation of biosynthesis clusters can be achieved by several methods that include gene disruption and heterologous expression. The main objective of gene disruption is the generation of non-producing mutants or mutants accumulating intermediates or novel products. In addition, gene disruption is used to establish the cluster boundaries. Gene or gene cluster expression in a heterologous host should result in the production of structures related to the studied compound or the final product.

Gene disruption

This approach involves the inactivation or deletion of single genes or groups of genes to confirm their involvement in a particular biosynthesis pathway.

Non-producing mutants

The inactivation of genes involved in biosynthesis of starter units and core precursors, condensation of precursor units or final cyclization steps and regulatory genes generally lead to generate non-producing mutants. These types of mutants have been obtained for the validation of gene clusters involved in the biosynthesis of many compounds. The inactivation of genes involved in regulatory processes was used for validating mithramycin (Lombó *et al.*, 1999), aclacinomycin (Räty *et al.*, 2000), neocarzinostatin (Liu *et al.*, 2005), tetronomycin (Demydchuk *et al.*, 2008) or sibiromycin (Li *et al.*, 2009a) biosynthesis clusters. The inactivation of genes involved in the biosynthesis of starter units was used for abrogating the production of different compounds such as avilamycin (Gaisser *et al.*, 1997), rifamycin (Kim *et al.*, 1998; Yu *et al.*, 2001), coumermycin (Wang *et al.*, 2000), ansamitocin (Yu *et al.*, 2002), candicidin (Campelo and Gil, 2002), borrelidin (Olano *et al.*, 2004a), chlorothricin (Jia *et al.*, 2006), oxazolomycin (Zhao *et al.*, 2006), tautomycetin (Choi *et al.*, 2007), tautomycin (Li *et al.*, 2006), thiostrepton (Kelly *et al.*, 2009) or zorbamycin (Galm *et al.*, 2009), thus confirming the involvement of these genes in the biosynthetic pathway.

For the condensation of initial precursors some biosynthetic pathways use enzyme complexes like PKSs, NRPSs or hybrid PKS-NRPS systems. Inactivation or deletion of PKS-I genes has been used for the validation of a number of clusters such as those involved in the biosynthesis of erythromycin (Weber *et al.*, 1990), niddamycin (Kakavas *et al.*, 1997), pikromycin-methymycin (Xue *et al.*, 1998a), pimaricin (Aparicio *et al.*, 1999 and 2000), nystatin (Brautaset *et al.*, 2000), amphotericin (Caffrey *et al.*, 2001), spinosyn

(Waldron *et al.*, 2001), ansamitocin (Yu *et al.*, 2002), nanchangmycin (Sun *et al.*, 2002), meilingmycin (Sun *et al.*, 2003a), salinomycin (Izumikawa *et al.*, 2003), borrelidin (Olano *et al.*, 2004a), hedamycin (Bililign *et al.*, 2004), neocarzilin (Otsuka *et al.*, 2004), concanamycin (Haydock *et al.*, 2005), neocarzinostatin (Liu *et al.*, 2005), chlorothricin (Jia *et al.*, 2006), meridamycin (Sun *et al.*, 2006), monensin (Harvey *et al.*, 2007), or tetrocarcin (Fang *et al.*, 2008). The same approach was used for the inactivation of NRPSs involved in the biosynthesis of enduracidin (Yin and Zabriskie, 2006), α-lipomycin (Bihlmaier *et al.*, 2006), meridamycin (He *et al.*, 2006), thiocoraline (Lombó *et al.*, 2006), anthramycin (Hu *et al.*, 2007), azinomycin (Zhao *et al.*, 2008), saframycin (Li *et al.*, 2008), tomaymycin (Li *et al.*, 2009b) or zorbamycin (Galm *et al.*, 2009). Hybrid PKS-NRPS enzymes were inactivated to validate clusters involved in the biosynthesis of leinamycin (Cheng *et al.*, 2002) and oxazolomycin (Song *et al.*, 2008a). Also, PKS-II have been frequent targets for validating aromatic polyketide clusters such as those involved in the biosynthesis of urdamycin (Decker and Haag, 1995), mithramycin (Lombó *et al.*, 1996), napthocyclinone (Brünker *et al.*, 1999), simocyclinone (Trefzer *et al.*, 2002), hedamycin (Bililign *et al.*, 2004), oviedomycin (Lombó *et al.*, 2004), fredericamycin (Wendt-Pienkowski *et al.*, 2005), cosmomycin (Garrido *et al.*, 2006), or pradimicins (Kim *et al.*, 2007a).

Inactivation or deletion of genes encoding other type of enzymes have been also used to generate non-producing mutants thus confirming their involvement in the biosynthesis of the corresponding compound: cyclases such as those involved in the biosynthesis of monensin (Oliynyk *et al.*, 2003), desferrioxamine (Barona-Gómez *et al.*, 2004), validamycin (Yu *et al.*, 2005), phenalinolactone (Dürr *et al.*, 2006), or brasilicardin (Hayashi *et al.*, 2008); isomerases involved in the biosynthesis of monensin (Oliynyk *et al.*, 2003); condensing enzymes like NonS involved in ester linkage of four enantiomeric nonactic acid building blocks during nonactin biosynthesis (Smith *et al.*, 2000); monooxygenases as cytochromes P450 involved in the biosynthesis of halstoctacosanolides (Tohyama *et al.*, 2006) or an epoxidase for phenalinolactone biosynthesis (Dürr *et al.*, 2006); a putative aldolase gene from clorobiocin cluster (Pojer *et al.*, 2002); prenyltransferases from phenalinolactone (Dürr *et al.*, 2006) or furanonaphthoquinone (Haagen *et al.*, 2006) clusters; a neocarzilin halogenase (Otsuka *et al.*, 2004); or a methyltransferase, a putative AMP-binding enzyme and a oxidoreductase from thienamycin pathway (Núñez *et al.*, 2003).

In some cases, the whole or most of the cluster was deleted to verify its involvement in the biosynthesis of a particular compound. This approach was used for the validation of monensin (Oliynyk *et al.*, 2003), coelichelin and desferrioxamine (Barona-Gómez *et al.*, 2006), endophenazine (Haagen *et al.*, 2006), pentalenolactone (Tetzlaff *et al.*, 2006), and nigericin (Harvey *et al.*, 2007) clusters.

A good example for the need of gene cluster validation is the biosynthesis of neocarzinostatin. Inactivation experiments showed that this pathway is encoded by three copies of the cluster in *S. carzinostaticus*. This made more difficult any attempt to disrupt the pathway since total abolishment of enediyne production required the inactivation of all copies of the cluster (Liu *et al.*, 2005).

Mutants accumulating intermediates or novel derivatives

The inactivation of genes involved in tailoring steps for decorating the main core of natural products usually generates mutants that accumulate biosynthetic intermediates or novel derivatives. In addition, inactivation or in-frame deletion of specific genes directing the biosynthesis of core structures sometimes led to similar results.

Several genes involved in tailoring modification have been targets for validating biosynthesis clusters, with inactivation of genes encoding enzymes participating in oxygenation reactions like cytochrome P450s, FAD-dependent oxygenases, dioxygenases, epoxidases or hydroxylases being one of the most prolific sources for obtaining novel compounds in the biosynthesis of erythromycin (Weber *et al.*, 1989), pikromycin (Xue *et al.*, 1998a,b), ansamitocin (Spiteller *et al.*, 2003), borrelidin (Olano *et al.*, 2004b), chartreusin (Xu *et al.*, 2005), halstoctacosanolides

(Tohyama *et al.*, 2006), phenalinolactone (Dürr *et al.*, 2006), geldanamycin (Shin *et al.*, 2008), or pladienolide (Machida *et al.*, 2008b). Inactivation of methyltransferase genes used for validating erythromycin (Weber *et al.*, 1989), ansamitocin (Carroll *et al.*, 2002; Spiteller *et al.*, 2003), furanonaphthoquinone (Haagen *et al.*, 2006), or mitomycin (Sitachitta *et al.*, 2007; Grüschow *et al.*, 2007) clusters led to the accumulation of novel derivatives. Inactivation of acyltransferases and carbamoyltransferases led to the formation of novel products derived from ansamitocin and geldanamycin pathways (Moss *et al.*, 2002; Spiteller *et al.*, 2003; Shin *et al.*, 2008). Other tailoring modification enzymes targeted in validating gene clusters and rendering novel derivatives were dehydrogenases from ansamitocin (Carroll *et al.*, 2002) and geldanamycin (Shin *et al.*, 2008) clusters; an ansamitocin halogenase gene (Spiteller *et al.*, 2003); a prephenate dehydratase from salinosporamide pathway (McGlinchey *et al.*, 2008); a borrelidin aminotransferase (Olano *et al.*, 2004b); or a thiostrepton amidotransferase (Kelly *et al.*, 2009).

Inactivation of genes involved in the production of deoxysugars or in glycosylation also leads to mutants that accumulate compounds with a different pattern of glycosylation. Inactivation of different genes involved in erythromycin (Dhillon *et al.*, 1989; Weber *et al.*, 1990), mithramycin (Lombó *et al.*, 1997; Lozano *et al.*, 2000), pikromycin-methymycin (Xue *et al.*, 1998a), novobiocin (Steffensky *et al.*, 2000), simocyclinone (Trefzer *et al.*, 2002a), hygromycin A (Palaniappan *et al.*, 2006), phenalinolactone (Dürr *et al.*, 2006) or gentamicin (Kim *et al.*, 2008) deoxysugar biosynthesis led to mutant strains producing the corresponding intermediate compounds. Of outstanding interest was the inactivation of L-rhamnose and D-forosamine biosynthesis genes from *Sac. spinosa* (Madduri *et al.*, 2001) and L-rhamnose biosynthesis cluster from *S. olivaceus* (Ramos *et al.*, 2008) involved in spinosyn and elloramycin pathways respectively, since these genes are in both cases unlinked to the rest of the cluster. Inactivation of genes encoding glycosyltransferases involved in the attachment of deoxysugar moieties to their respective aglycones has been also a great source of novel compounds during validation of clusters involved in the biosynthesis of oleandomycin (Olano *et al.*, 1998), mithramycin (Fernández *et al.*, 1998; Blanco *et al.*, 2000; Nur-e-Alam *et al.*, 2005), chromomycin (Menéndez *et al.*, 2006), α-lipomycin (Bihlmaier *et al.*, 2006), cosmomycin (Garrido *et al.*, 2006) or lactonamycin Z (Zhang *et al.*, 2008a).

Targeting genes involved in the biosynthesis of precursors different from deoxysugars has, in some cases, led to mutants accumulating novel products such as those derived from simocyclinone (Galm *et al.*, 2002) and geldanamycin (Rascher *et al.*, 2003) pathways. In addition, targeting genes encoding different PKS systems has led occasionally to accumulation of several shunt products, as happened by disrupting rifamycin PKS-I genes (August *et al.*, 1998), or to novel products such those generated by deleting a ketoreductase (KR6) domain from nanchangmycin PKS-I (Sun *et al.*, 2003b) and KR1 domain from meridamycin PKS-I (He *et al.*, 2006). Inactivation of a mithramycin KR gene *mtmW* led to a mutant accumulating three novel products (Remsing *et al.*, 2003). The inactivation of the chromomycin homologous gene *cmmWI* resulted also in the accumulation of chromomycin equivalent derivatives (Menéndez *et al.*, 2004). Finally, inactivation of the benastatin KSIII gene *benQ* led to the accumulation of five novel benastatin derivatives (Xu *et al.*, 2007).

Gene expression

A parallel approach for the validation of gene clusters is the expression of single genes or a group of genes in a heterologous host and the detection of the desired compound. *S. lividans* was used as host for expressing megalomycin PKS-I genes leading to the production of its aglycone (Volchegursky *et al.*, 2000). In addition, it was used for expressing cosmid clones containing part of the daunorubicin cluster (Stutzman-Engwall *et al.*, 1989; Otten *et al.*, 1990), the daunorubicin PKS-II (Grimm *et al.*, 1994) and genes involved in the biosynthesis and attachment of L-daunosamine (Olano *et al.*, 1999) that resulted in the production of several biosynthetic intermediates. The same host was used for the expression of cosmid 16F4 carrying elloramycin PKS-II that conferred the ability to produce a tetracenomycin compound (Decker *et al.*, 1995). Mithramycin PKS-II genes were expressed in different combinations in *S. lividans* and *S. coelicolor* producing different shunt products (Künzel *et al.*, 1997; Blanco *et al.*, 1996). The avilamycin iterative PKS-I gene was also expressed in both heterologous hosts conferring on them the capability for producing the aromatic precursor orsellinic acid (Gaisser *et al.*, 1997). *S. coelicolor* was used, in addition, for the expression of napthocyclinone PKS-II genes directing the production of several shunt products, and when co-expressed with the actinorhodin KR products accumulated (Brünker *et al.*, 1999) that were previously described to be produced by the minimal actinorhodin PKS (McDaniel *et al.*, 1993). The same host was used for the expression of rifamycin biosynthesis genes involved in the production of AHBA (Yu *et al.*, 2001). These genes were later on co-expressed with some ansamitocin PKS-I genes, which resulted in the accumulation of a triketide compound (Yu *et al.*, 2002).

In addition to the usual hosts *S. lividans* and *S. coelicolor*, other strains have been used for heterologous expression. The urdamycin PKS genes were expressed into tetracenomycin mutants complementing their mutations and restoring tetracenomycin production (Decker and Haag, 1995). The chalcomycin PKS genes were expressed in a *S. fradiae* derived strain, from which the tylosin PKS genes were deleted, resulting in the production of a hybrid compound that demonstrated the predicted domain organization of chalcomycin PKS (Ward *et al.*, 2004). The megosamine biosynthetic genes were expressed in the erythromycin producer leading to the accumulation of megalomycins (Volchegursky *et al.*, 2000). *S. albus* was used to express steffimycin PKS genes leading to production of an intermediate compound (Gullón *et al.*, 2006), and for expressing azinomycin biosynthetic genes involved in the generation of the NRPS starter unit (Zhao *et al.*, 2008).

The biochemical approach for validating enzymatic activities has been frequently used by expressing genes in *E. coli*. Briefly, it was used for characterizing mithramycin NGS MtmD and NGDH MtmE (Lombó *et al.*, 1997); vicenistatin glycosyltransferase VinC (Ogasawara *et al.*, 2004); validamycin 2-epi-5-epi-valiolone synthase ValA (Yu *et al.*, 2005); pentalenolactone pentalenene synthase PtlA and glyceraldehyde-3-phosphate dehydrogenases Gap1 and Gap2 (Tetzlaff *et al.*, 2006); or oleandomycin L-oleandrosyl-3-O-methyltransferase OleY (Rodríguez *et al.*, 2001).

Definition of cluster boundaries

Another important issue regarding the validation of gene clusters is the definition of the limits of the cluster. This has been addressed by gene disruption thus locating genes whose inactivation had no effect on metabolite production, or by expressing the whole biosynthesis gene cluster in a heterologous hosts once it was sequenced and annotated, leading to the production of the metabolite. In both cases, these approaches have been supported by *in silico* analysis.

Gene disruption

In the last years, the determination of the cluster boundaries has relied, in most cases, in inactivation or deletion of genes located at both ends of the cluster as was accomplished in thienamycin (Núñez *et al.*, 2003), leinamycin (Tang *et al.*, 2004), fredericamycin (Wendt-Pienkowski *et al.*, 2005), neocarzinostatin (Liu *et al.*, 2005), anthramycin (Hu *et al.*, 2007), pradimicin (Kim *et al.*, 2007a), sibiromycin (Li *et al.*, 2009a), tomaymycin (Li *et al.*, 2009b) or zorbamycin (Galm *et al.*, 2009) gene clusters. In the case of benastatin cluster, identified by homology-based approaches and expression in *S. lividans*, the cluster was narrowed to 17 genes necessary for benastatin production by excision of non essential flanking regions and by expression of the rest of the genes in *S. albus* yielding a benastatin-producing strain (Xu *et al.*, 2007).

Gene cluster expression

In this case, for validating the gene cluster it is necessary to produce heterologously the desired compound using different hosts and genetic tools to mobilize large DNA fragments (Kieser *et al.*, 2000; Martínez *et al.*, 2004; Zhang, *et al.*, 2008b; Fujii, 2009; Gust, 2009). *S. lividans* was used as a heterologous host for the production oxytetracycline (Binnie *et al.*, 1989), nonactin (Kwon *et al.*, 2000), griseorhodin (Li and Piel, 2002), aureothin (He and Hertweck, 2003), leptomycin (Hu *et al.*, 2005), furaquinocin (Kawasaki *et al.*, 2006), pentalenolactone (Tetzlaff *et al.*, 2006), elloramycin (Ramos *et al.*, 2008), anthramycin (Hu *et al.*, 2008) or meridamycin (Liu *et al.*, 2009) upon expression of the corresponding cluster. Medermycin (Ichinose *et al.*, 2003), non-chlorinated derivatives of neocarzilin and their respective keto forms (Otsuka *et al.*, 2004) or lincomycin (Koberská *et al.*, 2008) were produced in *S. coelicolor*. Novobiocin and clorobiocin (Eustáquio *et al.*, 2005) or phenalinolactone (Binz *et al.*, 2008) clusters were expressed in *S. lividans* and *S. coelicolor*. Fredericamycin (Wendt-Pienkowski *et al.*, 2005) and steffimycin (Gullón *et al.*, 2006) were produced in *S. albus*, while thiocoraline cluster was expressed in *S. albus* and *S. lividans* (Lombó *et al.*, 2006). *S. venezuelae* YJ003 was used for expression of kanamycin (Thapa *et al.*, 2007) and gentamicin A2 (Park *et al.*, 2008) clusters. In some cases cluster validation by heterologous expression led, in addition, to produce novel hybrid compounds as in the case of aranciamycin cluster expressed in *S. fradiae* A0 and *S. diastatochromogenes* Tü6028 that accumulated eight novel glycosylated compounds (Luzhetskyy *et al.*, 2007b and 2008a).

Combinatorial biosynthesis for the generation of novel compounds

Combinatorial biosynthesis can be defined as the application of genetic engineering to modify biosynthetic pathways in order to produce novel compounds (Fig. 11.3). After isolation of gene clusters the different approaches used for their validation are already part of combinatorial biosynthesis, since in most cases novel compounds are produced. The strategies used in combinatorial biosynthesis (Fig. 11.4) involve targeting gene disruption; expression of genes or combination of genes involved in precursor condensing steps or in tailoring modification in different hosts including mutant strains or producers of related structures; and mutasynthesis.

Precursor condensing enzymes

PKSs and NRPSs are the condensing enzymes most widely studied since they are responsible for generating a large number of the bioactive natural products in use. These enzymes are involved in the biosynthesis of polyketides by step-wise decarboxylative Claisen-type condensation of acyl-CoA precursors, PKS-I, -II and -III (Van Lanen and Shen, 2008a), or in the biosynthesis of small peptide molecules by condensation of proteinogenic and non-proteinogenic amino acids, including D-amino acids and other carboxylic acids, NRPSs (Fischbach and Walsh, 2006), and have been frequently employ for combinatorial biosynthesis. Traditionally, the main system used was PKS-II, since it is composed of discrete proteins that conform the minimal PKS (ACP, KSα, KSβ) for producing a linear polyketide that later is reduced by KRs, cyclized and further decorated (Hertweck *et al.*, 2007). In particular the most frequently used aromatic PKSs are those involved in the biosynthesis of actinorhodin, daunorubicin and tetracenomycin (Shen, 2000; Wohlert *et al.*, 2001; Kalaitzis and Moore, 2004). However, other PKS-II have been also used like those involved in the biosynthesis of aclacinomycin, enterocin, jadomycin, nogalamycin and pradimicin (Shen, 2000; Kantola *et al.*, 2003; Metsä-Ketelä *et al.*, 2003; Lee *et al.*, 2005; Das and Khosla, 2009). These studies led to production of a great number of novel compounds and to determine the requirements for the generation of naphthacenequinone and benz[a]anthraquinone ring systems characteristic of anthracyclines or tetracenomycins and angucyclines, respectively. PKS-I have also been used for combinatorial biosynthesis leading to produce a large number of novel compounds (Weissman and Leadlay, 2005; Menzella and Reeves, 2007). These systems are composed by large multifunctional enzymes organized into modules in the order in which they act, each harbouring a set of distinct activities responsible for the catalysis of one elongation cycle (Van Lanen and Shen, 2008a). In addition, protein–protein interactions between terminal docking domains of successive multienzymes determine the order of proteins and promote their correct positioning within the assembly line (Weissman, 2006). Compounds containing

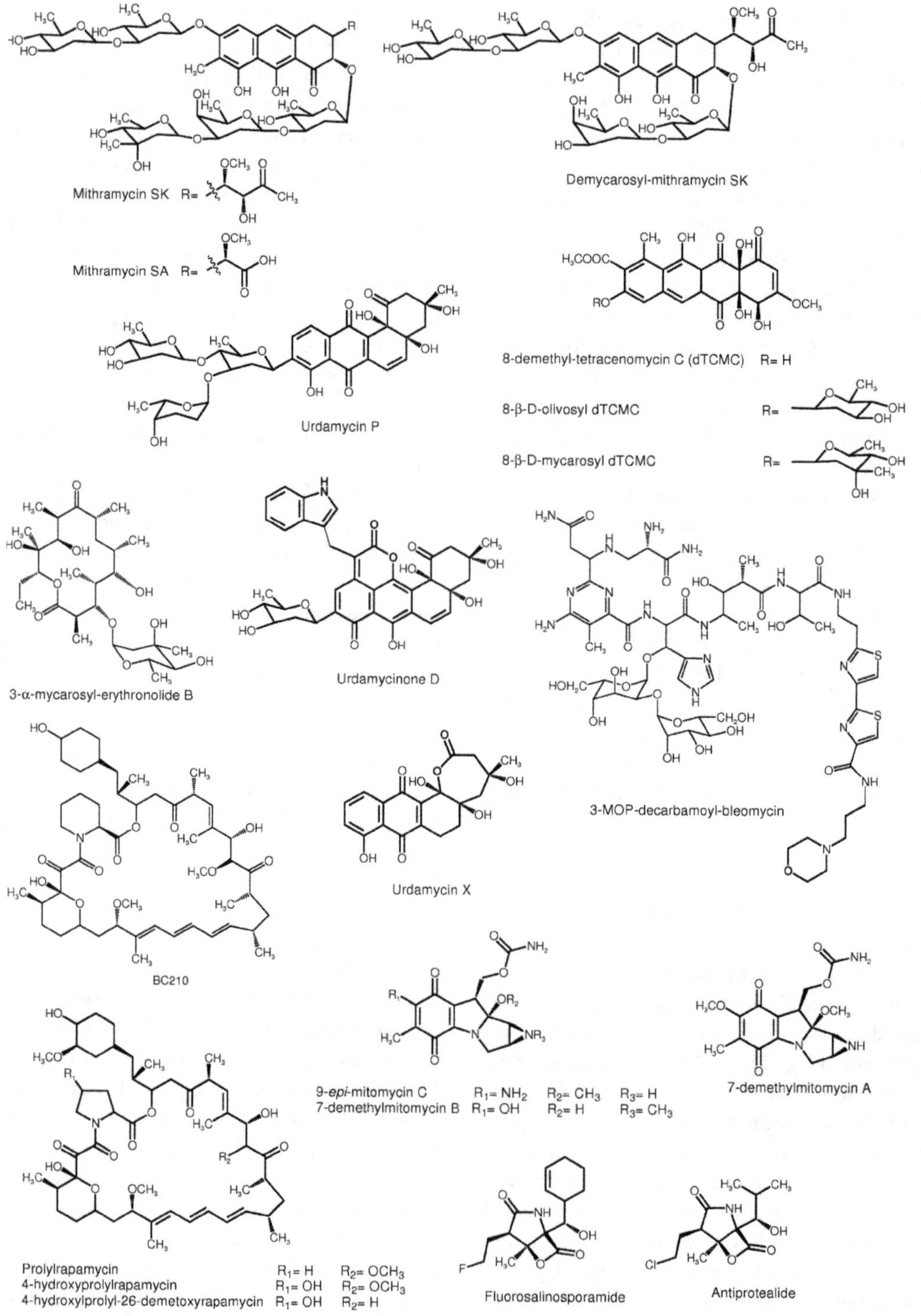

Figure 11.3 Structures of some novel compounds generated by combinatorial biosynthesis approaches.

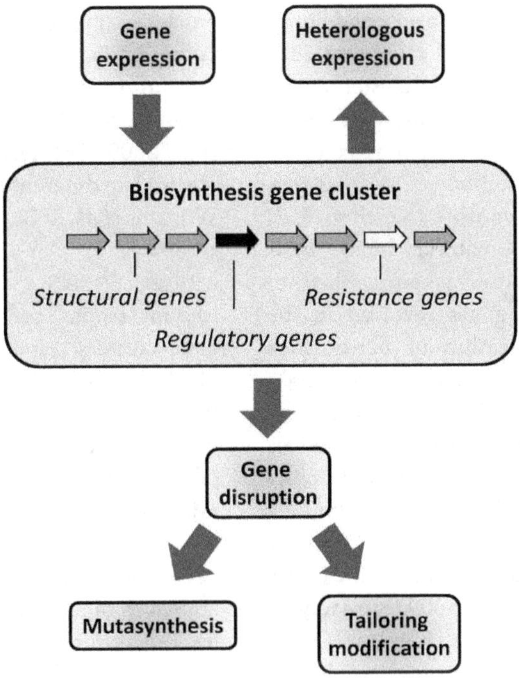

Figure 11.4 Combinatorial biosynthesis approaches for the generation of new compounds.

non-natural polyketide backbones have been generated through the combination of catalytic domains or modules from a variety of PKS-I such as those involved in avermectin, erythromycin, rapamycin, spinosyn and pikromycin biosynthesis (Oliynyk *et al.*, 1996; Rowe *et al.*, 2001; Weissman and Leadlay, 2005; Sheehan *et al.*, 2006; Menzella and Reeves, 2007; Kellenberger *et al.*, 2008). In other cases, large libraries of polyketides were generated by expression of combinations of PKS genes harbouring different mutations (Xue *et al.*, 1999). However, these approaches have not always been completely successful due to a poor understanding of the interactions between catalytic domains and modules (Kittendorf and Sherman, 2006) and to the presence of regions with extremely high sequence identity that leads to increase the instability of PKS-I genes in the expression plasmids used (Hu *et al.*, 2003). This has spurred on the investigation of the structure of PKS-I catalytic and docking domains and to use this information to improve combinatorial biosynthesis approaches (Broadhurst *et al.*, 2003; Weissman, 2006; Van Lanen and Shen, 2008a), to find approaches for circumventing recombination events that lead to large DNA deletions (Hu *et al.*, 2003) and to facilitate the cloning and expression of different combinations of genes leading to generation of large libraries of modified polyketide macrolactones (Xue *et al.*, 1999; Tang and McDaniel, 2001). Similar approaches have been applied to other modular enzymes like NRPS and hybrid PKS-NRPS (Keller and Schauwecker, 2003; Hahn and Stachelhaus, 2006; Menzella and Reeves, 2007) which has led to obtaining novel derivatives of daptomycin and related peptide antibiotics (Nguyen *et al.*, 2006; Baltz, 2008b). Other systems such as iterative PKS-I and PKS-III are attracting attention for their use in combinatorial biosynthesis in order to obtain enediyne and benastatin derivatives (Van Lanen and Shen, 2008b; Xu *et al.*, 2009).

Tailoring modification

The most prolific genes, in terms of novel compounds produced by combinatorial biosynthesis, are those encoding tailoring modification enzymes (Rix *et al.*, 2002; Olano *et al.*, 2009). Targeting disruption and expression experiments using tailoring modification genes has led to production of hundreds of novel compounds derived from different chemical structures such as those mentioned above in the cluster validation section. Similarly, novel compounds were generated derived from urdamycin (Rix *et al.*, 2003), oviedomycin (Lombó *et al.*, 2009a), rebeccamycin and staurosporine (Sánchez *et al.*, 2005 and 2006). Combinatorial biosynthesis has been especially fruitful when applied to glycosyltransferases and genes involved in the biosynthesis and modification of deoxysugars (Méndez *et al.*, 2008; Salas and Méndez, 2007 and 2009; Lombó *et al.*, 2009b) and has been used to obtain a great number of novel derivatives from different compounds including mithramycin (Trefzer *et al.*, 2002b; Pérez *et al.*, 2008; Baig *et al.*, 2008), pikromycin-methymycin (Hong *et al.*, 2004), rebeccamycin and staurosporine (Salas *et al.*, 2005; Sánchez *et al.*, 2009), urdamycin (Fedoryshyn *et al.*, 2007) and steffimycin (Olano *et al.*, 2008a). The utilization of glycosyltransferases for combinatorial biosynthesis has taken advantage of the elucidation of the structure of several of these enzymes (Hu and Walker, 2002; Luzhetskyy *et al.*, 2008b). This information has been used for the generation of hybrid or mutant versions of several glycosyltransferases with enhanced properties for obtaining glycosylated compounds (Xu *et al.*, 2008; Ramos *et al.*, 2009), for producing novel glycosylated derivatives (Hoffmeister *et al.*, 2001; Krauth *et al.*, 2009; Park *et al.*, 2009) or for generating enzymes capable to glycosylate different aglycones (Gantt *et al.*, 2008; Williams *et al.*, 2008).

Mutasynthesis

Mutant strains generated during validation of gene clusters, especially non-producing strains blocked in early biosynthetic steps involved in the generation of starter units or other precursors, can be used for precursor-directed biosynthesis of novel compounds. This approach couples the utilization of chemical synthesis with molecular biology (Weist and Süssmuth, 2005; Weissman, 2007; Kennedy, 2008). Mutasynthesis, also known as mutational biosynthesis, has been used for the generation of several non-natural compounds derived from different metabolites such as erythromycin (Toscano *et al.*, 1983), nikkomycin (Delzer *et al.*, 1984; Bormann *et al.*, 1999), enterocin (Kalaitzis *et al.*, 2003), rapamycin (Khaw *et al.*, 1998; Gregory *et al.*, 2005), borrelidin (Moss *et al.*, 2006), geldanamycin (Kim *et al.*, 2007b), aminocoumarin antibiotics (Heide *et al.*,

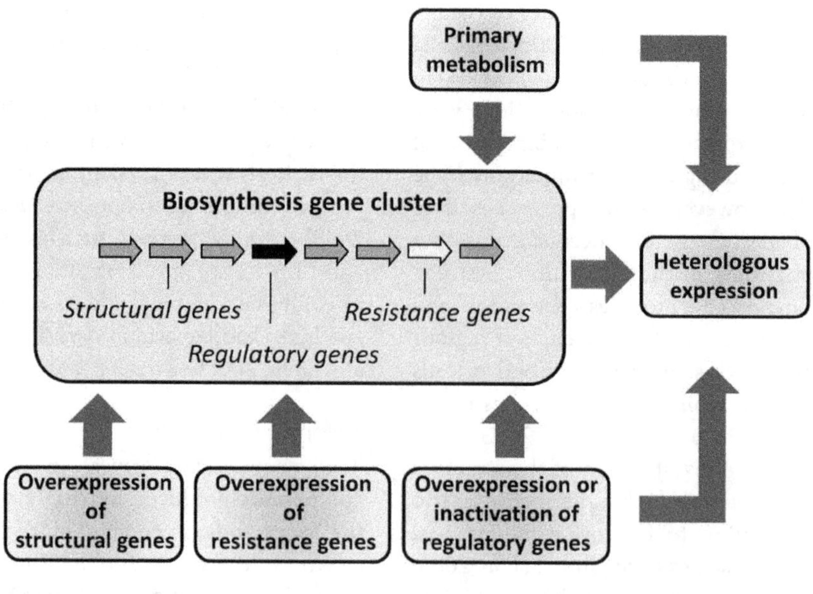

Figure 11.5 Methods used for improvement of secondary metabolite production.

2008), pikromycin (Gupta *et al.*, 2008) or salinosporamide (Eustáquio and Moore, 2008).

Improving metabolite production

Similar approaches to those used in combinatorial biosynthesis have been used, in addition, to improve the production of bioactive metabolites, including both natural and unnatural products (Olano *et al.*, 2008b). In general, overproduction of a particular compound can be achieved by engineering different aspects of cellular metabolism (Fig. 11.5). At the level of primary metabolism, precursor supply can be increased by altering the metabolic flux distribution (Reeves *et al.*, 2007) or by altering culture conditions in terms of pH or nutrients, in some cases searching for mutants showing the optimal employment of the resources (Song *et al.*, 2008b; Kim *et al.*, 2009). At the same level, protein synthesis rates can be increased by engineering RNA polymerase and ribosomes, which enhance the production of secondary metabolites (Wang *et al.*, 2008). The production of bioactive metabolites can also be improved by deregulating their specific biosynthetic pathways through pathway-specific (Smanski *et al.*, 2009) or pleiotropic regulatory genes (Duong *et al.*, 2009). Structural and self-resistance genes from biosynthesis clusters are also used for overexpression to derive an increase in production rates (Yanai *et al.*, 2006; Machida *et al.*, 2008b) or by generating mutations that improve the production of specific metabolites (Li *et al.*, 2008b). Other methods are the utilization of global genetic approaches such as genome shuffling and protoplast fusion (Chen *et al.*, 2009) or using decoy oligonucleotides for manipulating DNA–protein interactions that regulate specific genes (McArthur and Bibb, 2008). Finally, all these approaches can be used, in addition, by expressing the biosynthetic gene cluster in a heterologous host, streptomycete (Feng *et al.*, 2009) or *E. coli* (Harada *et al.*, 2009), previously engineered for producing bioactive compounds.

Concluding remarks

The development of recombinant DNA technology during the second part of the last century has set up the foundations for unravelling the information encoded in microbial genomes and to access to important information regarding one important aspect of the microbial world: the production of pharmacologically useful compounds. All the efforts executed for the isolation, characterization and manipulation of biosynthesis gene clusters have crystallized in obtaining a large number of unnatural compounds. Some of these compounds have improved biological properties, which make them lead compounds for their application in medicine, veterinary or agriculture fields. The future seems promising in terms of increasing the number and production titres of novel compounds from the already known gene clusters, and from those that will be discovered. This will be accomplished by applying traditional technologies and new ones developed in the '-omics' era: genomics, proteomics, metabolomics and transcriptomics that have created new opportunities to optimize microorganisms for the production of natural and unnatural compounds.

Acknowledgements

This research was supported by the Spanish Ministry of Science and Innovation (BFU2006–00404 to J.A.S. and BIO2005–04115 to C.M.), Red Temática de Investigación Cooperativa de Centros de Cáncer (Ministry of Health, ISCIII-RETIC RD06/0020/0026) and the EU FP6 (ActinoGen; Integrated project nº 005224). We thank Obra Social Cajastur for financial support to Carlos Olano.

References

Aguirrezabalaga, I., Olano, C., Allende, N., Rodríguez, L., Braña, A.F., Méndez, C., and Salas, J.A. (2000). Identification and expression of genes involved in biosynthesis of L-oleandrose and its intermediate L-olivose in the oleandomycin producer *Streptomyces antibioticus*. Antimicrob. Agents Chemother. *44*, 1266–1275.

Aparicio, J.F., Colina, A.J., Ceballos, E., and Martín, J.F. (1999). The biosynthetic gene cluster for the 26-membered ring polyene macrolide pimaricin. A new polyketide synthase organization encoded by two subclusters separated by functionalization genes. J. Biol. Chem. *274*, 10133–10139.

Aparicio, J.F., Fouces, R., Mendes, M.V., Olivera, N., and Martín, J.F. (2000). A complex multienzyme system encoded by five polyketide synthase genes is involved in the biosynthesis of the 26-membered polyene macrolide pimaricin in *Streptomyces natalensis*. Chem. Biol. *7*, 895–905.

August, P.R., Flickinger, M.C., and Sherman, D.H. (1994). Cloning and analysis of a locus (*mcr*) involved in mitomycin C resistance in *Streptomyces lavendulae*. J. Bacteriol. *176*, 4448–4454.

August, P.R., Tang, L., Yoon, Y.J., Ning, S., Müller, R., Yu, T.W., Taylor, M., Hoffmann, D., Kim, C.G., Zhang, X., *et al.* (1998). Biosynthesis of the ansamycin antibiotic rifamycin: deductions from the molecular analysis of the *rif* biosynthetic gene cluster of *Amycolatopsis mediterranei* S699. Chem. Biol. *5*, 69–79.

Ayuso-Sacido, A., and Genilloud, O. (2005). New PCR primers for the screening of NRPS and PKS-I systems in actinomycetes: detection and distribution of these biosynthetic gene sequences in major taxonomic groups. Microb. Ecol. *49*, 10–24.

Baig, I., Pérez, M., Braña, A.F., Gomathinayagam, R., Damodaran, C., Salas, J.A., Méndez, C., and Rohr, J. (2008). Mithramycin analogues generated by combinatorial biosynthesis show improved bioactivity. J. Nat. Prod. *71*, 199–207.

Bailey, C.R., Butler, M.J., Normansell, I.D., Rowlands, R.T., and Winstanley, D.J. (1984). Cloning a *Streptomyces clavuligerus* genetic locus involved in clavulanic acid biosynthesis. Biotechnology 2, 808–811.

Baltz, R.H. (2008a). Renaissance in antibacterial discovery from actinomycetes. Curr. Opin. Pharmacol. 8, 557- 563.

Baltz, R.H. (2008b). Biosynthesis and genetic engineering of lipopeptide antibiotics related to daptomycin. Curr. Top. Med. Chem. *8*, 618–638.

Barona-Gómez, F., Wong, U., Giannakopulos, A.E., Derrick, P.J., and Challis, G.L. (2004). Identification of a cluster of genes that directs desferrioxamine biosynthesis in *Streptomyces coelicolor* M145. J. Am. Chem. Soc. *126*, 16282–16283.

Barona-Gómez, F., Lautru, S., Francou, F.X., Leblond, P., Pernodet, J.L., and Challis, G.L. (2006). Multiple biosynthetic and uptake systems mediate siderophore-dependent iron acquisition in *Streptomyces coelicolor* A3(2) and *Streptomyces ambofaciens* ATCC23877. Microbiology *152*, 3355–3366.

Bauer, N.J., Kreuzman, A.J., Dotzlaf, J.E., and Yeh, W.K. (1988). Purification, characterization, and kinetic mechanism of S-adenosyl-L-methionine:macrocin *O*-methyltransferase from *Streptomyces fradiae*. J. Biol. Chem. *263*, 15619–15625.

Bentley, S.D., Chater, K.F., Cerdeño-Tárraga, A.M., Challis, G.L., Thomson, N.R., James, K.D., Harris, D.E., Quail, M.A., Kieser, H., Harper, D., *et al.* (2002). Complete genome sequence of the model actinomycete *Streptomyces coelicolor* A3(2). Nature *417*, 141–147.

Bérdy, J. (2005). Bioactive microbial metabolites. J. Antibiot. *58*, 1–26.

Bihlmaier, C., Welle, E., Hofmann, C., Welzel, K., Vente, A., Breitling, E., Müller, M., Glaser, S., and Bechthold, A. (2006). Biosynthetic gene cluster for the polyenoyltetramic acid α-lipomycin. Antimicrob. Agents Chemother. *50*, 2113–2121.

Bililign, T., Hyun, C.G., Williams, J.S., Czisny, A.M., and Thorson, J.S. (2004). The hedamycin locus implicates a novel aromatic PKS priming mechanism. Chem. Biol. *11*, 959–969.

Binnie, C., Warren, M., and Butler, M.J. (1989). Cloning and heterologous expression in *Streptomyces lividans* of *Streptomyces rimosus* genes involved in oxytetracycline biosynthesis. J. Bacteriol. *171*, 887–895.

Binz, T.M., Wenzel, S.C., Schnell, H.J., Bechthold, A., and Müller, R. (2008). Heterologous expression and genetic engineering of the phenalinolactone biosynthetic gene cluster by using Red/ET recombineering. Chembiochem *9*, 447–454.

Blanco, G., Fu, H., Méndez, C., Khosla, C., and Salas, J.A. (1996). Deciphering the biosynthetic origin of the aglycone of the aureolic acid group of anti-tumor agents. Chem. Biol. *3*, 193–196.

Blanco, G., Fernández, E., Fernández, M.J., Braña, A.F., Weissbach, U., Künzel, E., Rohr, J., Méndez, C., and Salas, J.A. (2000). Characterization of two glycosyltransferases involved in early glycosylation steps during biosynthesis of the antitumor polyketide mithramycin by *Streptomyces argillaceus*. Mol. Gen. Genet. *262*, 991–1000.

Blanco, G., Patallo, E.P., Braña, A.F., Trefzer, A., Bechthold, A., Rohr, J., Méndez, C., and Salas, J.A. (2001). Identification of a sugar flexible glycosyltransferase from *Streptomyces olivaceus*, the producer of the antitumor polyketide elloramycin. Chem. Biol. *8*, 253–263.

Bormann, C., Kálmánczhelyi, A., Süssmuth, R., and Jung, G. (1999). Production of nikkomycins Bx and Bz by mutasynthesis with genetically engineered *Streptomyces tendae* Tü901. J. Antibiot. *52*, 102–108.

Brautaset, T., Sekurova, O.N., Sletta, H., Ellingsen, T.E., Strøm, A.R., Valla, S., and Zotchev, S.B. (2000). Biosynthesis of the polyene antifungal antibiotic nystatin in *Streptomyces noursei* ATCC11455: analysis of the gene cluster and deduction of the biosynthetic pathway. Chem. Biol. *7*, 395–403.

Broadhurst, R.W., Nietlispach, D., Wheatcroft, M.P., Leadlay, P.F., and Weissman, K.J. (2003). The structure of docking domains in modular polyketide synthases. Chem. Biol. *10*, 723–731.

Brünker, P., McKinney, K., Sterner, O., Minas, W., and Bailey, J.E. (1999). Isolation and characterization of the naphthocyclinone gene cluster from *Streptomyces arenae* DSM40737 and heterologous expression of the polyketide synthase genes. Gene *227*, 125–135.

Butler, M.J., Friend, E.J., Hunter, I.S., Kaczmarek, F.S., Sugden, D.A., and Warren, M. (1989). Molecular cloning of resistance genes and architecture of a linked gene cluster involved in biosynthesis of oxytetracycline by *Streptomyces rimosus*. Mol. Gen. Genet. *215*, 231–238.

Caffrey, P., Lynch, S., Flood, E., Finnan, S., and Oliynyk, M. (2001). Amphotericin biosynthesis in *Streptomyces nodosus*: deductions from analysis of polyketide synthase and late genes. Chem. Biol. *8*, 713–723.

Campelo, A.B., and Gil, J.A. (2002). The candicidin gene cluster from *Streptomyces griseus* IMRU3570. Microbiology *148*, 51–59.

Carroll, B.J., Moss, S.J., Bai, L., Kato, Y., Toelzer, S., Yu, T.W., and Floss, H.G. (2002). Identification of a set of genes involved in the formation of the substrate for the incorporation of the unusual 'glycolate' chain extension unit in ansamitocin biosynthesis. J. Am. Chem. Soc. *124*, 4176–4177.

Challis, G.L., and Ravel, J. (2000). Coelichelin, a new peptide siderophore encoded by the *Streptomyces coelicolor* genome: structure prediction from the sequence of its non-ribosomal peptide synthetase. FEMS Microbiol. Lett. *187*, 111–114.

Challis, G.L. (2008). Mining microbial genomes for new natural products and biosynthetic pathways. Microbiology *154*, 1555–1569.

Chater, K.F., and Bruton, C.J. (1983). Mutational cloning in *Streptomyces* and the isolation of antibiotic production genes. Gene *26*, 67–78.

Chater, K.F., and Bruton, C.J. (1985). Resistance, regulatory and production genes for the antibiotic methylenomycin are clustered. EMBO J. *4*, 1893–1897.

Chater, K.F. (1998). Taking a genetic scalpel to the *Streptomyces* colony. Microbiology *144*, 1465–1478.

Chen, C.W., Lin, H.F., Kuo, C.L., Tsai, H.L., and Tsai, J.F.Y. (1988). Cloning and expression of a DNA sequence conferring cephamycin C production. Biotechnology *6*, 1222–1224.

Chen, X., Wei, P., Fan, L., Yang, D., Zhu, X., Shen, W., Xu, Z., and Cen, P. (2009). Generation of high-yield rapamycin-producing strains through protoplasts-related techniques. Appl. Microbiol. Biotechnol. *83*, 507–512.

Cheng, Y.Q., Tang, G.L., and Shen, B. (2002). Identification and localization of the gene cluster encoding biosynthesis of the antitumor macrolactam leinamycin in *Streptomyces atroolivaceus* S-140. J. Bacteriol. *184*, 7013–7024.

Choi, S.S., Hur, Y.A., Sherman, D.H., and Kim, E.S. (2007). Isolation of the biosynthetic gene cluster for tautomycetin, a linear polyketide T cell-specific immunomodulator from *Streptomyces* sp. CK4412. Microbiology *153*, 1095–1102.

Chuck, J.A., Dunn, C., Facultad, F.E., Nakazono, C., Nikodinovic, J., and Barrow, K.D. (2006). Amplification of DNA encoding entire type I polyketide synthase domains and linkers from *Streptomyces* species. Curr. Microbiol. *53*, 89–94.

Criado, L.M., Martín, J.F., and Gil, J.A. (1993). The *pab* gene from *Streptomyces griseus*, encoding *p*-aminobenzoic acid synthase, is located between genes possibly involved in candicidin biosynthesis. Gene *126*, 135–139.

Das, A., and Khosla, C. (2009). Biosynthesis of aromatic polyketides in bacteria. Acc. Chem. Res. *42*, 631–639.

Decker, H., and Haag, S. (1995). Cloning and characterization of a polyketide synthase gene from *Streptomyces fradiae* Tü2717, which carries the genes for biosynthesis of the angucycline antibiotic urdamycin A and a gene probably involved in its oxygenation. J. Bacteriol. *177*, 6126–6136.

Decker, H., Rohr, J., Motamedi, H., Zähner, H., and Hutchinson, C.R. (1995). Identification of *Streptomyces olivaceus* Tü2353 genes involved in the production of the polyketide elloramycin. Gene *166*, 121–126.

Decker, H., Gaisser, S., Pelzer, S., Schneider, P., Westrich, L., Wohleben, W., and Bechthold, A. (1996). A general approach for cloning and characterizing dNDP-glucose dehydratase genes from actinomycetes. FEMS Microbiol. Lett. *141*, 195–201.

Delzer, J., Fiedler, H.P., Müller, H., Zähner, H., Rathmann, R., Ernst, K., and König, W.A. (1984). New nikkomycins by mutasynthesis and directed fermentation. J. Antibiot. *37*, 80–82.

Demydchuk, J., Oliynyk, Z., and Fedorenko, V. (1998). Analysis of a kanamycin resistance gene (*kmr*) from *Streptomyces kanamyceticus* and a mutant with increased aminoglycoside resistance. J. Basic Microbiol. *38*, 231–239.

Demydchuk, Y., Sun, Y., Hong, H., Staunton, J., Spencer, J.B., and Leadlay, P.F. (2008). Analysis of the tetronomycin gene cluster: insights into the biosynthesis of a polyether tetronate antibiotic. Chembiochem *9*, 1136–1145.

Dhillon, N., Hale, R.S., Cortés, J., and Leadlay, P.F. (1989). Molecular characterization of a gene from *Saccharopolyspora erythraea* (*Streptomyces erythraeus*) which is involved in erythromycin biosynthesis. Mol. Microbiol. *3*, 1405–1414.

Distler, J., Mansouri, K., and Piepersberg, W. (1985). Streptomycin biosynthesis in *Streptomyces griseus* II. Adjacent genomic location of biosynthetic genes and one of two streptomycin resistance genes. FEMS Microbiol. Lett. *30*, 151–154.

Duong, C.T., Lee, H.N., Choi, S.S., Lee, S.Y., and Kim, E.S. (2009). Functional expression of SAV3818, a putative TetR-family transcriptional regulatory gene from *Streptomyces avermitilis*, stimulates antibiotic production in *Streptomyces* species. J. Microbiol. Biotechnol. *19*, 136–139.

Dürr, C., Schnell, H.J., Luzhetskyy, A., Murillo, R., Weber, M., Welzel, K., Vente, A., and Bechthold, A. (2006). Biosynthesis of the terpene phenalinolactone in *Streptomyces* sp. Tü6071: analysis of the gene cluster and generation of derivatives. Chem. Biol. *13*, 365–377.

Eliot, A.C., Griffin, B.M., Thomas, P.M., Johannes, T.W., Kelleher, N.L., Zhao, H., and Metcalf, W.W. (2008). Cloning, expression, and biochemical characterization of *Streptomyces rubellomurinus* genes required for biosynthesis of antimalarial compound FR900098. Chem. Biol. *15*, 765–770.

Epp, J.K., Burgett, S.G., and Schoner, B.E. (1987). Cloning and nucleotide sequence of a carbomycin-resistance gene from *Streptomyces thermotolerans*. Gene *53*, 73–83.

Eustáquio, A.S., Gust, B., Galm, U., Li, S.M., Chater, K.F., and Heide, L. (2005). Heterologous expression of novobiocin and clorobiocin biosynthetic gene clusters. Appl. Environ. Microbiol. *71*, 2452–2459.

Eustáquio, A.S., and Moore, B.S. (2008). Mutasynthesis of fluorosalinosporamide, a potent and reversible inhibitor of the proteasome. Angew. Chem. Int. Ed. Engl. *47*, 3936–3938.

Fang, J., Zhang, Y., Huang, L., Jia, X., Zhang, Q., Zhang, X., Tang, G., and Liu, W. (2008). Cloning and characterization of the tetrocarcin A gene cluster from *Micromonospora chalcea* NRRL11289 reveals a highly conserved strategy for tetronate biosynthesis in spirotetronate antibiotics. J. Bacteriol. *190*, 6014–6025.

Fedoryshyn, M., Nur-e-Alam, M., Zhu, L., Luzhetskyy, A., Rohr, J., and Bechthold, A. (2007). Surprising production of a new urdamycin derivative by *S. fradiae* Delta *urdQ/R*. J. Biotechnol. *130*, 32–38.

Feitelson, J.S., and Hopwood, D.A. (1983). Cloning of a *Streptomyces* gene for an *O*-methyltransferase involved in antibiotic biosynthesis. Mol. Gen. Genet. *190*, 394–398.

Feitelson, J.S., Malpartida, F., and Hopwood, D.A. (1985). Genetic and biochemical characterization of the *red* gene cluster of *Streptomyces coelicolor* A3(2). J. Gen. Microbiol. *131*, 2431–2441.

Feng, Z., Wang, L., Rajski, S.R., Xu, Z., Coeffet-LeGal, M.F., and Shen, B. (2009). Engineered production of iso-migrastatin in heterologous *Streptomyces* hosts. Bioorg. Med. Chem. *17*, 2147–2153.

Fernández, E., Lombó, F., Méndez, C., and Salas, J.A. (1996). An ABC transporter is essential for resistance to the antitumor agent mithramycin in the producer *Streptomyces argillaceus*. Mol. Gen. Genet. *251*, 692–698.

Fernández, E., Weissbach, U., Sánchez Reillo, C., Braña, A.F., Méndez, C., Rohr, J., and Salas, J.A. (1998). Identification of two genes from *Streptomyces argillaceus* encoding glycosyltransferases involved in transfer of a disaccharide during biosynthesis of the antitumor drug mithramycin. J. Bacteriol. *180*, 4929–4937.

Fischbach, M.A., and Walsh, C.T. (2006). Assembly-line enzymology for polyketide and nonribosomal peptide antibiotics: logic, machinery, and mechanisms. Chem. Rev. *106*, 3468–3496.

Fischer C., Lipata, F., and Rohr, J. (2003). The complete gene cluster of the antitumor agent gilvocarcin V and its implication for the biosynthesis of the gilvocarcins. J. Am. Chem. Soc. *125*, 7818–7819.

Fishman, S.E., Cox, K., Larson, J.L., Reynolds, P.A., Seno, E.T., Yeh, W.K., Van Frank, R., and Hershberger, C.L. (1987). Cloning genes for the biosynthesis of a macrolide antibiotic. Proc. Natl. Acad. Sci. U.S.A. *84*, 8248–8252.

Fouces, R., Mellado, E., Díez, B., and Barredo, J.L. (1999). The tylosin biosynthetic cluster from *Streptomyces fradiae*: genetic organization of the left region. Microbiology *145*, 855–868.

Fujii, I. (2009). Heterologous expression systems for polyketide synthases. Nat. Prod. Rep. *26*, 155–169.

Gaisser, S., Trefzer, A., Stockert, S., Kirschning, A., and Bechthold, A. (1997). Cloning of an avilamycin biosynthetic gene cluster from *Streptomyces viridochromogenes* Tü57. J. Bacteriol. *179*, 6271–6278.

Galm, U., Schimana, J., Fiedler, H.P., Schmidt, J., Li, S.M., and Heide, L. (2002). Cloning and analysis of the simocyclinone biosynthetic gene cluster of *Streptomyces antibioticus* Tü6040. Arch. Microbiol. *178*, 102–114.

Galm, U., Wendt-Pienkowski, E., Wang, L., George, N.P., Oh, T.J., Yi, F., Tao, M., Coughlin, J.M., and Shen, B. (2009). The biosynthetic gene cluster of zorbamycin, a member of the bleomycin family of antitumor antibiotics, from *Streptomyces flavoviridis* ATCC21892. Mol. Biosyst. *5*, 77–90.

Gantt, R.W., Goff, R.D., Williams, G.J., and Thorson, J.S. (2008). Probing the aglycon promiscuity of an engineered glycosyltransferase. Angew. Chem. Int. Ed. Engl. *47*, 8889–8892.

Garrido, L.M., Lombó, F., Baig, I., Nur-E.-Alam, M., Furlan, R.L., Borda, C.C., Braña, A.F., Méndez, C., Salas, J.A., Rohr, J., *et al.* (2006). Insights in the glycosylation steps during biosynthesis of the antitumor anthracycline cosmomycin: characterization of two glycosyltransferase genes. Appl. Microbiol. Biotechnol. 73, 122–131.

Gil, J.A., and Hopwood, D.A. (1983). Cloning and expression of a *p*-aminobenzoic acid synthetase gene of the candicidin-producing *Streptomyces griseus*. Gene *25*, 119–132.

Ginolhac, A., Jarrin, C., Gillet, B., Robe, P., Pujic, P., Tuphile, K., Bertrand, H., Vogel, T.M., Perrière, G., Simonet, P., *et al.* (2004). Phylogenetic analysis of polyketide synthase I domains from soil metagenomic libraries allows selection of promising clones. Appl. Environ. Microbiol. *70*, 5522–5527.

Goldberg, S.L., Romero, J.G., and Deo, Y.M. (1990). Cloning and characterization of the sisomicin-resistance gene from *Micromonospora inyoensis*. J. Antibiot. *43*, 992–999.

Gregory, M.A., Petkovic, H., Lill, R.E., Moss, S.J., Wilkinson, B., Gaisser, S., Leadlay, P.F., and Sheridan, R.M. (2005). Mutasynthesis of rapamycin analogues through the manipulation of a gene governing starter unit biosynthesis. Angew. Chem. Int. Ed. Engl. *44*, 4757–4760.

Grimm, A., Madduri, K., Ali, A., and Hutchinson, C.R. (1994). Characterization of the *Streptomyces peucetius* ATCC29050 genes encoding doxorubicin polyketide synthase. Gene *151*, 1–10.

Gross, H. (2007). Strategies to unravel the function of orphan biosynthesis pathways: recent examples and future prospects. Appl. Microbiol. Biotechnol. *75*, 267–277.

Grüschow, S., Chang, L.C., Mao, Y., and Sherman, D.H. (2007). Hydroxyquinone *O*-methylation in mitomycin biosynthesis. J. Am. Chem. Soc. *129*, 6470–6476.

Gullón, S., Olano, C., Abdelfattah, M.S., Braña, A.F., Rohr, J., Méndez, C., and Salas, J.A. (2006). Isolation, characterization, and heterologous expression of the biosynthesis gene cluster for the antitumor anthracycline steffimycin. Appl. Environ. Microbiol. *72*, 4172–4183.

Gupta, S., Lakshmanan, V., Kim, B.S., Fecik, R., and Reynolds, K.A. (2008). Generation of novel pikromycin antibiotic products through mutasynthesis. Chembiochem *9*, 1609–1616.

Gust, B., Challis, G.L., Fowler, K., Kieser, T., and Chater, K.F. (2003). PCR-targeted *Streptomyces* gene replacement identifies a protein domain needed for biosynthesis of the sesquiterpene soil odor geosmin. Proc. Natl. Acad. Sci. U.S.A. *100*, 1541–1546.

Gust, B. (2009). Cloning and analysis of natural product pathways. Methods Enzymol. *458*, 159–180.

Haagen, Y., Glück, K., Fay, K., Kammerer, B., Gust, B., and Heide, L. (2006). A gene cluster for prenylated naphthoquinone and prenylated phenazine biosynthesis in *Streptomyces cinnamonensis* DSM1042. Chembiochem *7*, 2016–2027.

Hahn, M., and Stachelhaus, T. (2006). Harnessing the potential of communication-mediating domains for the biocombinatorial synthesis of nonribosomal peptides. Proc. Natl. Acad. Sci. U.S.A. *103*, 275–280.

Harada, H., Yu, F., Okamoto, S., Kuzuyama, T., Utsumi, R., and Misawa, N. (2009). Efficient synthesis of functional isoprenoids from acetoacetate through metabolic pathway-engineered *Escherichia coli*. Appl. Microbiol. Biotechnol. *81*, 915–925.

Harvey, B.M., Mironenko, T., Sun, Y., Hong, H., Deng, Z., Leadlay, P.F., Weissman, K.J., and Haydock, S.F. (2007). Insights into polyether biosynthesis from analysis of the nigericin biosynthetic gene cluster in *Streptomyces* sp. DSM4137. Chem. Biol. *14*, 703–714.

Hayashi, Y., Matsuura, N., Toshima, H., Itoh, N., Ishikawa, J., Mikami, Y., and Dairi, T. (2008). Cloning of the gene cluster responsible for the biosynthesis of brasilicardin A, a unique diterpenoid. J. Antibiot. *61*, 164–174.

Haydock, S.F., Mironenko, T., Ghoorahoo, H.I., and Leadlay, P.F. (2004). The putative elaiophylin biosynthetic gene cluster in *Streptomyces* sp. DSM4137 is adjacent to genes encoding adenosylcobalamin-dependent methylmalonyl CoA mutase and to genes for synthesis of cobalamin. J. Biotechnol. *113*, 55–68.

Haydock, S.F., Appleyard, A.N., Mironenko, T., Lester, J., Scott, N., and Leadlay, P.F. (2005). Organization of the biosynthetic gene cluster for the macrolide concanamycin A in *Streptomyces neyagawaensis* ATCC27449. Microbiology *151*, 3161–3169.

He, J., and Hertweck, C. (2003). Iteration as programmed event during polyketide assembly; molecular analysis of the aureothin biosynthesis gene cluster. Chem. Biol. *10*, 1225–1232.

He, M., Haltli, B., Summers, M., Feng, X., and Hucul, J. (2006). Isolation and characterization of meridamycin biosynthetic gene cluster from *Streptomyces* sp. NRRL30748. Gene *377*, 109–118.

Heide, L., Gust, B., Anderle, C., and Li, S.M. (2008). Combinatorial biosynthesis, metabolic engineering and mutasynthesis for the generation of new aminocoumarin antibiotics. Curr. Top. Med. Chem. *8*, 667–679.

Hertweck, C., Luzhetskyy, A., Rebets, Y., and Bechthold, A. (2007). Type II polyketide synthases: gaining a deeper insight into enzymatic teamwork. Nat. Prod. Rep. *24*, 162–190.

Hoffmeister, D., Ichinose, K., and Bechthold, A. (2001). Two sequence elements of glycosyltransferases involved in urdamycin biosynthesis are responsible for substrate specificity and enzymatic activity. Chem. Biol. *8*, 557–567.

Hong, S.T., Carney, J.R., and Gould, S.J. (1997). Cloning and heterologous expression of the entire gene clusters for PD116740 from *Streptomyces* strain WP4669 and tetrangulol and tetrangomycin from *Streptomyces rimosus* NRRL3016. J. Bacteriol. *179*, 470–476.

Hong, J.S., Park, S.H., Choi, C.Y., Sohng, J.K., and Yoon, Y.J. (2004). New olivosyl-derivatives of methymycin/pikromycin from an engineered strain of *Streptomyces venezuelae*. FEMS Microbiol. Lett. *238*, 391–399.

Hong, W.R., Ge, M., Zeng, Z.H., Zhu, L., Luo, M.Y., Shao, L., and Chen, D.J. (2009). Molecular cloning and sequence analysis of the sisomicin biosynthetic gene cluster from *Micromonospora inyoensis*. Biotechnol. Lett. *31*, 449–455.

Hopwood, D.A. (1999). Forty years of genetics with *Streptomyces*: from *in vivo* through *in vitro* to *in silico*. Microbiology *145*, 2183–2202.

Hu, Y., and Walker, S. (2002). Remarkable structural similarities between diverse glycosyltransferases. Chem. Biol. *9*, 1287–1296.

Hu, Z., Desai, R.P., Volchegursky, Y., Leaf, T., Woo, E., Licari, P., Santi, D.V., Hutchinson, C.R., and McDaniel, R. (2003). Approaches to stabilization of inter-domain recombination in polyketide synthase gene expression plasmids. J. Ind. Microbiol. Biotechnol. *30*, 161–177.

Hu, Z., Reid, R., and Gramajo, H. (2005). The leptomycin gene cluster and its heterologous expression in *Streptomyces lividans*. J. Antibiot. *58*, 625–633.

Hu, Y., Phelan, V., Ntai, I., Farnet, C.M., Zazopoulos, E., and Bachmann, B.O. (2007). Benzodiazepine biosynthesis in *Streptomyces refuineus*. Chem. Biol. *14*, 691–701.

Hu, Y., Phelan, V.V., Farnet, C.M., Zazopoulos, E., and Bachmann, B.O. (2008). Reassembly of anthramycin biosynthetic gene cluster by using recombinogenic cassettes. Chembiochem *9*, 1603–1608.

Hyun, C., Kim, S.S., Sohng, J.K., Hahn, J., Kim, J., and Suh, J. (2000). An efficient approach for cloning the dNDP-glucose synthase gene from actinomycetes and its application in *Streptomyces spectabilis*, a spectinomycin producer. FEMS Microbiol. Lett. *183*, 183–189.

Ichinose, K., Bedford, D.J., Tornus, D., Bechthold, A., Bibb, M.J., Revill, W.P., Floss, H.G., and Hopwood, D.A. (1998). The granaticin biosynthetic gene cluster of *Streptomyces violaceoruber* Tü22: sequence analysis and expression in a heterologous host. Chem. Biol. *5*, 647–659.

Ichinose, K., Ozawa, M., Itou, K., Kunieda, K., and Ebizuka, Y. (2003). Cloning, sequencing and heterologous expression of the medermycin biosynthetic gene cluster of *Streptomyces* sp. AM-7161: towards comparative analysis of the benzoisochromanequinone gene clusters. Microbiology *149*, 1633–1645.

Ikeda, H., Wang, L.R., Ohta, T., Inokoshi, J., and Omura, S. (1998). Cloning of the gene encoding avermectin B 5-O-methyltransferase in avermectin-producing *Streptomyces avermitilis*. Gene *206*, 175–180.

Ikeda, H., Nonomiya, T., Usami, M., Ohta, T., and Omura, S. (1999). Organization of the biosynthetic gene cluster for the polyketide anthelmintic macrolide avermectin in *Streptomyces avermitilis*. Proc. Natl. Acad. Sci. U.S.A. *96*, 9509–9514.

Inouye, M., Morohoshi, T., Horinouchi, S., and Beppu, T. (1994). Cloning and sequences of two macrolide-resistance-encoding genes from mycinamicin-producing *Micromonospora griseorubida*. Gene *141*, 39–46.

Izumikawa, M., Murata, M., Tachibana, K., Ebizuka, Y., and Fujii, I. (2003). Cloning of modular type I polyketide synthase genes from salinomycin producing strain of *Streptomyces albus*. Bioorg. Med. Chem. *11*, 3401–3405.

Jensen, P.R., Williams, P.G., Oh, D.C., Zeigler, L., and Fenical, W. (2007). Species-specific secondary metabolite production in marine actinomycetes of the genus *Salinispora*. Appl. Environ. Microbiol. *73*, 1146–1152.

Jia, X.Y., Tian, Z.H., Shao, L., Qu, X.D., Zhao, Q.F., Tang, J., Tang, G.L., and Liu, W. (2006). Genetic characterization of the chlorothricin gene cluster as a model for spirotetronate antibiotic biosynthesis. Chem. Biol. *13*, 575–585.

Jones, G.H., and Hopwood, D.A. (1984). Molecular cloning and expression of the phenoxazinone synthase gene from *Streptomyces antibioticus*. J. Biol. Chem. *259*, 14151–14157.

Jones, G.H. (2000). Actinomycin production persists in a strain of *Streptomyces antibioticus* lacking phenoxazinone synthase. Antimicrob. Agents Chemother. *44*, 1322–1327.

Kakavas, S.J., Katz, L., and Stassi, D. (1997). Identification and characterization of the niddamycin polyketide synthase genes from *Streptomyces caelestis*. J. Bacteriol. *179*, 7515–7522.

Kalaitzis, J.A., Izumikawa, M., Xiang, L., Hertweck, C., and Moore, B.S. (2003). Mutasynthesis of enterocin and wailupemycin analogues. J. Am. Chem. Soc. *125*, 9290–9291.

Kalaitzis, J.A., and Moore, B.S. (2004). Heterologous biosynthesis of truncated hexaketides derived from the actinorhodin polyketide synthase. J. Nat. Prod. *67*, 1419–1422.

Kantola, J., Kunnari, T., Mäntsälä, P., and Ylihonko, K. (2003). Expanding the scope of aromatic polyketides by combinatorial biosynthesis. Comb. Chem. High Throughput Screen. *6*, 501–512.

Kawasaki, T., Hayashi, Y., Kuzuyama, T., Furihata, K., Itoh, N., Seto, H., and Dairi, T. (2006). Biosynthesis of a natural polyketide-isoprenoid hybrid compound, furaquinocin A: identification and heterologous expression of the gene cluster. J. Bacteriol. *188*, 1236–1244.

Kelemen, G.H., Cundliffe, E., and Financsek, I. (1991). Cloning and characterization of gentamicin-resistance genes from *Micromonospora purpurea* and *Micromonospora rosea*. Gene *98*, 53–60.

Kellenberger, L., Galloway, I.S., Sauter, G., Böhm, G., Hanefeld, U., Cortés, J., Staunton, J., and Leadlay, P.F. (2008). A polylinker approach to reductive loop swaps in modular polyketide synthases. Chembiochem *9*, 2740–2749.

Keller, U., and Schauwecker, F. (2003). Combinatorial biosynthesis of non-ribosomal peptides. Comb. Chem. High Throughput Screen. *6*, 527–540.

Kelly, W.L., Pan, L., and Li, C. (2009). Thiostrepton biosynthesis: prototype for a new family of bacteriocins. J. Am. Chem. Soc. *131*, 4327–4334.

Kennedy, J. (2008). Mutasynthesis, chemobiosynthesis, and back to semi-synthesis: combining synthetic chemistry and biosynthetic engineering for diversifying natural products. Nat. Prod. Rep. *25*, 25–34.

Khaw, L.E., Böhm, G.A., Metcalfe, S., Staunton, J., and Leadlay, P.F. (1998). Mutational biosynthesis of novel rapamycins by a strain of *Streptomyces hygroscopicus* NRRL5491 disrupted in *rapL*, encoding a putative lysine cyclodeaminase. J. Bacteriol. *180*, 809–814.

Kieser, T., Bibb, M.J., Buttner, M.J., Chater, K.F., and Hopwood, D.A. (2000). Practical *Streptomyces* Genetics (The John Innes Foundation, Norwich).

Kim, C.G., Yu, T.W., Fryhle, C.B., Handa, S., and Floss, H.G. (1998). 3-Amino-5-hydroxybenzoic acid synthase, the terminal enzyme in the formation of the precursor of mC7N units in rifamycin and related antibiotics. J. Biol. Chem. *273*, 6030–6040.

Kim, B.C., Lee, J.M., Ahn, J.S., and Kim, B.S. (2007a). Cloning, sequencing, and characterization of the pradimicin biosynthetic gene cluster of *Actinomadura hibisca* P157-2. J. Microbiol. Biotechnol. *17*, 830–839.

Kim, W., Lee, J.S., Lee, D., Cai, X.F., Shin, J.C., Lee, K., Lee, C.H., Ryu, S., Paik, S.G., Lee, J.J., *et al.* (2007b). Mutasynthesis of geldanamycin by the disruption of a gene producing starter unit: generation of structural diversity at the benzoquinone ring. Chembiochem *8*, 1491–1494.

Kim, J.Y., Suh, J.W., Kang, S.H., Phan, T.H., Park, S.H., and Kwon, H.J. (2008). Gene inactivation study of *gntE* reveals its role in the first step of pseudotrisaccharide modifications in gentamicin biosynthesis. Biochem. Biophys. Res. Commun. *372*, 730–734.

Kim, S.J., Kim, J.O., Shin, C.H., Park, H.W., and Kim, C.W. (2009). An approach to strain improvement and enhanced production of clavulanic acid in *Streptomyces clavuligerus*. Biosci. Biotechnol. Biochem. *73*, 160–164.

Kinashi, H., Shimaji, M., and Sakai, A. (1987). Giant linear plasmids in *Streptomyces* which code for antibiotic biosynthesis genes. Nature *328*, 454–456.

Kittendorf, J.D., and Sherman, D.H. (2006). Developing tools for engineering hybrid polyketide synthetic pathways. Curr. Opin. Biotechnol. *17*, 597–605.

Knirschová, R., Nováková, R., Fecková, L., Timko, J., Turna, J., Bistáková, J., and Kormanec, J. (2007). Multiple regulatory genes in the salinomycin biosynthetic gene cluster of *Streptomyces albus* CCM4719. Folia Microbiol. 52, 359–365.

Koberská, M., Kopecký, J., Olsovská, J., Jelínková, M., Ulanova, D., Man, P., Flieger, M., and Janata, J. (2008). Sequence analysis and heterologous expression of the lincomycin biosynthetic cluster of the type strain *Streptomyces lincolnensis* ATCC25466. Folia Microbiol. 53, 395–401.

Krauth, C., Fedoryshyn, M., Schleberger, C., Luzhetskyy, A., and Bechthold, A. (2009). Engineering a function into a glycosyltransferase. Chem. Biol. *16*, 28–35.

Kudo, F., Kasama, Y., Hirayama, T., and Eguchi, T. (2007). Cloning of the pactamycin biosynthetic gene cluster and characterization of a crucial glycosyltransferase prior to a unique cyclopentane ring formation. J. Antibiot. *60*, 492–503.

Künzel, E., Wohlert, S.E., Beninga, C., Haag, S., Decker, H., Hutchinson, C.R., Blanco, G., Méndez, C., Salas, J.A., and Rohr, J. (1997). Tetracenomycin M, a novel genetically engineered tetracenomycin resulting from a combination of mithramycin and tetracenomycin biosynthetic genes. Chem. Eur. J. 3, 1675–1678.

Kwon, H.J., Smith, W.C., Xiang, L., and Shen, B. (2000). Cloning and heterologous expression of the macrotetrolide biosynthetic gene cluster revealed a novel polyketide synthase that lacks an acyl carrier protein. J. Am. Chem. Soc. *123*, 3385–3386.

Lacalle, R.A., Tercero, J.A., and Jiménez, A. (1992). Cloning of the complete biosynthetic gene cluster for an aminonucleoside antibiotic, puromycin, and its regulated expression in heterologous hosts. EMBO J. *11*, 785–792.

Lee, T.S., Khosla, C., and Tang, Y. (2005). Engineered biosynthesis of aklanonic acid analogues. J. Am. Chem. Soc. *127*, 12254–12262.

Leskiw, B.K., Aharonowitz, Y., Mevarech, M., Wolfe, S., Vining, L.C., Westlake, D.W., and Jensen, S.E. (1988). Cloning and nucleotide sequence determination of the isopenicillin N synthetase gene from *Streptomyces clavuligerus*. Gene *62*, 187–196.

Li, A., and Piel, J. (2002). A gene cluster from a marine *Streptomyces* encoding the biosynthesis of the aromatic spiroketal polyketide griseorhodin A. Chem. Biol. *9*, 1017–1026.

Li, W., Ju, J., Osada, H., and Shen, B. (2006). Utilization of the methoxymalonyl-acyl carrier protein biosynthesis locus for cloning of the tautomycin biosynthetic gene cluster from *Streptomyces spiroverticillatus*. J. Bacteriol. *188*, 4148–4152.

Li, L., Deng, W., Song, J., Ding, W., Zhao, Q.F., Peng, C., Song, W.W., Tang, G.L., and Liu, W. (2008a). Characterization of the saframycin A gene cluster from *Streptomyces lavendulae* NRRL 11002 revealing a nonribosomal peptide synthetase system for assembling the unusual tetrapeptidyl skeleton in an iterative manner. J. Bacteriol. *190*, 251–263.

Li, M., Chen, Z., Lin, X., Zhang, X., Song, Y., Wen, Y., and Li, J. (2008b). Engineering of avermectin biosynthetic genes to improve production of ivermectin in *Streptomyces avermitilis*. Bioorg. Med. Chem. Lett. *18*, 5359–5363.

Li, W., Khullar, A., Chou, S., Sacramo, A., and Gerratana, B. (2009a). Biosynthesis of sibiromycin, a potent antitumor antibiotic. Appl. Environ. Microbiol. 75, 2869–2878.

Li, W., Chou, S., Khullar, A., and Gerratana, B. (2009b). Cloning and characterization of the biosynthetic gene cluster for tomaymycin, an SJG-136 monomeric analog. Appl. Environ. Microbiol. 75, 2958–2963.

Liu, W., Ahlert, J., Gao, Q., Wendt-Pienkowski, E., Shen, B., and Thorson, J.S. (2003). Rapid PCR amplification of minimal enediyne polyketide synthase cassettes leads to a predictive familial classification model. Proc. Natl. Acad. Sci. U.S.A. *100*, 11959–11963.

Liu, W., Nonaka, K., Nie, L., Zhang, J., Christenson, S.D., Bae, J., Van Lanen, S.G., Zazopoulos, E., Farnet, C.M., Yang, C.F., *et al.* (2005). The neocarzinostatin biosynthetic gene cluster from *Streptomyces carzinostaticus* ATCC15944 involving two iterative type I polyketide synthases. Chem. Biol. *12*, 293–302.

Liu, H., Jiang, H., Haltli, B., Kulowski, K., Muszynska, E., Feng, X., Summers, M., Young, M., Graziani, E., Koehn, F., *et al.* (2009). Rapid cloning and heterologous expression of the meridamycin biosynthetic gene cluster using a versatile *Escherichia coli-Streptomyces* artificial chromosome vector, pSBAC. J. Nat. Prod. 72, 389–395.

Locci, R. (1994). Actinomycetes as plant pathogens. Eur. J. Plant Pathol. *100*, 179–200.

Lombó, F., Blanco, G., Fernández, E., Méndez, C., and Salas, J.A. (1996). Characterization of *Streptomyces argillaceus* genes encoding a polyketide synthase involved in the biosynthesis of the antitumor mithramycin. Gene *172*, 87–91.

Lombó, F., Siems, K., Braña, A.F., Méndez, C., Bindseil, K., and Salas, J.A. (1997). Cloning and insertional inactivation of *Streptomyces argillaceus* genes involved in the earliest steps of biosynthesis of the sugar moieties of the antitumor polyketide mithramycin. J. Bacteriol. *179*, 3354–3357.

Lombó, F., Braña, A.F., Méndez, C., and Salas, J.A. (1999). The mithramycin gene cluster of *Streptomyces argillaceus* contains a positive regulatory gene and two repeated DNA sequences that are located at both ends of the cluster. J. Bacteriol. *181*, 642–647.

Lombó, F., Braña, A.F., Salas, J.A., and Méndez, C. (2004). Genetic organization of the biosynthetic gene cluster for the antitumor angucycline oviedomycin in *Streptomyces antibioticus* ATCC11891. Chembiochem *5*, 1181–1187.

Lombó, F., Velasco, A., Castro, A., de la Calle, F., Braña, A.F., Sánchez-Puelles, J.M., Méndez, C., and Salas, J.A. (2006). Deciphering the biosynthesis pathway of the antitumor thiocoraline from a marine actinomycete and its expression in two *Streptomyces* species. Chembiochem 7, 366–376.

Lombó, F., Abdelfattah, M.S., Braña, A.F., Salas, J.A., Rohr, J., and Méndez, C. (2009a). Elucidation of oxygenation steps during oviedomycin biosynthesis and generation of derivatives with increased antitumor activity. Chembiochem *10*, 296–303.

Lombó, F., Olano, C., Salas, J.A., and Méndez, C. (2009b). Sugar biosynthesis and modification. Methods Enzymol. *458*, 277–307.

Lozano, M.J., Remsing, L.L., Quirós, L.M., Braña, A.F., Fernández, E., Sánchez, C., Méndez, C., Rohr, J., and Salas, J.A. (2000). Characterization of two polyketide methyltransferases involved in the biosynthesis of the antitumor drug mithramycin by *Streptomyces argillaceus*. J. Biol. Chem. *275*, 3065–3074.

Luzhetskyy, A., Weiss, H., Charge, A., Welle, E., Linnenbrink, A., Vente, A., and Bechthold, A. (2007a). A strategy for cloning glycosyltransferase genes involved in natural product biosynthesis. Appl. Microbiol. Biotechnol. *75*, 1367–1375.

Luzhetskyy, A., Mayer, A., Hoffmann, J., Pelzer, S., Holzenkämper, M., Schmitt, B., Wohlert, S.E., Vente, A., and Bechthold, A. (2007b). Cloning and heterologous expression of the aranciamycin biosynthetic gene cluster revealed a new flexible glycosyltransferase. Chembiochem *8*, 599–602.

Luzhetskyy, A., Hoffmann, J., Pelzer, S., Wohlert, S.E., Vente, A., and Bechthold, A. (2008a). Aranciamycin analogs generated by combinatorial biosynthesis show improved antitumor activity. Appl. Microbiol. Biotechnol. *80*, 15–19.

Luzhetskyy, A., Méndez, C., Salas, J.A., and Bechthold, A. (2008b). Glycosyltransferases, important tools for drug design. Curr. Top. Med. Chem. *8*, 680–709.

Machida, K., Arisawa, A., Takeda, S., Tsuchida, T., Aritoku, Y., Yoshida, M., and Ikeda, H. (2008a). Organization of the biosynthetic gene cluster for the polyketide antitumor macrolide, pladienolide, in *Streptomyces platensis* Mer-11107. Biosci. Biotechnol. Biochem. *72*, 2946–2952.

Machida, K., Aritoku, Y., Nakashima, T., Arisawa, A., and Tsuchida, T. (2008b). Increase in pladienolide D production rate using a *Streptomyces* strain overexpressing a cytochrome P450 gene. J. Biosci. Bioeng. *105*, 649–654.

Madduri, K., Waldron, C., and Merlo, D.J. (2001). Rhamnose biosynthesis pathway supplies precursors for primary and secondary metabolism in *Saccharopolyspora spinosa*. J. Bacteriol. *183*, 5632–5638.

Malpartida, F., and Hopwood, D.A. (1984). Molecular cloning of the whole biosynthetic pathway of a *Streptomyces* antibiotic and its expression in a heterologous host. Nature *309*, 462–464.

Malpartida, F., Hallam, S.E., Kieser, H.M., Motamedi, H., Hutchinson, C.R., Butler, M.J., Sugden, D.A., Warren, M., McKillop, C., Bailey, C.R., *et al.* (1987). Homology between *Streptomyces* genes coding for synthesis of different polyketides used to clone antibiotic biosynthetic genes. Nature *325*, 818–821.

Malpartida, F., Niemi, J., Navarrete, R., and Hopwood, D.A. (1990). Cloning and expression in a heterologous host of the complete set of genes for biosynthesis of the *Streptomyces coelicolor* antibiotic undecylprodigiosin. Gene *93*, 91–99.

Mao, Y., Varoglu, M., and Sherman, D.H. (1999). Molecular characterization and analysis of the biosynthetic gene cluster for the antitumor antibiotic mitomycin C from *Streptomyces lavendulae* NRRL2564. Chem. Biol. *6*, 251–263.

Martínez, A., Kolvek, S.J., Yip, C.L., Hopke, J., Brown, K.A., MacNeil, I.A., and Osburne, M.S. (2004). Genetically modified bacterial strains and novel bacterial artificial chromosome shuttle vectors for constructing environmental libraries and detecting heterologous natural products in multiple expression hosts. Appl. Environ. Microbiol. *70*, 2452–2463.

McAlpine, J.B., Bachmann, B.O., Piraee, M., Tremblay, S., Alarco, A.M., Zazopoulos, E., and Farnet, C.M. (2005). Microbial genomics as a guide to drug discovery and structural elucidation: ECO-02301, a novel antifungal agent, as an example. J. Nat. Prod. *68*, 493–496.

McArthur, M., and Bibb, M.J. (2008). Manipulating and understanding antibiotic production in *Streptomyces coelicolor* A3(2) with decoy oligonucleotides. Proc. Natl. Acad. Sci. U.S.A. *105*, 1020–1025.

McCarthy, A.J., and Williams, S.T. (1992). Actinomycetes as agents of biodegradation in the environment-a review. Gene *115*, 189–192.

McDaniel, R., Ebert-Khosla, S., Hopwood, D.A., and Khosla, C. (1993). Engineered biosynthesis of novel polyketides. Science *262*, 1546–15450.

McGlinchey, R.P., Nett, M., Eustáquio, A.S., Asolkar, R.N., Fenical, W., and Moore, B.S. (2008). Engineered biosynthesis of antiprotealide and other unnatural salinosporamide proteasome inhibitors. J. Am. Chem. Soc. *130*, 7822–7823.

McNeil, M.M., and Brown, J.M. (1994). The medically important aerobic actinomycetes: epidemiology and microbiology. Clin. Microbiol. Rev. *7*, 357–417.

Méndez, C., Künzel, E., Lipata, F., Lombó, F., Cotham, W., Walla, M., Bearden, D.W., Braña, A.F., Salas, J.A., and Rohr, J. (2002). Oviedomycin, an unusual angucyclinone encoded by genes of the oleandomycin-producer *Streptomyces antibioticus* ATCC11891. J. Nat. Prod. *65*, 779–782.

Méndez, C., Luzhetskyy, A., Bechthold, A., and Salas, J.A. (2008). Deoxysugars in bioactive natural products: development of novel derivatives by altering the sugar pattern. Curr. Top. Med. Chem. *8*, 710–724.

Menéndez, N., Nur-e-Alam, M., Braña, A.F., Rohr, J., Salas, J.A., and Méndez, C. (2004). Biosynthesis of the antitumor chromomycin A3 in *Streptomyces griseus*: analysis of the gene cluster and rational design of novel chromomycin analogs. Chem. Biol. *11*, 21–32.

Menéndez, N., Nur-e-Alam, M., Fischer, C., Braña, A.F., Salas, J.A., Rohr, J., and Méndez, C. (2006). Deoxysugar transfer during chromomycin A3 biosynthesis in *Streptomyces griseus* subsp. *griseus*: new derivatives with antitumor activity. Appl. Environ. Microbiol. *72*, 167–177.

Menzella, H.G., and Reeves, C.D. (2007). Combinatorial biosynthesis for drug development. Curr. Opin. Microbiol. *10*, 238–245.

Metsä-Ketelä, M., Salo, V., Halo, L., Hautala, A., Hakala, J., Mäntsälä, P., and Ylihonko, K. (1999). An efficient approach for screening minimal PKS genes from *Streptomyces*. FEMS Microbiol. Lett. *180*, 1–6.

Metsä-Ketelä, M., Palmu, K., Kunnari, T., Ylihonko, K., and Mäntsälä, P. (2003). Engineering anthracycline biosynthesis toward angucyclines. Antimicrob. Agents Chemother. *47*, 1291–1296.

Miao, V., Coëffet-Legal, M.F., Brian, P., Brost, R., Penn, J., Whiting, A., Martin, S., Ford, R., Parr, I., Bouchard, M., *et al.* (2005). Daptomycin biosynthesis in *Streptomyces roseosporus*: cloning and analysis of the gene cluster and revision of peptide stereochemistry. Microbiology *151*, 1507–1523.

Migita, A., Watanabe, M., Hirose, Y., Watanabe, K., Tokiwano, T., Kinashi, H., and Oikawa, H. (2009). Identification of a gene cluster of polyether antibiotic lasalocid from *Streptomyces lasaliensis*. Biosci. Biotechnol. Biochem. *73*, 169–176.

Moss, S.J., Bai, L., Toelzer, S., Carroll, B.J., Mahmud, T., Yu, T.W., and Floss, H.G. (2002). Identification of *asm19* as an acyltransferase attaching the biologically essential ester side chain of ansamitocins using *N*-desmethyl-4, 5-desepoxymaytansinol, not maytansinol, as its substrate. J. Am. Chem. Soc. *124*, 6544–6545.

Moss, S.J., Carletti, I., Olano, C., Sheridan, R.M., Ward, M., Math, V., Nur-E.-Alam, M., Braña, A.F., Zhang, M.Q., Leadlay, P.F., Méndez, C., *et al.* (2006). Biosynthesis of the angiogenesis inhibitor borrelidin: directed biosynthesis of novel analogues. Chem. Commun. *22*, 2341–2343.

Motamedi, H., and Hutchinson, C.R. (1987). Cloning and heterologous expression of a gene cluster for the biosynthesis of tetracenomycin C, the anthracycline antitumor antibiotic of *Streptomyces glaucescens*. Proc. Natl. Acad. Sci. U.S.A. *84*, 4445–4449.

Murakami, T., Anzai, H., Imai, S., Satoh, A., Nagaoka, K., and Thompson, C.J. (1986). The bialaphos biosynthetic genes of *Streptomyces hygroscopicus*: Molecular cloning and characterization of the gene cluster. Mol. Gen. Genet. *205*, 42–50.

Nakata, K., Horinouchi, S., and Beppu, T. (1989). Cloning and characterization of the carbapenem biosynthetic genes from *Streptomyces fulvoviridis*. FEMS Microbiol. Lett. *48*, 51–55.

Nguyen, K.T., Ritz, D., Gu, J.Q., Alexander, D., Chu, M., Miao, V., Brian, P., and Baltz, R.H. (2006). Combinatorial biosynthesis of novel antibiotics related to daptomycin. Proc. Natl. Acad. Sci. U.S.A. *103*, 17462–17467.

Núñez, L.E., Méndez, C., Braña, A.F., Blanco, G., and Salas, J.A. (2003). The biosynthetic gene cluster for the beta-lactam carbapenem thienamycin in *Streptomyces cattleya*. Chem. Biol. *10*, 301–311.

Nur-e-Alam, M., Méndez, C., Salas, J.A., and Rohr, J. (2005). Elucidation of the glycosylation sequence of mithramycin biosynthesis: isolation of 3A-deolivosylpremithramycin B and its conversion to premithramycin B by glycosyltransferase MtmGII. Chembiochem *6*, 632–636.

Ogasawara, Y., Katayama, K., Minami, A., Otsuka, M., Eguchi, T., and Kakinuma, K. (2004). Cloning, sequencing, and functional analysis of the biosynthetic gene cluster of macrolactam antibiotic vicenistatin in *Streptomyces halstedii*. Chem. Biol. *11*, 79–86.

Ohnishi, Y., Ishikawa, J., Hara, H., Suzuki, H., Ikenoya, M., Ikeda, H., Yamashita, A., Hattori, M., and Horinouchi, S. (2008). Genome sequence of the streptomycin-producing microorganism *Streptomyces griseus* IFO13350. J. Bacteriol. *190*, 4050–4060.

Ohnuki, T., Imanaka, T., and Aiba, S. (1985). Self-cloning in *Streptomyces griseus* of an *str* gene cluster for streptomycin biosynthesis and streptomycin resistance. J. Bacteriol. *164*, 85–94.

Ohuchi, T., Ikeda-Araki, A., Watanabe-Sakamoto, A., Kojiri, K., Nagashima, M., Okanishi, M., and Suda, H. (2000). Cloning and expression of a gene encoding *N*-glycosyltransferase (*ngt*) from *Saccarothrix aerocolonigenes* ATCC39243. J. Antibiot. *53*, 393–403.

Olano, C., Rodríguez, A.M., Michel, J.M., Méndez, C., Raynal, M.C., and Salas, J.A. (1998). Analysis of a *Streptomyces antibioticus* chromosomal region involved in oleandomycin biosynthesis, which encodes two glycosyltransferases responsible for glycosylation of the macrolactone ring. Mol. Gen. Genet. *259*, 299–308.

Olano, C., Lomovskaya, N., Fonstein, L., Roll, J.T., and Hutchinson, C.R. (1999). A two-plasmid system for the glycosylation of polyketide antibiotics: bioconversion of epsilon-rhodomycinone to rhodomycin D. Chem. Biol. *6*, 845–855.

Olano, C., Wilkinson, B., Sánchez, C., Moss, S.J., Sheridan, R., Math, V., Weston, A.J., Braña, A.F., Martin, C.J., Oliynyk, M., *et al.* (2004a). Biosynthesis of the angiogenesis inhibitor borrelidin by *Streptomyces parvulus* Tü4055: cluster analysis and assignment of functions. Chem. Biol. *11*, 87–97.

Olano, C., Moss, S.J., Braña, A.F., Sheridan, R.M., Math, V., Weston, A.J., Méndez, C., Leadlay, P.F., Wilkinson, B., and Salas, J.A. (2004b). Biosynthesis of the angiogenesis inhibitor borrelidin by *Streptomyces parvulus* Tü4055: insights into nitrile formation. Mol. Microbiol. *52*, 1745–1756.

Olano, C., Abdelfattah, M.S., Gullón, S., Braña, A.F., Rohr, J., Méndez, C., and Salas, J.A. (2008a). Glycosylated derivatives of steffimycin: insights into the role of the sugar moieties for the biological activity. Chembiochem *9*, 624–633.

Olano, C., Lombó, F., Méndez, C., and Salas, J.A. (2008b). Improving production of bioactive secondary metabolites in actinomycetes by metabolic engineering. Metab. Eng. *10*, 281–292.

Olano, C., Méndez, C., and Salas, J.A. (2009). Antitumor compounds from actinomycetes: from gene clusters to new derivatives by combinatorial biosynthesis. Nat. Prod. Rep. *26*, 628–660.

Oliynyk, M., Brown, M.J., Cortés, J., Staunton, J., and Leadlay, P.F. (1996). A hybrid modular polyketide synthase obtained by domain swapping. Chem. Biol. *3*, 833–839.

Oliynyk, M., Stark, C.B., Bhatt, A., Jones, M.A., Hughes-Thomas, Z.A., Wilkinson, C., Oliynyk, Z., Demydchuk, Y., Staunton, J., and Leadlay, P.F. (2003). Analysis of the biosynthetic gene cluster for the polyether antibiotic monensin in *Streptomyces cinnamonensis* and evidence for the role of *monB* and *monC* genes in oxidative cyclization. Mol. Microbiol. *49*, 1179–1190.

Oliynyk, M., Samborskyy, M., Lester, J.B., Mironenko, T., Scott, N., Dickens, S., Haydock, S.F., and Leadlay, P.F. (2007). Complete genome sequence of the erythromycin-producing bacterium *Saccharopolyspora erythraea* NRRL23338. Nat. Biotechnol. *25*, 447–453.

Omura, S., Ikeda, H., Ishikawa, J., Hanamoto, A., Takahashi, C., Shinose, M., Takahashi, Y., Horikawa, H., Nakazawa, H., Osonoe, T., *et al.* (2001). Genome sequence of an industrial microorganism *Streptomyces avermitilis*: deducing the ability of producing secondary metabolites. Proc. Natl. Acad. Sci. U.S.A. *98*, 12215–12220.

Onaka, H., Taniguchi, S., Igarashi, Y., and Furumai, T. (2002). Cloning of the staurosporine biosynthetic gene cluster from *Streptomyces* sp. TP-A0274 and its heterologous expression in *Streptomyces lividans*. J. Antibiot. *55*, 1063–1071.

Otsuka, M., Ichinose, K., Fujii, I., and Ebizuka, Y. (2004). Cloning, sequencing, and functional analysis of an iterative type I polyketide synthase gene cluster for biosynthesis of the antitumor chlorinated polyenone neocarzilin in '*Streptomyces carzinostaticus*'. Antimicrob. Agents Chemother. *48*, 3468–3476.

Otten, S.L., Stutzman-Engwall, K.J., and Hutchinson, C.R. (1990). Cloning and expression of daunorubicin biosynthesis genes from *Streptomyces peucetius* and *S. peucetius* subsp. *caesius*. J. Bacteriol. *172*, 3427–3434.

Palaniappan, N., Ayers, S., Gupta, S., Habib, el-S., and Reynolds, K.A. (2006). Production of hygromycin A analogs in *Streptomyces hygroscopicus* NRRL2388 through identification and manipulation of the biosynthetic gene cluster. Chem. Biol. *13*, 753–764.

Park, J.W., Hong, J.S., Parajuli, N., Jung, W.S., Park, S.R., Lim, S.K., Sohng, J.K., and Yoon, Y.J. (2008). Genetic dissection of the biosynthetic route to gentamicin A2 by heterologous expression of its minimal gene set. Proc. Natl. Acad. Sci. U.S.A. *105*, 8399–8404.

Park, S.H., Park, H.Y., Sohng, J.K., Lee, H.C., Liou, K., Yoon, Y.J., and Kim, B.G. (2009). Expanding substrate specificity of GT-B fold glycosyltransferase via domain swapping and high-throughput screening. Biotechnol. Bioeng. *102*, 988–994.

Pérez, M., Baig, I., Braña, A.F., Salas, J.A., Rohr, J., and Méndez, C. (2008). Generation of new derivatives of the antitumor antibiotic mithramycin by altering the glycosylation pattern through combinatorial biosynthesis. Chembiochem *9*, 2295–2304.

Peschke, U., Schmidt, H., Zhang, H.Z., and Piepersberg, W. (1995). Molecular characterization of the lincomycin-production gene cluster of *Streptomyces lincolnensis* 78-11. Mol. Microbiol. *16*, 1137–1156.

Piel, J., Hoang, K., and Moore, B.S. (2000). Natural metabolic diversity encoded by the enterocin biosynthesis gene cluster. J. Am. Chem. Soc. *122*, 5415–5416.

Plater, R., and Robinson, J.A. (1992). Cloning and sequencing of a gene encoding macrotetrolide antibiotic resistance from *Streptomyces griseus*. Gene *112*, 117–122.

Pojer, F., Li, S.M., and Heide, L. (2002). Molecular cloning and sequence analysis of the clorobiocin biosynthetic gene cluster: new insights into the biosynthesis of aminocoumarin antibiotics. Microbiology *148*, 3901–3911.

Rafanan, Jr., E.R., Le, L., Zhao, L., Decker, H., and Shen, B. (2001). Cloning, sequencing, and heterologous expression of the *elmGHIJ* genes involved in the biosynthesis of the polyketide antibiotic elloramycin from *Streptomyces olivaceus* Tü2353. J. Nat. Prod. *64*, 444–449.

Ramos, A., Lombó, F., Braña, A.F., Rohr, J., Méndez, C., and Salas, J.A. (2008). Biosynthesis of elloramycin in *Streptomyces olivaceus* requires glycosylation by enzymes encoded outside the aglycon cluster. Microbiology *154*, 781–788.

Ramos, A., Olano, C., Braña, A.F., Méndez, C., and Salas, J.A. (2009). Modulation of deoxysugar transfer by the elloramycin glycosyltransferase ElmGT through site-directed mutagenesis. J. Bacteriol. *191*, 2871–2875.

Rascher, A., Hu, Z., Viswanathan, N., Schirmer, A., Reid, R., Nierman, W.C., Lewis, M., and Hutchinson, C.R. (2003). Cloning and characterization of a gene cluster for geldanamycin production in *Streptomyces hygroscopicus* NRRL3602. FEMS Microbiol. Lett. *218*, 223–230.

Räty, K., Kunnari, T., Hakala, J., Mäntsälä, P., and Ylihonko, K. (2000). A gene cluster from *Streptomyces galilaeus* involved in glycosylation of aclarubicin. Mol. Gen. Genet. *264*, 164–172.

Räty, K., Kantola, J., Hautala, A., Hakala, J., Ylihonko, K., and Mäntsälä, P. (2002). Cloning and characterization of *Streptomyces galilaeus* aclacinomycins polyketide synthase (PKS) cluster. Gene *293*, 115–122.

Reeves, A.R., Brikun, I.A., Cernota, W.H., Leach, B.I., González, M.C., Weber, J.M., (2007). Engineering of the methylmalonyl-CoA metabolite node of *Saccharopolyspora erythraea* for increased erythromycin production. Metab. Eng. *9*, 293–303.

Remsing, L.L., González, A.M., Nur-e-Alam, M., Fernández-Lozano, M.J., Braña, A.F., Rix, U., Oliveira, M.A., Méndez, C., Salas, J.A., and Rohr, J. (2003). Mithramycin SK, a novel antitumor drug with improved therapeutic index, mithramycin SA, and demycarosyl-mithramycin SK: three new products generated in the mithramycin producer *Streptomyces argillaceus* through combinatorial biosynthesis. J. Am. Chem. Soc. *125*, 5745–5753.

Rix, U., Fischer, C., Remsing, L.L., and Rohr, J. (2002). Modification of post-PKS tailoring steps through combinatorial biosynthesis. Nat. Prod. Rep. *19*, 542–580.

Rix, U., Remsing, L.L., Hoffmeister, D., Bechthold, A., and Rohr, J. (2003). Urdamycin L: a novel metabolic shunt product that provides evidence for the role of the *urdM* gene in the urdamycin A biosynthetic pathway of *Streptomyces fradiae* TU 2717. Chembiochem *4*, 109–111.

Rodríguez, A.M., Olano, C., Vilches, C., Méndez, C., and Salas, J.A. (1993). *Streptomyces antibioticus* contains at least three oleandomycin-resistance determinants, one of which shows similarity with proteins of the ABC-transporter superfamily. Mol. Microbiol. *8*, 571–582.

Rodríguez, L., Rodríguez, D., Olano, C., Braña, A.F., Méndez, C., and Salas, J.A. (2001). Functional analysis of OleY L-oleandrosyl 3-O-methyltransferase of the oleandomycin biosynthetic pathway in *Streptomyces antibioticus*. J. Bacteriol. *183*, 5358–5363.

Rowe, C.J., Böhm, I.U., Thomas, I.P., Wilkinson, B., Rudd, B.A., Foster, G., Blackaby, A.P., Sidebottom, P.J., Roddis, Y., Buss, A.D., *et al.* (2001). Engineering a polyketide with a longer chain by insertion of an extra module into the erythromycin-producing polyketide synthase. Chem. Biol. *8*, 475–485.

Salas, A.P., Zhu, L., Sánchez, C., Braña, A.F., Rohr, J., Méndez. C., and Salas, J.A. (2005). Deciphering the late steps in the biosynthesis of the anti-tumour indolocarbazole staurosporine: sugar donor substrate flexibility of the StaG glycosyltransferase. Mol. Microbiol. *58*, 17–27.

Salas, J.A., and Méndez, C. (2007). Engineering the glycosylation of natural products in actinomycetes. Trends Microbiol. *15*, 219–232.

Salas, J.A., and Méndez, C. (2009). Indolocarbazole antitumour compounds by combinatorial biosynthesis. Curr. Opin. Chem. Biol. *13*, 1–9.

Sánchez, C., Butovich, I.A., Braña, A.F., Rohr, J., Méndez, C., and Salas, J.A. (2002). The biosynthetic gene cluster for the antitumor rebeccamycin: characterization and generation of indolocarbazole derivatives. Chem. Biol. *9*, 519–531.

Sánchez, C., Zhu, L., Braña, A.F., Salas, A.P., Rohr, J., Méndez, C., and Salas, J.A. (2005). Combinatorial biosynthesis of antitumor indolocarbazole compounds. Proc. Natl. Acad. Sci. U.S.A. *102*, 461–466.

Sánchez, C., Méndez, C., and Salas, J.A. (2006). Engineering biosynthetic pathways to generate antitumor indolocarbazole derivatives. J. Ind. Microbiol. Biotechnol. *33*, 560–568.

Sánchez, C., Salas, A.P., Braña, A.F., Palomino, M., Pineda-Lucena, A., Carbajo, R.J., Méndez, C., Moris, F., and Salas, J.A. (2009). Generation of potent and selective kinase inhibitors by combinatorial biosynthesis of glycosylated indolocarbazoles. Chem. Commun. *27*, 4118–4120.

Savic, M., and Vasiljevic, B. (2006). Targeting polyketide synthase gene pool within actinomycetes: new degenerate primers. J. Ind. Microbiol. Biotechnol. *33*, 423–430.

Schauwecker, F., Pfennig, F., Schröder, W., and Keller, U. (1998). Molecular cloning of the actinomycin synthetase gene cluster from *Streptomyces chrysomallus* and functional heterologous expression of the gene encoding actinomycin synthetase II. J. Bacteriol. *180*, 2468–2474.

Schupp, T., Toupet, C., Engel, N., and Goff, S. (1998). Cloning and sequence analysis of the putative rifamycin polyketide synthase gene cluster from *Amycolatopsis mediterranei*. FEMS Microbiol. Lett. *159*, 201–207.

Schwecke, T., Aparicio, J.F., Molnar, I., Konig, A., Khaw, L.E., Haydock, S.F., Oliynyk, M., Caffrey, P., Cortes, J., Lester, J.B., *et al.* (1995). The biosynthetic gene cluster for the polyketide immunosuppressant rapamycin. Proc. Natl. Acad. Sci. U.S.A. *92*, 7839–7843.

Shao, L., Qu, X., Jia, X., Zhao, Q., Tian, Z., Wang, M., Tang, G., and Liu, W. (2006). Cloning and characterization of a bacterial iterative type I polyketide synthase gene encoding the 6-methylsalicyclic acid synthase. Biochem. Biophys. Res. Commun. *345*, 133–139.

Sheehan, L.S., Lill, R.E., Wilkinson, B., Sheridan, R.M., Vousden, W.A., Kaja, A.L., Crouse, G.D., Gifford, J., Graupner, P.R., Karr, L., *et al.* (2006). Engineering of the spinosyn PKS: directing starter unit incorporation. J. Nat. Prod. *69*, 1702–1710.

Shen, B. (2000). Biosynthesis of aromatic polyketides. Top. Curr. Chem. *209*, 1–51.

Sherman, D.H., Malpartida, F., Bibb, M.J., Kieser, H.M., Bibb, M.J., and Hopwood, D.A. (1989). Structure and deduced function of the granaticin-producing polyketide synthase gene cluster of *Streptomyces violaceoruber* Tü22. EMBO J. *8*, 2717–2725.

Shin, J.C., Na, Z., Lee, D.H., Kim, W.C., Lee, K., Shen, Y.M., Paik, S.G., Hong, Y.S., and Lee, J.J. (2008). Characterization of tailoring genes involved in the modification of geldanamycin polyketide in *Streptomyces hygroscopicus* JCM4427. J. Microbiol. Biotechnol. *18*, 1101–1108.

Sitachitta, N., Lopanik, N.B., Mao, Y., and Sherman, D.H. (2007). Analysis of a parallel branch in the mitomycin biosynthetic pathway involving the *mitN*-encoded aziridine *N*-methyltransferase. J. Biol. Chem. *282*, 20941–20947.

Smanski, M.J., Peterson, R.M., Rajski, S.R., and Shen, B. (2009). Engineered *Streptomyces platensis* strains that overproduce antibiotics platensimycin and platencin. Antimicrob. Agents Chemother. *53*, 1299–1304.

Smith, W.C., Xiang, L., and Shen, B. (2000). Genetic localization and molecular characterization of the *nonS* gene required for macrotetrolide biosynthesis in *Streptomyces griseus* DSM40695. Antimicrob. Agents Chemother. *44*, 1809–1817.

Song, D., Coughlin, J., Ju, J., Zhou, X., Shen, B., Zhao, C., and Deng, Z. (2008a). Alternative method for site-directed mutagenesis of complex polyketide synthase in *Streptomyces albus* JA3453. Acta Biochim. Biophys. Sin. *40*, 319–326.

Song, J.Y., Kim, Y.J., Hong, Y.S., and Chang, Y.K. (2008b). Enhancement of geldanamycin production by pH shock in batch culture of *Streptomyces hygroscopicus* subsp. *duamyceticus*. J. Microbiol. Biotechnol. *18*, 897–900.

Sosio, M., Bossi, E., Bianchi, A., and Donadio, S. (2000). Multiple peptide synthetase gene clusters in actinomycetes. Mol. Gen. Genet. *264*, 213–221.

Spiteller, P., Bai, L., Shang, G., Carroll, B.J., Yu, T.W., and Floss, H.G. (2003). The post-polyketide synthase modification steps in the biosynthesis of the antitumor agent ansamitocin by *Actinosynnema pretiosum*. J. Am. Chem. Soc. *125*, 14236–14237.

Stach, J.E., and Bull, A.T. (2005). Estimating and comparing the diversity of marine actinobacteria. Antonie van Leeuwenhoek *87*, 3–9.

Steffensky, M., Mühlenweg, A., Wang, Z.X., Li, S.M., and Heide, L. (2000). Identification of the novobiocin biosynthetic gene cluster of *Streptomyces spheroides* NCIB11891. Antimicrob. Agents Chemother. *44*, 1214–1222.

Streicher, S.L., Ruby, C.L., Paress, P.S., Danis, J.B., MacNeil, D.J., Gewain, K., MacNeil, T., Foor, F., Morin, N., Cimis, G., *et al.* (1989). Cloning genes for avermectin biosynthesis in *Streptomyces*. In Genetics and Molecular Biology of Industrial Microorganisms, Hershberger, C.L., Queener, S.W., and Hegeman, G., eds (American Society for Microbiology, Washington, D.C.), pp. 44–48.

Stutzman-Engwall, K.J., and Hutchinson, C.R. (1989). Multigene families for anthracycline antibiotic production in *Streptomyces peucetius*. Proc. Natl. Acad. Sci. U.S.A. *86*, 3135–3139.

Sugiyama, M., Thompson, C.J., Kumagai, T., Suzuki, K., Deblaere, R., Villarroel, R., and Davies, J. (1994). Characterisation by molecular cloning of two genes from *Streptomyces verticillus* encoding resistance to bleomycin. Gene *151*, 11–16.

Sun, Y., Zhou, X., Liu, J., Bao, K., Zhang, G., Tu, G., Kieser, T., and Deng, Z. (2002). *'Streptomyces nanchangensis'*, a producer of the insecticidal polyether antibiotic nanchangmycin and the antiparasitic macrolide meilingmycin, contains multiple polyketide gene clusters. Microbiology *148*, 361–371.

Sun, Y., Zhou, X., Tu, G., and Deng, Z. (2003a). Identification of a gene cluster encoding meilingmycin biosynthesis among multiple polyketide synthase contigs isolated from *Streptomyces nanchangensis* NS3226. Arch. Microbiol. *180*, 101–107.

Sun, Y., Zhou, X., Dong, H., Tu, G., Wang, M., Wang, B., and Deng, Z. (2003b). A complete gene cluster from *Streptomyces nanchangensis* NS3226 encoding biosynthesis of the polyether ionophore nanchangmycin. Chem. Biol. *10*, 431–441.

Sun, Y., Hong, H., Samborskyy, M., Mironenko, T., Leadlay, P.F., and Haydock, S.F. (2006). Organization of the biosynthetic gene cluster in *Streptomyces* sp. DSM4137 for the novel neuroprotectant polyketide meridamycin. Microbiology *152*, 3507–3515.

Swan, D.G., Rodríguez, A.M., Vilches, C., Méndez, C., and Salas, J.A. (1994). Characterisation of a *Streptomyces antibioticus* gene encoding a type I polyketide synthase which has an unusual coding sequence. Mol. Gen. Genet. *242*, 358–362.

Tang, L., and McDaniel, R. (2001). Construction of desosamine containing polyketide libraries using a glycosyltransferase with broad substrate specificity. Chem. Biol. *8*, 547–555.

Tang, G.L., Cheng, Y.Q., and Shen, B. (2004). Leinamycin biosynthesis revealing unprecedented architectural complexity for a hybrid polyketide synthase and nonribosomal peptide synthetase. Chem. Biol. *11*, 33–45.

Tetzlaff, C.N., You, Z., Cane, D.E., Takamatsu, S., Omura, S., and Ikeda, H. (2006). A gene cluster for biosynthesis of the sesquiterpenoid antibiotic pentalenolactone in *Streptomyces avermitilis*. Biochemistry *45*, 6179–6186.

Thapa, L.P., Oh, T.J., Lee, H.C., Liou, K., Park, J.W., Yoon, Y.J., and Sohng, J.K. (2007). Heterologous expression of the kanamycin biosynthetic gene cluster (pSKC2) in *Streptomyces venezuelae* YJ003. Appl. Microbiol. Biotechnol. *76*, 1357–1364.

Thiara, A.S., and Cundliffe, E. (1988). Cloning and characterization of a DNA gyrase B gene from *Streptomyces sphaeroides* that confers resistance to novobiocin. EMBO J. *7*, 2255–2259.

Thompson, C.J., Kieser, T., Ward, J.M., and Hopwood, D.A. (1982). Physical analysis of antibiotic resistance genes from *Streptomyces* and their use in vector construction. Gene *20*, 51–62.

Tohyama, S., Eguchi, T., Dhakal, R.P., Akashi, T., Otsuka, M., and Kakinuma, K. (2004). Genome-inspired search for new antibiotics. Isolation and structure determination of new 28-membered polyketide macrolactones, halstoctacosanolides A and B, from *Streptomyces halstedii* HC34. Tetrahedron *60*, 3999–4005.

Tohyama, S., Kakinuma, K., and Eguchi, T. (2006). The complete biosynthetic gene cluster of the 28-membered polyketide macrolactones, halstoctacosanolides, from *Streptomyces halstedii* HC34. J. Antibiot. *59*, 44–52.

Toscano, L., Fioriello, G., Spagnoli, R., Cappelletti, L., and Zanuso, G. (1983). New fluorinated erythromycins obtained by mutasynthesis. J. Antibiot. *36*, 1439–1450.

Trefzer, A., Pelzer, S., Schimana, J., Stockert, S., Bihlmaier, C., Fiedler, H.P., Welzel, K., Vente, A., and Bechthold, A. (2002a). Biosynthetic gene cluster of simocyclinone, a natural multihybrid antibiotic. Antimicrob. Agents Chemother. *46*, 1174–1182.

Trefzer, A., Blanco, G., Remsing, L., Künzel, E., Rix, U., Lipata, F., Braña, A.F., Méndez, C., Rohr, J., Bechthold, A., *et al.* (2002b). Rationally designed glycosylated premithramycins: hybrid aromatic polyketides using genes from three different biosynthetic pathways. J. Am. Chem. Soc. *124*, 6056–6062.

Udwary, D.W., Zeigler, L., Asolkar, R.N., Singan, V., Lapidus, A., Fenical, W., Jensen, P.R., and Moore, B.S. (2007). Genome sequencing reveals complex secondary metabolome in the marine actinomycete *Salinispora tropica*. Proc. Natl. Acad. Sci. U.S.A. *104*, 10376–10381.

Unwin, J., Standage, S., Alexander, D., Hosted, Jr., T., Horan, A.C., and Wellington, E.M. (2004). Gene cluster in *Micromonospora echinospora* ATCC15835 for the biosynthesis of the gentamicin C complex. J. Antibiot. *57*, 436–445.

Van Lanen, S.G., and Shen, B. (2008a). Advances in polyketide synthase structure and function. Curr. Opin. Drug Discov. Devel. *11*, 186–195.

Van Lanen, S.G., and Shen, B. (2008b). Biosynthesis of enediyne antitumor antibiotics. Curr. Top. Med. Chem. *8*, 448–459.

Vara, J., Malpartida, F., Hopwood, D.A., and Jiménez, A. (1985). Cloning and expression of a puromycin *N*-acetyltransferase gene from *Streptomyces alboniger* in *Streptomyces lividans* and *Escherichia coli*. Gene *33*, 197–206.

Vara, J., Lewandowska-Skarbek, M., Wang, Y.G., Donadio, S., and Hutchinson, C.R. (1989). Cloning of genes governing the deoxysugar portion of the erythromycin biosynthesis pathway in *Saccharopolyspora erythraea* (*Streptomyces erythreus*). J. Bacteriol. *171*, 5872–5881.

Ventura, M., Canchaya, C., Tauch, A., Chandra, G., Fitzgerald, G.F., Chater, K.F., and van Sinderen, D. (2007). Genomics of *Actinobacteria*: Tracing the evolutionary history of an ancient phylum. Microbiol. Mol. Biol. Rev. *71*, 495–548.

Volchegursky, Y., Hu, Z., Katz, L., and McDaniel, R. (2000). Biosynthesis of the anti-parasitic agent megalomicin: transformation of erythromycin to megalomicin in *Saccharopolyspora erythraea*. Mol. Microbiol. *37*, 752–762.

Waldron, C., Matsushima, P., Rosteck, Jr., P.R., Broughton, M.C., Turner, J., Madduri, K., Crawford, K.P., Merlo, D.J., and Baltz, R.H. (2001). Cloning and analysis of the spinosad biosynthetic gene cluster of *Saccharopolyspora spinosa*. Chem. Biol. *8*, 487–499.

Wang, Z.X., Li, S.M., and Heide, L. (2000). Identification of the coumermycin A_1 biosynthetic gene cluster of *Streptomyces rishiriensis* DSM 40489. Antimicrob. Agents Chemother. *44*, 3040–3048.

Wang, G., Hosaka, T., Ochi, K., (2008). Dramatic activation of antibiotic production in *Streptomyces coelicolor* by cumulative drug-resistance mutations. Appl. Environ. Microbiol. *74*, 2834–2840.

Ward, S.L., Hu, Z., Schirmer, A., Reid, R., Revill, W.P., Reeves, C.D., Petrakovsky, O.V., Dong, S.D., and Katz, L. (2004). Chalcomycin biosynthesis gene cluster from *Streptomyces bikiniensis*: novel features of an unusual ketolide produced through expression of the *chm* polyketide synthase in *Streptomyces fradiae*. Antimicrob. Agents Chemother. *48*, 4703–4712.

Wawrik, B., Kerkhof, L., Zylstra, G.J., and Kukor, J.J. (2005). Identification of unique type II polyketide synthase genes in soil. Appl. Environ. Microbiol. *71*, 2232–2238.

Weber, J.M., Schoner, B., and Losick, R. (1989). Identification of a gene required for the terminal step in erythromycin A biosynthesis in *Saccharopolyspora erythraea* (*Streptomyces erythreus*). Gene *75*, 235–241.

Weber, J.M., Leung, J.O., Maine, G.T., Potenz, R.H., Paulus, T.J., and DeWitt, J.P. (1990). Organization of a cluster of erythromycin genes in *Saccharopolyspora erythraea*. J. Bacteriol. *172*, 2372–2383.

Weissman, K.J., and Leadlay, P.F. (2005). Combinatorial biosynthesis of reduced polyketides. Nat. Rev. Microbiol. *3*, 925–936.

Weissman, K.J. (2006). The structural basis for docking in modular polyketide biosynthesis. Chembiochem *7*, 485–494.

Weissman, K.J. (2007). Mutasynthesis – uniting chemistry and genetics for drug discovery. Trends Biotechnol. *25*, 139–142.

Weist, S., and Süssmuth, R.D. (2005). Mutational biosynthesis-a tool for the generation of structural diversity in the biosynthesis of antibiotics. Appl. Microbiol. Biotechnol. *68*, 141–150.

Wendt-Pienkowski, E., Huang, Y., Zhang, J., Li, B., Jiang, H., Kwon, H., Hutchinson, C.R., and Shen, B. (2005). Cloning, sequencing, analysis, and heterologous expression of the fredericamycin biosynthetic gene cluster from *Streptomyces griseus*. J. Am. Chem. Soc. *127*, 16442–16452.

Westrich, L., Domann, S., Faust, B., Bedford, D., Hopwood, D.A., and Bechthold, A. (1999). Cloning and characterization of a gene cluster from *Streptomyces cyanogenus* S136 probably involved in landomycin biosynthesis. FEMS Microbiol. Lett. *170*, 381–387.

Williams, G.J., Goff, R.D., Zhang, C., and Thorson, J.S. (2008). Optimizing glycosyltransferase specificity via 'hot spot' saturation mutagenesis presents a catalyst for novobiocin glycorandomization. Chem. Biol. *15*, 393–401.

Wohlert, S.E., Wendt-Pienkowski, E., Bao, W., and Hutchinson, C.R. (2001). Production of aromatic minimal polyketides by the daunorubicin polyketide synthase genes reveals the incompatibility of the heterologous DpsY and JadI cyclases. J. Nat. Prod. *64*, 1077–1080.

Xu, Z., Jakobi, K., Welzel, K., and Hertweck, C. (2005). Biosynthesis of the antitumor agent chartreusin involves the oxidative rearrangement of an anthracyclic polyketide. Chem. Biol. *12*, 579–588.

Xu, Z., Schenk, A., and Hertweck, C. (2007). Molecular analysis of the benastatin biosynthetic pathway and genetic engineering of altered fatty acid-polyketide hybrids. J. Am. Chem. Soc. *129*, 6022–6030.

Xu, H., Minagawa, K., Bai, L., Deng, Z., and Mahmud, T. (2008). Catalytic analysis of the validamycin glycosyltransferase (ValG) and enzymatic production of 4"-epi-validamycin A. J. Nat. Prod. *71*, 1233–1236.

Xu, Z., Metsä-Ketelä, M., and Hertweck, C. (2009). Ketosynthase III as a gateway to engineering the biosynthesis of antitumoral benastatin derivatives. J. Biotechnol. *140*, 107–113.

Xue, Y., Zhao, L., Liu, H.W., and Sherman, D.H. (1998a). A gene cluster for macrolide antibiotic biosynthesis in *Streptomyces venezuelae*: architecture of metabolic diversity. Proc. Natl. Acad. Sci. U.S.A. *95*, 12111–12116.

Xue, Y., Wilson, D., Zhao, L., Liu, H., and Sherman, D.H. (1998b). Hydroxylation of macrolactones YC-17 and narbomycin is mediated by the *pikC*-encoded cytochrome P450 in *Streptomyces venezuelae*. Chem. Biol. *5*, 661–667.

Xue, Q., Ashley, G., Hutchinson, C.R., and Santi, D.V. (1999). A multiplasmid approach to preparing large libraries of polyketides. Proc. Natl. Acad. Sci. U.S.A. *96*, 11740–11745.

Yanai, K., and Murakami, T. (2004). The kanamycin biosynthetic gene cluster from *Streptomyces kanamyceticus*. J. Antibiot. *57*, 351–354.

Yanai, K., Murakami, T., and Bibb, M. (2006). Amplification of the entire kanamycin biosynthetic gene cluster during empirical strain improvement of *Streptomyces kanamyceticus*. Proc. Natl. Acad. Sci. U.S.A. *103*, 9661–9666.

Yin, X., and Zabriskie, T.M. (2006). The enduracidin biosynthetic gene cluster from *Streptomyces fungicidicus*. Microbiology *152*, 2969–2983.

Ylihonko, K., Tuikkanen, J., Jussila, S., Cong, L., and Mäntsälä, P. (1996). A gene cluster involved in nogalamycin biosynthesis from *Streptomyces nogalater*: sequence analysis and complementation of early-block mutations in the anthracycline pathway. Mol. Gen. Genet. *251*, 113–120.

Yu, T.W., Muller, R., Muller, M., Zhang, X., Draeger, G., Kim, C.G., Leistner, E., and Floss, H.G. (2001). Mutational analysis and reconstituted expression of the biosynthetic genes involved in the formation of 3-amino-5-hydroxybenzoic acid, the starter unit of rifamycin biosynthesis in *Amycolatopsis mediterranei* S699. J. Biol. Chem. *276*, 12546–11255.

Yu, T.W., Bai, L., Clade, D., Hoffmann, D., Toelzer, S., Trinh, K.Q., Xu, J., Moss, S.J., Leistner, E., and Floss, H.G. (2002). The biosynthetic gene cluster of the maytansinoid antitumor agent ansamitocin from *Actinosynnema pretiosum*. Proc. Natl. Acad. Sci. U.S.A. *99*, 7968–7973.

Yu, Y., Bai, L., Minagawa, K., Jian, X., Li, L., Li, J., Chen, S., Cao, E., Mahmud, T., Floss, H.G., *et al.* (2005). Gene cluster responsible for validamycin biosynthesis in *Streptomyces hygroscopicus* subsp. *jinggangensis* 5008. Appl. Environ. Microbiol. *71*, 5066–5076.

Zazopoulos, E., Huang, K., Staffa, A., Liu, W., Bachmann, B.O., Nonaka, K., Ahlert, J., Thorson, J.S., Shen, B., and Farnet, C.M. (2003). A genomics-guided approach for discovering and expressing cryptic metabolic pathways. Nat. Biotechnol. *21*, 187–190.

Zhang, H.Z., Schmidt, H., and Piepersberg, W. (1992). Molecular cloning and characterization of two lincomycin-resistance genes, *lmrA* and *lmrB*, from *Streptomyces lincolnensis* 78-11. Mol. Microbiol. *6*, 2147–2157.

Zhang, H., White-Phillip, J.A., Melançon, C.E., Kwon, H.J., Yu, W.L., and Liu, H.W. (2007). Elucidation of the kijanimicin gene cluster: insights into the biosynthesis of spirotetronate antibiotics and nitrosugars. J. Am. Chem. Soc. *129*, 14670–11483.

Zhang, X., Alemany, L.B., Fiedler, H.P., Goodfellow, M., and Parry, R.J. (2008a). Biosynthetic investigations of lactonamycin and lactonamycin Z: cloning of the biosynthetic gene clusters and discovery of an unusual starter unit. Antimicrob. Agents Chemother. *52*, 574–585.

Zhang, H., Wang, Y., and Pfeifer, B.A. (2008b). Bacterial hosts for natural product production. Mol. Pharm. *5*, 212–225.

Zhao, C., Ju, J., Christenson, S.D., Smith, W.C., Song, D., Zhou, X., Shen, B., and Deng, Z. (2006). Utilization of the methoxymalonyl-acyl carrier protein biosynthesis locus for cloning the oxazolomycin biosynthetic gene cluster from *Streptomyces albus* JA3453. J. Bacteriol. *188*, 4142–4147.

Zhao, Q., He, Q., Ding, W., Tang, M., Kang, Q., Yu, Y., Deng, W., Zhang, Q., Fang, J., Tang, G., *et al.* (2008). Characterization of the azinomycin B biosynthetic gene cluster revealing a different iterative type I polyketide synthase for naphthoate biosynthesis. Chem. Biol. *15*, 693–705.

Genomics of *Streptomyces* Cytochromes P450

12

David C. Lamb, Bin Zhao, F. Peter Guengerich, Steven L. Kelly and Michael R. Waterman

Abstract

In the present chapter, a review of the current status regarding the molecular biology and biochemistry of streptomycete cytochromes P450 (CYPs) will be presented. Emphasis will be placed upon CYPs with novel functions and identifying the roles of CYPs in the four sequenced and publicly available streptomycete genomes: *Streptomyces coelicolor, Streptomyces avermitilis, Streptomyces griseus,* and *Streptomyces scabies.* Additionally, the first detailed analysis of the CYP complement of a *Streptomyces* genome is being undertaken in the laboratories of the authors and colleagues. A summary of these findings will be presented. Finally, discussion on how new drugs might be generated from *Streptomyces* by modifying the structure of specific CYPs, as well as thoughts regarding actinomycete P450 research in the future, will be outlined.

Introduction

Cytochromes P450 (CYPs) are a superfamily of haem-thiolate containing enzymes, which catalyse the molecular scission of atmospheric oxygen (Ortiz de Montellano, 2005). Generally, one atom is incorporated into the substrate molecule and the other oxygen atom is reduced to a water molecule (see scheme I). For catalytic activity, CYPs have to be supplied with reducing equivalents from NADPH or NADH via channels from complementary redox partner proteins. Depending upon the microbe, CYP redox proteins are mainly divided into two classes (a) cytosolic ferredoxin (fd) and ferredoxin reductase (fpr) for prokaryotes and (b) endoplasmic reticulum associated NADPH-cytochrome P450 reductase (CPR) for eukaryotic microbes (Paine *et al.*, 2005).

Scheme I: $RH + O_2 + 2e^- + 2H^+ \rightarrow ROH + H_2O$

The reaction cycle of cytochrome P450 has been intensely studied and subjected to much experimentation. The monooxygenation reactions catalysed by P450s are diverse and include dehalogenation, C–C coupling, ring expansion, ring contraction, and others (Ortiz de Montellano, 2005; Fig. 12.1A). The reaction cycle is shown in Fig. 12.1B. In the resting state, the iron atom of the P450 haem is the ferric (III) low-spin state in which the iron is six coordinated by four nitrogen atoms in the haem pyrrole ring, the sulphur atom from cysteine and a water molecule in the distal position. The linkage of the iron to the cysteine thiolate gives rise to the characteristic Soret absorption at 450 nm when the ferrous form of the enzyme is complexed with carbon monoxide (Omura and Sato, 1964). Upon substrate binding, the water molecule is often expelled as the axial ligand, at least in some cases, to produce a five coordinated ferric high spin species that may have a concomitant increase of the redox potential. Next, an electron is transferred by an ancillary redox protein partner to the ferric haem generating the ferrous 5-coordinated haem. Molecular dioxygen binds to the ferrous enzyme resulting in the formation of a haem iron producing the ferric-dioxo species, which is subsequently protonated to produce the ferric peroxide complex termed Compound 0. A second solvent derived

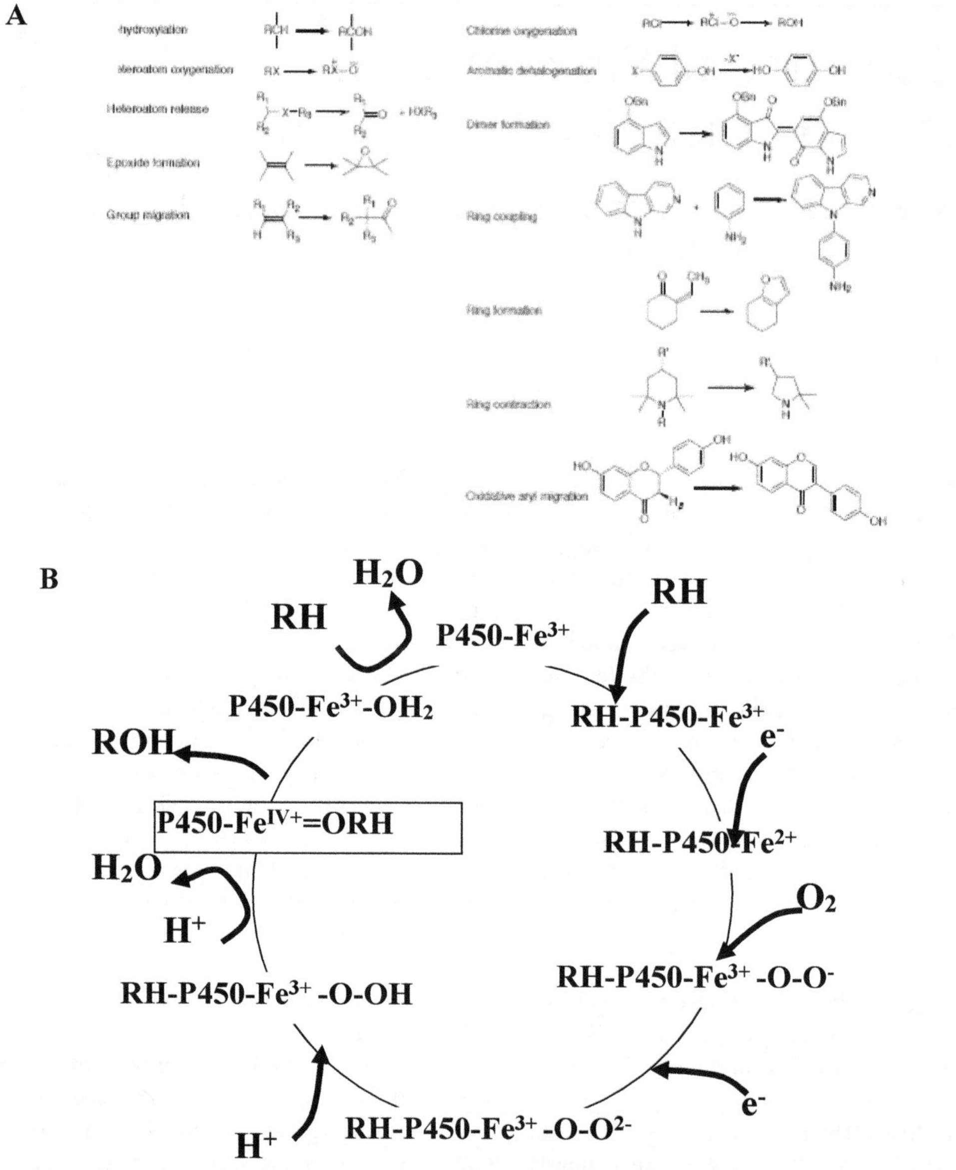

Figure 12.1 An overview of enzymatic reactions catalysed by cytochromes P450 and the P450 catalytic reaction cycle. (A) Examples of chemical reactions catalysed by CYP enzymes. Many are formal oxidations, but reductions and rearrangements are also known. (B) Schematic representation of the P450 catalytic cycle depicting oxygen activation, substrate hydroxylation and water production. For simplicity, the haem centre is represented as $P450\text{-}Fe^{3+}$ with the cysteine thiolate ligand omitted. Compound I is shown boxed.

proton leads to heterolytic cleavage of the O–O bond with loss of water generating the iron (IV) oxo-porphyrin π-radical (termed Compound I). Finally, Compound I extracts a proton from the substrate to form a ferryl-hydroxo species (Compound II) and a substrate-centred radical, followed by rebound of oxygen from the ferryl-hydroxo intermediate, generating the ferric haem and product (Schlichting *et al.*, 2000).

Given the huge diversity of P450 sequences generated particularly from microbial origin, a nomenclature system has been devised to categorize individual CYPs (Nelson *et al.*, 1996). This system is based upon sequence identity where P450s are grouped into families (1, 2, 3...), subfamilies (A, B, C...) and a number for each individual P450 (1, 2, 3, 4...), e.g. CYP105A1, CYP105B1, CYP105D1, CYP105D5. This nomenclature system is maintained by Dr. David Nelson, University of Tennessee, USA (http://drnelson.utmem.edu/CytochromeP450.html). It is the wide-spread belief that the ancestral CYP appeared some three billion years ago where it was thought to be involved in the detoxification of reactive oxygen species generated in the highly volatile atmosphere of the burgeoning Earth (Wickramasinghe and Ville, 1975). This event originally occurred within the domain of the microbes which where the first organisms to colonize the planet. Currently there are > 10,000 *CYP* gene sequences deposited in public genomic databases, and of these sequences > 30% are of microbial origin.

An overview of cytochromes P450 and actinomycetes

One of the surprising discoveries about the genomics of the CYP superfamily is the large number of *CYPs* in the bacterial class of actinomycetes. It had previously been imagined that bacteria have small numbers of *CYPs* or none at all. Most actinobacteria, particularly streptomycetes and mycobacteria, encode unusually large numbers of *CYP* genes (*Streptomyces coelicolor* A3(2), 18; *Streptomyces avermitilis*, 33; *Streptomyces scabies* 25; *Streptomyces peucetius*, 19; *Mycobacterium tuberculosis* 20; *Mycobacterium smegmatis*, 33; *Mycobacterium bovis*, 18; *Mycobacterium vanbaalenii*, 51; *Mycobacterium leprae*, 1 and *Nocardia farcinica*, 26). Additionally, the actinomycete *Frankia* spp., which are of importance in nitrogen fixation in a symbiotic process with certain non-leguminous plant, have many *CYP* genes. Of the three sequenced *Frankia* genomes, the frequency of *CYP* genes could be related exponentially to the size of each genome with the largest genome, 9.04 Mb, having 47 *CYP* genes in the *Frankia* sp. strain EAN1pec, the next largest, 7.5 Mb, having 22 in the *Frankia alni* strain ACN14a and the smallest, 5.4 Mb, having 13 in the *Frankia* sp. strain HFPCcI3 (Ventura *et al.*, 2007). Interestingly, scientific attention has now focused on the microbial populations of the Earth's seas and oceans. Many actinomycete strains have now been identified from these environments with a few species now having their complete genomes sequenced. These strains also have a preponderance of *CYP* genes. The first such ocean dwelling actinomycete to have its genome sequenced was that of *Salinispora tropica*, which has 19 annotated *CYP* genes with 11 linked to secondary metabolite producing gene clusters (Udwary *et al.*, 2007). However, some actinobacteria genomes possess no *CYP* genes including *Tropheryma whipplei* and *Corynebacterium diphtheriae* (http://sanger.ac.uk).

Interestingly, there is little overlap in the CYP contents of actinomycetes, with each actinobacterial family and species therein seemingly having its own unique complement of *CYP* genes. For example, the most conserved CYP family across different actinobacteria is CYP125, which is present only in some members of each genus and then in only certain strains (i.e. *S. avermitilis*, *N. farcinica*, *M. tuberculosis*, *M. smegmatis*, *S. tropica*) and in each instance is linked to one or more genes predicted to be involved in lipid degradation suggesting a role for this CYP in lipid metabolism. Many of the actinobacterial P450 genes are clustered with genes encoding ferredoxins (ancillary redox proteins required to provide reducing equivalent to the CYP for catalysis). Further, a substantial number of actinobacterial CYPs are associated with gene clusters for secondary metabolites, including many pathways which encode compounds which have been applied in human medicine (Table 12.1). Alternatively, many of the mycobacterial CYPs have been connected to the biosynthesis of the complex lipids that these organisms produce (Ouellet *et al.*, 2009).

Considering the actinomycete genomes sequenced so far, many have relevance to human and veterinary medicine, biotechnology, and ecology. This chapter will give an account of the distribution of cytochrome P450 enzymes in streptomycetes, their potential roles in secondary metabolism biosynthetic pathways and finally give an overview regarding some of the novel features of many of these P450s, particularly in

Table 12.1 Cytochromes P450 involved in streptomycete secondary metabolite biosynthetic gene clusters

Species	Secondary metabolite	CYP(s) involved	Therapeutic use
S. noursei	Nystatin	CYP161A1, CYP105H1	Antifungal
S. nodosus	Amphotericin	CYP161A3, CYP105H4	Antifungal
S. natalensis	Pimaricin	CYP164A2, CYP105H3	Antifungal
S. narbonensis	Marbomycin	CYP107L7P	Antifungal
S. venezuelae	Pikromycin	CYP107L1	Antibacterial
S. spheroides	Novobiocin	CYP163A1	Antibacterial
S. clavuligerus	Clavulanic acid	CYP105M1	Antibacterial
S. caelestis	Niddamycin	CYP113B2	Antibacterial
S. thermotolerans	Carbomycin	CYP107C1	Antibacterial
S. roseochromogenes	Chlorobiocin	CYP163A2	Antibacterial
S. hygroscopius	Rapamycin	CYP122A2, CYP107G1	Antibacterial
	Geldanamycin	CYP105U1	Antibacterial
S. fradiae	Tylosin	CYP113B1, CYP105L1, CYP154B1	Antibacterial
S. antibioticus	Oleandomycin	CYP107D1, CYP235A1	Antibacterial
	Simocyclinone	CYP163A3	Antifungal
S. avermitilis	Filipin	CYP105P1, CYP105D6	antifungal
	Avermectin	CYP171A1	Antiparasitic
S. peucetius	Doxorubicin	CYP131A1, CYP129A2	Antitumour
S. globisporus	C-1207	CYP211A1	Antitumour
S. sp strain C5	Daunorubicin	CYP131A2, CYP129A1	Antitumour
S. carzinostaticus	Neocarzinostatin	CYP154J1, CYP208A2	Antitumour
S. lavendulae	Mitomycin C	CYP107N1, CYP160A1, CYP105F1	Anti-HIV
	Complestatin	CYP165B5, CYP165E1	Cholesterol lowering
S. carbophilus	Pravastatin	CYP105A3	Cholesterol lowering
S. nanchangensis	Nanchangmycin	CYP124B2	Anticoccidial
	Meilingmycin	CYP171A2	Anticoccidial
S. tendae	Nikkomycin	CYP162A1, CYP105K1	Insecticidal
S. cinnamonesis	Monensin	CYP124B1	Anticoccidial
S. griseolus	7-Ethoxycoumarin	CYP105A1, CYP105B1	Antimicrobial

S. coelicolor A(3)2, were many important biochemical facets of the P450 superfamily have been discovered whilst characterising the 18 P450s of this prototypic actinomycete.

Application of *Streptomyces* spp. CYPs

Streptomycetes catalyse numerous transformations of xenobiotics of industrial and environmental importance. These oxidative transformations have been observed with coumarins (Sariaslani *et al.*, 1984), alkaloids (Sariaslani and Rosazza, 1983), and other complex xenobiotics (Taylor *et al.*, 1999). Some of the most significant of these biocatalytic reactions are aromatic and aliphatic hydroxylations, O- and N-dealkylations, N-oxidation, and C–C fission. Available experimental evidence has indicated that all of these oxidative reactions are catalysed by the related enzyme system known as cytochromes P450. Industrial application has been exploited in the synthesis of pravastatin, utilizing a streptomycete biotransformation step (Serizawa and Matsuoka, 1991) as well as in 16-hydroxylation of steroids (Berrie *et al.*, 1999) and in the preparation of drug metabolites for toxicological evaluation (Cannell *et al.*, 1997).

Biological roles of CYPs in *Streptomycetes*

The genus *Streptomyces* produces approximately two-thirds of naturally occurring antibiotics and a wide array of other biologically active molecules. These include numerous antibacterial agents (e.g. erythromycin, tetracycline), anticancer agents (doxorubicin, enediyenes), immunosuppressants (FK506, rapamycin), antiparasitic agents (avermectin, nemadectin), antifungals (amphotericin, griseofulvin, nystatin), cardiovascular agents (lovastatin, compactin) and veterinary products (monensin, tylosin) (for examples, see Fig. 12.2). Such compounds are often referred to as secondary metabolites, natural products such as polyketides containing multiple carbonyl and/or hydroxyl groups each separated by a methylene carbon. Although the final structures of the polyketides are diverse, these molecules share common features in their biosynthetic pathways, and in intermediates formed within these pathways. The impetus for the discovery of new polyketides and enhancing the activity of existing drugs has gathered momentum since infectious diseases continue to be a growing and serious health problem. The need is due to a number of reasons including the intrinsic power of infectious microbes to mutate and become drug resistant, clinical overuse of antibiotics, the changing diversity of the pathogens, immunocompromising treatments and immunodeficiency diseases.

An understanding of the molecular basis for the biosynthesis of many complex polyketide antibiotics has only been gained over the past decade and has provided an explanation for the basis of the extensive structural diversity seen amongst this class of naturally occurring molecules (Khosla and Harbury, 2001). Antibiotic diversity is generated during their biosynthesis by variations in (a) use of different starter and extender units, (b) degrees of reduction, (c) polyketide length, and (d) post-polyketide modification (e.g. hydroxylations, methylations).

Within *Streptomyces* spp., it is accepted that CYPs are involved in many polyketide biosynthetic pathways and catalyse the stereo- and regio-specific oxidative tailoring of the antibiotics (Table 12.1) e.g. as classically described for macrolides, such as in the biosynthesis of erythromycin, tylosin, mycinamicin, oleadomycin, and methymycin (Gaisser *et al.*, 2002). Such CYP reactions occur in the late stages of biosynthesis after formation of the macrolactone ring by the polyketide synthase. The biological importance of the resulting hydroxyl and/or epoxide substituents to the biological activity of the compounds is often represented by a significant increase in antibiotic potency. Although *CYP* genes are common in macrolide gene clusters, there are considerable variations in the substrates utilized and the regio- and stereochemistry of oxidation catalysed by these enzymes. This utility should come as no surprise given the known effect of even a single amino acid change in determining the position of steroid hydroxylation with mammalian CYP forms (Richardson and Johnson, 1994). The majority of the antibiotics produced by *Streptomyces* spp. are encoded by clustered genes located contiguously on the chromosome of the individual *Streptomyces* sp. CYPs have been described in many different species of *Streptomyces*, and in the majority of these, the individual CYP described is directly involved in the biosynthesis of an antibiotic produced by that specific strain. Examples of *Streptomyces* secondary metabolites having different biomedical functions and the CYP(s) involved in their biosynthesis are presented (Table 12.1).

The role of P450s in antibiotic gene clusters from *Streptomycetes*

In *Streptomyces* spp., P450s are primarily involved in the oxidative metabolism of polyketides, terpenes, and other secondary metabolites. Catalysis is often strictly stereo- and regio-specific as classically described for macrolide class of antibiotics, such as in the biosynthesis of pikromycin, tylosin, and oleadomycin. Such CYP reactions usually occur in the later stages of biosynthesis, after

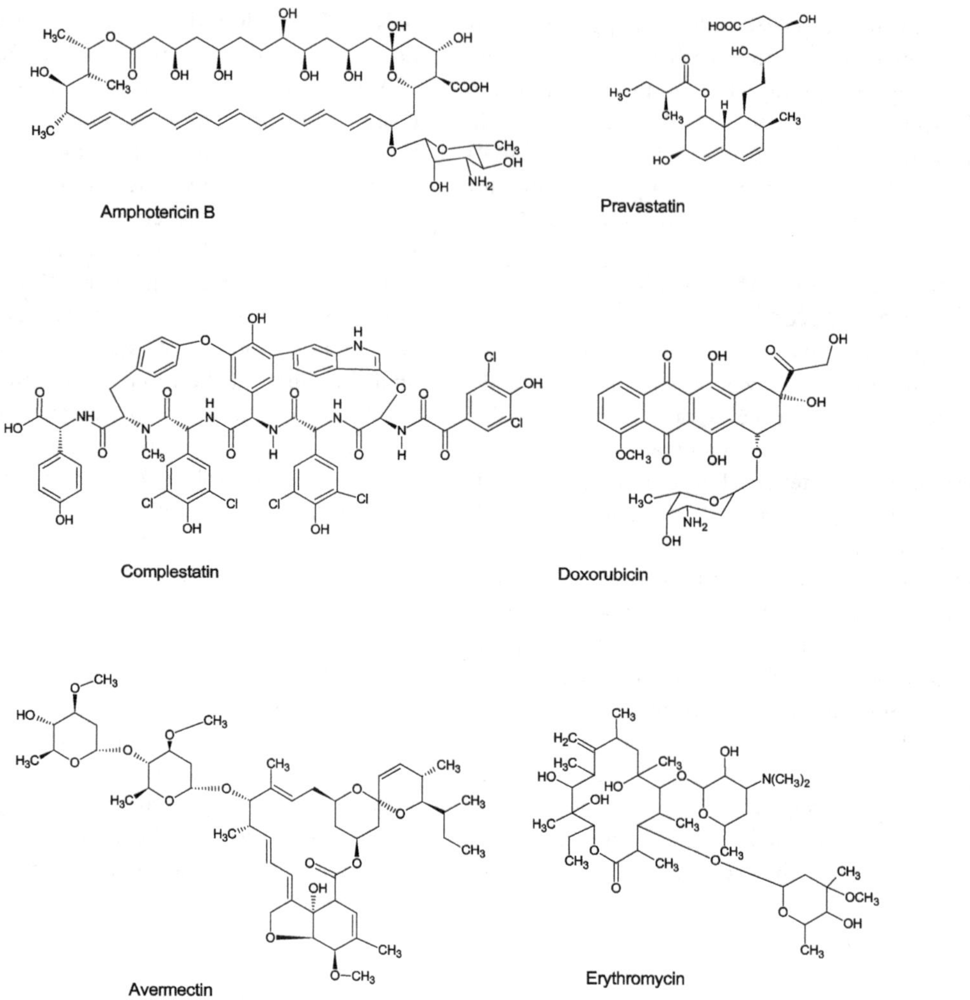

Figure 12.2 Examples of medically important *Streptomyces* secondary metabolites. The biosynthesis of each molecule is encoded by individual gene clusters within a specific streptomycete and each cluster contains one or more *CYP* gene(s).

formation of the secondary metabolite carbon skeleton by the polyketide synthase (PKS), non-ribosomal peptide synthetase (NRPS), and/or PKS–NRPS hybrid. However, examples contrary to this rule have been found recently and are described below. The majority of the antibiotics produced by *Streptomyces* are encoded by genes clustered contiguously on the chromosome of the individual Streptomyces strain. P450s have been described in numerous streptomycetes because they are directly involved in the biosynthesis of the antibiotic produced by that specific species (Table 12.2). The following sections will provide selected P450s of particular interest discovered from such pathways.

Novel P450 catalytic activities associated with antibiotic biosynthetic pathways

Following sequencing and gene identification of a streptomycete antibiotic-producing gene cluster, the task to unravel the biochemistry of individual steps takes prominence. Two approaches have been favoured to elucidate biochemical function; disruption of an individual gene and assessment of metabolite phenotype and/or isolation

Table 12.2 P450-containing secondary metabolite biosynthetic gene clusters revealed through genomic sequencing of streptomycetes

Streptomycete	P450-containing cluster and type	Product
S. coelicolor A(3)2	SCO1207 (CYP158A2)	HPQ melanin
	SCO0495 (CYP105N1)	Coelibactin
	SCO5223 (CYP170A1)	Albaflavenone
S. avermitilis	SAV412 (CYP105D6, CYP105P1)	Filipin
	SAV941 (CYP171A1)	Avermectin
	SAV2999 (CYP183A1)	Pentalenelactone
	SAV2894 (CYP107W1)	Oligomycin
	SAV3031 (CYP170A1)	Albaflavenone?
	SAV838 (CYP178A1)	Unknown
	SAV2377 (CYP107Y1, CYP181A1)	Unknown
	SAV7130 (CYP158A3)	HPQ melanin
	SAV7186 (CYP105R1)	Unknown
	SAV1987 (CYP107L2)	Unknown
S. griseus	SGR6619-SGR6620	HPQ melanin
	SGR574-SGR593, (NRPS)	Unknown
	SGR895-SGR901, (NRPS)	Unknown
	SGR604-SGR611	enediyne
	SGR810-SGR815, (PKS–NRPS) hybrid	Unknown
	SGR3239-SGR3288, type II PKS, NRPS	Unknown
	SGR527-SGR528	Melanin

and purification of enzyme components and *in vitro* reconstitution of enzymatic activity with previously isolated substrates to determine the chemistry of generated products. Such approaches with streptomycete P450s have led to exciting discoveries regarding their enzymology.

The staurosporine biosynthetic gene cluster in *Streptomyces* sp. TP-A0274 consists of 15 genes. Staurosporine, a protein kinase C inhibitor, consists of an indolocarbazole structure, aglycone linked to the sugar molecule deoxyaminosugar (Onaka *et al.*, 2002). These two molecules are linked together through a glycosidic bond at the anomeric position and an unusual glycosidic-like C–N bond between the C5 of the sugar moiety and the nitrogen of aglycone. Through gene disruption experiments it was shown that a cytochrome P450 enzyme (CYP245B1) was responsible for the formation of this unusual glycosidic-like bond (Fig. 12.3A). It is proposed that CYP245B1 catalyses hydroxylation at the 5′ position of the sugar (Onaka *et al.*, 2005). This hydroxyl group is subsequently protonated and then dehydrated to form an oxonium cation with the final step being nucleophilic addition of nitrogen at the N13 position to the oxonium ion forming the C–N bond. This is the first example of a P450 catalysing such a biochemical reaction.

Another example of the power of genetics to attribute a endogenous function to a P450 was described in the elucidation of griseorhodin A biosynthesis (Yunt *et al.*, 2009). Griseorhodin A belongs to the rubromycin class of antibiotics and has application as an anti HIV drug through inhibition of HIV reverse transcriptase. Griseorhodin A is made in *Streptomyces* sp. JP95 and is encoded by a 14-gene cluster. Specific disruption was made for each downstream gene and the chemical structure of the products produced from the expressed, mutant gene cluster examined. A cytochrome P450 gene encoded in this cluster was shown to catalyse an epoxidation

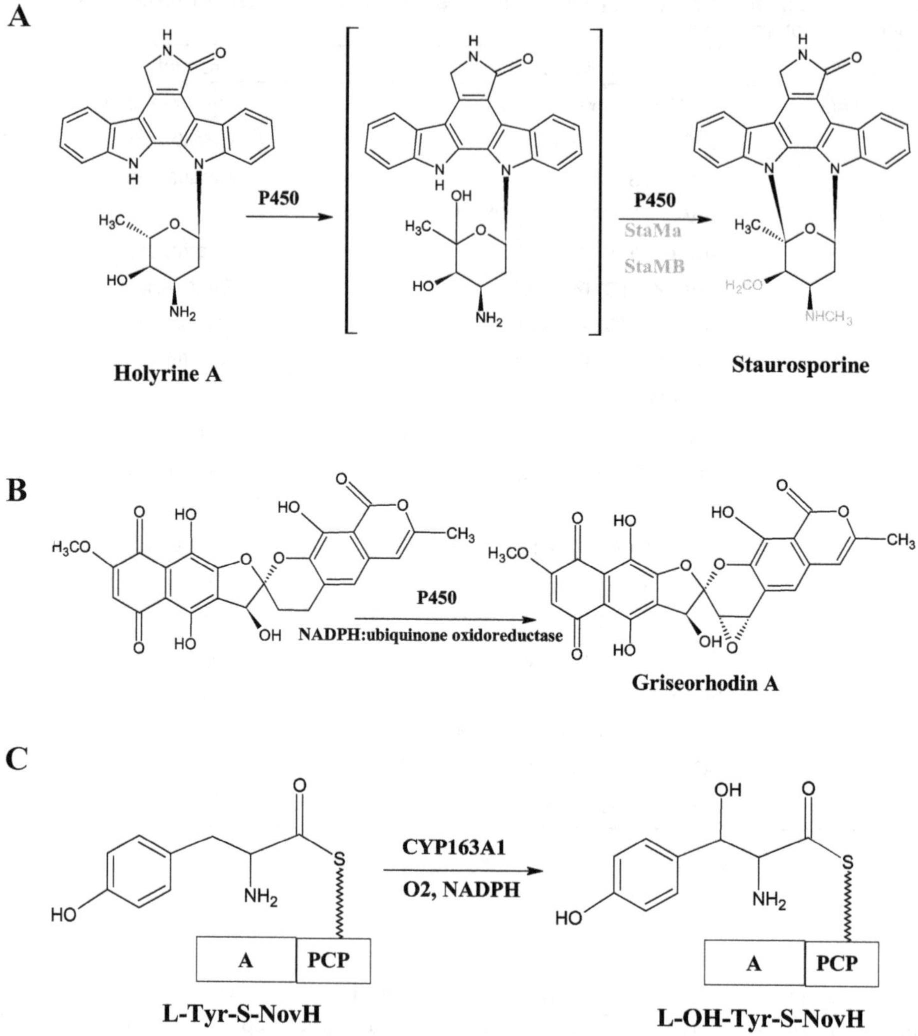

Figure 12.3 Examples of novel P450-catalysed reactions in *streptomycetes*. (A) The biosynthesis of staurosporine involves a P450 catalysing the formation of a novel C–N bond. (B) Introduction of an epoxide group to produce griseorhodin A requires a unique P450:NADPH ubiquinone oxidoreductase complex for catalysis. (C) Presentation of the substrate for CYP163A1 catalysis requires covalent tethering to a peptidyl carrier protein.

reaction, utilizing a saturated substrate to produce the final molecule (Yunt *et al.*, 2009; Fig. 12.3B). Interestingly, the P450 reaction was entirely dependent on a second gene product encoded within the cluster, NADPH:ubiquinone oxidoreductase. If the NADPH:ubiquinone oxidoreductase was omitted from the reaction then epoxidation did not occur suggesting that the two enzymes form a complex that is inactive if one component is removed (Yunt *et al.*, 2009). This is highly unusual for a P450 to be dependent on another protein for catalysis, other than a redox partner.

The antibacterial agent novobiocin belongs to the coumarin group of antibiotics which exerts its antibacterial activity by inhibition of the type II DNA topoisomerase DNA gyrase. The novobiocin gene cluster, isolated from *S.*

spheroides, revealed the presence of a single P450 gene (*CYP163A1*) which was predicted to catalyse the regio- and stereo-specific of L-tyrosine. Interestingly, CYP163A1 did not recognize free tyrosine (the starting material for the biosynthesis of novobiocin) or the soluble thioester surrogate L-tyrosyl-S-*N*-acetyl cysteamine. CYP163A1 only recognized the L-tyrosyl moiety presented on the pantetheinyl arm of a peptidyl carrier protein, which is also encoded in the gene cluster (Chen and Walsh, 2001; Fig. 12.3C). This is the first example of a substrate being presented to a P450 enzyme by another protein.

Amphotericin B is the last resort in the treatment of life-threatening systemic fungal infections but has also been shown to be effective against human immunodeficiency virus (HIV), *Leishmania* parasites, and prion diseases. However, the therapeutic use of this compound has been largely limited because of the toxic side effects associated with it. Carmody *et al.* have shown that by using targeted deletions of the *amphN* CYP in the amphotericin biosynthetic gene cluster of *S. nodusus* one can produce mutant strains producing chemically altered amphotericin molecules in which a methyl group replaces the exocyclic carboxyl group. This compound retained antifungal activity but had reduced toxic side effects, including reduced haemolytic activity (Carmody *et al.*, 2005).

Cytochromes P450 from *S. coelicolor* provide new insights into the superfamily

The completion of the genome sequence of *S. coelicolor* revealed the presence of 18 *CYP* genes (Bentley *et al.*, 2002; Lamb *et al.*, 2002). However, the function for all 18 CYPs remained unknown at that time. Consequently, in 2002 a project was initiated in the laboratories of Waterman, Guengerich, Kelly, and Lamb to determine the function of each CYP using genetic, biochemical and biophysical approaches and to gain an insight into their contribution to the life cycle of this organism. To date from this work many surprising and novel findings concerning P450 structure/function have been uncovered (Fig. 12.4) and summarized as follows.

Conserved P450 residues are reduced to a single cysteine

As sequencing projects advanced it was possible to examine and compare each P450 sequence and identify residues which could be considered as being essential to all P450 enzymes. It was universally thought the three residues were essential in all P450 sequences: the conserved cysteine which is the fifth ligand to the haem and the EXXR forming a charge pair in the K helix and which is involved in haem binding and overall P450 fold topology (Kalb and Loper, 1988; Ravichandran *et al.*, 1993). However, the genome of *S. coelicolor* revealed for the first time three CYPs—CYP156B1, CYP157A1 and CYP157C1—which did not contain this EXXR motif and thus challenged this dogma.

Examination of CYP157C1 revealed the presence of EQSLW in place of EXXR; however, the ferrous–CO complex gave a Soret maximum at 448 nm, typical of a 'normal' P450 (Rupasinghe *et al.*, 2006). Site directed mutagenesis to create mutant forms of CYP157C1 containing the EXXR motif were undertaken. However, all mutants did not result in P450 being produced, only the incorrectly folded P420 form (Rupasinghe *et al.*, 2006). These experiments irrevocably showed that EXXR motif is not required in all P450s for structural architecture and hence only a single residue is absolutely conserved in all P450 sequences, the cysteine which coordinates with the haem iron. x-ray structure analysis of a P450-lacking EXXR will provide new insights into P450 folding and architecture.

Insights into P450 haem incorporation are revealed

Until 2003, all P450 structures whose x-ray crystal structure had been resolved showed that following haem incorporation into the P450 molecule the haem adopted one specific orientation based upon the positions of the two protohaem vinyl groups. However, the crystal structure of CYP121A1 from *M. tuberculosis* showed that the haem group could assume two distinct orientations, the normal orientation and a small fraction in the opposite orientation which is inverted 180° (Leys *et al.*, 2003). CYP121A1 catalyses the formation of an intramolecular C–C bond between

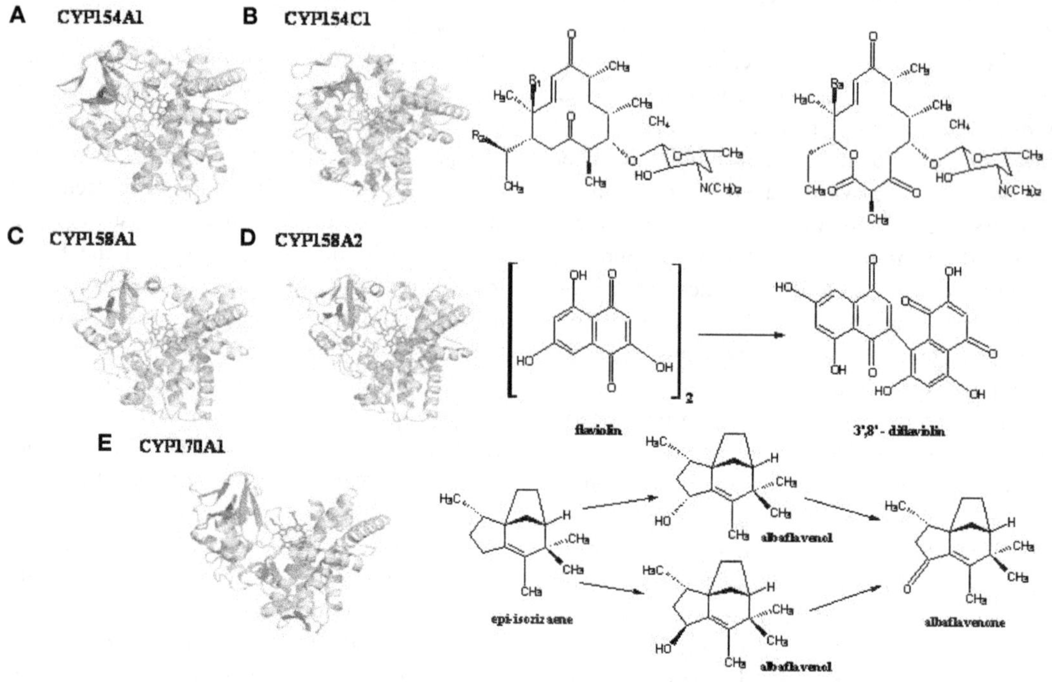

Figure 12.4 *S. coelicolor* cytochrome P450 crystal structures and their determined catalytic functions. (A) Structure of CYP154A1, an enzyme which shares 56% identity with CYP154C1 but does not share its catalytic function. (B) Structure of CYP154C1, an enzyme that can catalyse the final step in the biosynthesis of the antibacterial agents pikromycin, methymycin, and neomethymycin. (C) Structure of CYP158A1, an enzyme which catalyses the polymerization of flaviolin to dimers and a trimer. (D) Structure of CYP158A2, an enzyme which also catalyses the polymerization of flaviolin but with differing specificity compared with CYP158A1. (E) Structure of CYP170A1, an enzyme which catalyses two sequential allylic oxidations to convert epi-isozizaene to the antibacterial albaflavenone.

two tyrosyl carbon atoms of cyclodipeptide cyclo (*L*-Tyr-*L*-Tyr) (cYY) (Belin *et al.*, 2009). However, it is not known whether a particular orientation of the prosthetic haem is favoured during catalysis.

The crystal structure of *S. coelicolor* CYP154A1 revealed that the haem orientation is completely opposite to that of all reported CYP structures (Podust *et al.*, 2004; Fig. 12.4A). This is not a characteristic of the CYP154 family because the resolved structure of the other another family member, CYP154C1 (Podust *et al.*, 2003), contained all of its haem in the normal P450 orientation (Fig. 12.4B). The endogenous function of both CYP154A1 and CYP154C1 is unknown. However, using the substrates of *S. venezuleae* CYP107C1, a P450 involved in the biosynthesis of the antibacterial agents pikromycin, methymycin, and neomethymycin, CYP154C1 was shown to be active with both substrates, producing the final molecules (Podust *et al.*, 2003; Fig. 12.4B). Conversely CYP154A1 was shown to be inactive when these compounds were used as substrates. Thus it may be possible that the haem orientation may influence these enzymatic activities. It will be interesting to study the amino acids involved in haem incorporation in both P450s in order to gain further insight into P450 folding overall.

Elucidation of the function of two CYPs involved in C–C coupling

C–C coupling is a critical step in the biosynthesis of many important antibiotics. Vancomycin, the last line treatment for methicillin-resistant *Staphylococcus aureus* (MRSA), requires three such steps catalysed by P450s for its biosynthesis (Zerbe *et al.*, 2002). Further, both chloroeremomycin and balhimycin, antibiotics of the vancomycin family, contain P450-dependent biphenyl and biphenyl ether cross-links (Süssmuth *et al.*, 1999). *S. coelicolor* A3(2) contains a 3 gene operon containing a type III polyketide synthase, CYP158A2, and a quinione-forming monooxygenase (momA) (Cortes *et al.*, 2002). The PKSIII sequentially converts 5 molecules of malonyl CoA into 1,3,6,8-tetahydroxynapthalene which is oxidized to flaviolin by momA (Funa *et al.*, 2005). CYP158A2 subsequently catalyses C–C bond formation to polymerize flaviolin into the di- and trimer (Fig. 12.4C, Zhao *et al.*, 2005). The crystal structure of CYP158A2 has revealed important features of P450 coupling enzymes (Fig. 12.4C). CYP158A2 binds two molecules of flaviolin in its active site and the crystal structure predicts that the 2-OH group of flaviolin plays a role in anchoring the substrate in the active site while the 5-OH and 7-OH stabilize water molecules important for catalysis. The use of a substrate analogue, α-hydroxynapthalene (which is missing both the 5-OH and 7-OH), revealed 70-fold less activity in dimerization, thus suggesting the water molecules form a proton relay pathway to the bulk solvent.

The *S. coelicolor* genome contains a second *CYP158A* gene, *CYP158A1*, which shares 61% amino acid sequence identity with CYP158A2. It was found that CYP158A1 could catalyse the generation of the same dimers as CYP158A2 but with differing regio-specificity (Zhao *et al.*, 2007). The crystal structure of the flaviolin bound CYP158A1 complex (Fig. 12.4D) was different compared with CYP158A2 with the two molecules of flaviolin adopting different orientations. One molecule is positioned over the haem similar to CYP158A2 but the second molecule is bound at the entrance to the enzyme active site thus presenting difficulties in understanding how CYP158A1 catalyses dimerization compared with CYP158A2. It has been proposed that conformational changes in CYP158A1, possibly through binding ferredoxin during the electron transfer process, allows for the second flaviolin to move closer to the first in order to allow catalysis to occur.

Elucidation of the albaflavenone monooxygenase and associated moonlighting terpene active site

The *sco5223* gene encoding CYP170A1 in *S. coelicolor* A3(2) is part of a two-gene cluster with a sesquiterpene cyclase gene (*sco5222*) with which it shares a 4-nt ATGA transcriptional overlap at its 5′ end. The *sco5222* gene product has been shown to catalyse the synthesis of a novel sesquiterpene previously unseen in nature, epi-isozizaene, by cyclization of the universal sesquiterpene synthase substrate, farnesyl diphosphate (Lin and Cane, 2009; Fig. 12.4E). Using an *in vitro* assay it was established that purified CYP170A1 carries out two sequential allylic oxidations to convert epi-isozizaene to an epimeric mixture of albaflavenols and ultimately to the single ketone sesquiterpene, the antibiotic albaflavenone (Zhao *et al.*, 2008). Both epimers of albaflavenol as well as albaflavenone were produced by the wild-type strain, while none of these three metabolites could be detected in the extracts of the CYP170A1 knock-out stain. The *in vivo* observations verified the results of the *in vitro* experiments confirming the deduced biochemical role of CYP170A1 in the two-step oxidative conversion of epi-isozizaene to the sesquiterpene antibiotic albaflavenone. Although there are numerous examples of three-step oxidations of methyl groups to carboxylic acids that are catalysed by a single P450, the two-step oxidation of an allylic methylene to a conjugated ketone by a single P450 is highly unusual. Notably, in the biosynthesis of oxidized monoterpenes P450-catalysed allylic oxidation is normally followed by a conventional NAD(P)$^+$-dependent dehydrogenation, as for example in the conversion of (–)-limonene to (–)-isopiperitenone by way of (–)-*trans*-isopiperitenol in peppermint catalysed by successive P450 and dehydrogenase enzymes (Karp *et al.*, 1990).

During initial studies of the monooxygenase activity of CYP170A1 an unexpected additional activity was observed. Surprisingly, GC/MS analysis of the products of the CYP170A1 incubation mixture revealed the formation not only of both epimers of albaflavenol and the fully oxidized sesquiterpene antibiotic albaflavenone (Zhao *et al.*, 2008), but also the formation of (*E*)-β-farnesene (61%), (3*E*, 6*E*)-α-farnesene

(26%), (3*Z*, 6*E*)-α-farnesene (6.8%), nerolidol (4.9%), and farnesol (1.8%). The crystal structure of CYP170A1 revealed the presence of a novel terpene synthase active site which is moonlighting on the P450 structure (Zhao *et al.*, 2009; Fig. 12.4e). This includes signature sequences for divalent cation binding and an α-helical barrel. This barrel is unusual because it consists of only four helices rather than six found in all other terpene synthases. Mutagenesis established that this barrel is essential for the terpene synthase activity of CYP170A1 but not for the monooxygenase activity (Zhao *et al.*, 2009).

Elucidating the function of other P450s uncovered in genome sequenced *Streptomycetes*

Cytochromes P450 in S. avermitilis

The second streptomycete to have its genome sequenced was that of *S. avermitilis*, an industrial microorganism which produces the antiparasitic agent avermectin (Ikeda *et al.*, 2003). This potent antihelmintic compound is used commercially in animal healthcare and agriculture and is used for the control of human onchocerciasis. The fully sequenced genome of *S. avermitilis* revealed approximately 7600 ORFs with 33 putative CYP genes within the 9 Mb chromosome sequence (Ikeda *et al.*, 2003; Lamb *et al.*, 2003). Ten *S. avermitilis* P450s (CYP105D6, CYP105P1, CYP171A1, CYP178A1, CYP107Y1, CYP181A1, CYP107W1, CYP183A1, CYP158A3, and CYP105R1) are predicted to be involved in secondary metabolite production as determined by their genetic location in gene clusters known or postulated to be involved in secondary metabolite production. CYP105D6 and CYP105P1 (previous names PteC and PteD, respectively) are located in the filipin gene cluster of *S. avermitilis.* CYP171A1 plays a role in the biosynthesis of avermectin in *S. avermitilis*, being involved in formation of the furan ring at C6 to C8a of the avermectin algycone. Specifically, CYP171A1 catalyses two rounds of oxygen attack at the C8a position of the 6, 8a-seco-6, 8a-deoxyavermectin algycone (Ikeda *et al.*, 1999). CYP107W1 (previous name OlmB) catalyses the conversion of the macrolide oligomycin C (12-deoxyoligomycin A) to oligomycin A through a hydroxylation step at C12. *CYP178A1* is associated with a gene cluster containing the non ribosomal peptide synthetase Nrps7; *CYP107Y1* and *CYP181A1* are associated with a gene cluster – pks9, containing type II polyketide synthase; *CYP154D1* is adjacent to a gene encoding indole dioxygenase; *CYP107L2* is adjacent to a gene encoding for a phenylacetate degradation protein; *CYP183A1* lies in a gene cluster responsible for the biosynthesis of pentalenolactone; *CYP158A3*, is adjacent to a type III polyketide synthase; and *CYP105R1* lies in a polyketide synthase (pks4) gene cluster and is possibly involved in the formation of simple polyketides.

Only recently has biochemical analysis been applied to determine the function of the orphan *S. avermitilis* P450s, as described in the following examples:

The crystal structure of S. avermitilis CYP105P1 suggests a mechanism for substrate binding in macrolide metabolising P450s

Genetic experiments have shown that CYP105P1 and CYP105D6 catalyse hydroxylation at the C26 and C1′ positions of filipin respectively. The recently resolved crystal structure of CYP1051 revealed a novel feature which may contribute to this enzyme accepting a large hydrophobic molecule as a substrate (Xu *et al.*, 2009; Fig. 12.5A). The side chain of His72 in the BC loop ligates to the haem iron in the crystalline form. Further, the BC loop is partially disordered. Such a finding strongly suggests that the BC loop of CYP105P1 is highly flexible in solution which agrees with the hypothesis that to accept large macrolide substrates into their active site, the entrance to the active site should have conformational flexibility. Examination of the crystal structure of CYP107L1 (PicC) from *S. venezuelae* (Sherman *et al.*, 2006) shows disordering of the BC loop as well CYP165B3 from Amycolatopsis orientalis (Zerbe *et al.*, 2002), CYP51 from *M. tuberculosis* (Podust *et al.*, 2001), and CYP105A1 from *Streptomyces griseolus* (Sugimoto *et al.*, 2008). All these P450s accommodate large substrates in their active sites (Xu *et al.*, 2009).

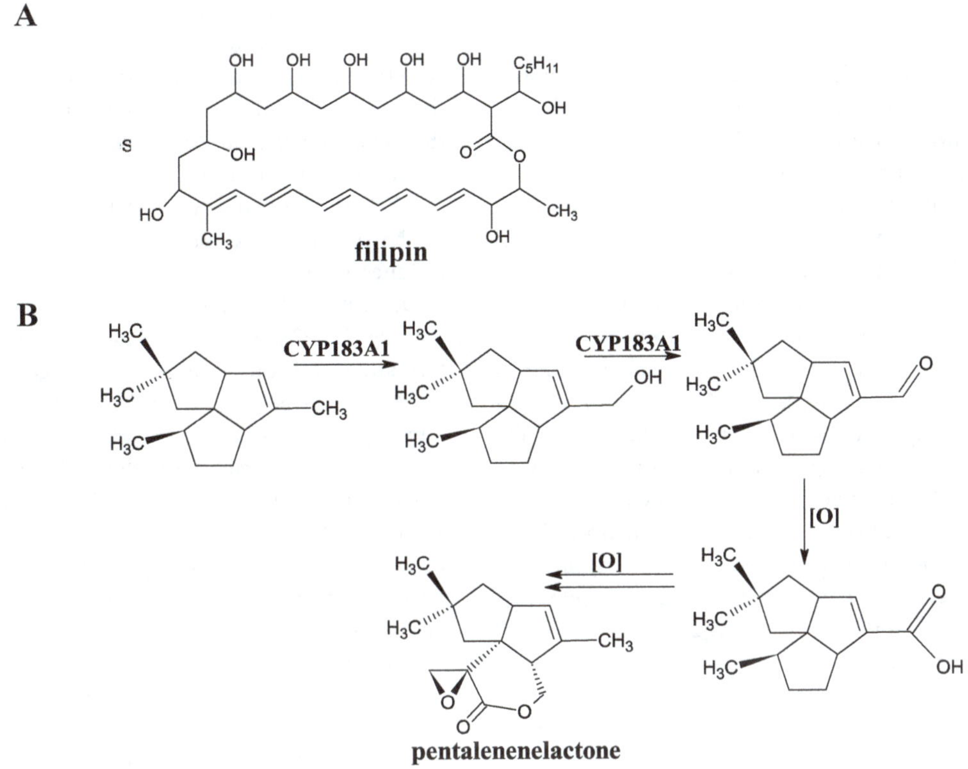

Figure 12.5 Biochemical analysis of P450s from *Streptomyces avermitilis*. (A) CYP105P1, a P450 which undergoes large conformational changes during catalysis in the biosynthesis of filipin. (B) The biosynthetic step of CYP183A1, involved in the production of pentalenelactone.

Determining the role of S. avermitilis CYP183A1 in pentalenolactone biosynthesis

S. avermitilis, as well as harbouring a gene cluster for the biosynthesis of avermectin, contains a 13 gene cluster for the biosynthesis of pentalenolactone, an antibacterial agent. Initiation of the biosynthesis of pentalenolactone involves the cyclization of farnesyl diphosphate by terpene synthase encoded within the cluster to make pentalenene. To generate the pentalenelactone antibiotic, pentalenene must be converted to a carboxylic acid. It has now been shown that the next step in this pathway involves a cytochrome P450 enzyme, CYP183A1, which catalyses the two- step oxidation of pentalenene to pentalen-13-al (Quaderer *et al.*, 2006; Fig. 12.5B). How pentalen-13-al is converted to the carboxylic acid remains unanswered. It has been speculated that CYP183A1 may perform this reaction. However, the remaining genes in the cluster encode for redox enzymes including non-haem iron dioxygenaseand six monooxygenases and dioxygenases, which may synthesize the carboxylic acid.

Structural genomics of streptomycete cytochromes P450

In addition to the five P450 crystal structures from *S. coelicolor* P450s (CYP154A1, CYP154C1, CYP158A1, CYP158A2, and CYP170A1) and the one P450 crystal structure from *S. avermitilis* (CYP105P1) (Xu *et al.*, 2009), only three additional streptomycete P450s have had its crystal structures resolved at the time of writing. These structures provide insights into substrate recognition and mechanisms of catalysis, such as to open routes for their manipulation to generate novel enzymatic products through mutagenesis.

One of the most studied streptomycete CYPs is CYP07L1 (PikC), a gene product from the pikromycin biosynthetic pathway from *Streptomyces venezuelae* (Betlach *et al.*, 1998). This biosynthetic pathway has revealed fundamental mechanisms of modular polyketide synthesis, aminodeoxysugar biosynthesis, and glycosyltransferase, as well as oxidative tailoring by P450 through analysis of PikC (Xue *et al.*, 1998). PikC is a unique P450 in streptomycetes in that it can produce a number of related macrolide products from the pathway. Considering the chemical structure of the PikC substrate, this P450 is able to catalyse monooxygenation at either C-10 or C-12 of the 12-membered ring macrolactone of YC17, giving rise to methymycin and neomethymycin respectively as well as at C-12 of the 14-membered ring of narbomycin, giving rise to pikromycin (Fig. 12.6A and B). Hydroxylation at C-14 of pikromycin give neopikromycin, dihydroxylation of YC-!7 gives rise to novamethymycin whilst dihydroxylation of narbomycin leads to novapikromycin (Fig. 12.6A and B). The crystal structure of PikC, both in the free structure or bound with each substrate, YC17 or narbomycin, revealed the structural basis for the activity observed (Sherman *et al.*, 2006). A unique anchoring system amongst P450s was revealed for PikC: a specific salt bridge of glutamate amino acid residues bind to the C3′ dimethylamino group of the deoxysugar substituent. These residues were confirmed by site-directed mutagenesis and they were shown to be essential for substrate binding. Interestingly, the desoamine group of each substrate can bind in two alternative binding pockets within the enzyme. Consequently, one can shift the product profile to a particular molecular form through mutagenesis of one of the binding sites. This strategy may be highly desirable in the future for the generation of novel bioactive molecules produced through manipulation of the pikromycin biosynthetic pathway.

As described above, staurosporine has attracted much attention because of its strong inhibitory activities against protein kinase C and topoisomerase, making it a potential anticancer agent (Onaka *et al.*, 2002). During the final step in the production of the indolocarbazole skeleton, CYP245A1 catalyses the intramolecular C–C coupling of chromopyrrolic acid (CPA) as well as oxidative decarboxylation. To understand the structural nature of this reaction, which will aid in the design of indolocarbazole biosynthesis system thus enabling the production of a wide variety indolocarbazole compounds, the crystal structure of CYP245A1 was resolved (Makino *et al.*, 2007). In the substrate free form of the enzyme, many residues in the substrate access channel cannot be identified due to disorder. Upon substrate binding, CYP245A1 adopts a more ordered conformation, and conformational rearrangements of residues in the active site are observed. Hydrogen bonding interactions of two carboxyl groups and T-shaped π–π interactions with indole rings hold the substrate in the substrate-binding cavity with a conformation perpendicular to the haem plane In such a position, it was proposed that C–C formation occurs through an indole cation radical intermediate.

Vitamin D3, a prohormone in mammals, requires activation by 25-hydroxylation in the liver and 1R-hydroxylation in the kidney by cytochrome P450 (CYP) enzymes. *Streptomyces griseolus* CYP105A1 has been shown to convert vitamin D3 into 1R,25-dihydroxyvitamin D3 (1R,25$(OH)_2D_3$) in two independent reactions, despite its low sequence identity with mammalian enzymes (<21% identity) (Sawada *et al.*, 2004). Structural analysis of CYP105A1 revealed that the location and orientation of 1R,25$(OH)_2D_3$ observed in a R84A mutant with high activity are similar to that of vitamin D_3 in human CYP2R1, suggesting a common substrate binding mechanism for 25-hydroxylation (Sugimoto *et al.*, 2008). R84A shows a 32-fold increase in 25-hydroxylation activity compared with the wild-type enzyme. A plausible explanation for this effect of the R84A mutant is that the loss of the interaction with the Arg84 side chain in B′ helix opens a direct channel to the active site and increases the adaptability of hydrophobic amino acids. Substrate-docking studies indicated that 1R$(OH)D_3$ and 25$(OH)D_3$ bind to the common site in two distinct orientations that present opposite ends of the sterol to the haem iron (Sugimoto *et al.*, 2008). Such results will be of use in the design of CYPs that can be used for the production of hormonally active forms of vitamin D3.

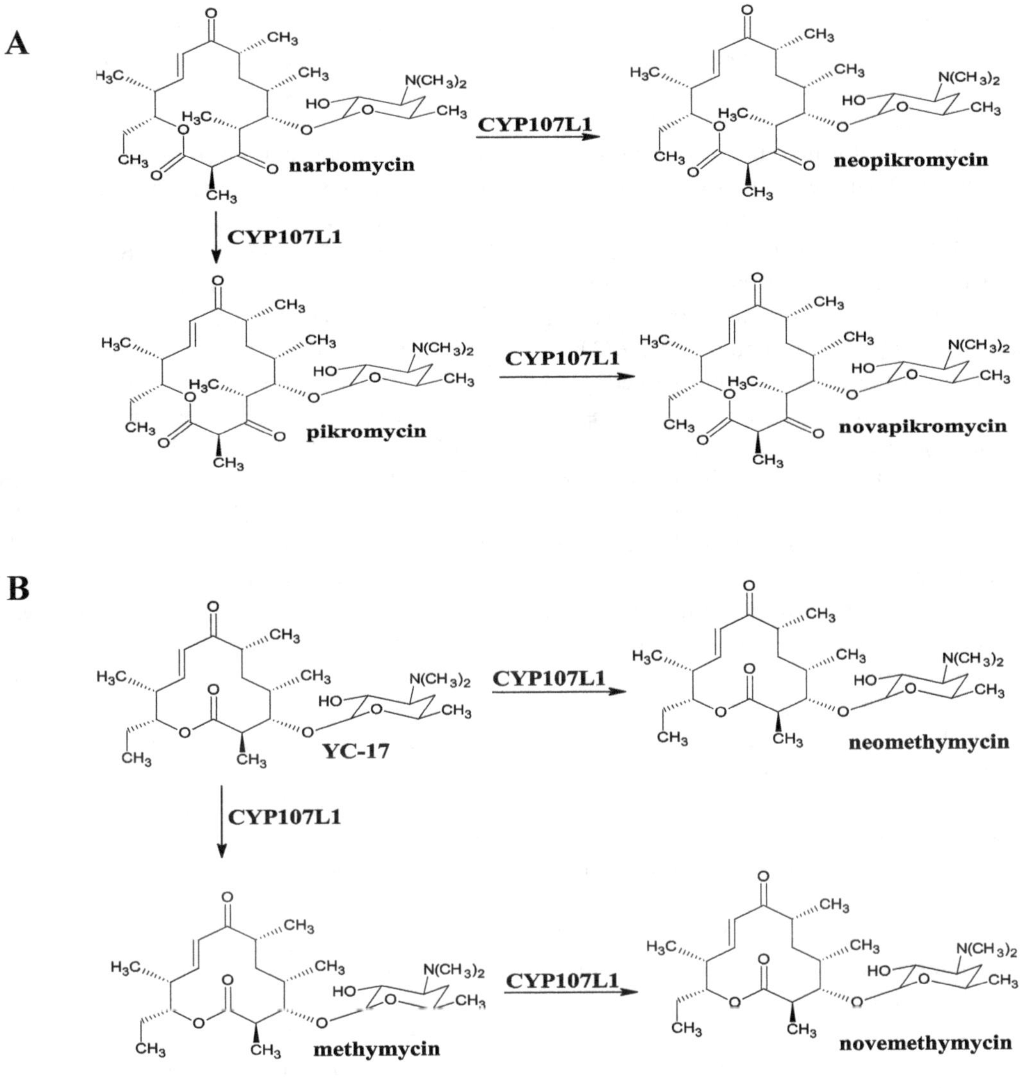

Figure 12.6 The biosynthetic steps of pikC (CYP107L1) involved in the production of the antibacterial agents pikromycin, methymycin and neomethmycin. (A) CYP107A1-mediated biosynthesis of pikromycin, neopikromycin and novapikromycin. (B) CYP107A1-mediated biosynthesis of methymycin, neomethymycin, and novemethymycin.

Continued sequencing of streptomycete genomes reveals an abundance of cytochrome P450 genes

Since the publication of the *S. coelicolor* A3(2) genome in 2002 and the *S. avermitilis* genome in 2003, three further streptomycete genomes (*Streptomyces peucetius*, *Streptomyces griseus* and *Streptomyces scabies*) have been sequenced of which only one, *S. griseus*, has so far been published (Ohnishi *et al.*, 2008). In the following section, an analysis of the P450s in each genome and potential biological functions will be briefly discussed.

The roles of P450s in S. peucetius

S. peucetius is an important streptomycete therapeutically because it produces the clinically important anthracycline chemotherapeutic agents, daunorubicin and doxorubicin. In 2004 the whole genome (8.7 Mb chromosome of strain *S. peucetius* ATCC 27952) was sequenced, but has yet to be published. The annotated sequence revealed the presence of 19 putative *CYP* genes with 15 functional genes and four pseudogenes (Parajuli *et al.*, 2004). Six of the 15 P450 genes are linked to secondary metabolite gene clusters: 4 (*CYP105P2, 105F2, 166B1,* and *107N3*) are clustered with putative modular polyketide synthase (PKS) of avermectin biosynthesis. The remaining two *CYP* genes (CYP131A1 and CYP129A2) are associated with the doxorubicin biosynthetic gene cluster, a drug used in the treatment of leukaemia. Interestingly, the exact biochemical reactions catalysed by CYP131A1 and CYP129A2 are still unclear but will be of biotechnological interest through genetic engineering in order to screen modified doxorubicin derivatives.

One interesting facet resulting from the sequencing of streptomycete genomes is the conservation of a four-gene cluster that was named 'conservon' (*cvn,* for conserved operon) in each genome. Each conservon encodes a predicted sensor kinase (*cvnA*) and-, an ATP-binding protein (*cvnD*), as well as two unknown ORFs (*cvnB* and *cvnC*), in which all CvnB contain the consensus signature roadblock/LC7 domain (Lamb *et al.,* 2002). In *S. coelicolor* A3(2) 13 such clusters were found (*cvnA, B, C, D, 1–13*) with unidirectional transcription and overlap of translational start and stop codons, suggesting an operon structure. Downstream of four conservons in *S. coelicolor* A3(2) the next ORF codes a CYP and in two conservons two CYPs are present. In *S. coelicolor* A3(2) *cvn10* the downstream CYPs are CYP157A1 and CYP154C1; in *cvn11* the downstream CYP is CYP157B1; in *cvn12* the downstream CYPs are CYP156A1 and CYP154A1; and in *cvn13* the downstream CYP is CYP157C1. Analysis of the genome sequence of *S. avermitilis* also revealed 11 conservons. Two conservons are followed downstream by genes encoding CYPs. In *cvnC9* the downstream CYPs are CYP157A2 and CYP154C2; and *cvnC10,* the downstream CYP is CYP157C2. Pull-down experiments in *S. coelicolor* has shown that the sensor kinase, ATP-binding protein and the two unknown ORFs encoded by *cvn9,* form a heterocomplex in the plasma membrane with specific interactions between each protein (Komatsu *et al.,* 2006). It has been suggested that this complex responds to possible environmental signals and/or hormone-like substances (Komatsu *et al.,* 2006). Presumably, given the genetic linkage of P450s to this gene cluster then the P450(s) has a role following on from the activation of the heterocomplex. Further work will be required to elucidate the role of these regulatory operons and that of the associated P450s.

Similarly, in *S. peucetius* many *CYPs* are clustered with regulatory elements: CYP251A1 clusters with a putative RoK-family transcriptional regulator; CYP252A1 clusters with ATP/GTP-binding protein, threonine dehydrogenase, and LysR-transcriptional regulator; CYP147F1 clusters with ferredoxin reductase, ATP/GTP-binding protein, and a putative transcriptional regulator; CYP154C4 clusters in a conservon; CYP154A3 clusters with a RNA polymerase sigma factor and MerR-family transcriptional regulator; CYP129A2 and 131A1 are associated with the pathway-specific regulatory gene *dnrI*; CYP107AH1 clusters with a putative dehydrogenase and expression regulator; finally, CYP107L9 clusters with a putative transcription regulator. Such gene linkages for these specific P450s linked to transcriptional factors and conservons in each genome strongly suggests a relationship between P450 regulatory genes and metabolism.

The roles of P450s in *S. griseus*

Streptomycin, the first aminoglycoside antibiotic, was discovered in the laboratory of Selman Waksman over 60 years go. This antibiotic was the first antituberculosis agent discovered and is responsible for saving millions of lives. This antibiotic is produced industrially through the fermentation of and extraction from *S. griseus*. Consequently, the 8.5 Mb genome of this organism was sequenced and completed in 2008 (Ohnishi *et al.,* 2008). In *S. griseus* the annotated genome contained 27 putative *CYP* genes, which have yet to be classified according to the CYP nomenclature convention and herein will be described by their SGR numbers. Eight of the 27 P450 genes are linked to secondary metabolite gene cluster with only one linked to a gene cluster which produces a known and characterized molecule. SGR6619 is part of a two-gene cluster with the second gene encoding a type III PKS; this operon produces HPQ melanin. This operon is conserved in *S. coelicolor* and *S. avermitilis* but not in *S. peucetius*. Additionally, six P450s are linked to secondary metabolism biosynthetic clusters which produce uncharacterized products. These P450s are linked to gene clusters that contain either a putative PKS and/or non-ribosomal peptide synthetase (NRPS) and/or PKS–NRPS hybrid genes: SGR582 and SGR895 are linked to NRPS gene clusters respectively; SGR810 is linked to a gene cluster containing a PKS–NRPS hybrid protein and two P450s, SGR3267 and SGR3274, are in a single cluster containing both a type II PKS and a NRPS. Finally SGR611 is in an eight-gene cluster which also contains an enediyne PKS, a gene organisation which is similar to that of many enediyne gene clusters found in several actinomycete strains. It will be important in the future to decipher these cryptic biosynthetic pathways in order to screen the therapeutic effects of these novel compounds.

The roles of P450s in *S. scabies*

A fourth streptomycete genome was recently sequenced, and although complete, has not yet been fully annotated (http://www.sanger.ac.uk/Projects/S_scabies/). *S. scabies* was first described in 1891 and has a global distribution (Lambert and Loria, 1989). *S. scabies* in an economically important plant pathogen responsible for potato scab resulting in the loss of millions of dollars annually through destroyed crops. Infection occurs through the lenticels of developing tubers and initially results in reddish-brown water-soaked lesions on the tuber periderm. As the pathogen continues to colonize the tuber, the host develops wound periderm, resulting in slightly raised lesions composed of rough corky tissue (Hooker, 1981). Consequently, there is much interest in identifying new pesticide targets from the completed genome in order to control this disease. Our analysis of the completed sequence revealed the presence of 25 *CYP* genes in *S. scabies*. Interestingly, many of these genes (5; CYP154A4, CYP156C1, CYP156D1, CYP157C5, CYP156B2) are in 'conservons' consisting of a predicted sensor kinase, an ATP-binding protein and two unknown ORFs, possibly reflecting the role of these P450s in pathways activated by environmental signals. Analysis of clusters to determine the possible function of the remaining *CYP* gene products remains difficult at present. This genome also contains a *CYP102B* homologue which is similar to the *S. coelicolor* CYP102B1 in that it predicted to encode a protein which is solely the haem domain. Furthermore, *S. scabies* contains a second *CYP102* gene, CYP102G1, which is a bifunctional P450 consisting of a haem domain fused to an FMN/FAD containing eukaryotic-like reductase. Presumably both CYP102s are involved in fatty acid metabolism. S. *scabies* has 13 *P450* genes which are unique to this organism CYP105Z1, CYP07AK1, CYP07AL1, CYP107AM1, CYP143B1, CYP143C1, CYP152D1, CYP154L1, CYP180B1, CYP182B1, CYP246A1, CYP282A1, and CYP283A1. This pattern is not unique for streptomycetes in that each sequence retains its own unique complement of *P450* genes, which probably reflects the biosynthetic needs of the organism in its ecological niche. Only one P450 family is conserved in all four sequenced genomes, CYP107U. This gene is not linked to any gene clusters and, thus, predicting function at this time is not possible.

A note on the biodiversity of marine actinomycete cytochrome P450s and their biosynthetic products

Intense interest has recently focused on marine actinomycetes as a source of novel secondary metabolites with potential therapeutic qualities. Many *Streptomyces* spp. have been identified from the marine environment but to date no genome of a marine streptomycete has been sequenced. However, the first marine water actinomycete, *Salinospora tropica*, had its genome sequenced in 2007 and provides a platform for evaluating the biosynthetic potential of marine actinomycete as a source of natural products (Udwary *et al.*, 2007). *S. tropica* contains a varied array of molecules including the potent proteasome inhibitor salinosporamide A, (which is currently in phase I human clinical trials for the treatment of cancer), halogenated macrolides sporolides

A and B, lymphostin, and salinilactam. Analysis for the presence of cytochrome P450 genes and their links to secondary-metabolites gene clusters was evaluated. Surprisingly, *S. tropica* has 16 P450 genes with at least 8 linked to the production of secondary metabolites: CYP163B1 is linked to a 32 gene cluster containing a hybrid PKS-NRPS and which biosynthesizes salinosporamide; CYP107E3 is in a cluster with a type I PKS involved in biosynthesising salinilactam; CYP154M1 is linked to an enediyne PKS. However, the remaining 5 P450s are in pathways whose products are yet to be determined: CYP113E1 is linked to an enediyne PKS, CYP107AY1 is linked to a cluster containing an NRPS, CYP125G1 is linked to cluster containing a hybrid PKS-NRPS proposed to make a siderophore, and two P450s – CYP3004A1 and CYP3004B1 – are in the same gene cluster and also proposed to synthesize a siderophore. The linkage of cytochromes P450 to secondary metabolite biosynthetic pathways in marine actinomycetes (and streptomycetes) will provide a new research direction in natural product biosynthesis in the future. Indeed, just from the shotgun sequencing of marine bacteria in the Sargasso Sea >3500 P450 sequences have been uncovered (Venter *et al.*, 2004). Such genetic information will no-doubt lead to the development of therapeutics for medicinal and veterinary applications

Conclusions and future prospects

Streptomycete CYPs are providing an exciting opportunity for the discovery of their roles in existing and in new secondary metabolite gene clusters, for their genetic engineering in such pathways to afford the generation of modified natural molecules as well as providing fundamental insights into the biochemistry of the P450 superfamily as a whole. The discovery of streptomycete gene clusters which encode therapeutic agents show that they invariably contain one or more P450(s) in the gene clusters. Additionally, many cryptic pathways have been uncovered from the streptomycete genome sequencing projects, many containing P450(s) in such pathways, and it is a priority to identify the end products of such pathways. Only then will such pathways be amenable to manipulation, and those P450(s) will no doubt be a target for mutagenesis experiments. Extension of such approaches to marine streptomycetes, as seen for the related actinomycete *S. tropica*, will be a goal for the future. Finally, examination of the *S. coelicolor* P450s have revealed fundamental discoveries for the P450 field as a whole: information regarding different P450 enzymatic reactions (C–C coupling, hydroxylation, allylic oxidation), different modes of substrate binding, recognition of redox partners, and structural architecture have all been uncovered. Clearly, such analysis of further streptomycete P450s will provide rewarding challenges and exhilarating results in the years ahead.

Acknowledgements

Support was provided by National Institutes of Health Grants GM69970 (M.R.W.), ES00267 (M.R.W.) and by a Leverhulme Trust Fellowship and the Royal Society (D.C.L.).

References

Belin, P., Le Du, M.H., Fielding, A., Lequin, O., Jacquet, M., Charbonnier, J.B., Lecoq, A., Thai, R., Courcon, M., Masson, C., *et al.* (2009). Identification and structural basis of the reaction catalyzed by CYP121, an essential cytochrome P450 in *Mycobacterium tuberculosis*. Proc. Natl. Acad. Sci. U.S.A. *106*, 7426–7431.

Bentley, S.D., Chater, K.F., Cerdeño-Tárraga, A.M., Challis, G.L., Thomson, N.R., James, K.D., Harris, D.E., Quail, M.A., Kieser, H., Harper, D., *et al.* (2002). Complete genome sequence of the model actinomycete *Streptomyces coelicolor* A3(2). Nature *417*, 141–147.

Berrie, J.R., Williams, R.A.D., and Smith, K.E. (1999). Microbial transformations of steroids-XI. Progesterone transformation by *Streptomyces roseochromogenes*-purification and characterisation of the 16 alpha-hydroxylase system. J. Steroid Biochem. *71*, 153–165.

Betlach, M.C., Kealey, J.T., Ashley, G.W., and McDaniel, R. (1998). Characterization of the macrolide P-450 hydroxylase from *Streptomyces venezuelae* which converts narbomycin to picromycin. Biochemistry *37*, 14937–14942.

Cannell, R.J., Rashid, T., Ismail, I.M., Sidebottom, P.J., Knaggs, A.R., and Marshall, P.S. (1997). Novel metabolites of warfarin produced by *Beauveria bassiana* and *Streptomyces rimosus*: a novel application of hplc-nmr. Xenobiotica *27*, 147–157.

Carmody, M., Murphy, B., Byrne, B., Power, P., Rai, D., Rawlings, B., and Caffrey, P. (2005). Biosynthesis of amphotericin derivatives lacking exocyclic carboxyl groups. J. Biol. Chem. *280*, 34420–34426.

Chen, H., and Walsh, C.T. (2001). Coumarin formation in novobiocin biosynthesis: beta-hydroxylation of the aminoacyl enzyme tyrosyl-S-NovH by a cytochrome P450 NovI. Chem. Biol. *8*, 301–312.

Cortes, J., Velasco, J., Foster, G., Blackaby, A.P., Rudd, B.A., and Wilkinson, B. (2002). Identification and cloning of a type III polyketide synthase required for diffusible pigment biosynthesis in *Saccharopolyspora erythraea*. Mol. Microbiol. *44*, 1213–1224.

Funa, N., Funabashi, M., Yoshimur, E., and Horinouchi, S. (2005). A novel quinine-forming monooxygenase family involved in modification of aromatic polyketides. J. Biol. Chem. *280*, 14514–14523.

Gaisser, S., Lill, R., Staunton, J., Mendez, C., Salas, J., and Leadlay, P.F. (2002). Parallel pathways for oxidation of 14-membered polyketide macrolactones in *Saccharopolyspora erythraea*. Mol. Microbiol. *44*, 771–781.

Hooker, W.J. (1981). Common scab. In Compendium of Potato Diseases, Hooker, W.J., ed. (American Phytopathological Society, St. Paul, MN), pp. 33–34.

Ikeda, H., Nonomiya, T., Usami, M., Ohta, T., and Omura, S. (1999). Organization of the biosynthetic gene cluster for the polyketide anthelmintic macrolide avermectin in *Streptomyces avermitilis*. Proc. Natl. Acad. Sci. U.S.A. *96*, 9509–9514.

Ikeda, H., Ishikawa, J., Hanamoto, A., Shinose, M., Kikuchi, H., Shiba, T., Sakaki, Y., Hattori, M., and Omura, S. (2003). Complete genome sequence and comparative analysis of the industrial microorganism *Streptomyces avermitilis*. Nat. Biotechnol. *21*, 526–531.

Kalb, V.F., and Loper, J.C. (1988). Proteins from eight eukaryotic cytochrome P-450 families share a segmented region of sequence similarity. Proc. Natl. Acad. Sci. U.S.A. *85*, 7221–7225.

Karp, F., Mihaliak, C.A., Harris, J.L., and Croteau, R. (1990). Monoterpene biosynthesis: specificity of the hydroxylations of (-)-limonene by enzyme preparations from peppermint (*Mentha piperita*), spearmint (*Mentha spicata*), and perilla (*Perilla frutescens*) leaves. Arch. Biochem. Biophys. *276*, 219–226.

Khosla, C., and Harbury, P.B. (2001). Modular enzymes. Nature *409*, 247–252.

Komatsu, M., Takano, H., Hiratsuka, T., Ishigaki, Y., Shimada, K., Beppu, T., and Ueda, K. (2006). Proteins encoded by the conservon of *Streptomyces coelicolor* A3(2) comprise a membrane associated heterocomplex that resembles eukaryotic G protein-coupled regulatory system. Mol. Microbiol. *62*, 1534–1546.

Lamb, D.C., Skaug, T., Song, H.-L., Jackson, C.J., Podust, L.M., Waterman, M.R., Kell, D.B., Kelly, D.E., and Kelly, S.L. (2002). The cytochrome P450 complement (CYPome) of *Streptomyces coelicolor* A3(2). J. Biol. Chem. 277, 24000–24005.

Lamb, D.C., Ikeda, H., Nelson, D.R., Ishikawa, J., Skaug, T., Jackson, C., Omura, S., Waterman, M.R., and Kelly, S.L. (2003). Cytochrome p450 complement (CYPome) of the avermectin-producer *Streptomyces avermitilis* and comparison to that of *Streptomyces coelicolor* A3(2). Biochem. Biophys. Res. Commun. *307*, 610–619.

Lambert, D.H., and Loria, R. (1989). *Streptomyces scabies* Sp-Nov, Nom-Rev. Int. J. Syst. Bacteriol. *39*, 387–392.

Leys, D., Mowat, C.G., McLean, K.J., Richmond, A., Chapman, S.K., Walkinshaw, M.D., and Munro, A.W. (2003). Atomic structure of *Mycobacterium tuberculosis* CYP121 to 1.06 A reveals novel features of cytochrome P450. J. Biol. Chem. *278*, 5141–5147.

Lin, X., and Cane, D.E. (2009). Biosynthesis of the sesquiterpene antibiotic albaflavenone in *Streptomyces coelicolor*. Mechanism and stereochemistry of the enzymatic formation of epi-isozizaene. J. Am. Chem. Soc. *131*, 6332–6333.

Makino, M., Sugimoto, H., Shiro, Y., Asamizu, S., Onaka, H., and Nagano, S. (2007). Crystal structures and catalytic mechanism of cytochrome P450 StaP that produces the indolocarbazole skeleton. Proc. Natl. Acad. Sci. U.S.A. *104*, 11591–11596.

Nelson, D.R., Koymans, L., Kamataki, T., Stegman, J.J., Feyereisen, R., Waxman, D.J., Waterman, M.R., Gotoh, O., Coon, M.J., Estabrook, R.W., *et al.* (1996). P450 superfamily: update on new sequences, gene mapping, accession numbers and nomenclature. Pharmacogenetics *6*, 1–42.

Ohnishi, Y., Ishikawa, J., Hara, H., Suzuki, H., Ikenoya, M., Ikeda, H., Yamashita, A., Hattori, M., and Horinouchi, S. (2008). Genome sequence of the streptomycin-producing microorganism *Streptomyces griseus* IFO 13350. J. Bacteriol. *190*, 4050–4060.

Omura, T., and Sato, R. (1964). The carbon monoxide binding pigment of liver microsomes. I. Evidence for its hemoprotein nature. J. Biol. Chem. *239*, 2379–2385.

Onaka, H., Taniguchi, S., Igarashi, Y., and Furumai, T. (2002). Cloning of the staurosporine biosynthetic gene cluster from *Streptomyces* sp. TP-A0274 and its heterologous expression in *Streptomyces lividans*. J. Antibiot. *55*, 1063–1071.

Onaka, H., Asamizu, S., Igarashi, Y., Yoshida, R., and Furumai, T. (2005). Cytochrome P450 homolog is responsible for C-N bond formation between aglycone and deoxysugar in thestaurosporine biosynthesis of *Streptomyces* sp. TP-A0274. Biosci. Biotechnol. Biochem. *69*, 1753–1759.

Ortiz de Montellano P.R. (2005). Cytochrome P450, Structure, Mechanism and Biochemistry (Kluwer Academic/Plenum Publishers, New York).

Ouellet, H., Johnston, J.B., and Ortiz de Montellano, P.R. (2009). The *Mycobacterium tuberculosis* cytochrome P450 system. Arch. Biochem. Biophys. *493*, 82–95.

Paine, M.J.I., Scrutton, N.S., Munro, A.W., Guiterrez, A., Roberts, G.C.K., and Wolf, C.R. (2005). In Cytochrome P450, Structure, Mechanism and Biochemistry, Ortiz de Montellano, P.R. ed. (Kluwer Academic/Plenum Publishers; New York), pp.115–148.

Parajuli, N., Basnet, D.B., Chan Lee, H., Sohng, J.K., and Liou, K. (2004). Genome analyses of *Streptomyces peucetius* ATCC 27952 for the identification and comparison of cytochrome P450 complement with other *Streptomyces*. Arch. Biochem. Biophys. *425*, 233–241.

Podust, L.M., Poulos, T.L., and Waterman, M.R. (2001). Crystal structure of cytochrome P450, 14alpha-sterol demethylase (CYP51) from *Mycobacterium tuberculosis* in complex with azole inhibitors. Proc. Natl. Acad. Sci. U.S.A. *98*, 3068–3073.

Podust, L.M., Kim, Y., Arase, M., Neely, B.A., Beck, B.J., Bach, H., Sherman, D.H., Lamb, D.C., Kelly, S.L., and Waterman, M.R. (2003). The 1.92-A structure of *Streptomyces coelicolor* A3(2) CYP154C1. A new monooxygenase that functionalizes macrolide ring systems. J. Biol. Chem. *278*, 12214–12221.

Podust, L.M., Bach, H., Kim, Y., Lamb, D.C., Arase, M., Sherman, D.H., Kelly, S.L., and Waterman, M.R. (2004). Comparison of the 1.85 A structure of CYP154A1 from *Streptomyces coelicolor* A3(2) with the closely related CYP154C1 and CYPs from antibiotic biosynthetic pathways. Protein Sci. *13*, 255–268.

Quaderer, R., Omura, S., Ikeda, H., and Cane, D.E. (2006). Pentalenolactone biosynthesis. Molecular cloning and assignment of biochemical function to PtlI, a cytochrome P450 of *Streptomyces avermitilis*. J. Am. Chem. Soc. *128*, 13036–13037.

Ravichandran, K.G., Boddupalli, S.S., Hasermann, C.A., Peterson, J.A., and Deisenhofer, J. (1993). Crystal structure of hemoprotein domain of P450BM-3, a prototype for microsomal P450's. Science *261*, 731–736.

Richardson, T.H., and Johnson, E.F. (1994). Alterations of the regiospecificity of progesterone metabolism by the mutagenesis of two key amino acid residues in rabbit cytochrome P450, 2C3v. J. Biol. Chem. *269*, 23937–23943.

Rupasinghe, S., Schuler, M.A., Kagawa, N., Yuan, H., Lei, L., Zhao, B., Kelly, S.L., Waterman, M.R., and Lamb, D.C. (2006). The cytochrome P450 gene family CYP157 does not contain EXXR in the K-helix reducing the absolute conserved P450 residues to a single cysteine. FEBS Lett. *580*, 6338–6342.

Sariaslani, F.S., and Rosazza, J.P. (1983). Novel biotransformations of 7-ethoxycoumarin by *Streptomyces griseus*. Appl. Environ. Microbiol. *46*, 468–474.

Sariaslani, F.S., Eckenrode, F.M., Beale, J.M., and Rosazza, J.P. (1984). Formation of a reactive iminium derivative by enzymatic and chemical oxidations of 16-O-acetylvindoline. J. Med. Chem. *27*, 749–754.

Sawada, N., Sakaki, T., Yoneda, S., Kusudo, T., Shinkyo, R., Ohta, M., and Inouye, K. (2004). Conversion of vitamin D3 to 1R, 25-dihydroxyvitamin D3 by *Streptomyces griseolus* cytochrome P450SU-1. Biochem. Biophys. Res. Commun. *320*, 156–164.

Schlichting, I., Berendzen, J., Chu, K., Stock, A.M., Maves, S.A., Benson, D.E., Sweet, R.M., Ringe, D., Petsko, G.A., and Sligar, S.G. (2000). The catalytic pathway of cytochrome p450cam at atomic resolution. Science *287*, 1615–1622.

Serizawa, N., and Matsuoka, T. (1991). A two component type cytochrome P450 monooxygenase system in a prokaryote that catalyses hydroxylation of ML-236B to pravastatin, a tissue selective inhibitor of 3-hydroxy-3-methylglutaryl coenzyme A reductase. Biochim. Biophys. Acta *1084*, 35–40.

Sherman, D.H., Li, S., Yermalitskaya, L.V., Kim, Y., Smith, J.A., Waterman, M.R., and Podust, L.M. (2006). The structural basis for substrate anchoring, active site selectivity, and product formation by P450 PikC from *Streptomyces venezuelae*. J. Biol. Chem. *281*, 26289–26297.

Sugimoto, H., Shinkyo, R., Hayashi, K., Yoneda, S., Yamada, M., Kamakura, M., Ikushiro, S., Shiro, Y., and Sakaki, T. (2008). Crystal structure of CYP105A1 (P450SU-1) in complex with 1alpha, 25-dihydroxyvitamin D3. Biochemistry *47*, 4017–4027.

Süssmuth, R.D., Pelzer, S., Nicholson, G., Walk, T., Wohlleben, W., and Ung, G. (1999). New advances in the biosynthesis of glycopeptide antibiotics of the vancomycin type from *Amycolatopsis mediterranei*. Angew. Chem. Int. Ed. *38*, 1976–1979.

Taylor, M., Lamb, D.C., Cannell, R., Dawson, M., and Kelly, S.L. (1999). Cytochrome P450105D1 (CYP105D1) from *Streptomyces griseus*: heterologous expression, activity and activation effects of multiple xenobiotics. Biochem. Biophys. Res. Commun. *263*, 838–842.

Udwary, D.W., Zeigler, L., Asolkar, R.N., Singan, V., Lapidus, A., Fenical, W., Jensen, P.R., and Moore, B.S. (2007). Genome sequencing reveals complex secondary metabolome in the marine actinomycete *Salinispora tropica*. Proc. Natl. Acad. Sci. U.S.A. *104*, 10376–10381.

Venter, J.C., Remington, K., Heidelberg, J.F., Halpern, A.L., Rusch, D., Eisen, J.A., Wu, D., Paulsen, I., Nelson, K.E., Nelson, W., *et al.* (2004). Environmental genome shotgun sequencing of the Sargasso Sea. Science *304*, 66–74.

Ventura, M., Canchaya, C., Tauch, A., Chandra, G., Fitzgerald, G.F., Gerald, K.F., and van Sinderen, D. (2007). Genomics of *Actinobacteria*: tracing the evolutionary history of an ancient phylum. Microbiol. Mol. Biol. Rev. *71*, 495–548.

Wickramasinghe, R.H., and Ville, C.A. (1975). Early role during chemical evolution for cytochrome P450 in oxygen detoxification. Nature *256*, 509–511.

Xu, L.H., Fushinobu, S., Ikeda, H., Wakagi, T., and Shoun, H. (2009). Crystal structures of cytochrome P450, 105P1 from *Streptomyces avermitilis*: conformational flexibility and histidine ligation state. J. Bacteriol. *191*, 1211–1219.

Xue, Y., Zhao, L., Liu, H.W., and Sherman, D.H. (1998). A gene cluster for macrolide antibiotic biosynthesis in *Streptomyces venezuelae*: architecture of metabolic diversity. Proc. Natl. Acad. Sci. U.S.A. *95*, 12111–12116.

Yunt, Z., Reinhardt, K., Li, A., Engeser, M., Dahse, H.M., Gütschow, M., Bruhn, T., Bringmann, G., and Piel, J. (2009). Cleavage of four carbon–carbon bonds during biosynthesis of the griseorhodin a spiroketal pharmacophore. J. Am. Chem. Soc. *131*, 2297–2305.

Zerbe, K., Pylypenko, O., Vitali, F., Zhang, W., Rouset, S., Heck, M., Vrijbloed, J.W., Bischoff, D., Bister, B., Süssmuth, R.D., *et al.* (2002). Crystal structure of

OxyB, a cytochrome P450 implicated in an oxidative phenol coupling reaction during vancomycin biosynthesis. J. Biol. Chem. 277, 47476–47485.

Zhao, B., Guengerich, F.P., Bellamine, A., Lamb, D.C., Izumikawa, M., Funa, N., Lei, L., Podust, L.M., Sundamoorthy, M., Reddy, L.M., *et al.* (2005). Bindings of two flaviolin substrate molecules, oxidative coupling, and crystal structure of *Streptomyces coelicolor* A3(2) cytochrome P450, 158A2. J. Biol. Chem. *280*, 11599–11607.

Zhao, B., Lamb, D.C., Lei, L., Kelly, S.L., Yuan, H., Hachey, D.L., and Waterman, M.R. (2007). Different binding modes of two flaviolin substrate molecules in cytochrome P450, 158A1 (CYP158A1) compared to CYP158A2. Biochemistry *46*, 8725–8733.

Zhao, B., Lin, X., Lei, L., Lamb, D.C., Kelly, S.L., Waterman, M.R., and Cane, D.E. (2008). Biosynthesis of the sesquiterpene antibiotic albaflavenone in *Streptomyces coelicolor* A3(2). J. Biol. Chem. *283*, 8183–8189.

Zhao, B., Lei, L., Vassylyev, D.G., Lin, X., Cane, D.E., Kelly, S.L., Yuan, H., Lamb, D.C., and Waterman, M.R. (2009). A Terpene Synthase Active Site Moonlighting on the Albaflavenone Monooxygenase. J. Biol. Chem. *284*, 36711–36719.

Index

www.ingramcontent.com/pod-product-compliance
Lightning Source LLC
LaVergne TN
LVHW062208200926
848484LV00006B/1354